T0259811

Lecture Notes in Computer Science   11801

Dimitris Fotakis · Evangelos Markakis (Eds.)

Algorithmic Game Theory

12th International Symposium, SAGT 2019
Athens, Greece, September 30 – October 3, 2019
Proceedings

 Springer

Editors
Dimitris Fotakis
National Technical
University of Athens
Athens, Greece

Evangelos Markakis
Athens University
of Economics and Business
Athens, Greece

ISSN 0302-9743   ISSN 1611-3349   (electronic)
Lecture Notes in Computer Science
ISBN 978-3-030-30472-0   ISBN 978-3-030-30473-7   (eBook)
https://doi.org/10.1007/978-3-030-30473-7

LNCS Sublibrary: SL3 – Information Systems and Applications, incl. Internet/Web, and HCI

This Springer imprint is published by the registered company Springer Nature Switzerland AG
The registered company address is: Gewerbestrasse 11, 6330 Cham, Switzerland

Preface

This volume contains the papers and extended abstracts presented at the 12th International Symposium on Algorithmic Game Theory (SAGT 2019), held during September 30–October 3, 2019, at the National Technical University of Athens, Greece.

This year, SAGT 2019 received 55 submissions, which were all rigorously peer-reviewed by at least 3 PC members, and evaluated on the basis of originality, significance, and exposition. The PC eventually decided to accept 26 papers to be presented at the conference. To accommodate the publishing traditions of different fields, authors of accepted papers could ask that only a one-page abstract of the paper appeared in the proceedings. Among the 26 accepted papers, the authors of 1 paper selected this option. The program also included three invited talks by distinguished researchers in Algorithmic Game Theory, namely Maria-Florina Balcan (Carnegie Mellon University, USA), Shahar Dobzinski (Weizmann Institute of Science, Israel), and Herve Moulin (University of Glasgow, UK, and Higher School of Economics, Russia). In addition, SAGT 2019 featured a tutorial on "Learning Theory in Algorithmic Economics," by Georgios Piliouras (Singapore University of Technology and Design, Singapore) and Vasilis Syrgkanis (Microsoft Research, USA).

The works accepted for publication in this volume cover most of the major aspects of Algorithmic Game Theory, including auction theory, mechanism design, two-sided markets, computational aspects of games, congestion games, resource allocation problems, and computational social choice. Furthermore, due to the general support by Springer, we were able to provide a best paper award. The PC decided to give the award to the paper "The Declining Price Anomaly is not Universal in Multi-buyer Sequential Auctions (but almost is)," by Vishnu V. Narayan, Enguerrand Prebet, and Adrian Vetta.

We would like to thank all the authors for their interest in submitting their work to SAGT 2019, as well as the PC members and the external reviewers for their great work in evaluating the submissions. We also want to thank EATCS, Springer, Facebook, the COST Action GAMENET (CA16228), the Athens University of Economics and Business (AUEB), the National Technical University of Athens (NTUA), and the Institute of Communication and Computer Systems (ICCS), for their generous financial support. We are grateful to the National Technical University of Athens for hosting the event, and special thanks also go to Eleni Iskou for her excellent local arrangements work and to Antonis Antonopoulos for his help with the conference website.

Finally, we would also like to thank Alfred Hofmann and Anna Kramer at Springer for helping with the proceedings, and the EasyChair conference management system.

July 2019

Dimitris Fotakis
Evangelos Markakis

Organization

Program Committee

Siddharth Barman	Indian Institute of Science, India
Vittorio Bilo	University of Salento, Italy
Yang Cai	Yale University, USA
George Christodoulou	University of Liverpool, UK
Riccardo Colini Baldeschi	Facebook, UK
Edith Elkind	University of Oxford, UK
Piotr Faliszewski	AGH University of Science and Technology, Poland
Felix Fischer	Queen Mary University of London, UK
Michele Flammini	Gran Sasso Science Institute, and University of L'Aquila, Italy
Dimitris Fotakis	National Technical University of Athens, Greece
Laurent Gourves	CNRS, France
Tobias Harks	Augsburg University, Germany
Martin Hoefer	Goethe University Frankfurt/Main, Germany
Panagiotis Kanellopoulos	University of Patras, and CTI Diophantus, Greece
Thomas Kesselheim	University of Bonn, Germany
Piotr Krysta	University of Liverpool, UK
Pinyan Lu	Shanghai University of Finance and Economics, China
Brendan Lucier	MSR New England, USA
David Manlove	University of Glasgow, UK
Evangelos Markakis	Athens University of Economics and Business, Greece
Georgios Piliouras	Singapore University of Technology and Design, Singapore
Alexandros Psomas	Carnegie Mellon University, USA
Guido Schaefer	Centrum Wiskunde & Informatica, The Netherlands
Alkmini Sgouritsa	Max Planck Institute for Informatics, Germany
Christos Tzamos	University of Wisconsin-Madison, USA
Carmine Ventre	University of Essex, UK
Rakesh Vohra	University of Pennsylvania, USA

Additional Reviewers

Amanatidis, Yorgos
Anastasiadis, Eleftherios
Arunachaleswaran, Eshwar Ram
Banerjee, Siddhartha
Barrot, Nathanaël
Bei, Xiaohui
Brokkelkamp, Ruben
Cseh, Ágnes
de Keijzer, Bart
Deligkas, Argyrios
Esfandiari, Hossein
Fanelli, Angelo
Ferraioli, Diodato
Gairing, Martin
Ghalme, Ganesh
Hahn, Niklas
Huang, Chien-Chung
Kahng, Anson
Kleer, Pieter
Klimm, Max
Kocot, Maciej
Kolonko, Anna
Kontonis, Vasilis
Kovacs, Annamaria
Kumar, Rachitesh
Lesca, Julien
Li, Bo
Li, Minming
Meir, Reshef
Melissourgos, Themistoklis
Miyazaki, Shuichi

Moscardelli, Luca
Oosterwijk, Tim
Pountourakis, Emmanouil
Rathi, Nidhi
Ray Chaudhury, Bhaskar
Savani, Rahul
Schmand, Daniel
Schroder, Marc
Schvartzman, Ariel
Sekar, Shreyas
Serafino, Paolo
Skoulakis, Stratis
Skowron, Piotr
Slavkovik, Marija
Solomon, Shay
Sprenger, Jan
Suksompong, Warut
Szufa, Stanisław
Taggart, Sam
Telelis, Orestis
Telikepalli, Kavitha
Tönnis, Andreas
Vakaliou, Eftychia
Vinci, Cosimo
von Stengel, Bernhard
Voudouris, Alexandros
Wang, Hongao
Wang, Zihe
Wilczynski, Anaëlle
Xu, Haifeng
Zhao, Mingfei

Invited Talks

Machine Learning for Mechanism Design

Maria-Florina Balcan

Carnegie Mellon University, Pittsburgh, PA 15213, USA
ninamf@cs.cmu.edu
http://www.cs.cmu.edu/~ninamf

Mechanism design is a field of game theory with significant real-world impact, encompassing areas such as pricing and auction design. A powerful and prominent approach in this field is automated mechanism design, which uses optimization and machine learning to design mechanisms based on data. In this talk I will discuss how machine learning theory tools can be adapted and extended to analyze important aspects of automated mechanism design.

I will first discuss revenue maximization in the setting where the mechanism designer has access to samples from the distribution over buyers' values, not an explicit description thereof. I will present a general technique for providing sample complexity bounds, that is, bounds on the number of samples sufficient to ensure that if a mechanism has high average revenue over the set of samples, then that mechanism will have high revenue in expectation over the buyers' valuation distribution. This technique applies to mechanisms that satisfy linear delineability, a general structural property that we show is shared by a myriad of pricing and auction mechanisms. Roughly speaking, a mechanism is linearly delineable if for any set of buyers' values, the revenue function is piecewise linear in the mechanism's parameters. I will discuss numerous applications of this result to both pricing mechanisms (including posted-price mechanisms and multi-part tariffs), and auctions (including second price auctions with reserves and classes of affine maximizer auctions).

I will also discuss how we can estimate the degree of incentive-compatibility of potentially non-incentive-compatible mechanisms based on typical inputs, namely independent samples from the type distribution. Our estimate is based on an empirical variant of approximate incentive compatibility which measures the maximum utility an agent can gain by misreporting his type, on average over the samples. I will discuss how to bound the difference between our empirical incentive compatibility estimate and the true incentive compatibility approximation factor by using a subtle mixture of tools from learning theory. This question is of high interest since many real-world mechanisms are not incentive-compatible.

This talk is based on work joint with Tuomas Sandholm and Ellen Vitercik appearing in [1, 2].

References

1. Balcan, M.F., Sandholm, T., Vitercik, E.: A general theory of sample complexity for multi-item profit maximization. In: ACM Conference on Economics and Computation (2018)
2. Balcan, M.F., Sandholm, T., Vitercik, E.: Estimating approximate incentive compatibility. In: ACM Conference on Economics and Computation (2019)

From Cognitive Biases to the Communication Complexity of Local Search

Shahar Dobzinski

Weizmann Institute of Science
shahar.dobzinski@weizmann.ac.il

In this talk I will tell you how analyzing economic markets in which agents have cognitive biases has led to better understanding of the communication complexity of local search procedures.

We begin the talk with studying combinatorial auctions with bidders that exhibit endowment effect. In most of the previous work on cognitive biases in algorithmic game theory (e.g., [3] and its follow-ups) the focus was on analyzing their implications and mitigating the negative consequences. In contrast, we show how cognitive biases can sometimes be harnessed to improve the outcome.

Specifically, we study Walrasian equilibria in combinatorial markets. It is well known that a Walrasian equilibrium exists only in limited settings, e.g., when all valuations are gross substitutes, but fails to exist in more general settings, e.g., when the valuations are submodular. We consider combinatorial settings in which bidders exhibit the endowment effect, that is, their value for items increases with owner-ship. Our main result here shows that when the valuations are submodular even a mild level of endowment effect suffices to guarantee the existence of Walrasian equilibrium. In fact, we show that in contrast to Walrsian equilibria with standard utility maximizers bidders – in which the equilibrium allocation must be a global optimum – when bidders exhibit endowment effect any local optimum can be an equilibrium allocation.

This raises a natural question: what is the complexity of computing a local maximum in combinatorial markets? We reduce it to a communication variant of local search: there is some commonly known graph G. Alice holds f_A and Bob holds f_B, both are functions that specify a value for each vertex. The goal is to find a local maximum of $f_A + f_B$: a vertex v for which $f_A(v) + f_B(v) \geq f_A(u) + f_B(u)$ for every neighbor u of v. We prove that finding a local maximum requires polynomial (in the number of vertices) communication.

Based on joint works with Moshe Babaioff, Yakov Babichenko, Noam Nisan, and Sigal Oren [1, 2].

References

1. Babaioff, M., Dobzinski, S., Oren, S.: Combinatorial auctions with endowment effect. In: ACM EC (2018)
2. Babichenko, Y., Dobzinski, S., Nisan, N.: The communication complexity of local search. In: ACM STOC (2019)
3. Kleinberg, J.M., Oren, S.: Time-inconsistent planning: a computational problem in behavioral economics. In: ACM EC (2014)

Bidding for a Fair Share

Herve Moulin[1,2]

[1] University of Glasgow, Glasgow, UK
Herve.Moulin@glasgow.ac.uk
[2] Higher School of Economics, St. Petersburg, Russia

The Diminishing Share (DS) algorithm by Steinhaus (generalizing Divide and Choose), as well as the Moving Knife (MK) algorithm by Dubins and Spanier, guarantee to all participants a Fair Share of the manna while eliciting very little information from them. Hence their appeal: they bypass most of the cognitive effort to form full-fledged preferences; they also preserve the privacy of these preferences to a considerable degree.

The DS algorithm does not treat the agents symmetrically (namely, it fails the Anonymity test). The MK algorithm does not treat the manna symmetrically (it fails the Neutrality test) and severely limits the set of its feasible allocations.

In the classic cake division model with additive utilities, we propose a new family of division algorithm(s), dubbed the Bid & Choose rules, guaranteeing Fair Shares, maintaining the informational parsimony of DS and MK, and placing no restrictions on the allocations of the manna. The B&C rules are Anonymous, and each rule is defined by a specific interpretation of Neutrality. These properties are characteristic in the additive domain.

For general monotone preferences, each B&C rule, unlike DS and MK, offers reasonable guaranteed utility levels to each participant.

Contents

Network Games and Congestion Games

Social Choice

Matchings and Fair Division

Abstracts

Algorithmic Mechanism Design

Optimal On-Line Allocation
Rules with Verification

Markos Epitropou[1]([✉]) and Rakesh Vohra[1,2]

[1] Department of Electrical and Systems Engineering,
University of Pennsylvania, Philadelphia, USA
mep@seas.upenn.edu
[2] Department of Economics, University of Pennsylvania, Philadelphia, USA

Abstract. We consider a principal who allocates an indivisible object among a finite number of agents who arrive on-line, each of whom prefers to have the object than not. Each agent has access to private information about the principal's payoff if he receives the object. The decision to allocate the object to an agent must be made upon arrival of an agent and is irreversible. There are no monetary transfers but the principal can verify agents' reports at a cost and punish them. A novelty of this paper is a reformulation of this on-line problem as a compact linear program. Using the formulation we characterize the form of the optimal mechanism and reduce the on-line version of the verification problem with identical distributions to an instance of the secretary problem with one fewer secretary and a modified value distribution. This reduction also allows us to derive a prophet inequality for the on-line version of the verification problem.

Keywords: Stopping problems · Verification · Prophet inequalities

1 Introduction

In many large organizations scarce resources must be allocated internally without the benefit of prices. Examples include, the headquarters of a firm that must choose between multiple investment proposals from each of its division managers and funding agencies allocating a grant among researchers. In these settings the private information needed to determine the right allocation resides with the agents and the principal must rely on verification of agents' claims, which can be costly. We interpret verification as acquiring information (e.g., requesting documentation, interviewing an agent, or monitoring an agent at work), which can be costly. The headquarters of the diversified firm can hire an external firm to conduct an assessment of any division manager's claims, for example. The funding agency must allocate time to evaluate the claims of the researcher applying for a grant. Furthermore, in these settings, the principal can punish an agent if his claim is found to be false. For example, the head of personnel can reject an applicant, fire an employee or deny a promotion. Funding agencies can cut off funding.

Research supported in part by DARPA grant HR001118S0045.

D. Fotakis and E. Markakis (Eds.): SAGT 2019, LNCS 11801, pp. 3–17, 2019.
https://doi.org/10.1007/978-3-030-30473-7_1

Prior work considered an off-line version of this problem. Specifically, there is a principal who has to allocate one indivisible object among a finite number of agents all of whom are present. The value to the principal of assigning the object to a particular agent is the private information of the agent. Each agent prefers to possess the object than not. The principal would like to give the object to the agent who has the highest value to her. [4], the first to pose the question, assumes punishment is unlimited in the sense that an agent can be rejected and not receive the resource. Punishment can be limited because verification is imperfect or information arrives only after an agent has been hired for a while. In [20], verification is free, but punishment is limited. [17] generalizes both papers by incorporating costly verification *and* limited punishment.

This paper introduces and analyzes an on-line version of this problem in which the agents arrive and depart one at a time, and the decision to allocate the object to an agent must be made upon arrival of an agent. If the principal declines to allocate the object to an agent, the agent departs and cannot be recalled. If the principal allocates the object to an agent, the decision is irreversible. The problem is analogous to the problem of choosing a selling mechanism when facing a stream of buyers who arrive over time (see for example [11]) except we do not have access to monetary transfers.

If each agent were to truthfully report the value to the principal, the principal faces a *cardinal* version of the secretary problem [15, 16]. In this version, one is shown n non-negative numbers, sequentially, that are independent draws from known distributions (not necessarily identical). The goal is to select a single element (a 'secretary') with maximum value. An element of the sequence must be selected or discarded upon its arrival, and this decision is irrevocable. The solution involves a sequence of thresholds, indexed by the agent, and the principal allocates the object to the first agent whose reported value exceeds their corresponding threshold.

If the principal were to adopt such a policy in our setting it would encourage all agents to exaggerate their values. To discourage this, the principal can ration at the top of the distribution of values or verify an agent's claim and punish him if his claim is found to be false. The first reduces allocative efficiency while the second is costly. The goal of this paper is to find the optimal way to provide incentives via these two devices in an on-line setting. The contributions of this paper are as follows:

1. A reformulation of the on-line problem as a compact linear program that may be useful in other applications.
2. This reformulation allows us to derive a prophet inequality [21] for the on-line version of the verification problem.
3. Under the assumption of identical distributions, we reduce the on-line version of the verification problem with its incentive constraints to an instance of the cardinal secretary problem with one fewer secretary and a modified value distribution.

Our paper is related to three lines of work. The first is on costly verification that begins with [22]. This paper and others that followed such as [10], and [19], analyze off-line settings with transfers, which we rule out.

The second is on partial but costless verification, see for example [6] or [3], for example. In these models, verification is costless but imperfect. In our model verification is perfect but costly. At a high level the two are related because one can think of partial verification as being costly, but the cost is endogenous, depending on the nature of the realized allocation. In our case the cost is exogenous.

Our paper is also related to the extensive literature on versions of the secretary problem where the principal can rely on prices that was initiated in [7,12]. This was subsequently extended to include additional constraints such as cardinality constraints [1,12], matroids [14], matchings [2], and knapsack constraints [8,9]. The absence of money in our setting means that the results from these papers do not apply. However, our linear programming approach may be useful in analyzing problems when the principal has access to prices.

In Sect. 2 we introduce our setting and the linear programming formulation. In Sect. 3 we characterize the form of the optimal mechanism and provide a corresponding prophet inequality. In Sect. 4 we study the variation of the problem with limited punishment.

2 Model

There is a single indivisible good to allocate among a set of agents denoted by $I = \{1, \ldots, n\}$. The type of agent $i \in I$ is t_i which is the value to the principal of allocating the object to agent i. We assume that the agents' types are independently distributed. The distribution of agent's i type has strictly positive density f_i over the interval $T_i = [\underline{t}_i, \overline{t}_i]$. The preferences of the agents are simple: each prefers to possess the object to not. The actual private benefit enjoyed by an agent from receiving the object does not need to be specified.

Agents arrive one after the other and report their type, not necessarily truthfully. The principal can verify the reported type of agent i at cost $c > 0$ and determine perfectly if the agent has lied. In the event an agent is discovered to have lied, we withhold the object from them. This is the case of unlimited punishment. The case of limited punishment is considered later.

By the revelation principle we can restrict attention to direct mechanisms. Denote by $t_{\leq i}$ the profile of reported types made by all agents up to and including agent i. We write $t_{<i}$ to denote the profile of reported types made by all agents up to but not including i. A direct mechanism specifies for each profile of type reports, an allocation rule and an verification rule for each agent i. The allocation rule specifies the probability $q_i(t_i)$ he is allocated the good conditional on the event that the good is not already allocated. Specifically, $q_i(t_i) = Pr[\text{choose } t_i | 1, \ldots, i - 1 \text{ not allocated}]$. This fully captures the set of online allocation rules, since independence means there is no need to condition

the decision to allocate the good to agent i upon $t_{<i}$. The verification rule is the probability that agent i is assigned the good *and* inspected conditional on the event that the good is not already allocated and denoted $a_i(t_i)$. Therefore:

$$0 \leq a_i(t_i) \leq q_i(t_i) \leq 1 \quad \forall i \in I \quad \forall t_i \in T_i. \tag{1}$$

Definition 1. *A direct mechanism $\mathcal{M} = (T_1, \ldots, T_{|I|}, \{q_i(\cdot), a_i(\cdot)\}_{i \in I})$ restricts the strategy set of each agent i to T_i, and returns an allocation rule $q_i : T_i \to [0,1]$ and a verification rule $a_i : T_i \to [0,1]$ for each agent $i \in I$.*

Definition 2. *A direct mechanism $\mathcal{M} = (T_1, \ldots, T_{|I|}, \{q_i(\cdot), a_i(\cdot)\}_{i \in I})$ is incentive compatible if each agent i has an incentive to truthfully report her type, i.e.*

$$q_i(t_i) \geq q_i(t_i') - a_i(t_i') \quad \forall i \in I \quad \forall t_i, t_i' \in T_i. \tag{2}$$

The left hand side of (2) is the probability of receiving the good with a truthful report. The right hand side is the probability of receiving the good with a misreport adjusted downwards for the possibility of being inspected and punished for the misreport.

The principal would like to choose the allocation and verification probabilities q and a satisfying (1) and (2) to maximize:

$$\sum_{i \in I} \mathbb{E}_{t_{<i}} [\prod_{j<i} (1 - q_j(t_j))] \mathbb{E}_{t_i} [t_i q_i(t_i) - c a_i(t_i)].$$

2.1   Reduced Form Representation

We work with a reduced form representation of the allocation and verification rules (see for example [5,17,23]). Given an allocation and verification rule, (q, a), let $Q_i(t_i) = q_i(t_i) \mathbb{E}_{t_{<i}} [\prod_{j<i} (1 - q_j(t_j))]$ and $A_i(t_i) = a_i(t_i) \mathbb{E}_{t_{<i}} [\prod_{j<i} (1 - q_j(t_j))]$ be the interim allocation and verification probabilities respectively. The interim allocation and verification probabilities are related to the allocation and verification probabilities as follows:

Lemma 1. *Let Q, A, q, a be the interim as well as actual allocation and verification rules of a direct mechanism. Then the interim and actual rules are related as follows:*

$$q_i(t_i) = \frac{Q_i(t_i)}{1 - \sum\limits_{j<i} \mathbb{E}_{t_j} [Q_j(t_j)]} \tag{3}$$

$$a_i(t_i) = \frac{A_i(t_i)}{1 - \sum\limits_{j<i} \mathbb{E}_{t_j} [Q_j(t_j)]} \tag{4}$$

Proof. We prove (3). The proof of (4) is similar. Now, $Q_i(t_i) = q_i(t_i)\mathbb{E}_{t_{<i}}[\prod_{j<i}(1 - q_j(t_j))]$. Thus,

$$q_i(t_i) = \frac{Q_i(t_i)}{\mathbb{E}_{t_{<i}}[\prod_{j<i}(1 - q_j(t_j))]}.$$

It suffices to prove the following:

$$\mathbb{E}_{t_{\leq i}}[\prod_{j\leq i}(1 - q_j(t_j))] = 1 - \sum_{j\leq i}\mathbb{E}_{t_j}[Q_j(t_j)].$$

We do so by induction. For $i = 1$, the equality reduces to

$$\mathbb{E}_{t_1}[1 - q_1(t_1)] = 1 - \mathbb{E}_{t_1}[Q_1(t_1)]$$

which holds since $Q_1(t_1) = q_1(t_1)$. Let's now prove the equality for i. This holds since

$$\mathbb{E}_{t_{\leq i}}[\prod_{j\leq i}(1 - q_j(t_j))] = \mathbb{E}_{t_i}[(1 - q_i(t_i))]\mathbb{E}_{t_{<i}}[\prod_{j<i}(1 - q_j(t_j))]$$

$$= (1 - \mathbb{E}_{t_i}[q_i(t_i)])\mathbb{E}_{t_{<i}}[\prod_{j<i}(1 - q_j(t_j))]$$

$$= \mathbb{E}_{t_{<i}}[\prod_{j<i}(1 - q_j(t_j))]$$

$$- \mathbb{E}_{t_i}[q_i(t_i)]\mathbb{E}_{t_{<i}}[\prod_{j<i}(1 - q_j(t_j))]$$

$$= 1 - \sum_{j=1}^{i-1}\mathbb{E}_{t_j}[Q_j(t_j)] - \mathbb{E}_{t_i}[Q_i(t_i)]$$

$$= 1 - \sum_{j=1}^{i}\mathbb{E}_{t_j}[Q_j(t_j)]$$

where the first equality follows from independence, the second equality follows from linearity of expectations, and the fourth equality follows from the inductive step and the definition of the interim allocation.   □

It follows from Lemma 1 that the set of constraints (1) can be reduced to

$$Q_i(t_i) + \sum_{j<i}\mathbb{E}_{t_j}[Q_j(t_j)] \leq 1 \quad \forall i \in I \quad \forall t_i \in T_i$$

$$0 \leq A_i(t_i) \leq Q_i(t_i) \quad \forall i \in I \quad \forall t_i \in T_i$$

Using the reduced form representation we can formulate the principal's problem as the following linear program (denoted LP):

$$\max_{Q,A} \sum_{i \in I} \mathbb{E}_{t_i} [t_i Q_i(t_i) - c A_i(t_i)] \tag{LP}$$

$$s.t.\ Q_i(t_i) + \sum_{j < i} \mathbb{E}_{t_j} [Q_j(t_j)] \leq 1 \quad \forall i \quad \forall t_i \in T_i$$

$$Q_i(t_i) \geq Q_i(t_i') - A_i(t_i') \quad \forall i \in I \quad \forall t_i, t_i' \in T_i$$

$$0 \leq A_i(t_i) \leq Q_i(t_i) \quad \forall i \in I \quad \forall t_i \in T_i$$

2.2   Prophet Inequality Under Truthful Reporting

We first provide an alternative proof of a classic prophet inequality for the selection problem, when the agents truthfully report their types. A prophet inequality lower bounds the expected value of the number on which one stops with respect to the expected maximum value in hindsight. The maximum value in hindsight is the expected value of the nth order statistic. [16] obtained a tight bound of $1/2$ for this problem. In words, the optimal reward of the stopping problem is at least half the size of the expected value of the largest of the n random numbers. The study of prophet inequalities has attracted an enthusiastic following. [13] as well as [18] provide surveys.

This proof below will be replicated to show similar results when incentive constraints are present. Using the reduced form representation the problem can be restated as a linear program:

$$\max_{Q} \sum_{i \in I} \mathbb{E}_{t_i} [t_i Q_i(t_i)]$$

$$s.t.\ Q_i(t_i) + \sum_{j < i} \mathbb{E}_{t_i} [Q_i(t_i)] \leq 1 \quad \forall i \quad \forall t_i \in T_i$$

$$Q_i(t_i) \geq 0 \quad \forall i \quad \forall t_i \in T_i$$

Notice that the incentive compatibility constraints are absent.

We think of the problem of choosing the maximum value in hindsight as the off-line version of our problem where all agents are present at the same time. Here the principal can choose which agent should receive the object based on the reported types of *all* agents.

Theorem 1. *The optimal online algorithm achieves at least $1/2$ of the performance of the optimal offline algorithm in expectation.*

Proof. Let $Q_i^*(t_i)$ be the interim expected probability with which agent i with type t_i receives the item in the optimal off-line solution. The expected total value to the principal is given by

$$\sum_{i \in I} \mathbb{E}_{t_i} [Q_i^*(t_i) t_i].$$

Pick online values $Q_i(t_i) = \frac{1}{2}Q_i^*(t_i)$. It is clear that the objective function with respect to the reduced form for both problems is linear and coincides. Thus, a simple scaling approximates the optimal objective:

$$\sum_{i \in I} \mathbb{E}_{t_i}[Q_i(t_i)t_i] = \frac{1}{2}\sum_{i \in I} \mathbb{E}_{t_i}[Q_i^*(t_i)t_i].$$

The proposed solution is feasible for the online problem. In detail, $Q_i(t_i) + \sum_{j<i} \mathbb{E}_{t_j}[Q_j(t_j)] = \frac{1}{2}Q_i^*(t_i) + \frac{1}{2}\sum_{j<i}\mathbb{E}_{t_j}[Q_j^*(t_j)] \leq 1$. The last inequality holds since $Q_i^*(t_i) \leq 1$ and the expected offline allocation for the first $i - 1$ agents is also less than 1. □

3 The Optimal Mechanism

In this section we derive the optimal interim allocation and verification rules. The interim verification rule will be derived as a function of the optimal interim allocation rule. The optimal interim allocation rule will be given as a solution to a linear program. The actual allocation and verification rules can be obtained from the interim ones via Lemma 1.

Given the optimal interim allocation rule, the optimal interim verification rule can be deduced from the incentive constraints in (LP). They can be reduced to the following:

$$\min_{t_i} Q_i(t_i) \geq Q_i(t_i') - A_i(t_i') \quad \forall i \in I \quad \forall t_i' \in T_i \tag{5}$$

Therefore, at optimality,

$$A_i(t_i) = Q_i(t_i) - \min_{t_i'} Q_i(t_i'). \tag{6}$$

We use (6) to eliminate the verification variables from (LP). We also introduce a new set of variables $\{\phi_i\}_{i \in I}$ accounting for the minimum interim allocation per agent. For a given $\{\phi_i\}_{i \in I}$, the optimal interim allocation rule is given by the following linear program denoted LP (ϕ):

$$V(\phi) = \max_Q \sum_{i \in I} \mathbb{E}_{t_i}[(t_i - c)Q_i(t_i)] + c\phi_i \tag{7}$$

$$s.t. \ Q_i(t_i) + \sum_{j<i} \mathbb{E}_{t_j}[Q_j(t_j)] \leq 1 \quad \forall i \quad \forall t_i \in T_i$$

$$Q_i(t_i) \geq \phi_i \geq 0 \quad \forall i \in I \quad \forall t_i \in T_i$$

Whenever $\sum_i \phi_i \leq 1$, $V(\phi)$ is well defined, otherwise there is no feasible solution. This is because $1 \geq \sum_i \mathbb{E}_{t_i}[Q_i(t_i)] \geq \sum_i \phi_i$ should hold. Hence, the problem of finding the optimal mechanism reduces to

$$\max_{\phi: \sum_{i \in I} \phi_i \leq 1} V(\phi),$$

which is also a linear program. We now characterize the optimal interim allocation and verification rules given ϕ.

Lemma 2. *The optimal solution of LP(ϕ) is monotonic in type, i.e.*

$$Q_i(t_i) \leq Q_i(t_i') \quad \forall i \in I \quad \forall t_i \leq t_i'$$

Proof. Suppose not. Then, there is an i and pair (t_i, t_i') such that $Q_i(t_i) > Q_i(t_i')$. We pick an $\epsilon > 0$ such that

- $Q_i(t_i) - \frac{\epsilon}{f_i(t_i)} \geq Q_i(t_i')$,
- $Q_i(t_i') + \frac{\epsilon}{f_i(t_i')} \leq Q_i(t_i)$.

If we reduce $Q_i(t_i)$ by $\frac{\epsilon}{f_i(t_i)}$ and increase $Q_i(t_i')$ by $\frac{\epsilon}{f_i(t_i')}$, feasibility is preserved. The objective function value increases by $\epsilon(t_i' - t_i) > 0$, which is a contradiction.   □

Hence, there exists a threshold $\hat{t}_i$ for all i such that $Q_i(t_i) = \phi_i$ for $t_i \leq \hat{t}_i$ and $Q_i(t_i) \geq \phi_i$ otherwise.

We show the optimal strategy is a threshold strategy in each round. A transformation of variables will prove convenient:

$$Q_i(t_i) = \phi_i + x_i(t_i) \tag{8}$$

Given ϕ, we can find the optimal strategy by identifying the solution to the following linear program:

$$\max_x \sum_{i \in I} \mathbb{E}_{t_i}[x_i(t_i)(t_i - c)] \tag{XP}$$

$$s.t. \ x_i(t_i) + \sum_{j < i} \mathbb{E}_{t_j}[x_j(t_j)] \leq 1 - \sum_{j \leq i} \phi_j \quad \forall i \in I \quad \forall t_i \in T_i$$

$$x_i(t_i) \geq 0 \quad \forall i \in I \quad \forall t_i \in T_i$$

(XP) is a simplified version of LP (ϕ) given by the transformation defined in (8).

Lemma 3. *Suppose that Q is the optimal solution to LP (ϕ). Then, for each agent i, there exists a threshold $\hat{t}_i$, such that*

$$Q_i(t_i) = \begin{cases} 1 - \sum_{j < i} \mathbb{E}_{t_j}[Q_j(t_j)] & \text{if } t_i \geq \hat{t}_i \\ \phi_i & \text{otherwise} \end{cases} \tag{9}$$

Proof. Suppose we are interested in the allocation and verification rules when we reach agent i. Fix all other variables. We are interested in solving the following linear program

$$\max_{x_i} \mathbb{E}_{t_i}[x_i(t_i)(t_i - c)]$$

$$s.t. \ x_i(t_i) \leq 1 - \sum_{j \leq i} \phi_j - \sum_{j < i} \mathbb{E}_{t_j}[x_j(t_j)] \quad \forall i \in I \quad \forall t_i \in T_i$$

$$\mathbb{E}_{t_i}[x_i(t_i)] \leq 1 - \sum_{j \leq k} \phi_j - x_k(t_k) - \sum_{j < k, j \neq i} \mathbb{E}_{t_j}[x_j(t_j)] \quad \forall k > i \quad \forall t_k \in T_k$$

$$x_i(t_i) \geq 0 \quad \forall i \in I \quad \forall t_i \in T_i$$

Now, it is clear that the optimal solution can actually be characterized by a threshold. All high types will be assigned their upper limit till the constraint on the aggregate allocation binds. Thus, the optimal solution x is given by

$$x_i(t_i) = \begin{cases} 1 - \sum_{j \leq i} \phi_j - \sum_{j < i} \mathbb{E}_{t_j}[x_j(t_j)] & \text{if } t_i \geq \hat{t}_i \\ 0 & \text{otherwise} \end{cases}$$

Returning back to Q variables completes the proof.   □

Lemma 1 allows us to derive the actual allocation and verification rules given the interim ones. We also provide the form for the actual allocations, given the characterization of the optimal interim allocation in terms of parameters $\phi, \hat{t}$,

Corollary 1. *For each agent i there exists a threshold $\hat{t}_i$ and constant α_i, such that the optimal actual allocation can be written as follows:*

$$q_i(t_i) = \begin{cases} 1 & \text{if } t_i \geq \hat{t}_i \\ \alpha_i & \text{otherwise} \end{cases} \qquad a_i(t_i) = \begin{cases} 1 - \alpha_i & \text{if } t_i \geq \hat{t}_i \\ 0 & \text{otherwise} \end{cases}$$

Proof. We use Lemma 1 to get the form of the actual allocation:

$$q_i(t_i) = \frac{Q_i(t_i)}{1 - \sum_{j < i} \mathbb{E}_{t_j}[Q_j(t_j)]} = \begin{cases} \frac{1 - \sum_{j < i} \mathbb{E}_{t_j}[Q_j(t_j)]}{1 - \sum_{j < i} \mathbb{E}_{t_j}[Q_j(t_j)]} & \text{if } t_i \geq \hat{t}_i \\ \frac{\phi_i}{1 - \sum_{j < i} \mathbb{E}_{t_j}[Q_j(t_j)]} & \text{otherwise} \end{cases} = \begin{cases} 1 & \text{if } t_i \geq \hat{t}_i \\ \alpha_i & \text{otherwise} \end{cases}$$

where

$$\alpha_i = \frac{\phi_i}{1 - \sum_{j < i} \mathbb{E}_{t_j}[Q_j(t_j)]}.$$

The form for the actual verification rule follows by (6).   □

Before continuing, we summarize the roadmap for determining the optimal allocation and verification rules:

1. Solve the linear program $\max_{\phi: \sum_{i \in I} \phi_i \leq 1} V(\phi)$ to find the optimal interim allocation rule Q.
2. Derive the optimal interim verification rule A from Eq. (6).
3. Derive the optimal actual allocation and verification rules q, a from the interim ones Q, A, via Lemma 1.

3.1   Identical Distributions

We examine the case of identical distributions, i.e., $T_i = T$ and $f_i(t) = f(t)$ for all $i \in I, t_i \in T$. In this case, we can give a neat representation of the optimal strategy.

Let $\mu = \mathbb{E}_t[t]$. Now, LP (ϕ) can be written as

$$\max_x \sum_{i \in I} \mathbb{E}_t[x_i(t)(t-c)] + \mu \sum_i \phi_i$$

$$s.t. \ x_i(t) + \sum_{j<i} \mathbb{E}_t[x_j(t)] \leq 1 - \sum_{j \leq i} \phi_j \quad \forall i \in I \quad \forall t \in T$$

$$x_i(t) \geq 0 \quad \forall i \in I \quad \forall t \in T$$

Let ϕ^* be the vector ϕ that maximizes $V(\phi)$. We can set $\phi_i = 0$ for all $i < n$ and $\phi_n = \sum_i \phi_i^*$. The objective function does not change while the right hand side of all inequalities in the above LP increases, but the one for $i = n$, which remains the same. Thus, we can restrict our attention to $\phi_i = 0$ for $i < n$. In the optimal solution of the initial LP, the last agent's rule is constrained as follows:

$$Q_n(t_n) \leq 1 - \sum_{j<n} \mathbb{E}_{t_j}[Q_j(t_j)] \quad \forall t_n.$$

Since the right hand side coincides for all types, and the objective function is increasing in the allocation rule, the constraint binds for all types. This means that $Q_n^*(t_n) = \phi_n^*$, which implies that $x_n(t) = 0$ for all t.

We can now reduce LP (ϕ^*) to the following linear program to determine the strategies for the first $n-1$ agents:

$$\max_x \sum_{i \in I \setminus \{n\}} \mathbb{E}_t[x_i(t)(t-c)]$$

$$s.t. \ x_i(t) + \sum_{j<i} \mathbb{E}_t[x_j(t)] \leq 1 - \phi_n^* \quad \forall i \in I \setminus \{n\} \quad \forall t \in T$$

$$x_i(t) \geq 0 \quad \forall i \in I \setminus \{n\} \quad \forall t \in T$$

By normalizing the right hand sides of this linear program to 1, we can interpret it as arising from a cardinal secretary problem with $n-1$ secretaries, where the value of each 'secretary' is $t-c$, drawn according to a density function f. In case the object is still available in the last round it is given to the last agent.

3.2   Prophet Inequality

We derive a prophet inequality for the setting with verification using the reduced form.[1] It scales the optimal offline solution so as to make it a feasible solution for the online setting. This technique can also be used in the standard setting.

Theorem 2. *The optimal online algorithm achieves at least 1/2 of the performance of the optimal offline algorithm on expectation.*

---

[1] This result does not assume that the distribution of types is IID.

Proof. Let $Q_i^*(t_i)$ be the interim expected probability with which agent i with type t_i receives the item in the optimal off-line solution. Let $\phi_i^* = \inf_{t_i} Q_i^*(t_i)$ as proposed in [4]. The expected total value to the principal is given by

$$\sum_{i \in I} [\mathbb{E}_{t_i}[Q_i^*(t_i)(t_i - c)] + \phi_i^* c].$$

Pick online values $Q_i(t_i) = \frac{1}{2}Q_i^*(t_i)$ and $\phi_i = \frac{1}{2}\phi_i^*$. It is clear that the objective function with respect to the reduced form for both problems is linear and coincides. Thus, a simple scaling approximates the optimal objective:

$$\sum_{i \in I} [\mathbb{E}_{t_i}[Q_i(t_i)(t_i - c)] + \phi_i c] = \frac{1}{2} \sum_{i \in I} [\mathbb{E}_{t_i}[Q_i^*(t_i)(t_i - c)] + \phi_i^* c]$$

It suffices to prove that the proposed solution is feasible for the online problem.

- $Q_i(t_i) + \sum_{j<i} \mathbb{E}_{t_j}[Q_j(t_j)] = \frac{1}{2}Q_i^*(t_i) + \frac{1}{2}\sum_{j<i} \mathbb{E}_{t_j}[Q_j^*(t_j)] \leq 1$: This holds since $Q_i^*(t_i) \leq 1$ and the expected offline allocation for the first $i-1$ agents is also less than 1.
- $Q_i(t) \geq \phi_i$: The constraint coincides with the offline constraint. Nothing changes by scaling both sides of the inequality. □

4   Limited Punishment

We say that punishment is limited if the principal cannot reduce an agent's payoff to his outside option by punishing him. If we interpret verification as acquiring information, then punishment can be limited because information is imperfect.[2] We assume that punishment is proportional to the private benefit enjoyed by the agent from receiving the object. If v_i is the private benefit enjoyed by agent i, punishment is $k_i v_i$, where each $k_i \in [0, 1]$. These are the same assumptions as in [17]. As we show below, limited punishment will cause the principal to 'ration at the top' as well. All types above some threshold face the same probability of receiving the good.

By the Revelation Principle we can focus on direct mechanisms. In this case, if an agent is inspected, it is optimal to penalize him if and only if he is found to have lied. After the allocation is made, the planner will observe the agent's type and destroy a fraction k_i of the agent's payoff. A direct mechanism specifies for each profile of type reports the probability $q_i(t_i)$ that the good is assigned to agent i conditional on the event that it is not already assigned. These variables must satisfy the following feasibility conditions:

$$0 \leq q_i(t_i) \leq 1 \quad \forall i \in I \quad \forall t_i \in T_i \tag{10}$$

---

[2] We take verification cost and punishment level as exogenous but it is possible that the principal can get more precise information by incurring a higher information acquisition cost, which, in turn, leads to a more severe expected punishment. The results in this paper readily extend to the case where the principal can jointly optimize over verification cost and punishment level.

The incentive compatibility constraints are as follows:

$$v_i q_i(t_i) \geq (v_i - k_i v_i) q_i(t_i') \Rightarrow$$

$$q_i(t_i) \geq (1 - k_i) q_i(t_i') \quad \forall i \in I \quad \forall t_i, t_i' \in T_i \tag{11}$$

The principal would like to choose the allocation probabilities q to maximize:

$$\sum \mathbb{E}_{t_{<i}} [\prod_{j<i} (1 - q_j(t_j)) \mathbb{E}_{t_i} [t_i q_i(t_i)]].$$

As before we work with a reduced form representation. This allows us to formulate the optimal mechanism as the following linear program:

$$\max_Q \sum_{i \in I} \mathbb{E}_{t_i} [t_i Q_i(t_i)]$$

$$s.t. \ Q_i(t_i) + \sum_{j<i} \mathbb{E}_{t_i} [Q_i(t_i)] \leq 1 \quad \forall i \quad \forall t_i \in T_i$$

$$Q_i(t_i) \geq (1 - k_i) Q_i(t_i') \quad \forall i \quad \forall t_i \in T_i \quad \forall t_i' \in T_i$$

$$Q_i(t_i) \geq 0 \quad \forall i \quad \forall t_i \in T_i$$

4.1   The Optimal Mechanism

We simplify the incentive constraint, as in [20]. We include the proof for completeness.

Lemma 4. *An allocation rule satisfies incentive compatibility if and only if for all i there exists χ_i such that*

$$(1 - k_i) \chi_i \leq Q_i(t_i) \leq \chi_i \quad \forall t_i \in T_i \tag{12}$$

Proof. If IC holds then (12) holds with $\chi_i = \sup_{t_i} Q_i(t_i)$. Conversely, if (12) holds for some χ_i, then it also holds with $\chi_i' = \sup_{t_i} Q_i(t_i)$, which implies incentive compatibility. $\square$

We now write down a linear program which finds the optimal strategy. We know that for optimal χ this linear program is going to return the optimal strategy.

$$\max_{Q, \chi} \sum_{i \in I} \mathbb{E}_{t_i} [t_i Q_i(t_i)]$$

$$s.t. \ Q_i(t_i) + \sum_{j<i} \mathbb{E}_{t_i} [Q_i(t_i)] \leq 1 \quad \forall i \in I \quad \forall t_i \in T_i$$

$$(1 - k_i) \chi_i \leq Q_i(t_i) \leq \chi_i \quad \forall i \in I \quad \forall t_i \in T_i$$

$$Q_i(t_i) \geq 0 \quad \forall i \in I \quad \forall t_i \in T_i$$

We now describe the optimal strategy.

Lemma 5. *Suppose that Q is the optimal online solution. Let $\chi_i = \sup\limits_{t_i \in T_i} Q_i(t_i)$.*
Then for each agent i, there exists a threshold $\hat{t}_i$ such that

$$Q_i(t_i) = \begin{cases} \chi_i & \text{if } t_i \geq \hat{t}_i \\ (1 - k_i)\chi_i & \text{otherwise} \end{cases} \tag{13}$$

Proof. Suppose we are interested in the allocation rule when we reach agent i. Fix all other variables at their optimal value. We are interested in solving the following linear program:

$$\max_{Q_i} \mathbb{E}_{t_i}[t_i Q_i(t_i)]$$

$$s.t. \; Q_i(t_i) \leq 1 - \sum_{j<i} \mathbb{E}_{t_j}[Q_j(t_j)] \quad \forall i \in I \quad \forall t_i \in T_i$$

$$\mathbb{E}_{t_i}[Q_i(t_i)] \leq 1 - Q_k(t_k) - \sum_{j<k, j\neq i} \mathbb{E}_{t_j}[Q_j(t_j)] \quad \forall k > i \quad \forall t_k \in T_k$$

$$(1 - k_i)\chi_i \leq Q_i(t_i) \leq \chi_i \quad \forall t_i \in T_i$$
$$Q_i(t_i) \geq 0 \quad \forall t_i \in T_i$$

Now, it is clear that the optimal solution can be characterized by a threshold policy. All high types will be assigned their upper limit till a constraint for the aggregate allocation binds. The optimal online solution has the following form:

$$Q_i(t_i) = \begin{cases} \min\{\chi_i, 1 - \sum_{j<i} \mathbb{E}_{t_j}[Q_j(t_j)]\} & \text{if } t_i \geq \hat{t}_i \\ (1 - k_i)\chi_i & \text{otherwise} \end{cases} \tag{14}$$

The upper limit can be simplified. We prove that

$$\chi_i \leq 1 - \sum_{j<i} \mathbb{E}_{t_j}[Q_j(t_j)] \quad \forall i \in I.$$

Suppose otherwise. We pick $\chi' = 1 - \sum_{j<i} \mathbb{E}_{t_j}[Q_j(t_j)]$. This makes the constraints less strict since the upper bound remains the same but the lower bound reduces. Thus we can reduce the allocation for lower types and increase the allocation of higher types while holding the aggregate allocation steady. This is a contradiction since such a change will increase total welfare. □

In the limited penalties case the actual allocation will have a slightly different form.

Corollary 2. *For each agent i there exists a threshold $\hat{t}_i$, and constant β_i, such that the optimal actual allocation rule can be written as follows:*

$$q_i(t_i) = \begin{cases} \beta_i & \text{if } t_i \geq \hat{t}_i \\ (1 - k_i)\beta_i & \text{otherwise} \end{cases}$$

Proof. We use Lemma 1 to get the form of the actual allocation rule:

$$q_i(t_i) = \frac{Q_i(t_i)}{1 - \sum\limits_{j<i} \mathbb{E}_{t_j}[Q_j(t_j)]} = \begin{cases} \frac{\chi_i}{1 - \sum\limits_{j<i} \mathbb{E}_{t_j}[Q_j(t_j)]} & \text{if } t_i \geq \hat{t}_i \\ \frac{(1-k_i)\chi_i}{1 - \sum\limits_{j<i} \mathbb{E}_{t_j}[Q_j(t_j)]} & \text{otherwise} \end{cases} = \begin{cases} \beta_i & \text{if } t_i \geq \hat{t}_i \\ (1 - k_i)\beta_i & \text{otherwise} \end{cases}$$

where $\beta_i = \frac{\chi_i}{1 - \sum\limits_{j<i} \mathbb{E}_{t_j}[Q_j(t_j)]}$. □

4.2 Prophet Inequality

We use the same machinery as before to further illustrate that extra constraints that restrict the optimal solution in both offline and online cases, do not have an effect on the prophet inequality.

Theorem 3. *The optimal online algorithm achieves at least 1/2 of the performance of the optimal offline algorithm on expectation.*

Proof. Let $Q_i^*(t_i)$ be the interim probability with which agent i with type t_i receives the item in the optimal off-line solution. Let $\chi_i^* = \sup\limits_{t_i \in T_i} Q_i^*(t_i)$ as proposed in [20]. The expected total value to the principal is given by

$$\sum_{i \in I} \mathbb{E}_{t_i}[t_i Q_i^*(t_i)]$$

Pick online values $Q_i(t_i) = \frac{1}{2}Q_i^*(t_i)$ and $\chi_i = \frac{1}{2}\chi_i^*$ for all $i \in I$. It is clear that the objective function with respect to the reduced form for both problems is linear and coincides. Thus, a simple scaling approximates the optimal objective:

$$\sum_{i \in I} \mathbb{E}_{t_i}[t_i Q_i(t_i)] = \frac{1}{2} \sum_{i \in I} \mathbb{E}_{t_i}[t_i Q_i^*(t_i)]$$

It suffices to prove that the proposed solution is feasible for the online problem.

- $Q_i(t_i) + \sum\limits_{j<i} \mathbb{E}_{t_j}[Q_j(t_j)] = \frac{1}{2}Q_i^*(t_i) + \frac{1}{2}\sum\limits_{j<i} \mathbb{E}_{t_j}[Q_j^*(t_j)] \leq 1$: This holds since $Q_i^*(t_i) \leq 1$ and the expected offline allocation for the first $i-1$ agents is also less than 1.
- $(1 - k_i)\chi_i \leq Q_i(t) \leq \chi_i$: The constraint coincides with the offline constraint. Nothing changes by scaling both sides of the inequalities. □

References

1. Alaei, S.: Bayesian combinatorial auctions: expanding single buyer mechanisms to many buyers. In: Proceedings of the 52nd IEEE Annual Symposium on Foundations of Computer Science, pp. 512–521 (2011)

2. Alaei, S., Hajiaghayi, M., Liaghat, V.: Online prophet-inequality matching with applications to ad allocation. In: Proceedings of the 13th ACM Conference on Electronic Commerce, pp. 18–35 (2012)
3. Ball, I., Kattwinkel, D.: Probabilistic verification in mechanism design. In: Proceedings of the 20th ACM Conference on Economics and Computation, pp. 389–390 (2019)
4. Ben-Porath, E., Dekel, E., Lipman, B.L.: Optimal allocation with costly verification. Am. Econ. Rev. **104**(12), 3779–3813 (2014)
5. Border, K.C.: Implementation of reduced form auctions: a geometric approach. Econometrica **59**, 1175–1187 (1991)
6. Caragiannis, I., Elkind, E., Szegedy, M., Yu, L.: Mechanism design: from partial to probabilistic verification. In: Proceedings of the 13th ACM Conference on Electronic Commerce, pp. 266–283 (2012)
7. Chawla, S., Hartline, J.D., Malec, D.L., Sivan, B.: Multi-parameter mechanism design and sequential posted pricing. In: Proceedings of the 42nd ACM Symposium on Theory of Computing, pp. 311–320 (2010)
8. Duetting, P., Feldman, M., Kesselheim, T., Lucier, B.: Prophet inequalities made easy: stochastic optimization by pricing non-stochastic inputs. In: 58th IEEE Annual Symposium on Foundations of Computer Science, pp. 540–551 (2017)
9. Feldman, M., Svensson, O., Zenklusen, R.: A simple O(loglog(rank))-competitive algorithm for the matroid secretary problem. In: Proceedings of the 26th Annual ACM-SIAM Symposium on Discrete Algorithms, pp. 1189–1201 (2015)
10. Gale, D., Hellwig, M.: Incentive-compatible debt contracts: the one-period problem. Rev. Econ. Stud. **52**(4), 647–663 (1985)
11. Gershkov, A., Moldovanu, B.: Dynamic Allocation and Pricing: A Mechanism Design Approach. MIT Press, Cambridge (2014)
12. Hajiaghayi, M.T., Kleinberg, R., Sandholm, T.: Automated online mechanism design and prophet inequalities. In: Proceedings of the 22nd National Conference on Artificial Intelligence, pp. 58–65 (2007)
13. Hill, T.P., Kertz, R.P.: A survey of prophet inequalities in optimal stopping theory. Contemp. Math. **125**, 191–207 (1992)
14. Kleinberg, R., Weinberg, S.M.: Matroid prophet inequalities. In: Proceedings of the 44th Annual ACM Symposium on Theory of Computing, pp. 123–136 (2012)
15. Krengel, U., Sucheston, L.: Semiamarts and finite values. Bull. Am. Math. Soc. **83**(4), 745–747 (1977)
16. Krengel, U., Sucheston, L.: On semiamarts, amarts, and processes with finite value. Adv. Prob. Related Topics **4**, 197–266 (1978)
17. Li, Y.: Mechanism design with costly verification and limited punishments. Working Paper (2019)
18. Lucier, B.: An economic view of prophet inequalities. SIGecom Exch. **16**(1), 24–47 (2017)
19. Mookherjee, D., Png, I.: Optimal auditing, insurance, and redistribution. Q. J. Econ. **104**(2), 399–415 (1989)
20. Mylovanov, T., Zapechelnyuk, A.: Optimal allocation with ex post verification and limited penalties. Am. Econ. Rev. **107**(9), 2666–94 (2017)
21. Samuel-Cahn, E.: Comparison of threshold stop rules and maximum for independent nonnegative random variables. Ann. Probab. **12**(4), 1213–1216 (1984)
22. Townsend, R.M.: Optimal contracts and competitive markets with costly state verification. J. Econ. Theor. **21**(2), 265–293 (1979)
23. Vohra, R.V.: Dynamic mechanism design. Surv. Oper. Res. Manag. Sci. **17**(1), 60–68 (2012)

Strategyproof Facility Location
for Three Agents on a Circle

Reshef Meir[(✉)]

Technion—Israel Institute of Technology, Haifa, Israel
`reshefm@ie.technion.ac.il`

Abstract. We consider the facility location problem in a metric space, focusing on the case of three agents. We show that selecting the reported location of each agent with probability proportional to the distance between the other two agents results in a mechanism that is strategyproof in expectation, and dominates the random dictator mechanism in terms of utilitarian social welfare. We further improve the upper bound for three agents on a circle to $\frac{7}{6}$ (whereas random dictator obtains $\frac{4}{3}$); and provide the first lower bounds for randomized strategyproof facility location in any metric space, using linear programming.

1 Introduction

In a facility location problem, a central authority faces a set of agents who report their locations in some space, and needs to decide where to place a facility. It is typically assumed that each agent i wants the facility to be placed as close as possible to her own location a_i. We want a *strategyproof* mechanism, such that reporting the truthful location is a weakly dominant strategy for every agent. The designer may have additional goals, where the most common one is to minimize the utilitarian social cost—the sum of distances to agents' locations.

Strategyproof facility location mechanisms have been studied at least since the mid-20th century [5]. In 2009, the agenda of approximation mechanisms without money was made explicit in a paper by Procaccia and Tennenholtz [28, 29], who used facility location as their primary domain of demonstration due to its simplicity. Moreover, facility location is often a bridge between mechanism design and social choice [7,11,23] and has applications to transport [24], disaster relief [14] and more. Facility location is thus often used as a testbed for ideas and techniques in mechanism design and noncooperative multiagent systems.

Most problems that include a single facility are by now well understood. For example, all deterministic strategyproof mechanisms on continuous and on discrete lines have been characterized [9,30], and it is well known that selecting the median agent location is both strategyproof and optimal in terms of utilitarian social cost [25,28]. One strand of the literature seeks to characterize domains where median-like mechanisms exist [17,27].

For other domains, e.g. graphs that contain cycles, research following [28] has focused on the minimal social cost that can be guaranteed by strategyproof

© Springer Nature Switzerland AG 2019
D. Fotakis and E. Markakis (Eds.): SAGT 2019, LNCS 11801, pp. 18–33, 2019.
https://doi.org/10.1007/978-3-030-30473-7_2

mechanisms. For deterministic mechanisms even the existence of a single cycle in a graph entails that any strategyproof mechanism must be dictatorial on a subdomain, and thus has an approximation ratio that increases linearly with the number of agents n [9,30]. Many variations of the problem have since been explored in the AI and multiagent systems community, including multiple facilities [3,10,31], complex incentives and forms of strategic behavior [13,32–34], and alternative design goals [1,12,19]. The circle in particular has received much attention in the facility location literature [1,2,6,9,30], both because it is the simplest graph for which median-like mechanisms cannot work, and because it is an abstraction of actual problems like selecting a time-of-the-day or a server in a ring of computers.

Yet, the fundamental strategyproof facility location problem for *randomized mechanisms* remains almost unscathed. It is easy to show that the *random dictator* (RD) mechanism obtains an approximation ratio of $2 - \frac{2}{n}$ for any metric space [1,23], and of course that 1 is a lower bound. However except for lines and trees (where the deterministic Median mechanism is optimal), nothing else is known.

To the best of our knowledge, the literature does not mention mechanisms that approximate the optimal social cost better than RD even for specific spaces like the circle, nor is there any lower bound higher than 1.[1] The current paper focuses on narrowing this gap by proving tighter upper and lower bounds for three agents.

A variant of the problem on which there was more (negative) progress is when we allow arbitrary constraints on the location of the facility (e.g., where agents can be placed anywhere on a graph, but only 5 vertices are valid locations for the facility). In the constrained variant, the RD mechanism obtains $3 - \frac{2}{n}$ approximation and this is known to be tight for all strategyproof mechanisms. The upper bound holds for any metric space [23], whereas the lower bound requires specific constructions on the n-dimensional binary cube [11,22]. Anshelevich and Postl [4] show a smooth transition of the RD approximation ratio from $2 - \frac{2}{n}$ to $3 - \frac{2}{n}$ as the location of the facility becomes more constrained. See [20] (Section 5.3) for an overview of approximation results for a single facility.

1.1   Contribution

Our main contribution is the introduction of two randomized mechanisms that beat the random dictator (RD) mechanism on a circle: the *Proportional Circle Distance* (PCD) mechanism, which selects each reported location a_i with probability proportional to the length L_i of the arc facing agent i; and the q-Quadratic Circle Distance mechanism (q-QCD) where the probability of selecting a_i is proportional to $(\max\{(L_i)^2, q^2\})$.

---

[1] Alon et al. [1] proposed a randomized strategyproof mechanism specifically for circles, called the *hybrid mechanism*. They showed that it obtains the best possible approximation ratio for the *minimax cost*, yet for the social cost it achieves a poor approximation ratio of $\frac{n-1}{2}$.

We prove that PCD is strategyproof for any odd number of agents. For 3 agents, we show that PCD obtains an approximation ratio of $\frac{5}{4}$ on the circle (in contrast to $\frac{4}{3}$ by RD), and has a natural extension that is strategyproof and weakly dominates RD on any metric space. The $\frac{1}{4}$-QCD mechanism is also strategyproof for 3 agents, and obtains an approximation ratio of $\frac{7}{6}$ on the circle.

For any finite graph with m vertices, there is a linear program of polynomial size that can compute the optimal randomized strategyproof mechanism. We use such programs to obtain first (but non-tight) lower bounds on the approximation ratio of any strategyproof mechanism on circles and on general graphs. See Table 1 for a summary.

Some of our proofs use a combination of formal analysis and computer optimization. For most proofs we provide only a sketch due to space constraints, but all complete proofs appear in the full version on arXiv [21].

2 Preliminaries

A domain of facility location problems is given by $\langle \mathcal{X}, d \rangle$, where $\mathcal{X}$ is a set, and $d : \mathcal{X} \times \mathcal{X} \to \mathbb{R}_+$ is a distance metric. In this paper, $\mathcal{X}$ is a (discrete or continuous) graph, and $d(x, y)$ is the length of the shortest path between x and y. An *instance* in the domain $\langle \mathcal{X}, d \rangle$ is given by a profile $\boldsymbol{a} \in \mathcal{X}^n$, where n is the number of agents (implicit in the profile).

We denote by $\boldsymbol{a}_{-i}$ the partial profile that includes all entries in $\boldsymbol{a}$ except a_i.

A n-agent *facility location mechanism* in domain $\langle \mathcal{X}, d \rangle$ (or simply a *mechanism*) is a function $f : \mathcal{X}^n \to \Delta(\mathcal{X})$, where $\Delta(\mathcal{X})$ is the set of distributions over $\mathcal{X}$. We denote the resulting lottery of applying f to profile $\boldsymbol{a}$ by $f_{\boldsymbol{a}}$. Mechanism f is *deterministic* if $f_{\boldsymbol{a}}$ is degenerated for any profile $\boldsymbol{a}$, in which case we denote $f_{\boldsymbol{a}} \in \mathcal{X}$. We denote the probability that mechanism f selects z on profile $\boldsymbol{a}$ by $f_{\boldsymbol{a}}(z) \in [0, 1]$.

When placing a facility on $z \in \mathcal{X}$, an agent located at a_i suffers a cost of $d(a_i, z)$. We denote by $c_i(\boldsymbol{a}, h) = E_{z \sim h}[d(a_i, z)]$ the expected cost of agent i in profile $\boldsymbol{a}$, when the facility is placed according to lottery h.

The (utilitarian) *social cost* of lottery h in profile $\boldsymbol{a}$ is denoted by $SC(\boldsymbol{a}, h) = \sum_{i \leq n} c_i(\boldsymbol{a}, h) = E_{z \sim h}[\sum_{i \leq n} d(a_i, z)]$.

We omit the parameter $\boldsymbol{a}$ from the last two definitions when clear from context. We also abuse notation by writing $c_i(\boldsymbol{a}, z), SC(\boldsymbol{a}, z)$ for a specific location $z \in \mathcal{X}$ rather than a lottery.

We denote by $OPT(\boldsymbol{a}) = \inf_{z \in \mathcal{X}} SC(\boldsymbol{a}, z)$ the optimal social cost (note that this is w.l.o.g. obtained in a deterministic location).

2.1 Common Mechanism Properties

A mechanism f is *strategyproof* if for any profile $\boldsymbol{a} \in \mathcal{X}^n$, any agent i, and any alternative report $a_i' \in \mathcal{X}$, $c_i(\boldsymbol{a}, f_{\boldsymbol{a}}) \leq c_i(\boldsymbol{a}, f_{\boldsymbol{a}_{-i}, a_i'})$ (i.e., i does not gain in expectation).

A mechanism f is *ex-post strategyproof* if it is a lottery over strategyproof deterministic mechanisms. Note that ex-post strategyproofness implies strategyproofness, but not vice versa.

A mechanism f is *peaks-only* if $f_a(z) = 0$ for all $z \notin a$. That is, if the facility can only be realized on agents' locations.

Mechanism f *dominates* mechanism g, if for any profile a, $SC(a, f_a) \leq SC(a, g_a)$ and the inequality is strict for at least one profile.

Finally, a mechanism f has an approximation ratio of ϕ, if for any profile a, $SC(a, f_a) \leq \phi \cdot OPT(a)$.

Familiar mechanisms. The *Random Dictator (RD) mechanism* selects each agent i with equal probability, and places the facility on a_i. Clearly RD is ex-post strategyproof, and it is also known to be group-strategyproof [2] (that is, no subset of agents can gain by a joint deviation). Further, RD has an approximation ratio of $2 - \frac{2}{n}$ (i.e., $\frac{4}{3}$ for $n = 3$ agents), and this is tight for any metric space with at least two distinct locations [1].

On one-dimensional spaces, where agent locations can be sorted, the deterministic *median* mechanism simply picks the location of the median agent. The median mechanism is strategyproof and optimal [25]. The median mechanism also extends to trees, maintaining both properties [30].

3   Circles

A circle is the simplest graph for which there is no median. We denote by C_M the circle graph with M equi-distant vertices V. Assume w.l.o.g. that agents are indexed in clockwise order. For a profile $a \in V^n$, and two consequent agents $j, j+1$ (the addition is modulo n), we denote by $L_a(a_j, a_{j+1})$ (or just $L(a_j, a_{j+1})$ when the profile is clear from context) the length of the arc between these agents, that does not contain any other agent. When $L(a_j, a_{j+1})$ is not larger than a semicircle, then it also coincides with the distance $d(a_j, a_{j+1})$.

We also define $L_i = L(a_j, a_{j+1})$ where $j = i + \lfloor n/2 \rfloor$ (modulo n) to be the length of the arc that is "facing" agent i (although it may not be antipodal). For 3 agents this simply means that $L_1 = L(a_2, a_3), L_2 = L(a_3, a_1)$, and $L_3 = L(a_1, a_2)$. Also note that for 3 agents, the optimal location is always the agent facing the longest arc. See Fig. 1a.

3.1   Proportional Distance

Definition 1. *The* Proportional Circle Distance (PCD) mechanism *assigns the facility to each location a_i w.p.* $\frac{L_i}{\sum_{j \leq n} L_j}$.

Theorem 1. *PCD is strategyproof for any odd n.*

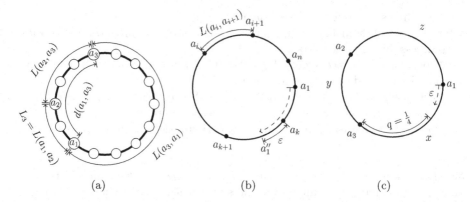

Fig. 1. Examples. (a) The circle C_{14}. Under PCD mechanism, the probabilities that the facility will be realized on a_1, a_2 and a_3, respectively, are $(\frac{3}{14}, \frac{9}{14}, \frac{2}{14})$. Under PD, the probabilities are $(\frac{3}{10}, \frac{5}{10}, \frac{2}{10})$. Under $\frac{1}{4}$-QCD, the probabilities are proportional to $((\frac{1}{4})^2, (\frac{9}{14})^2, (\frac{1}{4})^2)$, which gives us $(0.1161, 0.7677, 0.1161)$. The other two examples are used in the proof of Theorem 1 (b) and Case I of Theorem 2 (c).

Proof sketch. Suppose that a_1 tries the manipulate by moving (w.l.o.g.) clockwise to a_1'. Note that the probability of selecting agent 1 is not affected. Thus the agent's gain comes from increasing the selection probability of a closer agent at the expense of a farther agent. On the other hand, the agent's cost increases proportionally to her distance from her true location, and we show that this factor is more prominent. ☐

For 3 agents, the PCD mechanism guarantees an approximation ratio of $\frac{5}{4} = 1.25$. This is not hard to show, but will also follow from stronger results in Sect. 4. In Sect. 3.3 we further discuss what we know when $n > 3$.

3.2   The Quadratic Distance Mechanism

Since the optimal location with 3 agents is always the peak facing the longest arc, to improve the approximation ratio we must put more weight on peaks facing long arcs (at least in the "bad" instances).

Definition 2. *The* q-Quadratic Circle Distance (q-QCD) mechanism *considers the arc lengths* L_1, L_2, L_3. *It then assigns the facility to* a_i *w.p. proportional to*
$$s_i = \max\{(L_i)^2, q^2\}.$$

That is, q puts a lower bound on the probability that each agent is selected.

Theorem 2. *The* $\frac{1}{4}$-QCD *mechanism is strategyproof.*

Proof sketch. We denote $x = L_2, z = L_3$ and $y = L_1$. We denote by s_x, s_y, s_z the un-normalized weight assigned to the agent facing each respective arc, and by $p_i = \frac{s_i}{s}$ where $s = s_1 + s_2 + s_3$ the actual probability that i is selected. Note that $p_x + p_y + p_z = 1$. The notations are demonstrated on Fig. 1c.

The cost to agent 1 can be written as

$$c_1 = p_x z + p_z x = \frac{s_x z + s_z x}{s_x + s_y + s_z}.$$

Consider a step of size ε by agent 1 towards agent 3. Intuitively, moving towards the far agent only increases its probability of selection and is thus never beneficial for agent 1. Thus w.l.o.g. $z \geq x \geq \varepsilon$.

The move changes the arc lengths from (x, y, z) to $(x - \varepsilon, y, z + \varepsilon)$, and the cost changes accordingly to

$$c_1' = p_x' z + p_z' x + p_y' \varepsilon = \frac{s_{x-\varepsilon} z + s_{z+\varepsilon} x + s_y \varepsilon}{s_{x-\varepsilon} + s_y + s_{z+\varepsilon}}. \tag{1}$$

Our general strategy is to write the new cost c_1' as

$$c_1' = \frac{s_x z + s_z x + \varepsilon\gamma}{s_x + s_y + s_z + \varepsilon\theta} = \frac{c_1 s + \varepsilon\gamma}{s + \varepsilon\theta}, \tag{2}$$

where $\gamma, \theta \geq 0$. Then, we show that $\frac{\gamma}{\theta} \geq \frac{s_x z + s_z x}{s_x + s_y + s_z}(= c_1)$. This would conclude the proof, as it means that agent 1 does not gain:

$$c_1' = \frac{s_{x-\varepsilon} z + s_{z+\varepsilon} x + s_y \varepsilon}{s_{x-\varepsilon} + s_y + s_{z+\varepsilon}} \geq \frac{c_1 s + \varepsilon c_1 \theta}{s + \varepsilon\theta} = \frac{c_1 (s + \varepsilon\theta)}{s + \varepsilon\theta} = c_1. \tag{3}$$

The exact values of γ, θ depend on whether $x - \varepsilon \geq q$ (Case I, see Fig. 1c), $x \geq q > x - \varepsilon$ (Case II), or $q > x$ (Case III). We only show here Case I, which captures most of the proof's ideas. The proofs of the other cases are similar, with some caveats.

Suppose first that $y \geq q = \frac{1}{4}$ and that $z \leq \frac{1}{2}$ (we later show this does not matter). Then $s_x = x^2, s_y = y^2, s_z = z^2$, and

$$c_1 = p_x z + p_z x = \frac{x^2 z + z^2 x}{x^2 + z^2 + y^2}.$$

After the move, we have $s_x' = (x - \varepsilon)^2, s_z' = (z + \varepsilon)^2, s_y' = s_y = y^2$. Plugging into Eq. (1),

$$\begin{aligned} c_1' &= \frac{(x - \varepsilon)^2 z + (z + \varepsilon)^2 x + y^2 \varepsilon}{(x - \varepsilon)^2 + (z + \varepsilon)^2 + y^2} = \frac{x^2 z - 2\varepsilon x z + \varepsilon^2 z + z^2 x + 2\varepsilon z x + \varepsilon^2 x + y^2 \varepsilon}{x^2 - 2\varepsilon x + \varepsilon^2 + z^2 + 2\varepsilon z + \varepsilon^2 + y^2} \\ &= \frac{x^2 z + z^2 x + \varepsilon(y^2 + \varepsilon(z + x))}{x^2 + z^2 + y^2 + 2\varepsilon(z - x + \varepsilon)} = \frac{c_1 s + \varepsilon\gamma}{s + \varepsilon\theta}. \end{aligned}$$

It is worthwhile to take a step back and consider what we got so far. Note that γ in the nominator is always positive because the (linear) derivatives of the quadratic terms $s_x z, s_z x$ cancel out. This shows why using quadratic probabilities makes sense. However, this is not sufficient, since θ in the denominator is also positive, and when s_y is too small (specifically, smaller than $\frac{1}{16}$) then the nominator grows too slowly to counter the increase in the denominator. This explains why we need the parameter q—to make sure that the manipulator is selected with sufficient probability to counter the benefit of the increased probability of the agent that is closer to a_1.

Going back to the technical proof, we need to show that

$$\frac{\gamma}{\theta} = \frac{y^2 + \varepsilon(z + x)}{2(z - x + \varepsilon)} \geq \frac{x^2 z + z^2 x}{x^2 + z^2 + y^2}.$$

Rearranging, we should prove that

$$(y^2 + \varepsilon(z + x))(x^2 + z^2 + y^2) - (x^2\underline{z} + z^2 x)(2(z - x + \varepsilon)) \qquad (4)$$

is non-negative. It is easy to see that this expression is monotonically increasing in y (and $y \geq \frac{1}{4}$ in this case). It is a bit less easy to see (not shown here) that it is also monotonically increasing in ε. Thus it is sufficient to lower bound $(\frac{1}{16} + x(z + x))(x^2 + z^2 + \frac{1}{16}) - (x^2 z + z^2 x)2(z - x + x)$, or, equivalently,

$$(\frac{1}{16} + xz + x^2)(x^2 + z^2 + \frac{1}{16}) - 2z^2 x(x + z).$$

One can check that the minimum of this expression in the range $0 \leq x \leq z \leq \frac{1}{2}$ is exactly 0 (at $z = \frac{1}{2}, x = \frac{1}{4}$).[2] Thus $\frac{\gamma}{\theta} \geq c_1$, and we are done by Eq. (3).

Finally, suppose that $z > \frac{1}{2}$. The only change is that the underlined z in Eq. (4) would change to $x + y$ (which is smaller than z). This only increases the expression and would thus not make it negative.   □

Since the inequality we get in Eq. (4) is tight, the proof also shows that any q-QCD mechanism for $q < \frac{1}{4}$ would not be strategyproof.

Proposition 3. *The $\frac{1}{4}$-QCD mechanism has an approximation ratio of $\frac{7}{6} \cong 1.166$, and this is tight.*

Proof. Let $\boldsymbol{a} = (a_1, a_2, a_3)$ be a profile, and denote $x = d(a_1, a_2), y = d(a_2, a_3), z = d(a_1, a_3)$. We assume w.l.o.g. $z \geq y \geq x$, thus the optimal point is a_2. The optimal social cost is $x + y$.

We first argue that the approximation only becomes worse by moving a_2 to the mid point between her neighbors. By decreasing y to $y' = y - \varepsilon$ and increasing x to $x' = x + \varepsilon$, z remains the largest arc, so a_2 is still optimal and $x' + y' = x + y$ is still the optimal social cost. The social cost of the mechanism changes from $\frac{s_x(y+z)+s_y(x+z)+s_z(x+y)}{s_x+s_y+s_z}$ to $\frac{s'_x(y+z)+s'_y(x+z)+s_z(x+y)}{s'_x+s'_y+s_z}$. We have that $s'_x + s'_y \leq s_x + s_y$ since the new partition is more balanced. This means that the denominator weakly increases and the total weight p_z given to the optimal point a_2 can only decrease. Among the two non optimal points, note that a_3 has the higher cost ($z + y \geq z + x$). Now, $s'_x \geq s_x$ so the relative weight of the worst point

---

[2] We verified this with Wolfram Alpha.

a_3 only increases. Thus the social cost weakly increases and the approximation ratio becomes worse.

This means that we are left to find the worst instance among the instances with distances $(x, x, 1 - 2x)$ for some $x \leq \frac{1}{3}$. The optimum in such an instance is $2x$ whereas the social cost of $\frac{1}{4}$-QCD is:

- for $\frac{1}{3} \geq x \geq \frac{1}{4}$, we have in particular that $1 - 2x \geq \frac{1}{3} > \frac{1}{4}$. Then

$$SC = \frac{2x^2(x + (1 - 2x)) + (1 - 2x)^2 2x}{2x^2 + (1 - 2x)^2} = \frac{2x - 6x^2 + 6x^3}{1 - 4x + 6x^2}$$

and the approximation ratio is $\frac{1-3x+3x^2}{1-4x+6x^2}$. The derivative of this expression is negative for $x < \frac{1}{2}$ so it is maximized at the bottom of the range, at $x = \frac{1}{4}$.

- for $x \leq \frac{1}{4}$, we have that

$$SC = \frac{2(1/4)^2 3x + (1 - 2x)^2 2x}{2(1/4)^2 + (1 - 2x)^2},$$

and the approximation ratio is $\frac{3/16+(1-2x)^2}{2/16+(1-2x)^2}$, which is increasing in x, so once again we obtain the maximum at $x = \frac{1}{4}$.

Plugging $x = \frac{1}{4}$ to the expression of the approximation ratio above, we get that in the worst instance $\boldsymbol{a} = (0, \frac{1}{4}, \frac{1}{2})$, $\frac{1}{4}$-QCD obtains an approximation ratio of exactly $\frac{3/16+(1/2)^2}{2/16+(1/2)^2} = \frac{7}{6}$. □

3.3   Beyond 3 Agents

We already saw that the PCD mechanism is strategyproof for any odd n. However, calculating its worst-case approximation ratio is more tricky. In particular, the worst instance is *not* symmetric w.r.t. the optimal point (in contrast to 3 agents). In the limit, PCD and random dictator have the same approximation:

Proposition 4. *When n grows, the approximation ratio of PCD approaches* 2.

Proof. Let $n = 2k + 1$, and consider the profile in Fig. 2a, where $x = \frac{1}{4\sqrt{k}}$. The numbers inside the circle indicate the number of agents in each location.

The optimal point is the bottom concentration, with a social cost of $c_1 = kx + \frac{1}{2} - x \leq \frac{1}{4}\sqrt{k} + \frac{1}{2}$. The social cost of the left point is $c_2 = k(\frac{1}{2} - x) + k\frac{1}{2}$, and of the right point is $c_3 = kx + \frac{1}{2}$. Thus

$$SC(f^{PCD}(\boldsymbol{a})) = \frac{1}{2}c_1 + xc_2 + (\frac{1}{2} - x)c_3 = -2kx^2 + (2k - 1)x + \frac{1}{2}$$

$$= -\frac{1}{8} + \frac{1}{2}\sqrt{k} - \frac{1}{4\sqrt{k}} + \frac{1}{2} \geq \frac{1}{2}\sqrt{k},$$

and thus the approximation ratio is at least $\frac{\frac{1}{2}\sqrt{k}}{\frac{1}{4}\sqrt{k}+\frac{1}{2}} > 2 - \frac{8}{\sqrt{n}}$. □

It is an open question whether there is some mechanism (perhaps a variation of q-QCD) that strictly beats 2 approximation for any n. We believe that this is indeed the case but that would require simplifying the proof technique.

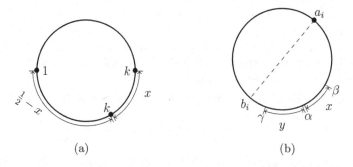

(a)   (b)

Fig. 2. Figures used in the proofs of Proposition 4 (a) and Proposition 6 (b).

Peak-only restrictions. We prove a weaker version of the following:

Conjecture 5. *For any n, the best strategyproof mechanism is peaks-only.*

Proposition 6. *For any n, the optimal strategyproof mechanism w.l.o.g. only places the facility either on peaks, or on points antipodal to peaks.*

Proof sketch. For a profile $\boldsymbol{a} = (a_1, \ldots, a_n)$, denote by b_i the point antipodal to a_i, and let $A = \{a_1, \ldots, a_n, b_1, \ldots, b_n\}$. Suppose that in some some profile $\boldsymbol{a}$, the mechanism f places the facility with some probability p on point $\alpha \notin A$. Denote by β, γ the nearest points to α from A clockwise and counterclockwise, respectively. Let $x = d(\alpha, \beta), y = d(\alpha, \gamma)$ (see Fig. 2b).

We define a mechanism f' that is identical to f, except that it "splits" the probability mass p of α between the adjacent points β, γ: it sets $f'_{\boldsymbol{a}}(\alpha) = 0$; $f'_{\boldsymbol{a}}(\beta) = f_{\boldsymbol{a}}(\beta) + p\frac{y}{x+y}$; and $f'_{\boldsymbol{a}}(\gamma) = f_{\boldsymbol{a}}(\gamma) + p\frac{x}{x+y}$.

We claim that for any agent i, $c_i(\boldsymbol{a}, f_{\boldsymbol{a}}) = c_i(\boldsymbol{a}, f'(\boldsymbol{a}))$. This would show both that f' is strategyproof and that $SC(\boldsymbol{a}, f_{\boldsymbol{a}}) = SC(\boldsymbol{a}, f'(\boldsymbol{a}))$ for all $\boldsymbol{a}$.

Indeed, consider some agent placed at a_i. From the three points α, β, γ, the one farthest from a_i cannot be α, since this would mean that b_i (the point antipodal to a_i) is strictly in the open interval (β, γ), whereas by construction there are no more points from A in this interval. Thus w.l.o.g. $d(a_i, \beta) < d(a_i, \alpha) < d(a_i, \gamma)$ (see figure). We omit the rest of the proof, which is not hard.   □

4   Beyond Circles

Definition 3. *The* Proportional Distance (PD) mechanism *for three agents selects each a_i ($i \in \{1, 2, 3\}$) with probability proportional to the distance between the other pair of agents.*

Note that for three agents on a circle, PD and PCD coincide when the agents are not all on the same semicircle, and otherwise PCD gives higher probability to the "middle" agent (which is optimal). Therefore PCD dominates PD. See Fig. 1a for an example. It is also not hard to show that PD dominates RD on any metric space. In particular, this means that $SC(f_{\boldsymbol{a}}^{PD}) \le \frac{4}{3}OPT(\boldsymbol{a})$ on any graph.

Theorem 7. *The PD mechanism is strategyproof in expectation for 3 agents in any metric space (in particular on any graph).*

In contrast to Theorem 1, the proof is rather technical and is thus omitted.

Observation 8. *The approximation ratio of any peaks-only mechanism (regardless of its incentive properties) on a general graph is at least $\frac{4}{3}$ $(2 - \frac{2}{n}$ for general $n)$.*

To see why, consider a star graph with n leafs, each containing one agent.

Proposition 9. *Let f be any peaks-only mechanism. Then for any profile $\mathbf{a} \in V^3$, we have that $SC(f_{\mathbf{a}}^{PD}) \leq \frac{5}{4} SC(f_{\mathbf{a}})$, and this bound is tight.*

Proof. Consider the distances between pairs $x \leq y \leq z$. W.l.o.g. we can denote $x + y = 1$. By triangle inequality, $z \leq x + y = 1$. The optimal peak location yields a cost of $x + y = 1$. The PD mechanism yields a cost of

$$SC(f^{PD}) = \frac{x(y+z)}{x+y+z} + \frac{y(x+z)}{x+y+z} + \frac{z(x+y)}{x+y+z} = \frac{2xy + xz + yz + z}{1+z}$$

$$= 2\frac{xy+z}{1+z} \leq 2\frac{xy+1}{1+1} = xy + 1 \leq (0.5)^2 + 1 = \frac{5}{4},$$

as required.

For tightness, consider any domain that contains three points a_1, a_2, a_3 such that a_2 is in the middle between a_1 and a_3 (e.g., a line). If there is one agent on each point then $x = d(a_1, a_2) = d(a_2, a_3) = y = 0.5$ whereas $z = d(a_1, a_3) = x + y = 1$. Then $SC(f^{PD}(\mathbf{a})) = 2\frac{xy+z}{1+z} = 1.25 = 1.25\,OPT(\mathbf{a})$, as the optimal peaks-only mechanism will select a_2.   □

Since the optimal point on a circle is always a peak, and since PCD dominates PD, we get the following.

Corollary 10. *For 3 agents on a circle, the PD and PCD mechanisms have an approximation ratio of $\frac{5}{4}$, and this is tight.*

Remark 1. Since $d(a_1, a_2) + d(a_2, a_3) + d(a_3, a_1)$ is a constant D given the profile, the PD mechanism selects each agent i with probability proportional to $D - SC(a_i)$. This allows us to easily extends the PD mechanism to any n, and it remains an open question whether the PD mechanism remains strategyproof. Recall however that already on the circle, PCD dominates PD, and is not asymptotically better than random dictator.

In [10], the authors suggest a randomized mechanism for placing $n-1$ facilities based on a similar idea: they place facilities on all agents except one (assuming all locations are distinct), where the placement omitted location a_i is selected with probability *inversely proportional* to the social cost of this placement (which in their case is just the distance to the closest agent $j \neq i$), and show it is strategyproof for any n and any metric space. Another mechanism that uses a similar proportional lottery (for two facilities) is in [18].

This suggests another possible direction for the single facility problem (or perhaps to more general problems), by considering various probabilities that are proportional to some decreasing function of the social cost.[3]

4.1   Lower Bounds via Linear Programming

An immediate corollary from Proposition 6, is that in that any upper bound on continuous circles implies the same upper bound on any finite circle with an even number of nodes, and thus any lower bound on a finite circle of any even size (or any size if Conjecture 5 is true) implies a lower bound for continuous circles.

It is well known that mechanism design problems for finite domains can be written as linear programs [8]. Automated mechanism design had also been applied to facility location problems, for one or more facilities on a line [15,26]. Due to the specifics of the problems they considered, they used advanced machine learning techniques rather than linear programming.

For a given graph (V, E), finding the optimal randomized strategyproof mechanism for three agents can be written as a simple linear optimization program as follows. There are $|V|^4 + 1$ variables: $(p_{a,z})_{a \in V^3, z \in V}$, where $p_{a,z} = f_a(z)$ is the probability that the facility is placed on z in profile a; and $\alpha \in \mathbb{R}$ which is the approximation factor. The optimization goal is simply to minimize α. There are four types of constraints:

1. Feasibility constraints: $p_{a,z} \geq 0$ for all $a \in V^3, z \in V$;
2. Probability constraints: $\sum_{z \in V} p_{a,z} = 1$ for all $a \in V^3$;
3. Incentive constraints: For every profile $a \in V^3$, any agent $i \in \{1,2,3\}$, and any alternative location $a_i' \in V$, we want to enforce the constraint $c_i(a, f_a) \leq c_i(a, f_{(a_{-i}, a_i')})$. This can be written as the following linear inequality over $2|V|$ variables: $\sum_{z \in V} d(z, a_i) p_{a,z} \leq \sum_{z \in V} d(z, a_i) p_{(a_{-i}, a_i'), z}$.
4. Approximation constraints: For every profile $a \in V^3$, we want to enforce the approximation $SC(a, f_a) \leq \alpha \cdot OPT(a)$. Since $OPT(a) = \min_{z \in V} \sum_{i \in \{1,2,3\}} d(z, a_i)$ can be computed once for each profile, the approximation constraint can also be written as a linear inequality: $\sum_{i \in \{1,2,3\}} \sum_{z \in V} d(z, a_i) p_{a,z} \leq \alpha \cdot \min_{z \in V} \sum_{i \in \{1,2,3\}} d(z, a_i)$.

In total, we get a bit more than $3|V|^4$ linear constraints. This is feasible for small graphs with commercial solvers, especially such that handle well sparse constraint matrices (we used Matlab's `linprog` function). By coding the graph in Fig. 3, we get the following:

Theorem 11. *There is no strategyproof mechanism for arbitrary graphs whose approximation ratio is better than $\frac{13}{12} \cong 1.0833$.*

---

[3] Note that using the reciprocal of the social cost (as in [10]) would lead to a poor outcome in the single facility problem.

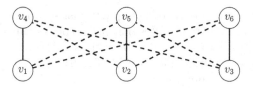

Fig. 3. A graph for which the best approximation ratio is $\frac{13}{12}$. The three solid edges have length 1, all dashed edges have length 2.

Small circles

Lemma 12. *For any strategyproof [peaks-only] mechanism f on the circle, there is a neutral and anonymous strategyproof [peaks-only] g,[4] such that $\max_a SC(g_a) \leq \max_{a'} SC(f_{a'})$.*

Proof. Mechanism g simply selects a permutation over agents uniformly at random, and direction+rotation for the circle uniformly at random, thereby mapping profile a to $\hat{a}$. Then, it runs f on $\hat{a}$ and maps back the outcome. Since this is a lottery over strategyproof mechanisms, it must also be strategyproof. It is also easy to see that if f is peaks-only then so is g. Finally, for any profile a, $SC(g_a)$ is averaging over several variations of $SC(f_{\hat{a}})$, all of which are bounded by $\max_{a'} SC(f_{a'})$. □

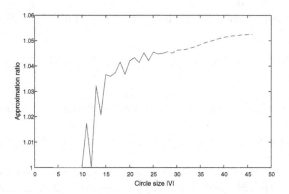

Fig. 4. The worst-case approximation ratio of the optimal 3-agent facility location mechanism on a circle with up to 44 vertices. The dashed part is computed only for peaks-only mechanisms on even M.

Theorem 13. *There is no strategyproof mechanism for circle graphs whose approximation ratio is better than 1.0456. If we add the peaks-only requirement, the lower bound is 1.0523.*

---

[4] A mechanism is *neutral* if it is invariant to renaming of vertices, and *anonymous* if it is invariant to renaming of agents.

To prove the theorem, we coded two linear programs: one that computes the optimal mechanism, and one that computes the optimal peaks-only mechanism. Since the number of variables for a circle with M vertices is M^4 (or $3M^3$ for peaks-only mechanisms) increases too fast for efficiently solving except for very small graphs, we applied the following improvements:

- By Lemma 12, it is sufficient to check mechanisms that are neutral. We thus fixed the location of the first agent, which reduces the number of variables by a factor of M.
- Also by Lemma 12, it is sufficient to check mechanisms that are anonymous. This allows us to add many symmetry constraints (both within profiles and between profiles) that effectively reduce the number of variables even more.
- By Proposition 6, it is sufficient to consider mechanisms that place the facility on one of the 6 peaks or anti-peaks.

This enables us to solve the obtained program for all mechanism on circles up to $M = 28$, and the program for peaks-only mechanisms for circles up to $M = 44$. We note that the worst-case approximation bounds in both programs are the same for any $|V| \leq 28$, which supports Conjecture 5, but leaves the proof as a challenge. The worst-case approximation ratios of the optimal mechanism for finite circles are shown in Fig. 4. It is non-monotone due to parity effects.

It remains an open question whether there is a better mechanism than the $\frac{1}{4}$-QCD mechanism for circles of arbitrary size, and what is the best approximation ratio that can be guaranteed. While we improved the upper bound from $\frac{4}{3}$ to $\frac{7}{6}$, and the lower bound from 1 to the bounds in Theorem 13, there is still a non-negligible gap.

5   Discussion

Table 1 summarizes our results for randomized mechanisms, and put them in the context of known bounds. It remains an open question whether the upper bound of $\frac{4}{3}$ ($2 - \frac{2}{n}$ for general n) is tight, and in particular whether general graphs are more difficult than circles.

The effect of the circle size on the available strategyproof mechanisms was evident in [9]. There, they showed (also using a computer search) a sharp dichotomy, where up to a certain size there are deterministic anonymous mechanisms, and above that size any strategyproof onto mechanism must be near-dictatorial. With randomized mechanisms, we see a more gradual effect.

The mechanisms we present seem quite specific to the problem at hand. Thus a natural question is what can be the takeaway messages for readers from the broader community of algorithmic game theory? We believe there are two.

Table 1. A summary of approximation bounds for 3-agent randomized mechanisms. (#) - obtains $\frac{5}{4} = 1.25$ approximation from best peak (Proposition 9).

Metric space	Any	Circle
Random dictator	$\frac{4}{3} \cong 1.333$ (from [1])	$\frac{4}{3} \cong 1.333$ (from [1])
Proportional [Circle] distance	$\frac{4}{3} \cong 1.333$ (#)	$\frac{5}{4} = 1.25$ (Corollary 10)
$\frac{1}{4}$-Quadratic circle distance	-	$\frac{7}{6} \cong 1.166$ (Theorem 2, Proposition 3)
best UB	1.333 (RD/PD)	1.166 ($\frac{1}{4}$-QCD)
LB (peaks-only)	1.333 (Observation 8)	1.0523 (Theorem 13)
LB	$\frac{13}{12} \cong 1.0833$ (Proposition 11)	1.0456 (Theorem 13)

First, the idea of focusing on the *derivative* of assignment probabilities as agents change their reported values. In the case of facility location, misreporting a value (say, by ε) causes the manipulator direct harm that is linear in ε, but may change the outcome probabilities in a way that still makes the manipulation beneficial. However, since the benefit is proportional to the *change* in probabilities (i.e., to their derivatives), using quadratic probabilities (whose derivatives are linear) puts the harm and benefit on the same scale. It is then left to the designer to tweak the parameters of the mechanism so as to make sure that the gain of a manipulator never exceeds the harm. Therefore, while the q-QCD mechanism seems more complicated than PCD and is more difficult to technically analyze, in a sense it is the result of a more structured and general approach to the problem, whereas PCD is a nice curiosity that happens to work.

The second idea is the combination of analytic and computational tools for solving a difficult design problem. While in some cases (e.g. in the analysis of our PD and PCD mechanisms) all the terms in the equations nicely cancel out to leave us with a clean proof, this is not always so. On the other hand, fully automated mechanism design [8] typically explodes with the size of the problem and leaves us with a solution that cannot be easily explained, modified or adapted to similar problems. This is true even for our linear programming approach in Sect. 4.1. However, one can come up with a specific or parametrized class of mechanisms, and use the computer capabilities to prove certain difficult inequalities, optimize parameters, or test various conjectures before setting out to prove them analytically. A similar combined approach has been applied e.g. in auctions [16], albeit with very different mechanisms.

We leave many open questions for future research. In particular, whether the PD and QCD mechanisms can be generalized for more agents, and whether there are classes of graphs that are inherently more difficult than circles.

Acknowledgments. This work was supported in part thanks to the Israeli Science Foundation grant number 773/16.

References

1. Alon, N., Feldman, M., Procaccia, A.D., Tennenholtz, M.: Strategyproof approximation of the minimax on networks. Math. Oper Res. **35**(3), 513–526 (2010)
2. Alon, N., Feldman, M., Procaccia, A.D., Tennenholtz, M.: Walking in circles. Discrete Math. **310**(23), 3432–3435 (2010)
3. Anastasiadis, E., Deligkas, A.: Heterogeneous facility location games. In: Proceedings of 17th AAMAS, pp. 623–631, 2018
4. Anshelevich, E., Postl, J.: Randomized social choice functions under metric preferences. J. Artif. Intell. Res. **58**, 797–827 (2017)
5. Black, D.: On the rationale of group decision-making. J. Polit. Econ. **56**(1), 23–34 (1948)
6. Cai, Q., Filos-Ratsikas, A., Tang, P.: Facility location with minimax envy. In: Proceedings of 25th IJCAI (2016)
7. Caragiannis, I., Kalaitzis, D., Markakis, E.: Approximation algorithms and mechanism design for minimax approval voting. In: Proceedings of 24th AAAI (2010)
8. Conitzer, V., Sandholm, T.: Complexity of mechanism design. In: Proceedings of the Eighteenth conference on Uncertainty in artificial intelligence, pp. 103–110. Morgan Kaufmann Publishers Inc. (2002)
9. Dokow, E., Feldman, M., Meir, R., Nehama, I.: Mechanism design on discrete lines and cycles. In: Proceedings of 13th ACM-EC, pp. 423–440 (2012)
10. Escoffier, B., Gourvès, L., Kim Thang, N., Pascual, F., Spanjaard, O.: Strategyproof mechanisms for facility location games with many facilities. In: Brafman, R.I., Roberts, F.S., Tsoukiàs, A. (eds.) ADT 2011. LNCS (LNAI), vol. 6992, pp. 67–81. Springer, Heidelberg (2011). https://doi.org/10.1007/978-3-642-24873-3_6
11. Feldman, M., Fiat, A., Golomb, I.: On voting and facility location. In: Proceedings of the 2016 ACM Conference on Economics and Computation, pp. 269–286. ACM (2016)
12. Feldman, M., Wilf, Y.: Strategyproof facility location and the least squares objective. In: Proceedings of 14th ACM-EC, pp. 873–890 (2013)
13. Filos-Ratsikas, A., Li, M., Zhang, J., Zhang, Q.: Facility location with double-peaked preferences. Auton. Agents Multi-agent Syst. **31**(6), 1209–1235 (2017)
14. Florez, J.V., Lauras, M., Okongwu, U., Dupont, L.: A decision support system for robust humanitarian facility location. Eng. Appl. Artif. Intell. **46**, 326–335 (2015)
15. Golowich, N., Narasimhan, H., Parkes, D.C.: Deep learning for multi-facility location mechanism design. In: IJCAI, pp 261–267 (2018)
16. Guo, M., Conitzer, V.: Computationally feasible automated mechanism design: general approach and case studies. In: AAAI, vol. 10, pp. 1676–1679 (2010)
17. Kalai, E., Muller, E.: Characterization of domains admitting nondictatorial social welfare functions and nonmanipulable voting procedures. J. Econ. Theory **16**, 457–469 (1977)
18. Lu, P., Sun, X., Wang, Y., Zhu, Z.A.: Asymptotically optimal strategy-proof mechanisms for two-facility games. In: Proceedings of the 11th ACM conference on Electronic commerce, pp. 315–324. ACM (2010)
19. Mei, L., Li, M., Ye, D., Zhang, G.: Strategy-proof mechanism design for facility location games: revisited. In: Proceedings of 15th AAMAS, pp. 1463–1464 (2016)
20. Meir, R.: Strategic voting. Synth. Lect. Artif. Intell. Mach. Learn. **13**(1), 1–167 (2018). Morgan & Claypool Publishers
21. Meir, R.: Strategyproof facility location for three agents on a circle. arXiv preprint arXiv:1902.08070 (2019)

22. Meir, R., Almagor, S., Michaely, A., Rosenschein, J.S.: Tight bounds for strategyproof classification. In: Proceedings of 10th AAMAS, pp. 319–326 (2011)
23. Meir, R., Procaccia, A.D., Rosenschein, J.S.: Algorithms for strategyproof classification. Artif. Intell. **186**, 123–156 (2012)
24. Moujahed, S., Simonin, O., Koukam, A., Ghédira, K.: A reactive agent based approach to facility location: application to transport. In: 4th Workshop on Agents in Traffic and Transportation, located at Autonomous Agents and Multiagent Systems (AAMAS 2006), pp. 63–69, Citeseer (2006)
25. Moulin, H.: On strategy-proofness and single-peakedness. Public Choice **35**, 437–455 (1980)
26. Narasimhan, H., Agarwal, S., Parkes, D.C.: Automated mechanism design without money via machine learning. In: Proceedings of the Twenty-Fifth International Joint Conference on Artificial Intelligence, pp. 433–439. AAAI Press (2016)
27. Nehring, K., Puppe, C.: The structure of strategy-proof social choice - part I: general characterization and possibility results on median spaces. J. Econ. Theory **135**(1), 269–305 (2007)
28. Procaccia, A.D., Tennenholtz, M.: Approximate mechanism design without money. In: Proceedings of the 10th ACM conference on Electronic commerce, pp. 177–186. ACM (2009)
29. Procaccia, A.D., Tennenholtz, M.: Approximate mechanism design without money. ACM Trans. Econ. Comput. (TEAC) **1**(4), 18 (2013)
30. Schummer, J., Vohra, R.V.: Strategy-proof location on a network. J. Econ. Theory **104**(2), 405–428 (2004)
31. Serafino, P., Ventre, C.: Truthful mechanisms without money for non-utilitarian heterogeneous facility location. In: Proceedings of 29th AAAI, pp. 1029–1035, (2015)
32. Sui, X., Boutilier, C.: Approximately strategy-proof mechanisms for (constrained) facility location. In: Proceedings of 14th AAMAS, pp. 605–613 (2015)
33. Todo, T., Iwasaki, A., Yokoo, M.: False-name-proof mechanism design without money. In: Proceedings of 10th AAMAS, pp. 651–658 (2011)
34. Zou, A., Li, M.: Facility location games with dual preference. In: Proceedings of 14th AAMAS, pp. 615–623 (2015)

Sharing Information with Competitors

Simina Brânzei[1(✉)], Claudio Orlandi[2(✉)], and Guang Yang[3,4(✉)]

[1] Purdue University, West Lafayette, USA
simina@purdue.edu
[2] Department of Computer Science, DIGIT, Aarhus University, Aarhus, Denmark
orlandi@cs.au.dk
[3] Institute of Computing Technology, Chinese Academy of Sciences, Beijing, China
[4] Conflux Technology Limited, Beijing, China
guang.research@gmail.com

Abstract. We study the mechanism design problem in the setting where agents are rewarded using information only, which is motivated by the increasing interest in secure multiparty computation. Specifically, we consider the setting of a joint computation where different agents have inputs of different quality and each agent is interested in learning as much as possible while maintaining exclusivity for information. Our high level question is how to design mechanisms that motivate all the agents (even those with high-quality inputs) to participate in the computation; we formally study problems such as set union, intersection, and average.

1 Introduction

Secure multiparty computation allows a set of parties to compute any functions on their private inputs. In recent years there has been a boom in the speed achieved by cryptographic protocols for secure multiparty computation (see e.g., [4,7,10,12,16,18] and references therein), to the point that start-ups and companies are beginning to offer products based on these technologies [1]. One question that has not been addressed in the cryptographic community so far is whether parties will have any *incentive* in participating in such protocols: In traditional multiparty computation tasks, multiple agents wish to evaluate some public function on their private inputs, where all agents are equal and the evaluated result is broadcasted to all of them or at least the honest ones. However, when viewing through the game-theoretic lens, the function evaluation process can be realized as the exchanging of private information among those agents, and hence the agents are *not equal*. For example, an agent with higher

The second author received support from: the Danish Independent Research Council under Grant-ID DFF-6108-00169 (FoCC); the European Union's under grant agreement No 731583 (SODA) and No 803096 (SPEC). Part of the work of the third author was done when working at Aarhus University. The third author was also supported in part by the National Natural Science Foundation of China Grants No. 61433014, 61602440, 61872334 and Shanghai Key Laboratory of Intelligent Information Processing, China. Grant No. IIPL-2016-006.

© Springer Nature Switzerland AG 2019
D. Fotakis and E. Markakis (Eds.): SAGT 2019, LNCS 11801, pp. 34–48, 2019.
https://doi.org/10.1007/978-3-030-30473-7_3

influence on the function tends to have a smaller incentive in the cooperation, and in the extreme case a "dictator" would have zero incentive; or even if the function is symmetric, an agent may still be less incentivized because of a high quality private input which provides a better prior than others. An example of a dictator is an agent with input zero when the function is *AND*; such an agent already knows the output of the computation and can learn nothing from others.

To this end we suggest to consider the procedure fairness (rather than the result fairness) in terms of *information benefit*, which measures how much an agent improves the quality of her own private information by participating. We believe this is a better characterization of the agent incentives. Also from the game-theoretic point of view, it makes sense to consider the agents as rational and self-motivated individuals rather than simply "good/bad" or "honest/semi-honest/malicious" as is typically done in cryptographic scenarios.

In this work, we study mechanisms for exchanging information without monetary transfer among rational agents. These agents are rational and self-motivated in the sense that they only care about maximizing their own utility defined in terms of information. More specifically, we focus on utility functions that capture the following properties about the behavior of the agents:

– *Correctness*: The agents wish to collect information from other agents.
– *Exclusivity*: The agents wish to have exclusive access to information.

The wish to collect information incentivizes cooperation, while the wish for exclusivity deters it. By unifying the above competing factors, agents aim to strike a balance between the two. The value of exclusivity is a concept studied in many areas of economics (e.g. labor economics, economics of the family, etc); see, e.g. [20] for a study on the role of exclusivity in contracts between buyers and sellers and [13] for platform-based information goods.

Utility functions that capture these competing factors are relevant in modeling situations where both cooperation and competition exist simultaneously, such as several companies wishing to exchange their private but probably overlapping information, e.g. training data for machine learning purpose, predictions for the stock market, etc.

We investigate specific information exchanging problems, such as Multiparty Set Union, as well as Set Intersection and Average. For example, in the set union problem there is a number of agents, each owning a set, and the goal is to find the union of the private sets held by all the agents. Set intersection is similarly defined except the goal is to find the intersection. Since for such problems the result is not Boolean and agents with different quality input should get different results, the value of result is measured by quality (accuracy) rather than by a Boolean indicator of whether it is the optimal one.

For the behavior of the agents, many of our results are for the "all-or-nothing model" where every agent either fully participates by truthfully submitting their input or refuses to participate. We also have several results for games with few agents in the richer model where agents can partially participate, by submitting some but not all of their information, as well as open questions.

The all-or-nothing model captures realistic scenarios where the inputs are authenticated by some trusted authority (e.g. using digital signatures), or where the inputs were already collected by some central entity, and the only choice of the agent is whether to allow their input to be used in the computation. Another motivation for the all-or-nothing model is when the inputs are later checked (e.g. in court or in future rounds of repeated games). The participants send their private input to the trusted mediator[1] (i.e. "principal") who runs a publicly known protocol (mechanism) to decide the payoff of each agent. Here the payoffs are customized pieces of information since we are studying the information exchanging mechanism without money.

As a simple example to motivate our work, consider the following scenario: Suppose there is a group of people and everyone is interested in finding a gold mine. The gold mine is situated in location t. Everyone has some estimate of where the gold mine is t_i and some uncertainty given by a radius d_i, i.e. each player i has an interval $[t_i - d_i, t_i + d_i]$. The players want to join their information to get a better approximation of the location of the gold mine and know that the gold mine lies in the intersection of all the estimates (sets). However if a player i knows that its radius d_i is much smaller than that of another player j, then player i knows that it won't learn much by interacting with player j. That is, in the worse case player i's interval is contained in player j's interval, so there is no information player i can infer from j. Since player i would rather not have player j gain free information without receiving anything in return, the problem is to design a mechanism that incentivizes the players to learn from each other (as much as possible). This is the problem that, later in the paper, we refer to as the *set intersection problem*.

1.1   Our Contribution

In this paper we propose a framework for non-monetary mechanism design with utility functions unifying both preferences of correctness and exclusivity. Let $N = \{1, \ldots, n\}$ be a set of agents. Suppose each agent has some piece of information, the details of which we intentionally leave informal for now. Given some mechanism M that the agents use to exchange information among themselves, we define the information benefit v_i of an agent i to be the additional "information" gained by i after participating in M. For example, in the case of the set union problem, where each agent owns a set of elements and tries to learn additional elements from other agents, this gain could be the number of additional elements learned by an agent compared to what that agent already knew.

The utility function will capture the tension between the wish to learn and the wish for exclusivity and the simple instantiation that we focus on is $u_i = v_i - \max_{j \in N \setminus \{i\}} v_j$. Thus each agent wishes to learn as much as possible while maintaining exclusivity over the information, which is captured by minimizing

---

[1] In the secure multiparty computation setting this trusted party is usually replaced by a cryptographic protocol. For the sake of simplicity, we do not further consider cryptographic protocols in this work.

the amount obtained by others. This definition is connected to the notion of envy-freeness; in particular, it captures the maximum "envy" that an agent i could have towards any another agent, and the goal is to reduce envy.

Our technical contribution is to design mechanisms for natural joint computation tasks such as *Multiparty Set Union*, as well as *Set Intersection* and *Average*. We focus on mechanisms that incentivize agents to submit the information they have as well as ensure properties such as Pareto efficiency[2] of the final allocation.

In the Multiparty Set Union Problem each agent owns a set x_i drawn from some universe $\mathcal{U}$. The utility functions are as described above. The strategy space of an agent consists of sets they can submit to the mechanism. We assume that agents can hide elements of their set, but not manufacture elements they don't have (i.e. there is a way to detect forgery). The question is to design a mechanism that incentivizes the agents to show their set of elements to others.

Theorem 1. *There is a truthful and Pareto efficient mechanism for set union among $n = 3$ agents. The mechanism runs in polynomial time.*

We leave open the mechanism design question for any number of agents.

Open Problem 1. *Is there a truthful polynomial time mechanism for set union for any number of agents? Are there randomized such mechanisms?*

However, we manage to solve this problem for the special case where each agent can either submit its whole set or the empty set, i.e. cooperate or not. We call this the "all-or-nothing" model.

Theorem 2. *There is a truthful, Pareto efficient, and welfare maximizing mechanism[3] for set union among any number n of all-or-nothing agents. The mechanism runs in polynomial time for any fixed n.*

We further show that this mechanism satisfies several other desirable properties, such as treating equal agents equally and rewarding more agents that contribute more.

Beyond multiparty set union, we also consider two case studies of problems with sets. The first is a set intersection problem, where each agent owns a connected set (interval) on the real line. The agents have to find an element in the intersection of all the sets and are promised that such an element exists. A high level example of this problem is when the agents are trying to find a gold mine as described in the introduction.

Theorem 3. *There is a truthful polynomial time mechanism for interval intersection among any number n of all-or-nothing agents.*

The second case study is a point average problem, where each agent has a point and the goal is to compute the average value of their inputs.

---

[2] Pareto effiency ensures no agent can be better off without making anyone worse off.

[3] Welfare maximization is achieved by maximizing over all Pareto efficient outcomes.

Theorem 4. *There is a truthful polynomial time mechanism for the point average problem for any number n of all-or-nothing agents.*

Finally, two more high-level remarks are in order.

Why not maximize social welfare? A trivial solution to problems such as set union can be to have everyone learn everything (i.e. maximize the sum of information gains). In traditional settings such as auctions or elections it is unlikely that every agent maximizes their information benefit simultaneously since their ideal outputs are usually conflicting, e.g. there is only one indivisible good that cannot be assigned to more than one agent. However, in the world where information replaces material goods, it becomes possible to duplicate the information at (nearly) zero cost such that every agent gets all information and hence maximizes their utility at the same time. This straightforward mechanism only works if all agents are selfless and choose to report truthfully. However, it is unfair in the sense that the more one agent contributes, the less benefit they could get (since the information benefit is bounded by the whole information minus their private information). Furthermore, the straightforward mechanism fails badly when agents take exclusivity into consideration: e.g. the dominant strategy would be "revealing nothing to the mechanism but combining the output with the private input afterward" and eventually the equilibrium becomes that no exchange happens at all (similar to the "rational secret sharing" problem discussed in [8,9,11]) when partial participation and strategic lies are allowed; and even in the all-or-nothing model an agent may prefer not participating according to their own utility function if their advantage over other competing agents would decrease.

A Note on Mechanism Design. The intuition behind our constructions is that every agent, when participating in the cooperation, should get a benefit no less than the loss they could cause to others by not participating. At first glance it might seem that the "loss to others" inflicted by a non-participating agent would be bounded by the exclusive information of that agent. However, it turns out that agents contribute much more to the mechanism than simply their private inputs. In particular, the participation of an agent may increase social welfare by giving incentives for participation to other agents with "better" inputs. An agent i with a high quality input might choose to join the computation, or reveal more of their private information, because, by doing so, they can reduce the information benefit of some other agent j (which is rational when it reduces i's own exclusivity loss).

Therefore, the key idea behind our constructions is to characterize the marginal contribution of every agent and assign information accordingly so that nobody prefers to leave the cooperation (and in the meanwhile we aim to maximize the social welfare among all stable allocations). For example, this idea is instantiated as a round-by-round exchange mechanism for the Three-Party Set Union problem (in Sect. 2.2), such that in every single "round" of exchange each agent gains more benefit than he offers to others.

1.2   Related Work

Our setting is reminiscent of cooperative game theory and the well-known solution of Shapley value [2,19,21], except that now the agents are rewarded with information instead of money. There are two main distinctions: (a) Information can be duplicated, for free or with negligible cost; (b) Every piece of information is unique whereas money is fungible, e.g. the same piece of information could have different values for different agents. The first property results in an unfixed total profit (sum of all agents' payoffs) and so breaks the intuition of "distribute the total surplus *proportionally* to each agent's contribution" used in Shapley value. The second property requires the mechanism to specify not only the *amount* of information but also the *details* of information allocation. In particular, the information already contained in an agent's input cannot be used to reward that agent. Such a property also leads to a subtle dilemma—the more an agent contributes, the less they can get as a reward from the mechanism— e.g. an agent with all information cannot get new information from other agents. Therefore, the mechanism must be able to motivate the most informed agents even though they may not benefit as much as those that know less (i.e. with lower quality inputs). A different line of work has studied the problem of sharing information when the inputs are substitutes or complements [6], which defined the value of information (and of a marginal unit of information) and instantiated it in the context of prediction markets.

Our model can be seen as an extension of the *non-cooperative computation* (NCC) framework and *informational mechanism design* (IMD) introduced in [14,22], where they characterize Boolean functions that are computable by rational agents with non-monetary utility functions defined in terms of information. In their model, the agents are trying to compute a public Boolean function on their private inputs with the help of a trusted center. Every agent claims their type (truthfully or not) to the center, and gets a response from the center (typically but not necessarily the Boolean function evaluated on claimed types). Agents may lie or refuse to participate, and they can apply any *interpretation function* (on the response from the center and their true input, so as to correct a wrong answer possibly caused by an earlier false declaration). In the setting of [22], the agents have a two-tiered preference of correctness preceding exclusivity[4], i.e. they are interested in misleading others only if this would not hurt their own correctness, whereas we generalize this lexicographic preference to a utility function incorporating both components (The lexicographic preference is a special case when one component is assigned a very small weight). Another extension is that we consider non-Boolean functions and allow distinct responses to different agents, which significantly enriches the space of candidate mechanisms.

The line of work [8,9,11] focuses on the cryptographic implementation of truthful mechanisms for secret sharing and multiparty computation by rational agents without a trusted mediator. In their setting there is an "issuer" who authenticates the initial shares of all agents so that the agents cannot forge a

---

[4] [14] considers other facets, such as privacy, but still in lexicographic ordering.

share (just as in the all-or-nothing model). Then the agents use simultaneous broadcast channels (non-simultaneous channels are also considered in [11]) to communicate in a round-by-round manner. Since all messages are broadcasted in this setting, a rational agent tends to keep silent so that they can receive others' information without revealing their own and hence possibly gain advantage in exclusivity. Therefore, much of the efforts and technical depth along this line is spent on catching dishonest agents (who do not broadcast their shares when they are supposed to), based on the key idea that in any given round the agents do not know whether this is just a test round designed to detect cheaters, or whether it is the final round for the actual information exchange. [9] achieve a fair, rational secure multiparty computation protocol which prevents coalitions and eliminates subliminal channels, despite the drawback of requiring special purpose hardware such as ideal envelopes and ballot boxes. However, all of these works assume the two-tiered preference of correctness and exclusivity as in [22], where in particular the correctness dimension is Boolean, i.e. either "correctly computed" or not. As a result, these works fall into the category of "implementing cryptographic protocols with rational agents" rather than the more game-theoretic topic "informational mechanism design" which we address in this paper.

There is another line of work [15, 17] on mechanism design with privacy-aware agents who care about their privacy rather than exclusivity. Considering privacy is relevant in many applications but technically orthogonal to what we study. (In our work, the privacy of the inputs is only a tool towards limiting the loss of utility due to the exclusivity preference, not a goal in itself).

The recent works of [5] and [3] investigate non-monetary mechanisms for cooperation among competing agents. However, an essential difference is that they consider a sequential delivery of outputs to different agents, such that the utility function is not merely in terms of information but also depends on the time or order when the output is delivered. For example, the "treasure hunting problem" in [5] is in particular very similar to the multiparty set intersection problem, except that in treasure-hunting only the first agent finding the common element gets positive utility while all others get zero.

2   Multiparty Set Union

Let $N = \{1, \ldots, n\}$ be a set of agents. There is a universe $\mathcal{U} = \{u_1, \ldots, u_m\}$ of possible numbers, from which each agent i owns a subset $S_i \subseteq \mathcal{U}$ that is private to the agent. The goal of the agents is to obtain more elements of the universe from other agents by sending elements from their own set in exchange.

We study the problem of designing mechanisms that incentivize the participants to share their information with each other. A mechanism $\mathcal{M}$ will take as input from each agent i a set $x_i \subseteq S_i$ and output a vector $\boldsymbol{y} = \mathcal{M}(\boldsymbol{x})$, so that the i-th entry of this vector contains the set received by agent i after the exchange.

Strategies. The strategy of an agent is the set it submits to the mechanism. Agents can hide elements (i.e. submit a strict subset of their true set), but not submit elements they don't actually have. A special case we will study in more depth is when the strategies of the agents are "all-or-nothing", i.e. $x_i \in \{\emptyset, S_i\}$. The input of each agent to the mechanism is sent through a private authenticated channel to the center.

Utility. We say the "information benefit" that agent i receives from sending their set S_i to the mechanism is the number of new elements that i obtains from the exchange: $v_i(\boldsymbol{x}) = |\mathcal{M}_i(\boldsymbol{x}) \setminus x_i|$. The *utility* of the agent is then defined as the minimum difference between their own information benefit and that of any other agent, formally given by $u_i(\boldsymbol{x}) = v_i(\boldsymbol{x}) - \max_{j \in N \setminus \{i\}} v_j(\boldsymbol{x})$.

The intuition is that each agent wishes to learn as much as possible while maintaining exclusivity, which is captured by minimizing the amount of information obtained by the other agents. This utility function is closely tied with the notion of envy as it compares the value for an agent with the maximum value of any other agent and the aim is to compute (approximately) envy-free outcomes.

Incentive Compatibility and Efficiency. We are interested in mechanisms that incentivize agents to share their information and will say that a mechanism is *truthful* if truth telling is a dominant strategy for each agent regardless of the strategies of the other agents. An allocation (outcome) is *Pareto efficient* (or *Pareto optimal*) if there is no other outcome where at least one agent is strictly better off and nobody is worse off.

Fairness. Some of our mechanisms also satisfy fairness and the fairness notions we consider are *symmetry* and *strong dominance*. Symmetry requires that if multiple agents report inputs of equivalent quality, then they get the same information benefit (and so the same utility). Strong dominance stipulates that if the information reported by an agent is inferior to the information reported by another agent under some partial order, then the result sent to the first agent is also (weakly) inferior to the result sent to the second agent under that order.

2.1   Two Agents

As a warm-up, we give a solution to the exchange problem for $n = 2$ agents.

Proposition 1. *There is a truthful polynomial time mechanism for the set union problem between two agents.*

Proof. W.l.o.g., the set owned by the second agent is larger: $|x_1| \leq |x_2|$. Let $y_2 = x_1 \cup x_2$ and $y_1 = x_1 \cup y_1'$, where y_1' is a set chosen so that $y_1' \subseteq x_2 \setminus x_1$ and $|y_1'| = |x_1 \setminus x_2|$. Then agents 1 and 2 can fairly exchange their exclusive elements until one of them has used up their exclusive elements. Note this type of exchange performed over multiple rounds can in fact be done in an atomic way by the principal. It is immediate that this mechanism ensures both agents get the same information benefit: $v_1 = v_2 = |x_1 \setminus x_2| \geq 0$ and it is weakly dominant for them to report their true information. $\square$

2.2   Three Agents

For three agents the problem becomes more subtle, as the mechanism must specify the order of pairwise exchanging, the number of exchanged elements, and, more importantly, which elements are exchanged.

Theorem 5. *There is a truthful polynomial time mechanism for set union among $n = 3$ agents.*

Proof. The theorem will follow from the construction in Mechanism 1. Mechanism 1 starts by removing the common elements among all three parties, since these elements will not affect the exchange; these elements are denoted by the set z_0. Then we consider the three pairwise intersections, from which the agents can exchange a number of elements bounded by the smallest intersection i.e. $s = \min\{|x_1 \cap x_2|, |x_2 \cap x_3|, |x_3 \cap x_1|\}$. Note that at the end of these exchanges at least one of these three intersections will be "used up". Thus we assume w.l.o.g. that after this step $x_2 \cap x_3 = \emptyset$ and $|x_2| \geq |x_3|$. Now we have reduced the original problem to a setting where there is no common intersection and only two pairwise intersections are non-empty, namely $x_1 \cap x_2$ and $x_1 \cap x_3$.

Let x_2, x_3 be partitioned into $x_2 = x_2' \cup x_2'', x_3 = x_3' \cup x_3''$ where $x_2' = x_2 \cap x_1, x_2'' = x_2 \backslash x_1$, and $x_3' = x_3 \cap x_1$, $x_3'' = x_3 \backslash x_1$. The intuition will be that elements in x_2' should be used to exchange elements in $x_3'' = x_3 \backslash (x_1 \cup x_2) = x_3 \cap \overline{x_1} \cap \overline{x_2}$, and similarly x_2'' for x_3'.

Next we discuss how exchanging occurs in several situations.

Case 1: $|x_2'| \geq |x_3''|$ and $|x_2''| \geq |x_3'|$. This is the simplest case, where we can simply make agent 3 exchange all elements in $x_3 = x_3'' \cup x_3'$ with both agents 1 and 2 for an equal amount of elements in $z \subseteq x_2'$ and $w \subseteq x_2''$ respectively. Then, agent 3 used up all its elements and the problem reduces to the two-party case between agents 1 and 2 with remaining elements in $(x_1 \backslash x_3', x_2 \backslash w)$.

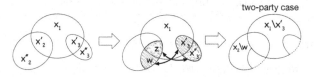

Case 2: $|x_3''| > |x_2'|$ and $|x_2''| > |x_3'|$. Then agent 2 uses $|x_3'|$ many elements in x_2'', denoted by w, to exchange all elements in x_3' with agents 1 and 3, and by symmetry agent 3 uses $|x_2'|$ many elements in z to exchange x_2' with agents 1 and 2. After this exchange all the three agents may have some elements left, but these are all exclusive elements, so the problem reduces to the easy case of three party with disjoint elements $(x_1 \backslash (x_2' \cup x_3'), x_2'' \backslash w, x_3'' \backslash z)$. Then the mechanism exchanges a number of elements equal to $\min\{|x_1 \backslash (x_2' \cup x_3')|, |x_2'' \backslash w|, |x_3'' \backslash z|\}$, further reducing the problem to the two-party case.

Mechanism 1: Three Party Set Union

Input: Set $x_i \subseteq \mathcal{U}$ for each player i
Output: Set $y_i \subseteq \mathcal{U}$ for each player i

1   $z_0 = x_1 \cap x_2 \cap x_3$

2   **foreach** *player* i **do**

3      $y_i = x_i$

4      $x_i = x_i \backslash z_0$

5   **end**

6   $s = \min\{|x_1 \cap x_2|, |x_2 \cap x_3|, |x_3 \cap x_1|\}$ /* *W.l.o.g.*, $|x_2| \geq |x_3|$ *and*
     $s = |x_2 \cap x_3|$ */

7   $z_1 = x_2 \cap x_3$

8   Select arbitrary sets $z_2 \subseteq x_3 \cap x_1$ and $z_3 \subseteq x_1 \cap x_2$ of sizes $|z_2| = |z_3| = s = |z_1|$

9   **foreach** *player* i **do**

10      $y_i = y_i \cup z_i$

11      $x_i = x_i \backslash (z_1 \cup z_2 \cup z_3)$

12   **end**

13   $x_2' = x_2 \cap x_1$; $x_2'' = x_2 \backslash x_1$

14   $x_3' = x_3 \cap x_1$; $x_3'' = x_3 \backslash x_1$

15   $(y_1', y_2', y_3') = (\emptyset, \emptyset, \emptyset)$ /* *Sets to store elements from recursive calls, if any.*   */

16   **if** $|x_2'| \geq |x_3''|$ *and* $|x_2''| \geq |x_3'|$ **then**
     /* *Case 1*   */

17   Select arbitrary sets $z \subseteq x_2'$ and $w \subseteq x_2''$ of sizes $|z| = |x_3''|$ and $|w| = |x_3'|$

18      $y_2 = y_2 \cup x_3$

19      $y_3 = y_3 \cup z \cup w$

20      $y_1 = y_1 \cup w \cup x_3''$

21      $(y_1', y_2') = \text{TwoPartySetUnion}(x_1 \backslash x_3'', x_2 \backslash w)$

22   **else if** $|x_3''| > |x_2'|$ *and* $|x_2''| > |x_3'|$ **then**
     /* *Case 2*   */

23   Select arbitrary sets $w \subseteq x_2''$ and $z \subseteq x_3''$ of sizes $|w| = |x_3'|$ and $|z| = |x_2'|$

24      $y_2 = y_2 \cup x_3' \cup z$

25      $y_3 = y_3 \cup x_2' \cup w$

26      $y_1 = y_1 \cup z \cup w$

27      $(y_1', y_2', y_3') = \text{ThreePartySetUnion}(x_1 \backslash (x_2' \cup x_3'), \ x_2'' \backslash w, x_3'' \backslash z)$
     /* *Recursive call with disjoint sets.*   */

28   **else**
     /* *Case 3:* $|x_3''| < |x_2'|$ *and* $|x_2''| < |x_3'|$ */

29   Select arbitrary sets $w \subseteq x_2''$ and $z \subseteq x_3''$ of sizes $|w| = |x_3''|$ and $|z| = |x_2''|$

30      $y_2 = y_2 \cup x_3'' \cup z$

31      $y_3 = y_3 \cup x_2'' \cup w$

32      $y_1 = y_1 \cup x_2'' \cup x_3''$

33      $(y_2', y_3') = \text{TwoPartySetUnion}(x_2' \backslash w, x_3' \backslash z)$

34   **foreach** *player* i **do**

35      $y_i = y_i \cup y_i'$ /* *Add elements obtained from recursive calls, if any, to the final set for each player.*   */

36   **end**

37   **return** (y_1, y_2, y_3)

disjoint three-party

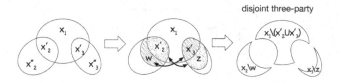

Case 3: $|x_3''| < |x_2'|$ and $|x_2''| < |x_3'|$. In this case agent 2 uses x_2'' in exchange for $|x_2''|$ many elements in $z \subseteq x_3'$, and, by symmetry, agent 3 uses x_3'' to exchange $|x_3''|$ many elements in $w \subseteq x_2'$. After such an exchange the problem reduces to three parties with $(x_1 \backslash (w \cup z), x_2' \backslash w, x_3' \backslash z)$.

two-party inside x_1

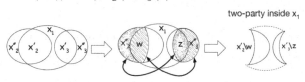

Finally, for improved welfare, agent 2 and agent 3 run a naïve two-agent exchange protocol with their remaining elements in $x_2' \backslash w$ and $x_3' \backslash z$. This is not optimal for agent 1, who has already collected full information and wants to end the exchange. However, agent 1 cannot prevent such exchange between agents 2 and 3 anyhow.

Mechanism 1 guarantees individual rationality because every round of exchange in its process is "fair" and "necessary". Every round is fair in the sense that all participants of that round get equal benefits—each of them gives out some elements in exchange for more new elements. Every round of such exchange is necessary because each element appears in at most one round, i.e. the mechanism does not reuse previously exchanged elements. Therefore, an agent that hides elements would suffer a loss lower bounded by the number of private elements that could have been traded, which is indeed a natural upper bound for the loss of others.   □

We note that Mechanism 1 is not Pareto efficient since the reduced problem (that is solved in the recursive call) is dealt with in a naïve way. For example, consider the last step in Case 1 i.e., after the problem has already been reduced to the two-party case. Here we could let agents 1 and 2 exchange their remaining elements without agent 3. This could be seen as fair, since agent 3 does not contribute new elements in those rounds. However, this procedure does not achieve Pareto efficiency, for that we can improve the social welfare by giving agent 3 some extra elements and, for sufficiently small number of elements, the utilities of agents 1 and 2 would not change and the solution would ensure that agents 1 and 2 remain truthful as before.

Theorem 6. *There is a truthful and Pareto efficient mechanism for set union among $n = 3$ agents.*

Proof. For case 1, consider the last step (of this case) in the execution of the mechanism. Mechanism 1 can be modified to assign randomly selected extra elements to player 3 so that $|y_3 \backslash x_3| = v_1 = v_2$ (recall that Mechanism 1 ensures $v_1 = v_2$). This modification achieves Pareto efficiency since any further improvement on social welfare will decrease the utility of player 1 or player 2, who already get all elements and cannot get more information benefit. Now we prove that the above modification also preserves truthfulness. This is immediate for players 1 and 2 but requires the following observations to see that it continues to hold from the point of view of player 3:

- each of player 3's exclusive elements in x_3'' leads to the same amount of marginal benefit to player 3 as to players 1 and 2, i.e. it is used to exchange for either one element in x_2', which will not be exchanged between players 1 and 2, or two elements when players 1 and 2 exchange elements in $x_2 \backslash w$ and $x_1 \backslash (x_2' \cup x_3')$, respectively.
- all of player 3's elements in x_3' do not affect others' information benefits; however, such elements can help player 3 since they might prevent the player from receiving some previously known element as the extra benefit.

We note that the same modification also works to ensure Pareto efficiency in case 2, while case 3 already ensures a Pareto efficient exchange. Thus there exists a truthful mechanism that is Pareto efficient. □

2.3   Any Number of Agents

We observe that the three agent mechanism above relies on a complex analysis that depends on the different intersection sets. The number of intersections increases exponentially as n grows and we leave open the question of whether it is possible to achieve an analogue of Mechanism 1 for more than three agents.

Open Problem 1. *Is there a truthful polynomial time mechanism for set union for any number of agents? Are there randomized such mechanisms?*

In Mechanism 2, we show a truthful mechanism for set union for any number of agents in the special case where each agent can either submit its whole set or the empty set, i.e. cooperate or not. We call this the "all-or-nothing" model and our main result is:

Theorem 7. There is a truthful, Pareto efficient, and welfare maximizing mechanism for set union among any number n of all-or-nothing agents. The mechanism runs in polynomial time for any fixed n.

Mechanism 2: Multiparty Set Union

Input: $(x_1, x_2, \ldots, x_n)$, where each set $x_i \subseteq \mathcal{U}$ is the input from player i.
Output: $(y_1, y_2, \ldots, y_n)$, where each set y_i is sent to player i.

1   Fix an ordering π of all elements in $\mathcal{U}$ /* π *will be used to specify the exchanged elements*   */

2   $u = \bigcup_{i=1}^{n} x_i$

3   $V = \texttt{ComputeV}(x_1, \ldots, x_n)$ /* *the function* $\texttt{ComputeV}$ *is defined below*   */

4   **foreach** *player* $i \in [n]$ **do**

5      $v_i = \max\{V, |u \backslash x_i|\}$

6   Let r_i be the set of first v_i elements in $z_{-i} \backslash x_i$ according to π

7      $y_i = x_i \cup r_i$

8   **end**

9   **return** $(y_1, y_2, \ldots, y_n)$.

10   **Function** $\texttt{ComputeV}(x_1, \ldots, x_n)$

11   **if** $n \leq 1$ **then**

12   **return** 0

13   **foreach** *player* $i \in [n]$ **do**

14        $z_{-i} = \bigcup_{j \neq i} x_j$

15        $V_{-i} = \texttt{ComputeV}(x_{-i})$

16   **end**

17     $V = \min\left\{\min_{k \in [n]}\left\{|z_{-k} \backslash x_k| + V_{-k}\right\}, \max_{k \in [n]}|z_{-k} \backslash x_k|\right\}$

18   **return** V.

---

3   Beyond Union: Intersection and Average

Moving beyond the multiparty set union problem, we suggest two other set problems where the agents own data points and wish to share them.

Intersection. The first problem is interval intersection, where each agent owns an interval in $\Re$ and the goal is to find a point in the intersection of all the sets. A high level scenario motivating this problem is the gold mine example from the introduction, where there is a group of people trying to find the location of a gold mine, and each person has an estimate of where the gold mine is, given by a center and a radius. The agents would like to merge their estimates to get a better idea of where the mine is situated, but the challenge is that agents with very good estimates (i.e. small radius) will not learn much from those with worse estimates (i.e. larger radius).

Theorem 8. *There is a truthful polynomial time mechanism for interval intersection among any number n of all-or-nothing agents.*

Set Average. The second problem is taking the average of a set that is distributed among the agents.

Theorem 9. *There is a truthful polynomial time mechanism for the average point problem among any number n of all-or-nothing agents.*

4 Discussion

Aside from our concrete open questions, the directions of generalizing the results to richer strategy spaces, allowing randomization, and more general utility functions are interesting.

References

1. Archer, D.W., et al.: From keys to databases - real-world applications of secure multi-party computation. Comput. J. **61**(12), 1749–1771 (2018)
2. Aumann, R.J.: Game theory. In: Game Theory, pp. 1–53. Springer (1989)
3. Azar, P.D., Goldwasser, S., Park, S.: How to incentivize data-driven collaboration among competing parties. In: ITCS, pp. 213–225 (2016)
4. Ben-Efraim, A., Lindell, Y., Omri, E.: Efficient scalable constant-round MPC via garbled circuits. In: Takagi, T., Peyrin, T. (eds.) ASIACRYPT 2017. LNCS, vol. 10625, pp. 471–498. Springer, Cham (2017). https://doi.org/10.1007/978-3-319-70697-9_17
5. Chen, Y., Nissim, K., Waggoner, B.: Fair information sharing for treasure hunting. In: AAAI, pp. 851–857 (2015)
6. Chen, Y., Waggoner, B.: Informational substitutes. In: FOCS, pp. 239–247 (2016)
7. Damgård, I., Pastro, V., Smart, N., Zakarias, S.: Multiparty computation from somewhat homomorphic encryption. In: Safavi-Naini, R., Canetti, R. (eds.) CRYPTO 2012. LNCS, vol. 7417, pp. 643–662. Springer, Heidelberg (2012). https://doi.org/10.1007/978-3-642-32009-5_38
8. Halpern, J., Teague, V.: Rational secret sharing and multiparty computation. In: STOC, pp. 623–632. ACM (2004)
9. Izmalkov, S., Micali, S., Lepinski, M.: Rational secure computation and ideal mechanism design. In: FOCS, pp. 585–594 (2005)
10. Keller, M., Orsini, E., Scholl, P.: MASCOT: faster malicious arithmetic secure computation with oblivious transfer. In: ACM SIGSAC, pp. 830–842 (2016)
11. Kol, G., Naor, M.: Cryptography and game theory: designing protocols for exchanging information. In: Canetti, R. (ed.) TCC 2008. LNCS, vol. 4948, pp. 320–339. Springer, Heidelberg (2008). https://doi.org/10.1007/978-3-540-78524-8_18
12. Lindell, Y., Pinkas, B., Smart, N.P., Yanai, A.: Efficient constant round multiparty computation combining BMR and SPDZ. In: Gennaro, R., Robshaw, M. (eds.) CRYPTO 2015. LNCS, vol. 9216, pp. 319–338. Springer, Heidelberg (2015). https://doi.org/10.1007/978-3-662-48000-7_16
13. Mantena, R., Sankaranarayanan, R., Viswanathan, S.: Platform-based information goods: the economics of exclusivity. Decis. Support Syst. **50**(1), 79–92 (2010)
14. McGrew, R., Porter, R., Shoham, Y.: Towards a general theory of non-cooperative computation. In: TARK, pp. 59–71. ACM (2003)
15. Miltersen, P.B., Nielsen, J.B., Triandopoulos, N.: Privacy-enhancing auctions using rational cryptography. In: Halevi, S. (ed.) CRYPTO 2009. LNCS, vol. 5677, pp. 541–558. Springer, Heidelberg (2009). https://doi.org/10.1007/978-3-642-03356-8_32

16. Nielsen, J.B., Nordholt, P.S., Orlandi, C., Burra, S.S.: A new approach to practical active-secure two-party computation. In: Safavi-Naini, R., Canetti, R. (eds.) CRYPTO 2012. LNCS, vol. 7417, pp. 681–700. Springer, Heidelberg (2012). https://doi.org/10.1007/978-3-642-32009-5_40
17. Nissim, K., Orlandi, C., Smorodinsky, R.: Privacy-aware mechanism design. In: ACM EC, pp. 774–789. ACM (2012)
18. Pinkas, B., Schneider, T., Zohner, M.: Scalable private set intersection based on OT extension. ACM Trans. Priv. Secur. 21(2), 7:1–7:35 (2018)
19. Roth, A.: The Shapley Value: Essays in Honor of Lloyd S. Shapley. Cambridge University Press, Cambridge (1988)
20. Segal, I., Whinston, M.: Exclusive contracts and protection of investments. RAND J. Econ. 31(4), 603–633 (2000)
21. Shapley, L.S.: A value for n-person games. The Shapley value, pp. 31–40 (1988)
22. Shoham, Y., Tennenholtz, M.: Non-cooperative computation: Boolean functions with correctness and exclusivity. Theor. Comput. Sci. 343(1), 97–113 (2005)

Mechanism Design for Locating a Facility Under Partial Information

Vijay Menon$^{(\boxtimes)}$ and Kate Larson

David R. Cheriton School of Computer Science,
University of Waterloo, Waterloo, Canada
{vijay.menon,kate.larson}@uwaterloo.ca

Abstract. We study the classic mechanism design problem of locating a public facility on a real line. In contrast to previous work, we assume that the agents are unable to fully specify where their preferred location lies, and instead only provide coarse information—namely, that their preferred location lies in some interval. Given such partial preference information, we explore the design of *robust* deterministic mechanisms, where by *robust* mechanisms we mean ones that perform well with respect to all the possible unknown true preferred locations of the agents. Towards this end, we consider two well-studied objective functions and look at implementing these under two natural solution concepts for our setting *(i)* very weak dominance and *(ii)* minimax dominance. We show that under the former solution concept, there are no mechanisms that do better than a naive mechanism which always, irrespective of the information provided by the agents, outputs the same location. However, when using the latter, weaker, solution concept, we show that one can do significantly better, and we provide upper and lower bounds on the performance of mechanisms for the objective functions of interest. Furthermore, we note that our mechanisms can be viewed as extensions to the classical optimal mechanisms in that they perform optimally when agents precisely know and specify their preferred locations.

1 Introduction

We consider the classic problem of locating a public facility on a real line or an interval, a canonical problem in *mechanism design without money*. In the standard version of this problem, there are n agents, denoted by the set $[n] = \{1, \cdots, n\}$, and each agent $i \in [n]$ has a preferred location x_i for the public facility. The cost of an agent for a facility located at p is given by $C(x_i, p) = |p - x_i|$, the distance from the facility to the agent's ideal location, and the task in general is to locate a facility that minimizes some objective function. The most commonly considered objective functions are *(a)* sum of costs for the agents and *(b)* the maximum cost for an agent. In the mechanism design version of the problem, the main question is to see if the objective under consideration can be implemented, either optimally or approximately, in (weakly) dominant strategies.

© Springer Nature Switzerland AG 2019
D. Fotakis and E. Markakis (Eds.): SAGT 2019, LNCS 11801, pp. 49–62, 2019.
https://doi.org/10.1007/978-3-030-30473-7_4

Table 1. Summary of our results. All the bounds are with respect to deterministic mechanisms.

	Average cost		Maximum cost	
	Upper bound	Lower bound	Upper bound	Lower bound
very weak dominance	$\frac{B}{2}$	$\frac{B}{2}$ [Theorem 1]	$\frac{B}{2}$	$\frac{B}{2}$ [Theorem 5]
minimax dominance	$\frac{3\delta}{4}$ [Theorem 3]	$\frac{\delta}{2}$ (only for mechanisms with finite range) [Theorem 4]	$\frac{B}{4} + \frac{3\delta}{8}$ [Theorem 6]	$\frac{B}{4}$ [12, Theorem 5]

While the standard version of the problem has received much attention, with several different variants like extensions to multiple facilities (e.g., [15,19]), looking at alternative objective functions (e.g., [3,10]) etc. being extensively studied, the common assumption in this literature is that the agents are always precisely aware of their preferred locations on the real line (or the concerned metric space). However, this might not always be the case and it is possible that the agents do not have accurate information about their ideal locations, or their preferences in general. To illustrate this, imagine a simple scenario where a city wants to build a school on a particular street (which we assume for simplicity is just a line) and aims to build one at a location that minimizes the maximum distance any of its residents have to travel to reach the school. While each of the residents is able to specify which block they would like the school to be located at, some of them are unable to precisely pinpoint where on the block they would like it because, for example, they do not currently have access to information (like infrastructure data) to better inform themselves, or they are simply unwilling to put in the cognitive effort to refine their preferences further. Therefore, instead of giving a specific location x, they end up giving an interval $[a, b]$, intending to say *"I know that I prefer the school to be built between the points a and b, but I am not exactly sure where I want it."*

The above described scenario is precisely the one we are concerned about in this paper. That is, in contrast to the standard setting of the facility location problem, we consider the setting in which the agents are uncertain (or partially informed) about their own preferred locations x_i and the only information they have is that their preferred location $x_i \in [a_i, b_i]$, where $b_i - a_i \leq \delta$ for some parameter δ which models the amount of inaccuracy. Now, given such partially informed agents, our task is to look at the problem from the perspective of a designer whose goal is to design *robust* mechanisms under this setting. Here by *robust* we mean that, for a given performance measure and when considering implementation under an appropriate solution concept, the mechanism should provide good guarantees with respect to this measure for all the possible underlying unknown true locations of the agents. The performance measure we use here is based on the minimax regret solution criterion, which, informally, for a given objective function, S, is an outcome that has the "best worst case", or one that induces the least amount of regret after one realizes the true input. More formally, if $\mathcal{P} = [0, B]$ denotes the set of all points where a facility can be located and $\mathcal{I} = [a_1, b_1] \times \cdots \times [a_n, b_n]$ denotes the set of all the possible vectors that correspond to the true ideal locations of the agents, then the minimax optimal solution, p_{opt}, for some objective function S is given by

$$p_{opt} = \arg\min_{p \in \mathcal{P}} \underbrace{\max_{I \in \mathcal{I}} \left(S(I, p) - \min_{p' \in \mathcal{P}} S(I, p') \right)}_{\text{maxRegret}(p, \mathcal{I})},$$

where $S(I, p)$ denotes the value of S when evaluated with respect to $I \in \mathcal{I}$ and a point p.

Thus, our aim is to design mechanisms that approximately implement the optimal minimax value (i.e., maxRegret$(p_{opt}, \mathcal{I})$) w.r.t. two objective functions—average cost and maximum cost—and under two solution concepts—very weak dominance and minimax dominance—that naturally extend to our setting (see Sect. 2 for definitions). In particular, we focus on deterministic and anonymous mechanisms that additively approximate the optimal minimax value, and our results are summarized in Table 1.

Before we move on to the rest of the paper, we anticipate that a reader might have some questions, especially w.r.t. our choice of performance measure and our decision to use additive as opposed to multiplicative approximations. We try to preemptively address these briefly in the section below.

1.1 Some Q & A

Why Regret? We argue below why this is a good measure by considering some alternatives.

1. Why not bound the ratio of the objective values of (a) the outcome that is returned by the mechanism and (b) the optimal outcome for that input? This, for instance, is the approach taken by Chiesa et al. [4]. In our case this is not a good measure because we can quickly see that this ratio is always unbounded in the worst-case.
2. Why not find a bound X such that for all $I \in \mathcal{I}, S(I, p) - S(I, p_I) \leq X$, where p is the outcome of the mechanism and p_I is the optimal solution associated with I? This, for instance, is the approach taken by Chiesa et al. [5]. Technically, this is essentially what we are doing when using max. regret. However, using regret is more informative because if we make a statement of the form maxRegret$(p, I) -$ maxRegret$(p_{opt}, I) \leq Y$, then this conveys two things: (a) for any p' there is at least one $I \in \mathcal{I}$ such that $S(I, p') - S(I, p_I) \geq Z$, where $Z =$ maxRegret(p_{opt}) (i.e. it gives us a sense on what is achievable at all—which in turn can be thought of as a natural lower bound) and (b) the point p chosen by the mechanism is at most $(Y + Z)$-far from the optimal objective value for any I. Hence, to convey these, we employ the notion of regret. We refer the reader to Appendix A in the full version of the paper[1] for a slightly more elaborate discussion.

---

[1] https://arxiv.org/pdf/1905.09230.pdf.

Why Additive Approximations? We use additive as opposed to multiplicative approximations because one can see that when using the latter and w.r.t. the max. cost objective function both the solution concepts that we consider in this paper—which we believe are natural ones to consider in this setting—do not provide any insight into the problem as there are no bounded mechanisms. Again, we refer the reader to Appendix A in the full version for a more elaborate discussion.

1.2   Related Work

There are two broad lines of research that are related to the topic of this paper. Below we discuss the most relevant papers with respect to each of them.

Designing Mechanisms with Incomplete Preferences. Among work in this space, the papers that are most relevant are the series of papers by Chiesa et al. [4–6], and the works of Hyafil and Boutilier [13,14]. The series of papers by Chiesa et al. [4–6] considers auction settings (single-item, combinatorial, and multi-unit auctions, respectively) where the agents are uncertain about their own types and the only information they have about their valuations is that it is contained in a set K, where K is any subset of the set of all possible valuations. The partial information model that we use in this paper is inspired by this series of papers. In particular, our prior-free and absolute worst-case approach under partial information is similar. However, our work is also different in that, unlike auctions, the problem we consider falls within the domain of *mechanism design without money* and so their results do not carry over to our setting.

Hyafil and Boutilier [13,14] consider the problem of designing mechanisms that have to make decisions using partial type information. While the overall theme in both their works is similar to ours, the questions they are concerned with and the model used are different. For instance, whereas in ours and Chiesa et al.'s models the agents do not know their true types and are therefore providing partial inputs, the assumption in the works of Hyafil and Boutilier [13,14] is that the mechanism has access to partial types, but agents are aware of their true type. This subtle change in turn leads to the focus being on solution concepts that are different from ours.

An alternative way to model uncertain agents is to assume that each of them has a probability distribution which tells them the probability of a point being their ideal location. For instance, this is the model that is used by Feige and Tennenholtz [8] in the context of task scheduling. However, in our model the agents do not have any more information than that they are within some interval, which we emphasize is not equivalent to assuming that, for a given agent, every point in the its interval is equally likely to be its true location.

The Facility Location Problem. Starting with the work of Moulin [17] there has been a flurry of research looking at designing strategyproof mechanisms for the facility location problem. These can be broadly divided into two branches. The first one consists of work, e.g., [2,7,16,17,20], that focuses on characterizing the class of strategyproof mechanisms in different settings (see [1] and

[18, Chapter 10] for surveys). The second branch consists of more recent papers which fall under the broad umbrella of *approximate mechanism design without money*, initially advocated by Procaccia and Tennenholtz [19], that focus on looking at how well a strategyproof mechanism can perform under different objective functions [9–11,15,19]. Our paper falls under this branch of the literature.

2 Preliminaries

Recall that in the standard (mechanism design) version of the facility location problem there are n agents, denoted by the set $[n] = \{1, \cdots, n\}$, and each agent $i \in [n]$ has a true preferred[2] location $\ell_i^* \in [0, B]$, for some fixed[3] constant $B \in \mathbb{R}$. A vector $I = (\ell_1, \cdots, \ell_n)$, where $\ell_i \in [0, B]$, is referred to as a location profile and the cost of agent i for a facility located at p is given by $C(\ell_i^*, p) = |p - \ell_i^*|$ (or equivalently, their utility is $-|p - \ell_i^*|$), the distance from the facility to the agent's location.[4] In general, the task in the facility location problem is to design mechanisms—which are, informally, functions that map location profiles to a point (or a distribution over points) in $[0, B]$—that (approximately) implement the outcome associated with a particular objective function.

In the version of the problem that we are considering, each agent i, although they have a true location $\ell_i^* \in [0, B]$, is currently unaware of their true location and instead only knows an interval $[a_i, b_i] \subseteq [0, B]$ such that $\ell_i^* \in [a_i, b_i]$. The interval $[a_i, b_i]$, which we denote by K_i, is referred to as the *candidate locations* of agent i, and we use $\mathbb{K}_i$ to denote the set of all possible candidate locations of agent i (succinctly referred to as the set of candidate locations). Now, given a profile of the set of candidate locations $(\mathbb{K}_1, \cdots, \mathbb{K}_n)$, we have the following definition.

Definition 1 (δ-uncertain-facility-location-game). *For all $n \geq 1$, $B > 0$, and $\delta \in [0, B]$, a profile of the set of candidate locations $(\mathbb{K}_1, \cdots, \mathbb{K}_n)$ is said to induce a δ-uncertain-facility-location-game if, for each i, $\mathbb{K}_i = \{[a_i, b_i] \mid b_i - a_i \leq \delta \text{ and } [a_i, b_i] \subseteq [0, B]\}$ (in words, for each i, their set of candidate locations can only have intervals of length at most δ).*

Remark: We refer to δ as the inaccuracy parameter. In general, when proving lower bounds we assume that the designer knows this δ as this only makes our results stronger, whereas for positive results we explicitly state what the designer knows about δ. Additionally, note that if $\delta = 0$, then we have the

---

[2] We often omit the term "preferred" and instead just say that ℓ_i^* is agent i's location.

[3] Here we make the assumption that the domain under consideration is bounded instead of assuming that the agents can be anywhere on the real line. This is necessary only because we are focusing on additive approximations. (For a slightly more elaborate explanation, see Sect. 1 in the paper by Golomb and Tzamos [12].)

[4] The particular utility function considered here is equivalent to the notion of symmetric single-peaked preferences that is often used in the economics literature (see, e.g., [16]).

standard facility location setting where the set of candidate locations associated with every agent is just a set of points in $[0, B]$. For a given profile of candidate locations $(K_1, \cdots, K_n)$, we say that "the reports are exact" when, for each agent i, K_i is a single point and not an interval.

2.1   Mechanisms, Solution Concepts, and Implementation

A (deterministic) mechanism $\mathcal{M} = (X, F)$ in our setting consists of an action space $X = (X_1, \cdots, X_n)$, where X_i is the action space associated with agent i, and an outcome function $F \colon X_1 \times \cdots \times X_n \to [0, B]$. A mechanism is said to be *direct* if, for all i, $X_i = \mathbb{K}_i$, where $\mathbb{K}_i$ is the set of all possible candidate locations of agent i. For every i, a strategy is a function $s_i \colon \mathbb{K}_i \to X_i$, and Σ_i and $\Delta(\Sigma_i)$ respectively denote the set of all pure and mixed strategies of i.

Since the outcome of a mechanism needs to be achieved in equilibrium, it remains to be defined what equilibrium solution concepts we consider in this paper. Below we define the two solution concepts that we use here. We note that the first (very weak dominance) was also used by Chiesa et al. [4].

Definition 2 (very weak dominance). *In a mechanism* $\mathcal{M} = (X, F)$, *an agent* i *with candidate locations* K_i *has a very weakly dominant strategy* $s_i \in \Sigma_i$ *if* $\forall s_i' \in \Sigma_i, \forall \ell_i \in K_i$, *and* $\forall s_{-i} \in \Sigma_{-i}$,

$$C\left(\ell_i, F(s_i(K_i), s_{-i}(K_{-i}))\right) \leq C\left(\ell_i, F(s_i'(K_i), s_{-i}(K_{-i}))\right).$$

In words, the above definition implies that for agent i with candidate locations K_i, it is always best for i to play the strategy s_i, irrespective of the actions of the other players and irrespective of which of the points in K_i is her true location.

Definition 3 (minimax dominance). *In a mechanism* $\mathcal{M} = (X, F)$, *an agent* i *with candidate locations* K_i *has a minimax dominant strategy* $s_i \in \Sigma_i$ *if* $\forall s_i' \in \Sigma_i$ *and* $\forall s_{-i} \in \Sigma_{-i}$,

$$\max_{\ell_i \in K_i} \max_{\sigma_i \in \Delta(\Sigma_i)} C(\ell_i, F(s_i(K_i), s_{-i}(K_{-i})))$$
$$- C(\ell_i, F(\sigma_i(K_i), s_{-i}(K_{-i})))$$
$$\leq \max_{\ell_i \in K_i} \max_{\sigma_i \in \Delta(\Sigma_i)} C(\ell_i, F(s_i'(K_i), s_{-i}(K_{-i}))$$
$$- C(\ell_i, F(\sigma_i(K_i), s_{-i}(K_{-i})).$$

Before explaining what the definition above implies, let $p = F(s_i(K_i), s_{-i}(K_{-i}))$ be the outcome of the mechanism when agent i plays strategy s_i and all the others play some s_{-i}. Now, consider the term $\mathrm{maxRegret}_i(p)$ $= \max_{\ell_i \in K_i} \max_{\sigma_i \in \Delta(\Sigma_i)} C(\ell_i, p) - C(\ell_i, F(\sigma_i(K_i), s_{-i}(K_{-i})))$, which calculates agent i's max. regret (i.e., the absolute worst case loss agent i will experience if and when she realizes her true location from her candidate locations) for playing s_i and for the output p. Then, what the above definition implies is that for a regret minimizing agent i with candidate locations K_i, it is always best for i to play s_i, irrespective of the actions of the other players, as any other strategy s_i' results in an outcome p' w.r.t. which i experiences at least as much max. regret as she experiences with p.

Remark: Note that both the solution concepts defined above can be seen as natural extensions of the classical (i.e., the usual mechanism design setting where the agents know their types exactly) weak dominance notion to our setting. That is, for all $i \in [n]$, if K_i is a single point, then both of them collapse to the classical weak dominance notion.

As stated in the introduction, given a profile of candidate locations $(K_1, \cdots, K_n)$, we want the mechanism to "perform well" against all the possible underlying true locations of the agents, i.e., with respect to all the location profiles $I = (\ell_1, \cdots, \ell_n)$ where $\ell_i \in K_i$. Hence, for a given objective function S, we aim to design mechanisms that achieve a good approximation of the optimal minimax value, which, for $\mathcal{I} = K_1 \times \cdots \times K_n$, is denoted by $\mathrm{OMV}_S(\mathcal{I})$ and is defined as

$$\mathrm{OMV}_S(\mathcal{I}) = \mathrm{maxRegret}(p_{opt}, \mathcal{I}), \tag{1}$$

where for a point $p \in [0, B]$, if $S(I, p)$ denotes the value of the function S when evaluated with respect to the vector I and p, then the maximum regret associated with p for the instance $\mathcal{I}$ is defined as

$$\mathrm{maxRegret}(p, \mathcal{I}) = \max_{I \in \mathcal{I}} \left(S(I, p) - \min_{p' \in [0, B]} S(I, p') \right), \tag{2}$$

and

$$p_{opt} = \arg\min_{p \in [0, B]} \mathrm{maxRegret}(p, \mathcal{I}). \tag{3}$$

Throughout, we refer to the point p_{opt} as the optimal minimax solution for the instance $\mathcal{I}$.

Finally, now that we have our performance measure, we define implementation in very weakly dominant and minimax dominant strategies.

Definition 4 (Implementation in very weakly dominant (minimax dominant) strategies). *For a δ-uncertain-facility-location-game, we say that a mechanism $\mathcal{M} = (X, F)$ implements α-OMV$_S$, for some $\alpha \geq 0$ and some objective function S, in very weakly dominant (minimax dominant) strategies, if for some $s = (s_1, \cdots, s_n)$, where s_i is a very weakly dominant (minimax dominant) strategy for agent i with candidate locations K_i,*

$$maxRegret(F(s_1(K_1), \cdots s_n(K_n)), \mathcal{I}) - OMV_S(\mathcal{I}) \leq \alpha.$$

3   Implementing the Average Cost Objective

In this section we consider the objective of locating a facility so as to minimize the average cost (sometimes succinctly referred to as avgCost and written as

AC). While the standard objective in the facility location setting is to minimize the sum of costs, here, like in work of Golomb and Tzamos [12], we use average cost because since we are approximating additively, it is easy to see that in many cases a deviation from the optimal solution results in a factor of order n coming up in the approximation bound. Hence, to avoid this, and to make comparisons with our second objective function, maximum cost, easier we use average cost.

In the standard setting where the agents know their true location, the average cost of locating a facility at a point p is defined as $\frac{1}{n} \sum_{i \in [n]} C(x_i, p)$, where x_i is the location of agent i. Designing even optimal strategyproof mechanisms in this case is easy since one can quickly see that the optimal location for the facility is the median of $x_1, \cdots, x_n$ and returning the same is strategyproof. In contrast to the standard setting, for some $\delta \in (0, B]$ and a corresponding δ-uncertain-facility-location-game, even computing what the minimax optimal solution for the average cost objective (see Eq. 3) is is non-trivial, let alone seeing if it can be implemented with any of the solution concepts discussed in Sect. 2.1. Although we will need some properties about the minimax optimal solution when proving properties about the mechanisms we design, we do not state them here, since due to space constraints we are unable to include any proofs. We refer the reader to the full version of the paper where we provide a complete discussion on the computing the minimax optimal solution and provide the proofs of the results in Sects. 3.1 and 3.2.

3.1   Implementation in Very Weakly Dominant Strategies

While very weak dominance is indeed a natural solution concept which extends the classical notion of weak dominance, we will see below that it is too strong as no deterministic mechanism can achieve a better approximation bound than $\frac{B}{2}$. This implies that, among deterministic mechanisms, the naive mechanism which always, irrespective of the reports of the agents, outputs the point $\frac{B}{2}$ is the best one can do.

Theorem 1. *Given a $\delta \in (0, B]$, let $\mathcal{M} = (X, F)$ be a deterministic mechanism that implements α-OMV_{AC} in very weakly dominant strategies for a δ-uncertain-facility-location-game. Then, $\alpha \geq \frac{B}{2}$.*

Although one could argue that this result is somewhat expected given how Chiesa et al. also observed similar poor performance for implementation with very weakly dominant strategies in the context of the single-item auctions [4, Theorem 1], we believe that it is still interesting because not only do we observe a similar result in a setting that is considerably different from theirs, but this observation also reinforces their view that one would likely have to look beyond very weakly dominant strategies in settings like ours. This brings us to our next section, where we consider an alternative, albeit weaker, but natural, extension to the classical notion of weakly dominant strategies.

3.2 Implementation in Minimax Dominant Strategies

In this section we move our focus to implementation in minimax dominant strategies. We first present a general result that applies to all mechanisms in our setting that are anonymous and minimax dominant, in particular showing that any such mechanism cannot be onto. The proof of this result, which can be found in the full version, is based on a characterization result for strategyproof, anonymous, and onto mechanisms when agents have symmetric single-peaked preferences [16, Corollary 2].

Remark: Note that in this section we focus only on direct mechanisms. This is w.l.o.g. since it turns out that the revelation principle holds in our setting for minimax dominant strategies. We refer the reader to Appendix B in the full version for the proof.

Theorem 2. *For all $\delta \in (0, B]$, let $\mathcal{M} = (X, F)$ be a deterministic mechanism that is anonymous and minimax dominant for a δ-uncertain-facility-location-game. Then, $\mathcal{M}$ cannot be onto.*

Given the fact that we cannot have an anonymous, minimax dominant, and onto mechanism, the natural question to consider is if we can find non-onto mechanisms that perform well. We answer this question in the next section.

Non-onto Mechanisms. In this section we consider non-onto mechanisms. We first show a positive result by presenting an anonymous mechanism that implements $\frac{3\delta}{4}$-OMV$_{AC}$ in minimax dominant strategies. Following this, we present a conditional lower bound that shows that one cannot achieve an bound better than $\frac{\delta}{2}$ when considering mechanisms that have a finite range.

An Anonymous and Minimax Dominant Mechanism. Consider the $\frac{\delta}{2}$-equispaced-median mechanism defined in Algorithm 1, which can be thought of as an extension to the standard median mechanism. The key assumption in this mechanism is that the designer knows a δ such that any agent's candidate locations has a length at most δ. Given this δ, the main idea is to divide the interval $[0, B]$ into a set of "grid points" and then map every profile of reports to one of these points, while at the same time ensuring that the mapping is minimax dominant. In particular, in the case of the $\frac{\delta}{2}$-equispaced-median mechanism, when $\delta > 0$, its range is restricted to the finite set of points $A = \{g_1, g_2, \cdots g_m\}$ such that, for $i \geq 1$, $g_{i+1} - g_i = \frac{\delta}{2}$, $g_1 = 0$, and $g_m \leq B$.

Below we state that the $\frac{\delta}{2}$-equispaced-median mechanism implements $\frac{3\delta}{4}$-OMV$_{AC}$ in minimax dominant strategies. The main idea in its proof, which can be found in the full version, is that for an agent i with candidate locations $[a_i, b_i]$ the ℓ_i associated with i in the mechanism is in fact the agent's "best alternative" among the alternatives in A. Once we have this, we then show the approximation bound.

Theorem 3. *For a given $\delta \in [0, B]$, the $\frac{\delta}{2}$-equispaced-median mechanism is anonymous and implements $\frac{3\delta}{4}$-OMV$_{AC}$ in minimax dominant strategies for a δ-uncertain-facility-location-game.*

Input: a $\delta \geq 0$ and for each agent i, their input interval $[a_i, b_i]$
Output: location of the facility p
1: $A \leftarrow \{g_1, \cdots, g_k\}$,where $g_1 = 0, g_k \leq B, g_{i+1} - g_i = \frac{\delta}{2}$
2: **for** each $i \in \{1, \cdots, n\}$ **do**
3:     $x_i \leftarrow$ point closest to a_i in A (in case of a tie, break in favour of the point in $[a_i, b_i]$ if there exists one, break in favour of point to the left otherwise)
4:     $y_i \leftarrow$ point closest to b_i in A (break ties as in line 3)
5:     **if** $|[x_i, y_i] \cap A| == 1$ **then**   ▷ the case when $x_i = y_i$
6:         $\ell_i \leftarrow x$
7:     **else if** $|[x_i, y_i] \cap A| == 2$ **then**
8:         **if** $|[x_i, y_i] \cap [a_i, b_i]| < 2$ **then**
9:             **if** $a_i + b_i \leq x_i + y_i$ **then**
10:                 $\ell_i \leftarrow x_i$
11:             **else**
12:                 $\ell_i \leftarrow y_i$
13:             **end if**
14:         **else**
15:             $\ell_i \leftarrow x_i$
16:         **end if**
17:     **else if** $|[x_i, y_i] \cap A| == 3$ **then**
18:         $\ell_i \leftarrow z_i$, where $z_i \in [x_i, y_i] \cap A, z_i \neq x_i, z_i \neq y_i$
19:     **end if**
20: **end for**
21: **return** median$(\ell_1, \cdots, \ell_n)$

Algorithm 1: $\frac{\delta}{2}$-equispaced-median mechanism

A Conditional Lower Bound. In the context of our motivating example from the introduction, it is possible, and in fact quite likely, that the city can only build the school at a finite set of locations on the street. Therefore, an interesting class of non-onto mechanisms to consider is ones which have a finite range. Furthermore, seeing our mechanism above, an inquisitive reader might wonder: "why $\frac{\delta}{2}$-equispaced? why not $\frac{\delta}{3}$-equispaced or something smaller than $\frac{\delta}{2}$?" First, one can easily construct counter-examples to show that any ϵ-equispaced-median mechanism is not minimax dominant for $\epsilon < \frac{\delta}{2}$. However, that still does not rule out mechanisms whose range is some finite set $\{g_1, \cdots, g_m\}$. Below we consider this question and we show that the approximation bound associated with any mechanism that is anonymous, minimax dominant, and has a finite range, is at least $\frac{\delta}{2}$. The key idea that is required in order to show this bound is the following lemma, which informally says that if the mechanism has a finite range, is minimax dominant, and achieves a bound less than $\frac{3\delta}{4}$, then there is "sufficient-gap" between four consecutive points in the range, A, of the mechanism. Once we have this observation it is then in turn used to construct profiles that will result in the stated bound. The proofs of both the lemma and theorem make use of a characterization result by Massó and De Barreda [16, Corollary 1], and they can be found in the full version. (Below we ignore mechanisms which have less than six points in their range as one can easily show that such mechanisms perform poorly.)

Lemma 1. *For all $\delta \in (0, \frac{B}{6}]$, let $\mathcal{M}$ be a deterministic mechanism that has a finite range A (of size at least six), is anonymous, and one that implements α-OMV_{AC} in minimax dominant strategies for a δ-uncertain-facility-location-game. Then, either $\alpha \geq \frac{3\delta}{4}$, or there exists four consecutive points $g_1, g_2, g_3, g_4 \in A$ such that $g_1 < g_2 < g_3 < g_4$ and $\frac{d_1}{2} + d_2 + \frac{d_3}{2} \geq \delta$, where, for $i \in [3], d_i = g_{i+1} - g_i$.*

Theorem 4. *For all $\delta \in (0, \frac{B}{6}]$, let $\mathcal{M}$ be a deterministic mechanism that has a finite range (of size at least six), is anonymous, and one that implements α-OMV_{AC} in minimax dominant strategies for a δ-uncertain-facility-location-game. Then, for any $\epsilon > 0$, $\alpha \geq \frac{\delta}{2} - \epsilon$.*

4 Implementing the Maximum Cost Objective

In this section we consider the objective of minimizing the maximum cost (succinctly referred to as maxCost and written as MC). In the standard setting where the reports are exact, the max. cost associated with locating a facility at p is defined as $\max_{i \in [n]} C(x_i, p)$ and if we assume w.l.o.g. that the x_i's are in sorted order, then it is easy to see that the optimal solution to this objective is to locate the facility at $p = \frac{x_1 + x_n}{2}$. However, unlike in the case of the avgCost objective that was considered in Sect. 3, one cannot design an optimal strategyproof mechanism even when the reports are exact, and it is known that the best one can do in terms of additive approximation is to achieve a bound of $\frac{B}{4}$ in the case of deterministic mechanisms [12, Theorem 5].

Now, coming to our setting, unlike in the case of the avgCost objective, calculating the minimax optimal solution is straightforward in this case. In fact, given the candidate locations $[a_i, b_i]$ for all i, if $L_1, \cdots, L_n$ and $R_1, \cdots, R_n$ denote the sorted order of the points in $\{a_i\}_{i \in [n]}$ and $\{b_i\}_{i \in [n]}$, respectively, then it is not too hard to show that the minimax optimal solution is the point $\frac{L_1 + R_1 + L_n + R_n}{4}$. Therefore, below we directly move on to implementation. We refer the reader to the full version for a complete discussion on the minimax optimal solution and for the proofs of the results that follow.

4.1 Implementation in Very Weakly Dominant Strategies

Here again see that very weak dominance is too strong a solution concept as we can show that one cannot do better than the naive mechanism which always outputs the point $\frac{B}{2}$. The following theorem, which can be proved by proceeding exactly like in the proof of Theorem 1, formalizes this statement.

Theorem 5. *For all $\delta \in (0, B]$, $\epsilon \in (0, \delta)$, let $\mathcal{M} = (X, F)$ be a deterministic mechanism that implements α-OMV_{MC} in very weakly dominant strategies for a δ-uncertain-facility-location-game. Then, $\alpha \geq \frac{B}{2} - \epsilon$.*

Given this, we consider minimax dominant strategies in the hope of getting an analogous result as Theorem 3.

4.2 Implementation in Minimax Dominant Strategies

When it comes to implementation in minimax dominant strategies, we again see that even in the case of the maxCost objective function one can do a lot better than under very weak dominance. Before we see the exact bounds, recall that Theorem 2 rules out the existence of mechanisms that are anonymous, minimax dominant, and onto. Hence, our focus will be on non-onto mechanisms. We note that the ideas in the following section are similar to the ones in Sect. 3.2 since here, too, we focus on "grid-based" mechanisms.

Non-onto Mechanisms. In this section we show that there exists a mechanism that implements $\left(\frac{B}{4} + \frac{3\delta}{8}\right)$-OMV$_{MC}$ in minimax dominant strategies. The mechanism is similar to the $\frac{\delta}{2}$-equispaced-median mechanism and can be considered as an extension to the phantom-half mechanism proposed by Golomb and Tzamos [12]. Hence, we only highlight the changes below.

$\frac{\delta}{2}$**-equispaced-phantom-half.** We need to make only two changes to Algorithm 1: *(a)* redefine A to be the set $\{g_1, \cdots, g_j, \cdots, g_m\}$, where $g_j = \frac{B}{2}, g_{i+1} - g_i = \frac{\delta}{2}$, for $1 \leq i \leq k - 1$, $g_0 \geq 0$, and $g_m \leq B$. *(b)* instead of returning the median of the l_is in line 21, we return the median of the points $\ell_{min}, \frac{B}{2}$, and ℓ_{max}, where $\ell_{min} = \min_i\{\ell_i\}$ and $\ell_{max} = \max_i\{\ell_i\}$.

Below, we state that the mechanism described above implements $\left(\frac{B}{4} + \frac{3\delta}{8}\right)$-OMV$_{MC}$ in minimax dominant strategies. The proof can be found in the full version.

Theorem 6. *For a given $\delta \in [0, \frac{2B}{3}]$, the $\frac{\delta}{2}$-equispaced-phantom-half mechanism is anonymous and one that implements $\left(\frac{B}{4} + \frac{3\delta}{8}\right)$-OMV$_{MC}$ in minimax dominant strategies for a δ-uncertain-facility-location-game.*

Finally, given this result, it is natural to ask if we have a lower bound like the one in Sect. 3.2. Unfortunately, the only answer we have is the obvious bound of $\frac{B}{4}$ that follows from the result of Golomb and Tzamos [12, Theorem 15] who showed that under exact reports, and when using deterministic mechanisms, one cannot achieve a bound lower than $\frac{B}{4}$.

5 Conclusions

The standard assumption in mechanism design that the agents are precisely aware of their preferences may not be realistic in many situations. Hence, we believe that there is a need to look at models that account for partially informed agents and, at the same time, design mechanisms that provide robust guarantees. In this paper we looked at such a model in the context of the classic single-facility location problem, where an agent specifies an interval instead of an exact location, and our focus was on designing robust mechanisms that perform well w.r.t. all the possible underlying true preferred locations of the agents. Towards this end, we looked at two solution concepts, very weak dominance and minimax dominance, and we showed that, with respect to both the objective functions we considered, while it was not possible to achieve any good mechanism in the context of the former solution concept, extensions to the classical optimal mechanisms—i.e., mechanisms that perform optimally in the classical setting where the agents exactly know their locations—performed significantly better under the latter, weaker, solution concept.

There are some immediate open questions in the context of the problem we considered like looking at randomized mechanisms, providing tighter bounds, and potentially even finding deterministic mechanisms that perform better than the ones we showed. More broadly, we believe that it will be interesting to revisit the classic problems in mechanism design, see if one can look at models which take into account partially informed agents, and design mechanisms where one can explicitly relate the performance of the mechanism with the quality of preference information.

References

1. Barberà, S.: An introduction to strategy-proof social choice functions. Soc. Choice Welf. **18**(4), 619–653 (2001)
2. Barberà, S., Jackson, M.: A characterization of strategy-proof social choice functions for economies with pure public goods. Soc. Choice Welf. **11**(3), 241–252 (1994)
3. Cai, Q., Filos-Ratsikas, A., Tang, P.: Facility location with minimax envy. In: Proceedings of the Twenty-Fifth International Joint Conference on Artificial Intelligence (IJCAI), pp. 137–143 (2016)
4. Chiesa, A., Micali, S., Zhu, Z.A:. Mechanism design with approximate valuations. In: Proceedings of the Third Innovations in Theoretical Computer Science (ITCS), pp. 34–38 (2012)
5. Chiesa, A., Micali, S., Zhu, Z.A.: Knightian self uncertainty in the VCG mechanism for unrestricted combinatorial auctions. In: Proceedings of the Fifteenth ACM Conference on Economics and Computation (EC), pp. 619–620 (2014)
6. Chiesa, A., Micali, S., Zhu, Z.A.: Knightian analysis of the vickrey mechanism. Econometrica **83**(5), 1727–1754 (2015)
7. Dokow, E., Feldman, M., Meir, R., Nehama, I.: Mechanism design on discrete lines and cycles. In: Proceedings of the Thirteenth ACM Conference on Electronic Commerce (EC), pp. 423–440. ACM (2012)
8. Feige, U., Tennenholtz, M.: Mechanism design with uncertain inputs: (to err is human, to forgive divine). In: Proceedings of the Forty-Third Annual ACM Symposium on Theory of Computing (STOC), pp. 549–558 (2011)
9. Feigenbaum, I., Sethuraman, J., Ye, C.: Approximately optimal mechanisms for strategyproof facility location: minimizing Lp norm of costs. Math. Oper. Res. **42**(2), 434–447 (2016)
10. Feldman, M., Wilf, Y.: Strategyproof facility location and the least squares objective. In: Proceedings of the Fourteenth ACM Conference on Electronic Commerce (EC), pp. 873–890 (2013)
11. Fotakis, D., Tzamos, C.: Strategyproof facility location for concave cost functions. Algorithmica **76**(1), 143–167 (2016)

12. Golomb, I., Tzamos, C.: Truthful facility location with additive errors (2017). arXiv preprint arXiv:1701.00529
13. Hyafil, N., Boutilier, C.: Partial revelation automated mechanism design. In: Proceedings of the Twenty-Second National Conference on Artificial Intelligence (AAAI), pp. 72–77 (2007)
14. Hyafil, N., Boutilier, C.: Mechanism design with partial revelation. In: Proceedings of the Twentieth International Joint Conference on Artificial Intelligence (IJCAI), pp. 1333–1340 (2007)
15. Lu, P., Sun, X., Wang, Y., Zhu, ZA.: Asymptotically optimal strategy-proof mechanisms for two-facility games. In: Proceedings of the Eleventh ACM Conference on Electronic Commerce (EC), pp. 315–324 (2010)
16. Massó, J., Barreda, I.M.D.: On strategy-proofness and symmetric single-peakedness. Games Econ. Behav. **72**(2), 467–484 (2011)
17. Moulin, H.: On strategy-proofness and single peakedness. Public Choice **35**(4), 437–455 (1980)
18. Nisan, N., Roughgarden, T., Tardos, E., Vazirani, V.V.: Algorithmic Game Theor. Cambridge University Press, Cambridge (2007)
19. Procaccia, A.D., Tennenholtz, M.: Approximate mechanism design without money. ACM Trans. Econ. Comput. **1**(4), 18:1–18:26 (2013)
20. Schummer, J., Vohra, R.V.: Strategy-proof location on a network. J. Econ. Theor. **104**(2), 405–428 (2002)

Mechanism Design for Constrained Heterogeneous Facility Location

Maria Kyropoulou[1], Carmine Ventre[2(✉)], and Xiaomeng Zhang[1]

[1] University of Essex, Colchester, UK
{maria.kyropoulou,xzhangao}@essex.ac.uk
[2] King's College London, London, UK
carmine.ventre@kcl.ac.uk

Abstract. The facility location problem has emerged as the benchmark problem in the study of the trade-off between incentive compatibility without transfers and approximation guarantee, a research area also known as approximate mechanism design without money. One limitation of the vast literature on the subject is the assumption that agents and facilities have to be located on the same physical space. We here initiate the study of constrained heterogeneous facility location problems, wherein selfish agents can either like or dislike the facility and facilities can be located on a given feasible region of the Euclidean plane. In our study, agents are assumed to be located on a real segment, and their location together with their preferences towards the facilities can be part of their private type. Our main result is a characterization of the feasible regions for which the optimum is incentive-compatible in the settings wherein agents can only lie about their preferences or about their locations. The stark contrast between the two findings is that in the former case any feasible region can be coupled with incentive compatibility, whilst in the second, this is only possible for feasible regions where the optimum is constant.

Keywords: Mechanism design without money · Facility location · Incentive compatibility

1 Introduction

Deciding where to locate a public facility, like a school, in order to serve a group of strategic agents, is a fundamental problem that has received a great deal of attention. Under such a setting, the city council, or some other public authority, needs to elicit private information from the concerned (local) people, or agents, without using money, and choose the location of the school based on that information. The authority defines the rules of choosing the location with the objective to maximize the *social welfare*, i.e., the total satisfaction of the agents. However, agents might misreport their private information in an attempt to maximize their own individual utility, which is usually captured using some

© Springer Nature Switzerland AG 2019
D. Fotakis and E. Markakis (Eds.): SAGT 2019, LNCS 11801, pp. 63–76, 2019.
https://doi.org/10.1007/978-3-030-30473-7_5

distance measure between their ideal location for the facility, commonly considered to be (part of) their private information (a.k.a., *type*), and the location of the facility itself. The absence of money makes it very challenging to align the incentives of the authority with those of the individual agents.

The field of mechanism design [12] focuses on the implementation of desired outcomes in strategic settings. A primary designer goal that has been extensively studied is that of truthfulness, which informally states that an agent should be able to optimize her own individual utility by reporting truthfully her private information. However, achieving this is not always compatible with maintaining a high social welfare [8,15]. Monetary compensations have been commonly used as a means towards aligning the incentives of the individuals with those of society, however, the use of payments is not always allowed due to ethical [12], legal (e.g., organ donations), or even just practical reasons. With this motivation in mind, researchers have started turning their attention to possible ways of achieving truthfulness without the use of payments, i.e., designing truthful (or *strategy-proof*, SP for short) mechanisms that do not use monetary transfers.

Mechanism design without money has been examined from the point of view of exact and approximate solutions. Exact mechanism design without money has a rich history in social choice literature (cf., e.g., [11]), while Procaccia and Tennenholtz [14] were the first to consider achieving truthfulness (or strategy-proofness) without using payments, by sacrificing the optimality of the solution and settling for just an approximation; their work has given rise to what is now known as approximate mechanism design without money. In a nutshell, the objective is that of finding the best approximation guarantee which guarantees strategy-proofness for a given optimization problem.

However, in many settings (such as the school location discussed above) the mechanism designer has some control on the set of feasible solutions (e.g., the area in the city where a school can be built) and would arguably be more interested in leveraging this power to marry strategy-proofness and optimality. In this paper, we initiate the investigation of this research direction and ask whether we can achieve strategy-proofness without using payments by restricting the feasibility of the solution space. As a case study, we consider a facility location problem similarly to Procaccia and Tennenholtz [14]. In our model, the agents are located on a single-dimensional space, just like in [14], while the facility can be located in a feasible region in $\mathbb{R}^2$. Contrary to their approach, we keep the requirement for an optimal solution and study how the shape of the feasible region for the facility location can impact the incentives. We are interested in the following general question:

What is the biggest feasible region for the facility that would allow for the optimal solution to be implemented in a truthful way?

We consider this question in the setting of *heterogeneous preferences* [1,18, 21], where the facility is not commonly believed to be desirable by the agents; some agents might find it attractive and wish to have it located as close to them as possible, but others might have different views and desire to be far away from them. This preference might also be part of the private information of each agent (in addition/place of their location). When the facility is a school, for example, it is reasonable to expect that families with small children will want to reside close to a school, yet others might prefer to live as far away as possible from it in order to avoid possible noise and traffic.

We define the utility of the agents to be quadratic in the distance between the agent and the facility. On the one hand, the literature on facility location in higher dimensional spaces has twists in the definition of distance in order to make this study feasible. On the other, in many problems, one dimension is not rich enough to fully describe preferences. For example, Barberá et al. [2] mention the city block metric, i.e., the shortest path between two points on a multidimentional grid, as a possible appropriate metric. Our model captures real-life scenarios wherein the agents are environmentally conscious or have resources that are depleted quadratically in the distance (as, e.g., power consumption in wireless communication [13]). In our city planner motivating example, quadratic costs align with environmentally conscious agents, i.e., agents who suffer quadratically in the distance to the facility due to the pollution caused by the travel to cover that distance.

1.1   Our Contribution

We examine whether restricting the solution space for the facility location problem can be used as a means of achieving the optimal social welfare in a strategy-proof way. Our findings show a dichotomy result in the sense that all or nothing can be done in the setting of heterogeneous facility location for exact optimal solutions.[1]

Specifically, we consider two different settings, where either the preferences of the agents are private information but their locations are publicly known (unknown preferences case), or the opposite, i.e., the locations of the agents are private information but their preferences are publicly known (unknown locations case).

In Sect. 3 we treat the case of unknown preferences and we show that the optimal mechanism (the one that maximizes the social welfare) is *group strategy-proof (GSP)* no matter the feasible region. GSP is a stronger requirement than strategy-proofness as it does not even allow for profitable deviations of coalitions of agents. Technically, this is proved by reducing the optimization problem of maximizing the social welfare to the geometric problem of selecting a point in the feasible region which is at maximum/minimum (depending on the shape of the instance) distance from a carefully defined point on the line where the agents reside. This point, which we call β, is a snapshot of the instance and is what the coalitions of agents can manipulate (together with the rule to choose the point

---

[1] Note that depending on the feasible region, the optimal solution might not be well defined, e.g. if the feasible region is $\mathbb{R}^2$ and all agents dislike the facility. Our results implicitly assume that the optimization problem is well-defined and focus on coupling it with incentive considerations.

in the feasible region). The proof identifies key properties that must be satisfied by a successful manipulation and then observes how those are incompatible with optimality.

We then handle the case of unknown locations in Sect. 4, where we need to distinguish between different cases depending on the majority of the preferences (which is, in this setting, public knowledge). In each of these cases, we show that in order to be able to implement the optimal solution in a strategy-proof way it would have to hold that the optimal social welfare is constant, i.e., there is a unique point in the feasible region that maximizes the sum of utilities. From the conceptual point of view, this result shows that in the unknown location setting the power coming from the restriction of the feasible region is null as to obtain a strategy-proof optimum, the incentives have to disappear altogether. This is a quite strong negative characterizing result, which paves the way for future research where the interplay between approximation guarantee and the feasible region is considered (see conclusions). From the technical point of view, the proof of this result adopts an iterative approach which identifies several instances showing that the optimum must be the same for both the minimum and the maximum possible value of β. However, while for the case wherein the preferences are homogenous, the argument uses SP and optimality constraints to establish the shape of the feasible region in the limit, the proof for heterogeneous instances requires a more careful step-by-step argument to prove that the optimum is constant.

We note that some of the proofs are deferred to the full version, due to lack of space.

1.2 Related Work

The facility location problem has been studied by many diverse research communities previously. We here discuss some of the most fundamental research directions that have been explored in the context of facility location.

Relevant research from a Social Choice perspective has mostly focused on the problem of locating a single facility on the line. In his seminal paper [11], Moulin characterizes the class of generalized median voter schemes as the only deterministic SP mechanisms for *single-peaked* agents on the line. Schummer and Vohra [17] extend the result of Moulin to trees and *continuous graphs*. Dokow et al. [3] prove that for small discrete graphs there are anonymous SP mechanisms, contrarily to the case of continuous cycles studied in [17]. They prove that SP mechanisms on *discrete large cycles* are nearly-dictatorial in that all agents can affect the outcome to a certain extent.

Facility location has also been one of the fundamental problems in the field of Mechanism Design without money. The work of Procaccia and Tennenholtz on facility location in [14] initiates the study of approximate mechanism design without money, where they suggest the idea of sacrificing a factor of the approximation guarantee as a means to obtain strategy-proofness. For the 2-facility location problem, they propose the Two-Extremes algorithm, that places the two facilities in the leftmost and rightmost location of the instance, and prove that it is group strategy-proof and $(n-2)$-approximate, where n is the number of agents. Furthermore, they provide a lower bound of $3/2$ on the approximation ratio of any SP algorithm for the facility location problem on the line and conjecture a lower bound of $\Omega(n)$. The latter conjecture has been proven by Fotakis et al. [7]. Their main result is the characterization of deterministic SP mechanisms with *bounded approximation ratio* for the 2-facility location problem on the line. They show that there exist only two such algorithms: (i) a mechanism that admits a unique dictator or (ii) the Two-Extremes mechanism proposed in [14]. The authors of [5] show how verification can be used to get truthful mechanisms with better approximation guarantees for the problem.

Lu et al. [10], improve several bounds studied in [14]. In particular, as regards deterministic algorithms they prove a better (w.r.t. [14]) lower bound of $2 - \mathcal{O}(\frac{1}{n})$. Furthermore, they prove a 1.045 lower bound for randomized mechanisms for the 2-facility location problem on the line and present a randomized $n/2$-approximate mechanism.

Our work falls under the category of exact (as opposed to approximate) mechanism design without money. We consider the restriction of the feasibility space so that optimality and strategy-proofness are not mutually exclusive. To the best of our knowledge, this work is the first in the facility location literature to distinguish the region of the agents' locations and the feasible region for the facility. However, similar restrictions have been studied in the judgement aggregation literature, see [4]. We study the case of heterogeneous preferences [1,18–21], and distinguish between cases where agents can only misreport their locations or their preferences (but not both). A similar distinction has been considered in [6].

In [9], Lu et al. consider general metric spaces for the 2-facility game. They give an $\Omega(n)$ lower bound for the approximation of deterministic strategy-proof mechanisms and prove that a constant approximation ratio can be achieved by a natural randomized mechanism, the so-called *Proportional* Mechanism.

2 Model and Preliminaries

We assume to have k agents, located on the segment $[0, \ell]$; we say that agent i is located at x_i in that segment. We let $\mathbf{x} = (x_1, \ldots, x_k)$. We need to locate a facility on a given *feasible region* $\Gamma \subseteq \mathbb{R}^2$. (We assume for simplicity that, in the larger space, the segment has the second coordinate equal to 0; our results hold no matter this choice, given that this coordinate is known in either the setting considered.) Each agent might like the facility or dislike it; we let $p_i \in \{-1, 1\}$ denote the preference of agent i with the meaning that if agent i likes (dislikes, respectively) the facility then $p_i = 1$ ($p_i = -1$, respectively). In this sense, our model is heterogeneous, in that not all the agents will have the same opinion of the usefulness of the facility. We let $\mathbf{p} = (p_1, \ldots, p_k)$, let m denote the number of agents whose preference is 1, and let n the number of agents whose preference is -1, so that $k = m + n$. Agent i has a *utility* which depends on her location x_i and preference p_i, and the location $f = (x_f, y_f)$ of the facility in Γ, that is,

$$u_i((x_i, p_i), f) = \begin{cases} (x_i - x_f)^2 + y_f^2 & \text{if } p_i = -1 \\ \lambda - (x_i - x_f)^2 - y_f^2 & \text{if } p_i = 1 \end{cases},$$

where λ is a constant which guarantees that the utilities are not negative. Intuitively, an agent who likes the facility wants to be close to f, while an agent who dislikes it wants to be far from it. Our definition of utility captures that and is similar in spirit to the one in [1]; in our definition, however, the utility is quadratic in the distance between x_i and f (see Sect. 1 for a relevant discussion).

We study this problem from a mechanism design perspective. That is, we assume that the agents have a private *type* t_i, and we consider the two extreme cases of type being either the preference or the location of each agent, i.e. $t_i \in \{x_i, p_i\}$. A mechanism $\mathcal{M}$ collects reports from the agents, which are potentially different *bids* b_i, and on this input returns a location for the facility in Γ. With a slight abuse of notation, we assume that the bid of agent i to the mechanism is completed with the public part of $\{x_i, p_i\}$. Our objective is to design a *truthful mechanism* (a.k.a., strategy-proof, SP for short) $\mathcal{M}$, i.e., a mechanism such that for any t_i, b_i and $\mathbf{b}_{-i} = (b_j)_{j \neq i}$,

$$u_i(t_i, \mathcal{M}(t_i, \mathbf{b}_{-i})) \geq u_i(t_i, \mathcal{M}(b_i, \mathbf{b}_{-i})).$$

A stronger requirement is for the mechanism to be *group strategy-proof* (GSP, for short). A mechanism $\mathcal{M}$ is GSP if for any profile $\mathbf{b}$ and any coalition $C \subseteq [k]$, there is no joint deviation $\mathbf{b}'_C = (b'_i)_{i \in C}$ of the agents in C such that no agent in C loses and at least one gains, that is, for all $\mathbf{b}$, for all $C \subseteq [k]$ and for all $\mathbf{b}'_C$ there exists $i \in C$ such that

$$u_i(b_i, \mathcal{M}(\mathbf{b})) > u_i(b_i, \mathcal{M}(\mathbf{b}'_C, \mathbf{b}_{-C}))$$

or for all $i \in C$,

$$u_i(b_i, \mathcal{M}(\mathbf{b})) \geq u_i(b_i, \mathcal{M}(\mathbf{b}'_C, \mathbf{b}_{-C})),$$

where $\mathbf{b}_{-C} = (b_i)_{i \notin C}$. We restrict the focus in this work on optimal mechanisms for the *social welfare*, that is, we want $\mathcal{M}$ to find, on input an instance $\mathbf{b} = (b_1, \ldots, b_k)$, the point $f^\star = \arg\max_{f \in \Gamma} SW(\mathbf{b}, f)$, where $SW(\mathbf{b}, f) = \sum_{i=1}^{k} u_i(b_i, f)$. Clearly, optimality depends on the choice of the feasible region Γ, but we omit this dependence when referring to optimal mechanisms for clarity of exposition.

3   Unknown Preferences

We begin by introducing some notations, that allow a more useful formulation of the social welfare and ultimately a geometric characterization of the optimum.

Fix a profile $\mathbf{b}$. Recall that m denotes the number of agents whose preference is 1 and n the number of agents whose preference is -1. Note that in the case of unknown preferences, m and n depend on the agents' strategies. We let $\gamma = m - n$, $s_p = \sum_{i:p_i=p} x_i$, for $p \in \{-1,1\}$ and $\delta = s_1 - s_{-1}$. Tedious calculations can verify that we can rewrite $SW(\mathbf{b},(x_f,y_f))$ as follows:

$$-\gamma\left(x_f - \frac{\delta}{\gamma}\right)^2 - \gamma y_f^2 + \frac{\delta^2}{\gamma} + m\lambda - \left(\sum_{i:p_i=1} x_i^2 - \sum_{i:p_i=-1} x_i^2\right) \qquad \text{if } \gamma \neq 0;$$

$$2\delta x_f + m\lambda - \left(\sum_{i:p_i=1} x_i^2 - \sum_{i:p_i=-1} x_i^2\right) \qquad \text{if } \gamma = 0.$$

Therefore, for a given instance $\mathbf{b}$, optimizing the social welfare is equivalent to finding a point (i.e., the x_f and y_f) in Γ that maximizes the equations above, in the respective cases. This amounts to choosing a point in Γ with maximum/minimum x-coordinate (depending on the sign of δ) when the number of players with preference 1 is equal to the number of players whose preference is -1. In the case in which $m \neq n$ ($\gamma \neq 0$), however, we need to maximize the quadratic equation. This is equivalent to finding the point in Γ that either maximizes (when $\gamma < 0$) or minimizes (when $\gamma > 0$) the (square of the) distance from the point $\beta = \left(\frac{\delta}{\gamma}, 0\right)$.[2]

Below, we will let $d(\cdot,\cdot)$ denote the distance between two points. Moreover, we will let $m,n,s_1,s_{-1},p_i,\delta$ denote the parameters for the instance in which agent i is truthful and $f = (x_f,y_f)$ be the output of the mechanism. $x_\beta = \delta/\gamma$ naturally corresponds to the x-coordinate of point β defined above. We add a prime symbol to denote aspects of the instance where agents misreport their type.

We are now ready to prove the first part of our dichotomy.

Theorem 1. *For all $\Gamma \subseteq \mathbb{R}^2$, the optimum mechanism is GSP.*

Proof. We focus on the optimum mechanism that breaks ties between solutions in a bid-independent way, that is, if there are $f, f' \in \Gamma$ that are optimal for two instances of the problem, then the mechanism will consistently return the same (e.g., the one with minimum y-coordinate and, in case of further ties, with the minimum x-coordinate).

Assume that there exists $\Gamma \subseteq \mathbb{R}^2$ such that this optimum mechanism is not GSP. This means that there is a coalition C that by joint deviation manages to change the outcome from f to f' while no agent in C loses and at least one gains. We denote with C_l the subset of C containing the agents in the coalition who lie. We extend the notations above with a c and $-c$ symbol to restrict the respective quantities to the agents inside and outside C, respectively; so, for example, γ_c (γ_{-c}, resp.) denotes the difference between the number of agents inside

---

[2] This geometric characterization of the optimum is the only aspect where the quadratic distances play a fundamental role; with Euclidean distances the optimum is less well behaved. For the agents' utilities and the optimum, maximizing/minimizing distances is equivalent to maximizing/minimizing the square of the distances.

(outside, resp.) C with preference 1 and those with preference -1. Furthermore, we use subscripts c, l and c, nl to differentiate the quantities calculated on the agents in C who lie and do not lie, respectively. So for example, $\gamma_{c,l}$ and $\gamma_{c,nl}$ correspond to the difference between the numbers of preferences for members of the coalition that lie and do not lie, respectively.

We begin by showing that if there is a profitable deviation for the coalition, then $x_f \neq x_{f'}$. First, consider the case in which the agents in C have heterogeneous preferences, that is, there are agents in C with either preference. Since the mechanism is not GSP then there exist agents i and j in C such that $p_i = 1$ and $p_j = -1$ and it holds that:

$$d^2(x_i, f) > d^2(x_i, f')$$
$$d^2(x_j, f) \leq d^2(x_j, f')$$

where we assumed w.l.o.g. that i is the agent in C for whom the inequality is strict (at least one such agent must be in C). By simple algebraic manipulations, we conclude that

$$(x_j - x_i)(x_f - x_{f'}) > 0.$$

This implies that $x_f \neq x_{f'}$.

Consider now the case in which all the agents in C have the same preference, so that $\gamma_c \neq 0$. Assume for a contradiction that $x_f = x_{f'}$. We shall prove that:

$$\gamma_c(\gamma_{c,l} + \gamma_{c,nl} + \gamma_{-c}) \leq 0, \tag{1}$$
$$\gamma_c(-\gamma_{c,l} + \gamma_{c,nl} + \gamma_{-c}) \geq 0. \tag{2}$$

First observe that $\gamma = \gamma_{c,l} + \gamma_{c,nl} + \gamma_{-c}$ and $\gamma' = -\gamma_{c,l} + \gamma_{c,nl} + \gamma_{-c}$, so it suffices to prove that $\gamma_c \cdot \gamma \leq 0$ and that $\gamma_c \cdot \gamma' \geq 0$. Consider the case in which $\gamma_c > 0$ (the case $\gamma_c < 0$ is symmetric). Since the coalition finds it profitable to change the output from f to f' then, since $x_f = x_{f'}$, it must be the case that $|y_{f'}| < |y_f|$. But then since the optimum for the original instance chooses f but not f' it cannot be that $\gamma > 0$. We then have that $\gamma_c \cdot \gamma \leq 0$, thus proving (1). Similarly, given that the optimum for the modified instance (in which the agents in C lie) returns f' and not f it cannot be that $\gamma' < 0$, hence $\gamma_c \cdot \gamma' \geq 0$. Summing up (1) and (2), we have $2\gamma_c\gamma_{c,l} \leq 0$, which is a contradiction since γ_c and $\gamma_{c,l}$ have the same sign and $\gamma_c \neq 0$.

It now remains to argue only about the case $x_f \neq x_{f'}$. Contradicting this case as well will prove that there is no profitable deviation for a coalition, as desired. Let ϵ be

$$\epsilon = \frac{x_f + x_{f'}}{2} + \frac{x_f + x_{f'}}{2} \frac{y_f - y_{f'}}{x_f - x_{f'}}. \tag{3}$$

Intuitively, the point $(\epsilon, 0)$ is the intersection of the x-axis with the perpendicular crossing the middle of the line segment connecting f and f'. As $x_f \neq x_{f'}$, this intersection must exist. Note that ϵ is an important parameter to determine where β and β' are in the cases in which $\gamma \neq 0$ and $\gamma' \neq 0$. In fact, $(\epsilon, 0)$ partitions the points on the x-axis according to the facility they are closer to. So, for example, by definition of the optimum, β and f must be on the same side of $(\epsilon, 0)$ for $\gamma > 0$.

We continue with two observations (inequalities (4) and (5)) that will be useful later on. When $\gamma_{c,l} = 0$, we have

$$\delta_{c,l}(x_f - x_{f'}) \leq 0. \tag{4}$$

Indeed, since $\gamma_{c,l} = 0$ then the coalition is heterogeneous and $m_{c,l} = n_{c,l}$. Assume that $x_{f'} < x_f$ (the opposite case being symmetric). Since the coalition prefers f' over f, then it must be that $x_j \leq \epsilon \leq x_l$, for every $j, l \in C$ such that $p_j = 1$ and $p_l = -1$. But then since $\gamma_{c,l} = 0$ (i.e., there is a bijection between agents lying in either direction) we can conclude that $\delta_{c,l} \leq 0$.

When $\gamma_{c,l} \neq 0$, we can prove:

$$\gamma_{c,l}(x_{\beta_{c,l}} - \epsilon)(x_f - x_{f'}) \leq 0. \tag{5}$$

Indeed, consider the case $\gamma_{c,l} > 0$, i.e., $m_{c,l} > n_{c,l}$. Assume that $x_{f'} < x_f$; as argued above, all the agents in the coalition with preference 1 (-1, resp.) must be to the left (right, resp.) of point $(\epsilon, 0)$ and at least one must have a location different from ϵ. If $n_{c,l} = 0$, then we can conclude that $x_{\beta_{c,l}} \leq \epsilon$ and prove (5). If $n_{c,l} > 0$ then choose a subset S of $n_{c,l}$ agents in the coalition with preference 1 who lie, including one with location not ϵ (if any). We can then conclude that $\sum_{j \in S} x_j - \sum_{l \in C_l, p_l = -1} x_l < 0$. But then, observing that $\sum_{j \in C_l \setminus S, p_j = 1} x_j \leq (m_{c,l} - n_{c,l})\epsilon$, we can conclude that $x_{\beta_{c,l}} < \epsilon$ and prove (5). (The remaining cases can be proved with the same argument mutatis mutandis.)

In order to conclude the proof of the theorem, we will now show a contradiction with (either) (4) and (5); we will consider three different cases depending on the values of γ and γ'. Note that $\delta' = -\delta_{c,l} + \delta_{c,nl} + \delta_{-c}$ and so:

$$x_\beta = \frac{\delta_{c,l} + \delta_-}{\gamma_{c,l} + \gamma_-} \text{ and } x'_\beta = \frac{-\delta_{c,l} + \delta_-}{-\gamma_{c,l} + \gamma_-},$$

where the $-$ as a subscript denotes the parameters of the instance that do not change because of the lies (i.e., the sum of c, nl and $-c$ components).

Case $\gamma \neq 0$ and $\gamma' \neq 0$. Let us only discuss here $\gamma > 0, \gamma' > 0$; the other cases can be proved with the same argument. By the definition of optimum for positive values of γ and γ', we have:

$$(x_\beta - \epsilon)(x_f - x_{f'}) \geq 0 \text{ and } (x_{\beta'} - \epsilon)(x_f - x_{f'}) \leq 0.$$

Observe that since the optimum uses a fixed-tie breaking rule it cannot be the case that both the inequalities above are actually equality. In fact, were this the case, then f and f' would be optimal locations for the facility in both instances;

this would contradict that $f \neq f'$. We assume without loss of generality that the first is true with a strict sign. Therefore, for $x_f > x_{f'}$ we have

$$\frac{\delta_{c,l} + \delta_-}{\gamma_{c,l} + \gamma_-} > \epsilon \text{ and } \frac{-\delta_{c,l} + \delta_-}{-\gamma_{c,l} + \gamma_-} \leq \epsilon.$$

By simple algebraic manipulations, we can conclude that $\delta_{c,l} - \epsilon\gamma_{c,l} > 0$. Similarly, we can show that $x_f < x_{f'}$ yields $\delta_{c,l} - \epsilon\gamma_{c,l} < 0$. But this contradicts (5), when $\gamma_{c,l} \neq 0$ and (4) in the case $\gamma_{c,l} = 0$.

Case $\gamma = 0$ and $\gamma' = 0$. We begin by observing that, by summing up the conditions on $\gamma = 0, \gamma' = 0$, we get that $\gamma_{c,l} = 0$ (and then (5) does not hold in this case). By the definition of optimum, we have :

$$(\delta_{c,l} + \delta_-)(x_f - x_{f'}) > 0$$
$$(-\delta_{c,l} + \delta_-)(x_f - x_{f'}) \leq 0,$$

where, as above, we assume without loss of generality that the first is strictly true (one has to be strict by the tie-breaking rule of the optimum algorithm). From the two inequalities, we get $\delta_{c,l}(x_f - x_{f'}) > 0$, which contradicts (4).

Case $\gamma = 0$ (exclusive) or $\gamma' = 0$. We here discuss only $\gamma > 0, \gamma' = 0$; the remaining cases can be proved in the same manner. By the definition of optimum, we have:

$$(x_\beta - \epsilon)(x_f - x_{f'}) > 0$$
$$(-\delta_{c,l} + \delta_-)(x_f - x_{f'}) \leq 0.$$

Again, the former is assumed to be strict by the tie-breaking rule adopted by the optimum. As $\gamma' = -\gamma_{c,l} + \gamma_- = 0$, we have $\gamma_{c,l} = \gamma_-$ and then since $\gamma = \gamma_{c,l} + \gamma_- > 0$ we can conclude that $\gamma_{c,l} > 0$ (and so (4) does not hold here). Thus, from the first inequality, we have:

$$\frac{1}{2\gamma_{c,l}}(\delta_{c,l} + \delta_- - 2\epsilon\gamma_{c,l})(x_f - x_{f'}) > 0$$
$$\Rightarrow (\delta_{c,l} + \delta_- - 2\epsilon\gamma_{c,l})(x_f - x_{f'}) > 0.$$

Combined with the second inequality above, we have

$$(\delta_{c,l} - \epsilon\gamma_{c,l})(x_f - x_{f'}) > 0,$$

which contradicts (5).

4   Unknown Locations

We now prove the second part of our dichotomy result. Given that the value of $\gamma = m - n$ changes combinatorially the optimum and that, in the case of unknown locations, γ is known to the designer, our analysis needs to differentiate all the three cases about the relative order between m and n.

4.1   Case $m > n$

We begin with all the instances where the number of 1's is bigger than the number of -1's. For two points α and ζ on the plane, we let $C(\alpha, \zeta)$ denote the points in the interior of the circle centred in α of radius $d(\alpha, \zeta)$; formally,

$$C(\alpha, \zeta) = \left\{ (x, y) \middle| (x - x_\alpha)^2 + (y - y_\alpha)^2 < d^2(\alpha, \zeta) \right\}.$$

We denote by $\bar{C}(\alpha, \zeta)$ the points of the circle including those on the circumference.

Theorem 2. *When $m > n = 0$, if the optimum mechanism is strategy-proof, then Γ is such that the optimum is constant.*

Proof. We are going to show, through a sequence of instances, that for the optimum to be strategy-proof on those instances, Γ must have a certain shape. We can then observe that given such a shape, the optimum is constant no matter the instance. Specifically, we will prove that there is $f \in \Gamma$ such that

$$\left[C(\beta_{\min}, f) \cup C(\beta_{\max}, f) \right] \cap \Gamma = \emptyset, \tag{6}$$

where $\beta_{\max} = (\ell, 0)$ and $\beta_{\min} = (0, 0)$.

We let f be the optimum of the instance in which $m - 2$ agents are on $\ell/2$, one agent i is on $x_L = \frac{\ell}{2}\xi$ and the last one, named j, is on $x_R = \frac{\ell}{2}\zeta$, where $\xi = \frac{m}{m+1} < 1$ and $\zeta = \frac{m+2}{m+1} > 1$. We call this instance $\mathbf{x}$ and note that $x_\beta = \ell/2$. Assume agent i lies and reports 0 instead of x_L. Then the optimum will be computed according to point β', which is such that $x_{\beta'} = x_L$. By strategy-proofness, the outcome cannot be closer to i's true location (which is the same as β'), i.e., $C((x_L, 0), f) \cap \Gamma = \emptyset$.

We now will argue that there is a sequence of instances that prove in the limit that $C(\beta_{\min}, f) \cap \Gamma = \emptyset$. We set $\mathbf{x}_1$ to be as follows: $m - 2$ agents are on $\frac{\ell}{2}\xi$, x_{L_1} is $\xi^2(\ell/2)$ and x_{R_1} is such that the point used to calculate the optimum, denoted β_1, satisfies $x_{\beta_1} = x_L$ (that is, $x_{R_1} = \frac{\ell}{2}\frac{m(m+2)}{(m+1)^2} < \frac{\ell}{2}$). From the previous step we know that Γ has no intersection with $C(\beta_1, f)$ and then the optimum in this case will again be f. Now, consider the case in which the true type of i is x_{L_1} and she misreports to 0. Then the optimum f_1' will be computed according to point β_1', which satisfies $x_{\beta_1'} = x_{L_1}$. By strategy-proofness, we must then have that $C((x_{L_1}, 0), f) \cap \Gamma = \emptyset$.

We can now iterate the reasoning above and define instance $\mathbf{x}_r$, as follows: $x_{L_r} = (\ell/2)\xi^{r+1}$, $m - 2$ agents are on $x_{\beta_{r-1}}$ and $x_{R_r} \leq \ell$ is such that the point used to calculate the optimum, denoted β_r satisfies $x_{\beta_r} = x_{L_{r-1}}$. Using similar reasoning, we conclude that $C(\beta_r, f) \cap \Gamma = \emptyset$. Given that $\xi < 1$, in the limit we have that $C(\beta_{\min}, f) \cap \Gamma = \emptyset$, as desired.

We can use the same argument now on the right side of β to conclude that $C(\beta_{\max}, f) \cap \Gamma = \emptyset$.

We adopt a different argument to account for the case in which there are agents who dislike the facility.

Theorem 3. *When $m > n > 0$, if the optimum mechanism is strategy-proof, then Γ is such that the optimum is constant.*

4.2   Case $m < n$

The arguments used for the case $m > n$ can be used in a very similar manner to prove the following claim. The only change is in the definition of optimum (maximum distance from Γ as opposed to minimum distance as for $m > n$) and, consequently, the constraints on the shape of the feasible region Γ. We omit the details.

Theorem 4. *When $m < n$, if the optimum mechanism is strategy-proof, then Γ is such that the optimum is constant.*

4.3   Case $m = n$

We now complete our proof for the case in which $m = n > 1$; we leave the arguably less interesting case of two agents open for future research.

Theorem 5. *When $m = n > 1$, if the optimum mechanism is strategy-proof, then Γ is such that the optimum is constant.*

Proof. Recall that when $m = n$ ($\gamma = 0$), the optimum requires to choose the point in Γ with maximum/minimum x-coordinate (depending on the sign of δ). We will then prove that for all $f, f' \in \Gamma$, it holds $x_{f'} = x_f$.

Consider the instance $\mathbf{x}$ comprised of four agents, where

$$0 < x_1 < x_2 < x_3 < x_4 < \ell$$
$$p_1 = p_3 = 1, \quad p_2 = p_4 = -1,$$
$$x_1 + \ell - x_2 - x_4 > 0,$$
$$x_1 + x_3 - x_4 > 0.$$

The above can be satisfied for example if $x_1 = 2\xi, x_2 = 3\xi, x_3 = 4\xi, x_4 = 5\xi$, and $\xi = \ell/7$. Let f be the optimum for instance $\mathbf{x}$. Observe that δ has negative sign and therefore f has minimum x-coordinate in Γ. Assume by contradiction that the claim is not true and let f' be one of the rightmost points in Γ (a point with the maximum x-coordinate), such that $x_{f'} > x_f$. Let ϵ be the quantity defined in (3) for f and f' and recall that $(\epsilon, 0)$ partitions the points on the x-axis according to the facility they are closer to.

If $\epsilon < x_3$, then the third agent has an incentive to declare $x'_3 = \ell$ so that $\delta' > 0$ (this is guaranteed by the definition of the instance) and the optimum becomes f', which is closer to x_3 than f. Similarly, when $\epsilon > x_2$, the second agent has an incentive to declare $x'_2 = 0$ to change the sign of δ and move the (undesirable) facility further from her location. Observe that one of these two conditions on ϵ must be true, since $x_3 \neq x_2$. Thus, wherever ϵ is, there exists at least one agent who has an incentive to lie – a contradiction.

5 Conclusions

We have introduced a new perspective in the research on mechanism design without money. Whereas the quality of the solutions, in terms of their approximation guarantee, has been used as a way to obtain truthfulness (or strategy-proofness), we propose here to use the feasibility of the solution space as a way to get incentive-compatibility. The former is usually detrimental to the designer, who instead might well be in charge of defining feasibility. Just as one aims at the best possible approximation, here we would aim at having the largest possible set of feasible solutions.

In addition to this conceptual contribution, our work has given a set of involved technical contributions showing a dichotomy in the case study of heterogeneous facility location problem. Whilst any feasible region can be used to design optimal GSP mechanisms when agents can lie about their preferences, very little can be done for SP mechanisms facing agents who can misreport their location.

Our work leaves a number of compelling open questions. Even only for the variant of facility location considered, one might wonder to what extent the two sides of our dichotomy generalize. For the positive side of the coin of unknown preferences, we wonder whether a similar theorem holds in the case in which the agents are located on a bidimensional subset of $\mathbb{R}^2$ rather than a segment; possible interesting case studies include agents located on the boundary of a circle, a region expressed by a quadratic function, or even an arbitrary shape. The negative part of our dichotomy for unknown locations is reminiscent of the known characterization for collusion-resistant mechanisms with money [16]. Differently from that, our constant-outcome characterization is qualified by optimality. The natural next question is then to relax the requirement of optimality to (constant) approximations. One question of interest could be: What approximation guarantee allows us to achieve truthfulness for every possible combination of positive and negative preferences? More generally, our research agenda can be applied to other mechanism design optimization problems studied in the literature; how many feasible solutions can we allow to get truthfulness?

References

1. Anastasiadis, E., Deligkas, A.: Heterogeneous facility location games. In: AAMAS, pp. 623–631 (2018)
2. Barberá, S., Gul, F., Stacchetti, E.: Generalized median voter schemes and committees. J. Econ. Theor. **61**(2), 262–289 (1993)
3. Dokow, E., Feldman, M., Meir, R., Nehama, I.: Mechanism design on discrete lines and cycles. In: ACM EC, pp. 423–440 (2012)
4. Endriss, U.: Judgment aggregation. In: Brandt, F., Conitzer, V., Endriss, U., Lang, J., Procaccia, A.D. (eds.) Handbook of Computational Social Choice, Chapter 17. Cambridge University Press, Cambridge (2016)
5. Ferraioli, D., Serafino, P., Ventre, C.: What to verify for optimal truthful mechanisms without money. In: AAMAS, pp. 68–76 (2016)

6. Fong, C.K.K., Li, M., Lu, P., Todo, T., Yokoo, M.: Facility location games with fractional preferences. In: AAAI (2018)
7. Fotakis, D., Tzamos, C.: On the power of deterministic mechanisms for facility location games. In: ICALP, pp. 449–460 (2013)
8. Gibbard, A.: Manipulation of voting schemes: a general result. Econometrica **41**(4), 587–601 (1973)
9. Lu, P., Sun, X., Wang, Y., Zhu, Z.A.: Asymptotically optimal strategy-proof mechanisms for two-facility games. In: ACM EC, pp. 315–324 (2010)
10. Lu, P., Wang, Y., Zhou, Y.: Tighter bounds for facility games. In: WINE, pp. 137–148 (2009)
11. Moulin, H.: On strategy-proofness and single-peakedness. Public Choice **35**, 437–455 (1980)
12. Nisan, N., Roughgarden, T., Tardos, E., Vazirani, V.V.: Algorithmic Game Theor. Cambridge University Press, Cambridge (2007)
13. Penna, P., Ventre, C.: Sharing the cost of multicast transmissions in wireless networks. In: SIROCCO, pp. 255–266 (2004)
14. Procaccia, A.D., Tennenholtz, M.: Approximate mechanism design without money. ACM Trans. Econ. Comput. **1**(4), 18 (2013)
15. Satterthwaite, M.A.: Strategy-proofness and arrow's conditions: existence and correspondence theorems for voting procedures and social welfare functions. J. Econ. Theor. **10**(2), 187–217 (1975)
16. Schummer, J.: Manipulation through bribes. J. Econ. Theor. **91**(2), 180–198 (2000)
17. Schummer, J., Vohra, R.V.: Strategy-proof location on a network. J. Econ. Theor. **104**, 405–428 (2002)
18. Serafino, P., Ventre, C.: Heterogeneous facility location without money on the line. In: ECAI - Including PAIS, pp. 807–812 (2014)
19. Serafino, P., Ventre, C.: Truthful mechanisms without money for non-utilitarian heterogeneous facility location. In: AAAI, pp. 1029–1035 (2015)
20. Serafino, P., Ventre, C.: Heterogeneous facility location without money. Theor. Comput. Sci. **636**, 27–46 (2016)
21. Zou, S., Li, M.: Facility location games with dual preference. In: AAMAS, pp. 615–623 (2015)

Obvious Strategyproofness, Bounded Rationality and Approximation
The Case of Machine Scheduling

Diodato Ferraioli[1]([⊠])([iD]) and Carmine Ventre[2]([iD])

[1] Università degli Studi di Salerno, Fisciano, SA, Italy
dferraioli@unisa.it
[2] King's College London, London, UK
carmine.ventre@kcl.ac.uk

Abstract. Obvious strategyproofness (OSP) has recently emerged as the solution concept of interest to study incentive compatibility in presence of agents with a specific form of bounded rationality, i.e., those who have *no* contingent reasoning skill whatsoever. We here want to study the relationship between the approximation guarantee of incentive-compatible mechanisms and the *degree* of rationality of the agents, intuitively measured in terms of the number of contingencies that they can handle in their reasoning. We weaken the definition of OSP to accommodate for cleverer agents and study the trade-off between approximation and agents' rationality for the paradigmatic machine scheduling problem. We prove that, at least for the classical machine scheduling problem, "good" approximations are possible if and only if the agents' rationality allows for a significant number of contingencies to be considered, thus showing that OSP is not too restrictive a notion of bounded rationality from the point of view of approximation.

Keywords: Mechanism design · Machine scheduling ·
Simple mechanisms · Bounded rationality · Lookahead

1 Introduction

Mechanism design is an established research field, by now rooted in a number of academic disciplines including theoretical computer science and AI. Its main objective is that of computing in presence of selfish agents who might misguide the designer's algorithm if it is profitable for them to do so. The concept of *strategyproofness* (SP-ness) (a.k.a., *truthfulness*) ensures that the algorithm and the agents' incentives are compatible and computation is indeed viable.

SP is based on the assumption of full rationality: agents are able to consider all possible strategies and their combinations to reason about their incentives.

An extended abstract of the paper appeared as [15].

D. Ferraioli—Partially supported by GNCS-INdAM and by the Italian MIUR PRIN 2017 Project ALGADIMAR "Algorithms, Games, and Digital Markets".

D. Fotakis and E. Markakis (Eds.): SAGT 2019, LNCS 11801, pp. 77–91, 2019.
https://doi.org/10.1007/978-3-030-30473-7_6

Nevertheless, this assumption is seldom true in reality and it is often the case that people strategize against mechanisms that are known to be truthful [4]. One then needs a different notion to compute in the presence of agents with bounded rationality. The problem here is twofold: how can we formalize strategyproofness for agents with (some kind of) bounded rationality? If so, can we *quantify* this bounded rationality and relate that to the *performances* of the mechanisms?

The first question has been recently addressed by Li [18], who defines the concept of *obvious strategyproofness* (OSP-ness); this notion has attracted quite a lot of interest in the community [3,6,12–14,17,19,21,24]. Here, the mechanism is seen as an extensive-form game; when a decision upon the strategy to play has to be made, it is assumed that the reasoning of each agent i is as simple as the following: the *worst* possible outcome that she can get when behaving well (this typically corresponds to playing the game according to the so-called agent's true *type*) must be at least as good as the *best* outcome when misbehaving (that is, following a different strategy). Best/Worst are quantified over *all* the possible strategies that the players playing in the game after i can adopt. Li [18] proves that this is the right solution concept for a model of bounded rationality wherein agents have *no* contingent reasoning skills; rather than thinking about the possible cases of if-then-else's, an agent is guaranteed that honesty is the best strategy to follow no matter all the contingencies.

Given the OSP formalization of bounded rationality, we focus, in this work, on the second question. On the one hand, OSP is too restrictive in that people might be able, within their computational limitations, to consider *some* contingent reasoning, that is, a few cases of if-then-else's. On the other hand, OSP mechanisms appear to be quite limited, with respect to SP ones, in terms of their approximation guarantee [12,13]. The question then becomes:

> *Can we quantify the trade-off between the "degree" of bounded rationality of the agents and the approximation guarantee of the mechanisms incentivizing them?*

Our Contribution. The concept of *lookahead* is discussed in the literature in the context of (strategies to play) games, and agents with limited computational capabilities. De Groot [9] found that all chess players (of whatever standard) used essentially the same thought process – one based upon a lookahead heuristic. Shannon [23] formally proposed the lookahead method and considered it a practical way for machines to tackle complex problems, whilst, in his classical book on heuristic search, Pearl [20] described lookahead as the technique being used by "almost all game-playing programs".

We propose to consider lookahead as a way *to quantify bounded rationality*, in relation to OSP. Whilst in OSP the players have no lookahead at all, we here consider the case in which the agents have lookahead k, k going from 0 (OSP) to $n-1$ (SP). Intuitively, k measures the number of players upon which each player reasons about in her decision making. We allow the set of k "lookahead" players to be player and time specific (that is, different players can reason about different competitors, and the set of players is not fixed but may change at different time steps of the mechanism). So when agent i has to decide upon the strategy to play, she will consider all the possible cases (strategies) for these k agents at that time (à la SP) and a no-contingent reasoning (à la OSP) for the others. This definition, which is somewhat different from that of the next k moves in the game, is dictated by different subtleties of extensive-form mechanisms. In particular, these k agents can be chosen in different ways to cover diverse angles. (A more technical discussion is deferred to Sect. 2.) In absence of other formal definitions of incentive compatibility for different degrees of rationality, we regard our definition of *OSP with k-lookahead* (k-OSP, for short) as a major conceptual contribution of our work.

We then look at the trade-off between the value of k and the approximation guarantee of k-OSP mechanisms. We focus of the well-studied problem of machine scheduling, where n agents control related machines and the objective is to schedule a set of m (identical) jobs to the machines so to minimize the *makespan* (i.e., the latest machine's completion time). In our main technical contribution, we prove a lower bound on approximation guarantee of $\tau_k(n) = \frac{\sqrt{k^2+4n}-k}{2}$, thus providing a smooth transition function between the known approximation factors of $\sqrt{n}$ for OSP mechanisms [12] and 1 for SP mechanisms [2]. We also show that this bound is tight, at least for three-values domains. (Such a restriction is common to the state of the art of OSP mechanisms [12].) Our lower and upper bounds significantly extend and generalize to k-OSP the analysis done in [12] for OSP mechanisms. Specifically, the lower bound needs to identify some basic properties of the function $\tau_k(n)$ and prove what features the *implementation tree* of a mechanism (i.e., extensive-form game induced by it) with good approximation guarantee must have. Our upper bound instead defines a mechanism (algorithm, implementation tree and payment function) which combines a descending auction phase, to identify a certain number of slowest machines, with an ascending auction to find out the $k + 1$ fastest machines. The analysis of the approximation guarantee of our k-OSP mechanism is significantly more involved than the one used in [12] for $k = 0$.

The main message of our work is that having more rational agents only slightly improves the approximation guarantee of incentive-compatible mechanisms, at least in the case of machine scheduling. In fact, to have a constant approximation of the optimum makespan one would need agents with $\omega(1)$-lookahead. We can then conclude that, in the cases in which the agents are not that rational, OSP is not that restrictive a solution concept to study the approximation of mechanisms for agents with bounded rationality.

Related Work. Recent research in algorithmic mechanism design has suggested to focus on "simple" mechanisms to deal with bounded rationality [7,16,22]. OSP provides a formal definition for simple mechanisms, by focusing on a specific aspect of bounded rationality (see references above for the body of work on this concept). However, different concepts of simple mechanisms have been recently adopted in literature, most prominently posted-price mechanisms have received great attention and have been applied to many different settings [1,5,8,10,11].

2   The Definition

We have a set N of n agents; each agent i has a domain D_i of possible *types* – encoding some feature of theirs (e.g., their speed). The actual type of agent i is her private knowledge.

An extensive-form mechanism $\mathcal{M}$ is a triple $(f, p, \mathcal{T})$, where f is an algorithm that takes as input bid profiles and returns a feasible solution, $p = (p_1, \ldots, p_n)$ is the payment function, one for each agent, and $\mathcal{T}$ is an extensive-form game, that we call *implementation tree*[1]. Intuitively, $\mathcal{T}$ represents the steps that the mechanism will take to determine its outcome. More formally, each internal node u of $\mathcal{T}$ is labelled with a player $S(u)$, called the *divergent agent* at u, and the outgoing edges from u are labelled with types in the domain of $S(u)$ that are compatible with the history leading to u; the edge labels denote a partition of the compatible types. We denote by $D_i(u)$ the types in the domain of i that are compatible with the history leading to node $u \in \mathcal{T}$. The tree models how $\mathcal{M}$ interacts with the agents: at node u the agent $S(u)$ is queried and asked to choose an action, that corresponds to selecting one of u's outgoing edges. The chosen action *signals* that the type of $S(u)$ is in the set of types labeling the corresponding edge. The leaves of the tree will then be linked to (a set of) bid profiles; the mechanism will return (f, p) accordingly; in other words, each leaf corresponds to an outcome of the mechanism. (Observe that this means that the domain of f and p is effectively given by the leaves of $\mathcal{T}$.)

We use $\mathbf{b}$ to denote bid profiles, so that b_i stands for the type that i signalled to the mechanism. For simplicity, we use $f(\mathbf{b})$ and $p_1(\mathbf{b}), \ldots, p_n(\mathbf{b})$ to denote the outcome of (f, p) for the leaf of $\mathcal{T}$ to which $\mathbf{b}$ belongs. We assume that agents have quasi-linear utilities, that is, agent i of type t who signals (i.e., plays the game $\mathcal{T}$ according to) b has utility $u_i(b, \mathbf{b}_{-i}) = p_i(\mathbf{b}) - t(f(\mathbf{b}))$, where, with a slight abuse of notation, $t(f(\mathbf{b}))$ is the cost that player i pays to implement the outcome $f(\mathbf{b})$ when her type is t, and $\mathbf{b}_{-i}$ is the declaration vector of (i.e. types signalled by) all agents except i. (In general, we let $\mathbf{b}_A = (b_j)_{j \in A}$ for $A \subset N$.)

Figure 1 gives an example of an implementation tree where three players have a two-value domain $\{L, H\}$. The root partitions the domain of machine 1 into L and H. If we let v denote the left child of the root, then $D_1(v) = \{L\}$ as type H is no longer compatible with the history of v.

We now define *OSP with k-lookahead*. OSP informally implies that whenever an agent is asked to diverge, she is better off acting according to her true type in *any* possible future scenario: the worst possible outcome after selecting her true type is at least as good as the best possible outcome after misreporting her type, at that particular point in the implementation tree. This models agents with no contingent reasoning, i.e., those unable to think through hypothetical

---

[1] The literature on mechanism design usually omits $\mathcal{T}$ from the definition of mechanism, since it often focuses only on specific classes of mechanisms defined by a given implementation tree (e.g., direct revelation mechanisms, posted price mechanisms). However, it turns out that for OSP (and k-OSP) the design of the extensive-form implementation is essential to define the incentive constraints.

scenarios such as "if player 2 will play L and player 3 will play L, then I prefer L; if they will play L and H respectively, then I prefer L, too; ans so on". In OSP, agent thinking is gross-grained: "If I play L, then the outcome will correspond to leaves $l_1, \ldots, l_4$, otherwise it will correspond to leaves $l_5, \ldots, l_8$".

However, it would be possible that agents have some limited ability of doing contingent reasoning: they can think through hypothetical scenarios corresponding to the action profiles of some players, but not all of them. Specifically, we would like to model a player able to reason as follows: "If player 2 will play L, I know that by choosing L I will finish either in l_1 or in l_2, otherwise I will finish in l_5 or l_6; if player 2 will play R, then my choice will be between the outcomes corresponding to l_3 and l_4 and the one corresponding to l_7 and l_8". That is, we here consider a more finely grained partition of the leaves of the tree, allowing for some steps of contingent reasoning by the divergent agent. Intuitively, our definition will allow the agent to reason about the moves of k agents; informally, OSP with k-lookahead then implies that whenever an agent is asked to diverge, she is better off acting according to her true type for *any fixed* choice of strategies of the k agents she reasons about (just like truthfulness) and *any* possible future scenario of the actions of the remaining $n - k - 1$ agents.

For the formal definition, we need to introduce some more notation. We call a bid profile $\mathbf{b}$ *compatible with* u if $\mathbf{b}$ is compatible with the history of u for all agents. We furthermore say that $(t, \mathbf{b}_{-i})$ and $(b, \mathbf{b}'_{-i})$ diverge at u if $i = S(u)$ and t and b are labels of different edges outgoing u (we sometimes will abuse notation and we also say that t and b diverge at u). E.g., (L, H, H) and (L, L, H) are compatible with node v on Fig. 1 and diverge at that node, whilst (L, L, H) and (L, L, L) are compatible with v but do not diverge at v.

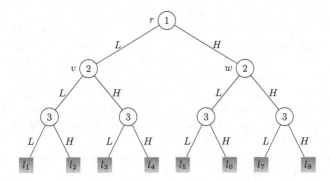

Fig. 1. An implementation tree with three players with two-value domains $\{L, H\}$; each player separates the domain types upon playing; at each leaf l_i the mechanism computes $f(\mathbf{b})$ and $p(\mathbf{b})$, $\mathbf{b}$ being the bid vector at l_i.

For every agent i and types $t, b \in D_i$, we let $u^i_{t,b}$ denote a vertex u in the implementation tree $\mathcal{T}$, such that $(t, \mathbf{b}_{-i})$ and $(b, \mathbf{b}'_{-i})$ are compatible with u, but diverge at u for some $\mathbf{b}_{-i}, \mathbf{b}'_{-i} \in D_{-i}(u) = \times_{j \neq i} D_j(u)$. Note that such a vertex might not be unique as agent i will be asked to separate t from b in different paths from the root (but only once for every such path). We call these vertices of $\mathcal{T}$ *tb-separating* for agent i. For example, the node r in the tree in Fig. 1 is a LH-separating node for agent 1; while v and w are two LH-separating node for agent 2. These nodes are crucial, as at any point in which an agent distinguishes two different types we will need to add a (set of) constraints to account for her incentives. We finally denote i's lookahead at $u^i_{t,b}$ as $\mathcal{L}_k(u^i_{t,b})$, that is, a set of (at most) k agents that move in $\mathcal{T}$ after i. (When k is clear from the context, we simply let $\mathcal{L}(u)$ be the lookahead of agent $S(u)$ at u.)

Definition 1 (OSP with k-lookahead). *An extensive-form mechanism* $\mathcal{M} = (f, \mathcal{T}, p)$ *is OSP with k-lookahead (k-OSP, for short) given* $\mathcal{L}_k(u^i_{t,b})$, *if for all* i, $t, b \in D_i$, t *being i's true type,* $u^i_{t,b} \in \mathcal{T}$, $\mathbf{b}_K \in D_K(u^i_{t,b})$ *and* $\mathbf{b}_T, \mathbf{b}'_T \in D_T(u^i_{t,b})$, *it holds that*

$$u_i(t, \mathbf{b}_K, \mathbf{b}_T) \geq u_i(b, \mathbf{b}_K, \mathbf{b}'_T),$$

where $K = \mathcal{L}_k(u^i_{t,b})$, $T = N \setminus (K \cup \{i\})$ *and* $D_A(u) = \times_{j \in A \subset N} D_j(u)$.

In words, a mechanism is OSP with lookahead if each agent is willing to behave truthfully at each node of the tree in which she interacts with the mechanism, provided that she exactly knows the types of agents in K ($\mathbf{b}_K$ is the same either side of the inequality) but has no information about agents in T, except that their types are compatible with the history.

We remark that with $k = 0$ we get the definition of OSP – wherein K is empty – and with $k = n - 1$ we have truthfulness, T being empty.

Discussion. The set $\mathcal{L}_k(u)$ in the definition above crucially captures our notion of lookahead. We highlight the following features of our definition. The size of set $\mathcal{L}_k(u)$ tells us how many players, agent $S(u)$ can contingently reason about. This means that the boundaries of k indeed go from 0, which corresponds to OSP, to $n - 1$, which is equivalent to strategyproofness. In this sense, our definition represents a smooth transition between the two notions, measuring the degree of rationality of the players. For example, consider Fig. 1 and focus on player 1; when $k = 0$ then our notion is exactly OSP and the constraints require to compare the utility of 1 in the leaves $l_1, \ldots, l_4$ with her utility in $l_5, \ldots, l_8$; when, instead, $k = 1$ and $\mathcal{L}_1(r) = \{2\}$ then the constraints compare the utility of 1 in the leaves l_1, l_2 with that in l_5, l_6 (this corresponds to the case in which 2 plays L) *and* the utility of 1 in the leaves l_3, l_4 with that in l_7, l_8 (this corresponds to the case in which 2 plays H); finally, for $k = 2$ we get truthfulness as we need to compare the utility of 1 in l_j and l_{4+j} for $j = 1, \ldots, 4$. We note that intermediate values of k are consistent with the vast literature stating that human reasoning only has limited depth: for example, it is known that in chess most professional players are usually able to think ahead few steps only [9]. We remark that k-OSP differs from k-level reasoning: the latter considers a Nash equilibrium in which an agent plays a best response to what happens in the next k steps; the former considers a(n obviously) dominant strategy.

The set $\mathcal{L}_k(u)$ depends on u; this means that the number and the identities of players on which $S(u)$ can reason about can (in principle) adaptively depend on the actual position in the implementation tree. This in particular allows us to also capture extensive-form games where the choice of the players to query is adaptive and a definition of lookahead where the players on which $S(u)$ can reason about are (a subset of) those who move next: this is for example the case in many multi-player board games in which the player can take actions that change who is the next player to play, e.g., by blocking some opponents or reversing the order of play.

Note that whenever $\mathcal{L}_k(u) = \mathcal{L}_k(v)$ for $S(u) = S(v)$ then we model the case in which the lookahead is independent from the actual implementation tree and only depends on $S(u)$'s prior knowledge of the other agents.

Differently from the examples of chess and multi-player board games in which a player only looks ahead to opponents that play in the next rounds, our definition of $\mathcal{L}_k(u)$ allows this set to contain also players that will play far away in the future. This clearly makes our definition more general.

Moreover, we observe that this definition of $\mathcal{L}_k(u)$ also allows us to overcome a paradox that would arise if one defines the set of opponents that one looks ahead only with respect to the implementation tree. For the sake of argument, let us fix $k = 1$. Consider an adaptive implementation tree, where at node u different actions taken by agent $S(u)$ correspond to different players taking the next move. As a limit case, one can imagine that $S(u)$ has $n-1$ different available actions and each of them enables a different opponent to react (e.g., this is the case for those board games where each player can decide who plays next). Hence, assuming that $S(u)$ can look ahead to players moving in the next step means that $S(u)$ has the ability to look ahead to all of them. Hence, in this setting limited look-ahead is not limiting at all the ability of contingent reasoning of $S(u)$ (that is, in this setting every mechanism that is 1-OSP according to this tree-only definition of lookahead is actually SP).

This is not surprising, since in this setting we are giving each agent i the chance to "reason about" each opponent regardless of the action that i takes. A more realistic alternative would be to assume that the agent exactly knows the actions of an opponent j only when i takes an action that enables j to be the next player to play (e.g., in the board game example described above, the current player i is assumed to know which actions player j will take when i chooses j as the next player to play, but i has no hint about the actions of j if she chooses $k \neq j$ as the next player to play). However, in this case i would have to reason about all the possible action combinations of all the different players that move after her; this might not weaken OSP and indeed means that the agent is not more rational at all. In fact, a careful inspection shows that, in this case, 1-OSP according to this alternative definition of tree-only lookahead has the same constraints of OSP.

Anyway, it must be highlighted that in non-adaptive trees, i.e., trees where the identity of the next player to move after $S(u)$ is the same irrespectively of $S(u)$'s action, tree-only lookahead would indeed weaken OSP and effectively capture a more rational agent capable of one step of contingent reasoning. Since this is a special case of our notion, our lower bound continues to hold.

Our definition requires that an agent with k-lookahead is capable of exactly pinpointing the type of the agents in K. This is in fact the same assumption that is implicitly done in the classical definition of truthfulness. Moreover, this makes our definition of k-OSP mechanism a special case of mechanisms implementable with *partition dominant strategy* as defined in [24]. Consequently, our definition satisfies a natural generalization of the standard decision theory axioms of monotonicity, continuity and independence, necessary to model the reasoning of agents with a knowledge of the state of nature (e.g., the type profiles) limited only to partitions of the set of these states (e.g., the type profiles that are compatible with the history of the mechanism). We also observe that this requirement only reinforces our lower bound below (even if they were so rational to do that, still the approximation guarantee would be a constant only for non-constant values of k). However, we leave open the problem of understanding whether our upper bound is tight even for a weaker notion of rationality where the types of the agents in K are not fully known but only have further restricted domains (e.g., an agent with k-lookahead only knows the next ℓ actions, for some $\ell > 0$, that will be taken by the agents in K).

3   The Case of Machine Scheduling

We now study the relationship between lookahead and approximation for the well-studied problem of machine scheduling. Here, we are given a set of m identical jobs to execute and the n agents control related machines. Agent i's type is a job-independent processing time t_i per unit of job (equivalently, an execution speed $1/t_i$ that is independent from the actual jobs). The algorithm f must choose a possible schedule $f(\mathbf{b}) = (f_1(\mathbf{b}), \ldots, f_n(\mathbf{b}))$ of jobs to the machines, where $f_i(\mathbf{b})$ denotes the job load assigned to machine i when agents take actions signalling $\mathbf{b}$. The cost that agent i faces for the schedule $f(\mathbf{b})$ is $t_i(f(\mathbf{b})) = t_i \cdot f_i(\mathbf{b})$. We focus on algorithms f^* minimizing the *makespan*, i.e., $f^*(\mathbf{b}) \in \arg\min_{\mathbf{x}} \max_{i=1}^{n} b_i(\mathbf{x})$; f is α-approximate if it returns a solution with cost at most α times the optimum.

3.1   Lower Bound

Let $\tau_k(n) = \frac{\sqrt{k^2 + 4n} - k}{2}$. That is, τ_k is a function of n such that $n = \tau_k(n)(\tau_k(n) + k)$. Observe that $\tau_0(n) = \sqrt{n}$ and $\tau_{n-1}(n) = 1$. In this section, we prove the following theorem, that states the main result of our work. Henceforth, for sake of readability, let us denote $\tau := \tau_k(n)$.

Theorem 1. *For the machine scheduling problem, no k-OSP mechanism can be better than τ-approximate, regardless of the value of the sets $\mathcal{L}_k(\cdot)$. This even holds for homogeneous three-value domains, i.e., $D_i = \{L, M, H\}$ for each i.*

Proof. Consider $m = n$. Moreover, consider a domain $D_i = \{L, M, H\}$ for every i, with $M \geq \tau \left\lceil \frac{m}{\lfloor \tau \rfloor} \right\rceil L$ and $H \geq \tau \cdot mM$.

The proof will work in three steps. First, we prove some algebraic property of τ (cf. Lemma 1). We then characterize implementation tree and algorithm of a k-OSP mechanism with approximation better than τ (cf. Lemma 2). Finally, we identify an instance for which any such mechanism cannot return an approximation better than τ – a contradiction.

Lemma 1. $\tau = c + \delta$, with $\delta \in \left[0, \frac{k}{\tau+k-1}\right]$, where $c = \max\left\{\alpha \in \mathbb{N}: k \le \frac{n-c^2}{c}\right\}$.

Suppose now that a mechanism $\mathcal{M}$ with approximation ratio $\rho < \tau$ exists for the setting at the hand, and let $\mathcal{T}$ be its implementation tree. Let us rename the agents as follows: Agent 1 is the 1st distinct agent that diverges in $\mathcal{T}$; because of its approximation guarantee, the mechanism must have at least one divergent agent for our domain. We now call agent 2, the 2nd distinct agent that diverges in the subtree of $\mathcal{T}$ defined by agent 1 taking an action signalling type H; if no agent diverges in this subtree of $\mathcal{T}$ we simply call 2 an arbitrary agent different from 1. More generally, agent i is the ith distinct agent that diverges, if any, in the subtree of $\mathcal{T}$ that corresponds to the case that the actions taken by agents that previously diverged are signalling their type being H. As above, if no agent diverges in the subtree of interest, we just let i denote an arbitrary agent different from $1, 2, \ldots, i-1$. We denote with u_i the node in which i diverges in the subtree in which all the other agents have taken actions signalling H; if i got her id arbitrarily, then we denote with u_i a dummy node. We then have the following lemma.

Lemma 2. Any k-OSP $\mathcal{M}$ which is ρ-approximate, with $\rho < \tau$, must satisfy the following conditions:

1. For every $i \le n+1 - \lceil \tau \rceil - k$, if agent i diverges at node u_i, it must diverge on M and H.
2. For every $i \le n - \lfloor \tau \rfloor - k$, if agent i diverges at node u_i and takes an action signalling type H, then $\mathcal{M}$ does not assign any job to i whenever the action of agents in $\mathcal{L}(u_i)$ are all signalling H.

Proof. Let us first prove part 1. Suppose that there is $i \le n+1 - \lceil \tau \rceil - k$ such that at node u_i i does not diverge on M and H (i.e., any action signalling M is signalling also H). Then it must diverge on L and M, since u_i must have at least two outgoing edges (since i is assumed to diverge at u_i), and the remaining edges can only be labeled with L. Consider the type profile $\mathbf{x}$ such that $x_i = M$, and $x_j = H$ for every $j \ne i$. Observe that, by definition of u_i, $x_j \in D_j(u_i)$ for every agent j. The optimal allocation for the type profile $\mathbf{x}$ assigns all jobs to machine i, with cost $OPT(\mathbf{x}) = mM$. Since $\mathcal{M}$ is ρ-approximate, then it also assigns all jobs to machine i. Indeed, if a job is assigned to a machine $j \ne i$, then the cost of the mechanism would be at least $H \ge \tau \cdot mM > \rho \cdot OPT(\mathbf{x})$, that contradicts the approximation bound.

Consider now the profile $\mathbf{y}$ such that $y_i = L$, $y_j = H$ for every $j < i$ and $j \in \mathcal{L}(u_i)$, and $y_j = L$ for every $j > i$ such that $j \notin \mathcal{L}(u_i)$. (We stress that our lower bound holds no matter the definition of the sets $\mathcal{L}(u_i)$.)

Observe that, as for $\mathbf{x}$, we have that $y_j \in D_j(u_i)$ for every agent j. It is not hard to see that $OPT(\mathbf{y}) \leq \left\lceil \frac{m}{n-i-k+1} \right\rceil L$. Let μ be the number of jobs that $\mathcal{M}$ assigns to machine i in this case. Since $\mathcal{M}$ is ρ-approximate, then $\mu < m$. Indeed, if $\mu = m$, then the cost of the mechanism contradicts the approximation bound, since $mL \geq \tau \left\lceil \frac{m}{n-i-k+1} \right\rceil L > \rho \cdot OPT(\mathbf{y})$, where we used that

$$\tau \left\lceil \frac{m}{n-i-k+1} \right\rceil \leq \tau \left\lceil \frac{n}{\lceil \tau \rceil} \right\rceil = \tau \left\lceil \frac{\tau(\tau+k)}{\tau+1-\delta} \right\rceil$$
$$\leq \tau \frac{\tau(\tau+k) + (\tau-\delta)}{\tau+1-\delta} \leq \tau(\tau+k) = m,$$

where the last inequality follows from $\delta \leq \frac{k}{\tau+k-1}$ by Lemma 1.

Hence, for the mechanism to be OSP with k-lookahead we need that both the following conditions are satisfied: (i) $p_i(\mathbf{x}) - mM \geq p_i(\mathbf{y}) - \mu M$, and (ii) $p_i(\mathbf{y}) - \mu L \geq p_i(\mathbf{x}) - mL$, where $p_i(\mathbf{x})$ and $p_i(\mathbf{y})$ denote the payment that i receives from the mechanism $\mathcal{M}$ when agents' actions are signalling $\mathbf{x}$ and $\mathbf{y}$, respectively. However, this leads to the contradiction that $L \geq M$.

Let us now prove part 2. Suppose that there is $i \leq n - \lfloor \tau \rfloor - k$ and $\mathbf{x}_{-i}$, with $x_j \in D_j(u_i)$ for every agent j and $x_j = H$ for every $j \in \mathcal{L}(u_i)$, such that if i takes an action signalling type H, then $\mathcal{M}$ assigns at least a job to i. According to part 2, machine i diverges at node u_i on H and M.

Consider then the profile $\mathbf{y}$ such that $y_i = M$, $y_j = H$ for $j \leq i + k$ with $i \neq j$, and $y_j = L$ for $j > i + k$. Observe that $OPT(\mathbf{y}) = \left\lceil \frac{m}{n-i-k} \right\rceil \cdot L$. Since $\mathcal{M}$ is ρ-approximate, then it does not assign any job to machine i, otherwise its cost would be at least $M \geq \tau \left\lceil \frac{m}{\lceil \tau \rceil} \right\rceil L \geq \tau \left\lceil \frac{m}{n-i-k} \right\rceil L > \rho \cdot OPT(\mathbf{x})$.

Hence, for the mechanism to be OSP with k-lookahead we need that both the following conditions are satisfied: (i) $p_i(\mathbf{x}) - H \geq p_i(\mathbf{y}) - 0$, and (ii) $p_i(\mathbf{y}) - 0 \geq p_i(\mathbf{x}) - M$. However, this leads to the contradiction that $H \leq M$.   □

Roughly speaking, Lemma 2 states that any k-OSP mechanism must have an implementation tree such that the first $n - \lfloor \tau \rfloor - k$ agents interacting with the mechanism, must be asked if their type is H, and, in the case of affirmative answer, they must not receive any job.

We next observe that such a mechanism cannot have approximation lower than τ, contradicting our hypothesis that M was k-OSP and ρ-approximate.

To this aim, assume first that for each agent $i \leq n - \lfloor \tau \rfloor - k$ diverges at u_i. We consider the profile $\mathbf{x}$ such that $x_i = H$ for every i. The optimal allocation consists in assigning a job to each machine, and has cost $OPT(\mathbf{x}) = H$. According to Part 2 of Lemma 2, since $\mathcal{M}$ is supposed to be k-OSP, if machines take actions that signal $\mathbf{x}$, then the mechanism $\mathcal{M}$ does not assign any job to machine i, for every $i \leq n - \lfloor \tau \rfloor - k$. Hence, the best outcome that $\mathcal{M}$ can return for $\mathbf{x}$ consists in fairly assigning the m jobs to the remaining $\lfloor \tau \rfloor + k$ machines. Observe that, if $\delta = 0$, i.e., τ is an integer, then each machine receives τ job, and thus the cost of $\mathcal{M}$ is at least $\tau H > \rho OPT(\mathbf{x})$, which contradicts the approximation ratio of $\mathcal{M}$. Otherwise, there is at least one machine that receives at least $\lceil \tau \rceil$ jobs, since $\lfloor \tau \rfloor (\lfloor \tau \rfloor + k) < \tau (\tau + k) = m$. In this case, the cost of $\mathcal{M}$ is at least $\lceil \tau \rceil H > \tau H = \tau OPT(\mathbf{x})$, contradicting again the approximation ratio of $\mathcal{M}$.

Consider now the case that there is $1 < i \leq n - \lfloor \tau \rfloor - k$ that does not diverge at u_i. It is not hard to see that this would contradict the approximation of $\mathcal{M}$ given that it would be unaware of the type of too many machines. $\qquad \square$

3.2 Upper Bound

We next show that for every k and every possible choice of lookahead sets $\{\mathcal{L}_k(u)\}_{u \in T}$, the bound above is tight, for three-values domains, i.e., $D_i = \{L_i, M_i, H_i\}$ for every i. To this aim, consider the following mechanism $\mathcal{M}_k$, that consists of a Descending Phase (Algorithm 1) followed by an Ascending Phase (Algorithm 2). The algorithmic output is augmented with a payment, to agent i, of M_i for each unit of job load received.

---

1 Set $A = [n]$, and $t_i = \max\{d \in D_i\}$
2 **while** $|A| > \lceil \tau \rceil + k$ **do**
3   Set $p = \max_{a \in A}\{t_a\}$ and $i = \min\{a \in A : t_a = p\}$
4   Ask machine i if her type is equal to p
5   **if** *yes* **then** remove i from A, and set $t_i = p$
6   **else** set $t_i = \max\{t \in D_i : t < p\}$

Algorithm 1: The descending phase keeps in A the machines that are still *alive* and in t_i the maximum non-discarded type for each agent; then proceeds by removing from A the slowest machines, until there are only $\lceil \tau \rceil + k$ left.

---

1 Set $s_i = \min\{d \in D_i\}$
2 Set $B = \emptyset$
3 **while** $|B| \leq k$ **do**
4   Set $p = \min_{a \in A \setminus B}\{s_a\}$ and $i = \min\{a \in A \setminus B : s_a = p\}$
5   Ask machine i if her type is equal to p
6   **if** *yes* **then** Set $t_i = p$ and insert i in B
7   **else** set $s_i = \min\{d \in D_i : d > p\}$
8 Consider the profile $\hat{z}$ with $\hat{z}_i = t_i$ for $i \in B$ and $\hat{z}_j = \min_{w \notin A} t_w$ for $j \in A \setminus B$
9 Let $f^*(\hat{z}) = (f_i^*(\hat{z}))_{i \in A}$ be the optimal assignment of jobs on input profile $\hat{z}$
10 Assign $f_j^*(\hat{z})$ jobs to each machine $j \in A$

Algorithm 2: The ascending phase adds to B the k fastest machines; then it computes the optimal assignment by using the revealed type for machines in B and a suitably chosen placeholder type for the remaining machines.

In case of multiple optimal assignments in line 9 of Algorithm 2, we assume that the mechanism returns the one that maximizes the number of jobs assigned to machines in B. This is exactly the solution returned by the optimal greedy algorithm, and thus can be computed in polynomial time.

Roughly speaking, mechanism $\mathcal{M}_k$ works by discovering in the descending phase the $n - \lfloor \tau \rfloor - k$ slowest machines and discarding them (i.e., no job will be assigned to these machines). (Our mechanism satisfies the conditions of Lemma 2 thus showing that our analysis is tight for both approximation and design of the mechanism.) The ascending phase then serves to select a good assignment to the non-discarded machines. To this aim, the mechanism discovers in the ascending phase the $k + 1$ fastest machines. The assignment that is returned is then the optimal assignment to the non-discarded machines in the case that the type of the $k+1$ fastest machines is as revealed, whereas the type of the remaining non-discarded machines is supposed to be as high as possible, namely equivalent to the type of the last discarded machine (i.e., the fastest among the slow machines).

Proposition 1. *Mechanism $\mathcal{M}_k$ is k-OSP if $D_i = \{L_i, M_i, H_i\}$ for each i.*

Proof. We prove that $M_i \cdot f_i(\mathcal{M}_k(\mathbf{x})) - x_i \cdot f_i(\mathcal{M}_k(\mathbf{x})) \geq M_i \cdot f_i(\mathcal{M}_k(\mathbf{y})) - x_i \cdot f_i(\mathcal{M}_k(\mathbf{y}))$ for each machine i, for each node u in which the mechanism makes a query to i, for every $\mathbf{z}_{\mathcal{L}(u)}$ such that $z_j \in D_j(u)$ for $j \in \mathcal{L}(u)$, for every x_i and y_i that diverge at u, for each pair of type profiles $\mathbf{x}, \mathbf{y}$ such that $x_j \in D_j(u)$, $y_j \in D_j(u)$ for every agent j and $x_j = y_j = z_j$ for every $j \in \mathcal{L}(u)$.

This is obvious for $x_i = M_i$. We next prove that $x_i = H_i$ implies $f_i(\mathcal{M}_k(\mathbf{x})) \leq f_i(\mathcal{M}_k(\mathbf{y}))$, that immediately implies the desired claim. Let us first consider a node u corresponding to the descending phase of the mechanism. In this case, $x_i = p$, where p is as at node u. Moreover, in all profiles as described above there are at least $\lceil \tau \rceil + k$ machines that either have a type lower than p, or they have type p but are queried after i. However, for every $\mathbf{x}_{-i}$ satisfying this property, we have that $f_i(\mathcal{M}_k(\mathbf{x})) = 0 \leq f_i(\mathcal{M}_k(\mathbf{y}))$ for every alternative profile $\mathbf{y}$.

Suppose now that node u corresponds to the ascending phase of the mechanism. In this case, $y_i = p$, where p is as at node u. Observe that $f_i(\mathcal{M}_k(\mathbf{y})) = f_i^\star(y_i, \mathbf{z}_{\mathcal{L}(u)}, \hat{\mathbf{z}}_{-i,\mathcal{L}(u)})$, where $f_i^\star(y_i, \mathbf{z}_{\mathcal{L}(u)}, \hat{\mathbf{z}}_{-i,\mathcal{L}(u)})$ is the number of jobs assigned to machine i by the optimal outcome on input profile $(y_i, \mathbf{z}_{\mathcal{L}(u)}, \hat{\mathbf{z}}_{-i,\mathcal{L}(u)})$, $\hat{\mathbf{z}}_{-i,\mathcal{L}(u)}$ being such that $\hat{z}_j = \max_{k \in A} t_k$ for every $j \in A \setminus (\{i\} \cup \mathcal{L}(u))$.

Observe that for every $\mathbf{x}$ as described above, it must be the case that $x_j \geq y_i$ for every $j \in A \setminus \mathcal{L}(u)$. Hence, we distinguish two cases: if $\min_{j \in A \setminus \mathcal{L}(u)} x_j = x_i$, then $f_i(\mathcal{M}_k(\mathbf{x})) = f_i^\star(x_i, \mathbf{z}_{\mathcal{L}(u)}, \hat{\mathbf{z}}_{-i,\mathcal{L}(u)}) \leq f_i^\star(y_i, \mathbf{z}_{\mathcal{L}(u)}, \hat{\mathbf{z}}_{-i,\mathcal{L}(u)}) = f_i(\mathcal{M}_k(\mathbf{y}))$; if instead $\min_{j \in A \setminus \mathcal{L}(u)} x_j = x_k$, for some $k \neq i$, then

$$f_i(\mathcal{M}_k(\mathbf{x})) = f_i^\star(x_k, \mathbf{z}_{\mathcal{L}(u)}, \hat{\mathbf{z}}_{-k,\mathcal{L}(u)}) \leq f_k^\star(x_k, \mathbf{z}_{\mathcal{L}(u)}, \hat{\mathbf{z}}_{-k,\mathcal{L}(u)})$$
$$\leq f_i^\star(y_i, \mathbf{z}_{\mathcal{L}(u)}, \hat{\mathbf{z}}_{-i,\mathcal{L}(u)}) = f_i(\mathcal{M}_k(\mathbf{y})),$$

where we used that $\hat{\mathbf{z}}_{-k,\mathcal{L}(u)} = \hat{\mathbf{z}}_{-i,\mathcal{L}(u)}$ and the inequalities follow since: (i) in the optimal outcome the fastest machine must receive at least as many jobs as slower machines; (ii) in the optimal outcome, given the speeds of other machines, the number of jobs assigned to machine i decreases as its speeds decreases. □

Proposition 2. *Mechanism $\mathcal{M}_k$ is $\left(\frac{m+k+\lceil \tau \rceil - 1}{m} \lceil \tau \rceil \right)$-approximate.*

Proof (Sketch). We denote with $OPT(\mathbf{x})$ the makespan of the optimal assignment when machines have type profile $\mathbf{x}$. We will use the same notation both if the optimal assignment is computed on n machines and if it is computed and on $\lceil \tau \rceil + k$ machines, since these cases are distinguished through the input profile.

Fix a type profile $\mathbf{x}$. Let A and B as at the end of the mechanism when agents behave according to $\mathbf{x}$. Let β be the smallest multiple of $|A|$ such that $\beta \geq \sum_{i \in A} OPT_i(\mathbf{x})$. Moreover, let $t = \min_{j \notin A} t_j$. We define the profile $\mathbf{y}$ as follows: $y_i = w$ for every $i \in A$ and $y_i = t$ otherwise, where w is chosen so that $\frac{\beta}{|A|} \cdot w = \max_{j \in A}(x_j \cdot OPT_j(\mathbf{x}))$. Consider then the assignment $\mathbf{a}$ that assigns β jobs equally split among agents in A and $m - \beta$ jobs equally split among agents not in A. It is immediate to see that $OPT(\mathbf{x}) \geq MS(\mathbf{a}, \mathbf{y})$, where $MS(\mathbf{a}, \mathbf{y})$ is the makespan of the assignment $\mathbf{a}$ with respect to the type profile $\mathbf{y}$.

Let $\mathcal{M}(\mathbf{x})$ be the makespan of the assignment returned by our mechanism if agents behave according to $\mathbf{x}$. Then, $\mathcal{M}(\mathbf{x})$ is equivalent to $OPT(\hat{\mathbf{z}})$, where $\hat{\mathbf{z}}$ is such that $\hat{z}_j = x_j$ for $j \in B$ and $\hat{z}_j = t$ for $j \in A \setminus B$. Let α be the smallest multiple of $|B|$ such that $\alpha \geq \sum_{i \in B} OPT_i(\hat{\mathbf{z}})$. We define the profile $\hat{\mathbf{y}}$ as follows: $\hat{y}_i = \hat{w}$ for every $i \in B$ and $y_i = t$ otherwise, where $\hat{w}$ is chosen so that $\frac{\alpha}{|B|} \cdot \hat{w} = \max_{j \in B}(x_j \cdot OPT_j(\hat{\mathbf{z}}))$. Consider then the assignment $\hat{\mathbf{a}}$ that assigns α jobs equally split among agents in B and $m - \alpha$ jobs equally split among agents in $A \setminus B$. It is immediate to see then $\mathcal{M}(\mathbf{x}) = OPT(\hat{\mathbf{z}}) = MS(\hat{\mathbf{a}}, \hat{\mathbf{y}})$. The theorem then follows, since it occurs that $\frac{OPT(\hat{\mathbf{y}})}{MS(\mathbf{a}, \mathbf{y})} \leq \frac{m+k+\lceil \tau \rceil - 1}{m} \lceil \tau \rceil$. $\square$

The next corollary follows by simple algebraic manipulations.

Corollary 1. *Mechanism $\mathcal{M}_k$ is $(\lceil \tau \rceil + 1)$-approximate for $m > \lceil \tau \rceil (k + \lceil \tau \rceil)$ and the approximation tends to $\lceil \tau \rceil$ as m increases.*

4 Conclusions

We have studied the relationship between the bounded rationality of the agents and the approximation guarantee of mechanisms incentivizing these agents. We have relaxed the popular notion of OSP [18] to allow for more fine grained notions of rationality. For machine scheduling, we proved that more rational agents do not help in getting close to the optimum, unless the level of rationality is significant to a point where the meaning of bounded becomes questionable. On one hand, our findings motivate the focus on OSP for future work on the approximation guarantee of mechanisms for agents with bounded rationality. On the other hand, one might wonder whether similar results hold also for different optimization problems. To this aim, we observe that the techniques that we use in our proof have a resemblance with the ones used in [13] for proving the inapproximability of OSP mechanisms for the facility location problem (with money). Hence, we believe that results similar to the ones we give for machine scheduling may be proved for facility location. As for other problems, we highlight that no approximation result is known even for OSP mechanisms. In particular, for binary allocation problems (that have been considered already in [18]), only a characterization of optimal OSP mechanism is known.

References

1. Adamczyk, M., Borodin, A., Ferraioli, D., de Keijzer, B., Leonardi, S.: Sequential posted price mechanisms with correlated valuations. In: Markakis, E., Schäfer, G. (eds.) WINE 2015. LNCS, vol. 9470, pp. 1–15. Springer, Heidelberg (2015). https://doi.org/10.1007/978-3-662-48995-6_1

2. Archer, A., Tardos, É.: Truthful mechanisms for one-parameter agents. In: FOCS 2001, pp. 482–491 (2001)

3. Ashlagi, I., Gonczarowski, Y.A.: Stable matching mechanisms are not obviously strategy-proof. J. Econ. Theor. **177**, 405–425 (2018)

4. Ausubel, L.M.: An efficient ascending-bid auction for multiple objects. Am. Econ. Rev. **94**(5), 1452–1475 (2004)

5. Babaioff, M., Immorlica, N., Lucier, B., Weinberg, S.M.: A simple and approximately optimal mechanism for an additive buyer. In: FOCS 2014, pp. 21–30 (2014)

6. Bade, S., Gonczarowski, Y.A.: Gibbard-Satterthwaite success stories and obvious strategyproofness. In: EC 2017, p. 565 (2017)

7. Chawla, S., Hartline, J., Malec, D., Sivan, B.: Multi-parameter mechanism design and sequential posted pricing. In: STOC 2010, pp. 311–320 (2010)

8. Correa, J., Foncea, P., Hoeksma, R., Oosterwijk, T., Vredeveld, T.: Posted price mechanisms for a random stream of customers. In: EC 2017, pp. 169–186 (2017)

9. De Groot, A.: Thought and Choice in Chess. Mouton, Oxford (1978)

10. Eden, A., Feldman, M., Friedler, O., Talgam-Cohen, I., Weinberg, S.M.: A simple and approximately optimal mechanism for a buyer with complements. In: EC 2017, pp. 323–323 (2017)

11. Feldman, M., Fiat, A., Roytman, A.: Makespan minimization via posted prices. In: EC 2017, pp. 405–422 (2017)

12. Ferraioli, D., Meier, A., Penna, P., Ventre, C.: Obviously strategyproof mechanisms for machine scheduling. In: ESA 2019 (2019)

13. Ferraioli, D., Ventre, C.: Obvious strategyproofness needs monitoring for good approximations. In: AAAI 2017, pp. 516–522 (2017)

14. Ferraioli, D., Ventre, C.: Probabilistic verification for obviously strategyproof mechanisms. In: IJCAI 2018 (2018)

15. Ferraioli, D., Ventre, C.: Obvious strategyproofness, bounded rationality and approximation. In: AAMAS 2019 (2019)

16. Hartline, J., Roughgarden, T.: Simple versus optimal mechanisms. In: EC 2009, pp. 225–234 (2009)

17. Kyropoulou, M., Ventre, C.: Obviously strategyproof mechanisms without money for scheduling. In: AAMAS 2019 (2019)

18. Li, S.: Obviously strategy-proof mechanisms. Am. Econ. Rev. **107**(11), 3257–87 (2017)

19. Mackenzie, A.: A revelation principle for obviously strategy-proof implementation. Research Memorandum 014, (GSBE) (2017)

20. Pearl, J.: Heuristics: Intelligent Search Strategies for Computer Problem Solving. Addison-Wesley, Reading (1984)

21. Pycia, M., Troyan, P.: Obvious dominance and random priority. In: EC 2019 (2019)
22. Sandholm, T., Gilpin, A.: Sequences of take-it-or-leave-it offers: near-optimal auctions without full valuation revelation. In: Faratin, P., Parkes, D.C., Rodríguez-Aguilar, J.A., Walsh, W.E. (eds.) AMEC 2003. LNCS (LNAI), vol. 3048, pp. 73–91. Springer, Heidelberg (2004). https://doi.org/10.1007/978-3-540-25947-3_5
23. Shannon, C.: Programming a computer for playing chess. Philos. Mag. **41**(314), 256–275 (1950)
24. Zhang, L., Levin, D.: Bounded rationality and robust mechanism design: an axiomatic approach. Am. Econ. Rev. **107**(5), 235–39 (2017)

Auctions and Markets

Risk Robust Mechanism Design for a Prospect Theoretic Buyer

Siqi Liu[1], J. Benjamin Miller[2], and Alexandros Psomas[3(✉)]

[1] UC Berkeley, Berkeley, USA
[2] UW–Madison, Madison, USA
[3] Carnegie Mellon University, Pittsburgh, USA
cpsomas@cs.cmu.edu

Abstract. Consider the revenue maximization problem of a risk-neutral seller with m heterogeneous items for sale to a single additive buyer, whose values for the items are drawn from known distributions. If the buyer is also risk-neutral, it is known that a simple and natural mechanism, namely the better of selling separately or pricing only the grand bundle, gives a constant-factor approximation to the optimal revenue. In this paper we study revenue maximization without risk-neutral buyers. Specifically, we adopt cumulative prospect theory, a well established generalization of expected utility theory.

Our starting observation is that such preferences give rise to a very rich space of mechanisms, allowing the seller to extract arbitrary revenue. Specifically, a seller can construct extreme lotteries that look attractive to a mildly optimistic buyer, but have arbitrarily negative true expectation. Therefore, giving the seller absolute freedom over the design space results in absurd conclusions; competing with the optimal mechanism is hopeless. Instead, in this paper we study four broad classes of mechanisms, each characterized by a distinct use of randomness. Our goal is twofold: to explore the power of randomness when the buyer is not risk-neutral, and to design simple and attitude-agnostic mechanisms—mechanisms that do not depend on details of the buyer's risk attitude—which are good approximations of the optimal in-class mechanism, tailored to a specific risk attitude. Our main result is that the same simple and risk-agnostic mechanism (the better of selling separately or pricing only the grand bundle) is a good approximation to the optimal non-agnostic mechanism within three of the mechanism classes we study.

1 Introduction

Expected utility theory (EUT) has long reigned as the prevailing model of decision making under uncertainty. However, a substantial body of evidence, including the famous Allais paradox [1], shows that most people make choices that violate this theory. Cumulative prospect theory [27] is arguably the most prominent alternative. A key element of this theory is a non-linear transformation of cumulative probabilities by a *probability weighting function*. This transformation

© Springer Nature Switzerland AG 2019
D. Fotakis and E. Markakis (Eds.): SAGT 2019, LNCS 11801, pp. 95–108, 2019.
https://doi.org/10.1007/978-3-030-30473-7_7

can model a person's tendency towards optimism or pessimism.[1] On the other hand, as mechanism designers we use randomization as an important tool in optimizing our objective, typically (and crucially) assuming that agents make choices according to the tenets of expected utility theory. While we have vastly deepened our understanding of mechanism design under this assumption, it is essential to study empirically validated models of human decision-making. In this paper we study the revenue-maximization problem of a risk-neutral seller with m heterogeneous items for sale to a single, additive buyer with cumulative prospect theory preferences. Our goal is to design simple mechanisms that are agnostic to the underlying probability weighting function of the buyer, yet achieve a good approximation to the revenue of the optimal mechanism *tailored* to this weighting function. To understand our results in context, we begin by briefly reviewing cumulative prospect theory.

1.1   Prospect Theory Basics

In full generality, cumulative prospect theory (CPT) asserts that preferences are parameterized by a reference point (or status quo) r, a value function U that maps (deterministic, i.e. certain) outcomes into *utils* (or dollars), and two probability weighting functions, w^+ and w^-, for weighting the cumulative probabilities of positive and negative outcomes (relative to r). By taking $r = 0$ and the weighting functions w^+ and w^- to be the identity function, one recovers expected utility theory; thus, CPT generalizes EUT. However, like most works in mechanism design, we assume linear utility for money: $U(x) = x$. That is, our agents have value 1 for \$1 and value 1000 for \$1000. What remains, then, are the weighting functions w^+ and w^- and the reference point r.

For intuition, consider first a simple event E which occurs with probability $\frac{1}{2}$, and assume that $r = 0$. Suppose that E corresponds to an agent receiving value 10; if E does not occur, the agent receives nothing. A risk-neutral agent would value this potential income at $10 \cdot \Pr[E] = 5$. An optimistic agent, over-estimating the possibility of receiving 10, might value E at slightly more than 5, whereas a pessimistic agent might value it at slightly less. CPT uses a weighting function w^+ which modifies probabilities of positive outcomes: the agent values event E at $10 \cdot w^+(\Pr[E])$. Then $w^+(x) > x$ corresponds to optimism, and $w^+(x) < x$ corresponds to pessimism. CPT captures much more complex behavior than merely optimism and pessimism. For example, in experiments (e.g. [6,27]), subjects tend to overweight extreme events: in a sense, people are optimistic about very good outcomes and pessimistic about very bad outcomes. This sort of behavior can be readily captured by CPT.

In general, the event of interest might correspond to a positive or negative outcome. For example, E might correspond to the agent *losing* value 10. In that case, we expect the optimistic agent to *underweight* the probability of E occurring. For this reason, CPT models probability weighting for gains and losses

---

[1] As we discuss below, real-world attitudes are not merely "optimistic" or "pessimistic", but such simplistic attitudes are easily and naturally captured by this model.

with functions w^+ and w^-, respectively. When the random variable is supported on multiple non-zero values, applying w^+ (or w^-) directly to the probability of each event leads to violations of first-order stochastic dominance. For this reason, [23] proposed to weight the cumulative distribution function, rather than the probability mass function; hence *cumulative* prospect theory.

Our interest here is highlighting the effects of nonlinear probability weighting. We will therefore focus on a special case of cumulative prospect theory, namely *rank dependent utility theory* (RDUT). This theory is rich enough to explain a number of known violations of expected utility theory, e.g., the Allais paradox [24], general enough to include expected utility theory as a special case, while at the same time simple enough to be mathematically tractable. This theory is equivalent to the following assumption.

Assumption 1 *([23]). For all $p \in [0,1]$, $w^-(p) = 1 - w^+(1-p)$.*

Assumption 1 allows us to rank all the outcomes from worst to best, independent of whether they are gains or losses, and weight their probabilities with a single weighting function $w(x)$. Furthermore, it makes the reference point r irrelevant. [7] have previously studied the same model, giving a class of mechanisms which optimally sell a single item to a pessimistic buyer. However, they restrict themselves to convex weighting functions. Here we study general weighting functions and multi-item auctions. We postpone more details about rank dependent utility theory until Sect. 2, and refer the reader to the full version of this paper for what the expected utility of a general CPT agent (that is, without Assumption 1) for even a simple lottery looks like.

1.2   Our Results

Our starting point is the observation that even very mild probability weighting gives rise to rich seller behavior, which allows the seller to extract unbounded revenue. Specifically, we show that under assumptions satisfied by most weighting functions in the literature, the seller can design a bet that has arbitrarily negative (risk-neutral) expectation, but looks attractive to a RDUT buyer. This bet can be easily turned into an auction for selling any number of items by giving the items for free if and only if the buyer takes the bet. Similar behavior has been observed before this work for more general models, e.g. [2,12].

In light of these negative results for arbitrary buyer-seller interaction, we focus our attention to specific classes of mechanisms, imposing various restrictions on the mechanism's description and implementation. These restrictions are not onerous: when offered to a risk-neutral buyer, two of the classes are equivalent to the class of all mechanisms, and another is equivalent to all deterministic mechanisms. Our restrictions thus serve to isolate particular uses of randomization and to illustrate the various effects RDUT preferences have on mechanism design.

The first class we consider is that of *deterministic price* mechanisms, which we denote C_{dp}. Here, the seller offers a menu of (possibly correlated) distributions over the items, each at a fixed price. The buyer may pay the price for a distribution, after which she receives a draw from the distribution. To bypass some technical barriers, we also consider a special case of this class, *nested deterministic price* mechanisms, or $\mathcal{C}_{ndp}$, which impose certain constraints on the distributions over items in a menu. These constraints are very mild (for example they are always satisfied by independent distributions) and are without loss of generality for a risk-neutral buyer. Next, we consider the class of *deterministic allocation* mechanisms, $\mathcal{C}_{da}$, where the mechanism deterministically allocates a bundle of items for a possibly randomized, non-negative payment. $\mathcal{C}_{da}$ is equivalent to deterministic mechanisms for a risk neutral buyer. Finally, we consider a multi-item generalization of the single-item class of mechanisms that is optimal for convex weighting functions (as shown by [7]). We call this class *binary-lottery* mechanisms and denote it by $\mathcal{C}_b$.

Our main result is that, for classes $\mathcal{C}_{ndp}$, $\mathcal{C}_{da}$ and $\mathcal{C}_b$, a single simple mechanism, agnostic to the underlying weighting function, gives a good approximation on the revenue of the *optimal* in-class mechanism *tailored* to w. That mechanism is the better of selling every item separately at a fixed price (henceforth SREV) and selling the grand bundle as a single item at a fixed price (henceforth BREV), which is a valid mechanism in all classes considered. Furthermore, this mechanism is deterministic, which implies that its expected revenue is the same for all weighting functions w, and only depends on the buyer's value distribution $\mathcal{D}$. Our proof is by relating the revenue of each class of mechanisms to the revenue obtainable from a risk-neutral buyer via any mechanism, combined with a result of [3], which shows that max{SREV, BREV} is a constant approximation to this risk-neutral revenue. For $\mathcal{C}_{dp}$ our understanding is partial; we show that max{SREV, BREV} approximates the optimal, risk non-agnostic $\mathcal{C}_{dp}$ auction within a factor that is doubly exponential in the number of items. This implies a constant approximation for a constant number of items (in fact, for two items we can show an approximation factor of 2 for just SREV), but we leave it as an open problem whether a constant approximation is possible for the general case. All our results can be extended to a unit-demand and additive up to a downward closed constraint buyer by paying an extra factor of 4 and 31.1, respectively, using the results of [8].

Intuitively, the difficulty with analyzing mechanisms for RDUT buyers (and especially optimal mechanisms) is that, given a mechanism, we cannot generally argue about how much a buyer type t values the menu item purchased by a type t'. This is especially the case for general deterministic price mechanisms, where allocations over items could be arbitrarily correlated. This, in turn, prevents us from using basic "simulation arguments": starting from an auction $\mathcal{M}$, manipulate the allocation rule and pricing rule to get a different auction $\mathcal{M}'$. Such arguments are very useful in getting meaningful upper bounds on the optimal revenue. For example, [17] upper bound the optimal revenue from a product distribution, $\text{REV}(\mathcal{D} \times \mathcal{D}')$, by $\text{REV}(\mathcal{D}) + \text{VAL}(\mathcal{D}')^2$ using such an argument, where they give a concrete auction for $\mathcal{D}$ by manipulating the allocation and payment rule of the optimal auction for $\mathcal{D} \times \mathcal{D}'$. Similar "marginal mechanism"

---

[2] $\mathcal{D}$ and $\mathcal{D}'$ here are distribution over m_1 and m_2 items, respectively. $\text{VAL}(\mathcal{D}') = \sum_{j \in [m_2]} \mathbb{E}[\mathcal{D}'_j]$, i.e. the total expected sum of values from items in $\mathcal{D}'$.

arguments are crucial in many works that give simple and approximately optimal mechanisms for additive buyers, e.g. [3, 19, 31]; for example, the so-called core-tail decomposition technique depends on such arguments. On the other hand, the recently developed Lagrangian duality based approach ([4, 5, 9, 13, 14, 20]) also seems to fail here. This technique has been successful in getting benchmarks in a number of settings, by giving a solution to the dual of the mathematical program that computes the optimal auction. To the best of our knowledge, all works that use this technique start from a linear program. Here, the mathematical program for the optimal, risk non-agnostic auction is not even convex. Even though in theory only weak duality is necessary for this technique to work, we haven't been successful in applying it to our problem.

1.3 Related Work and Roadmap

Prospect theory was originally defined by [18] but, though successful in explaining experimentally observed behavior, it suffered from a number of weaknesses, namely violations of first-order stochastic dominance between random variables. Several works ([23, 26, 29, 30]) proposed solutions to these issues, resulting in cumulative prospect theory ([27]). Next to expected utility theory, cumulative prospect theory is likely the best studied theory of decision-making under uncertainty. We refer the reader to the book of [28] for a thorough exposition of the model. Also see [21] for a survey of non-EUT models. Although widely studied in behavioral economics, prospect theory has received much less attention in the game theory and mechanism design literature. Our work is most closely related to that of [7], who study optimal and robust mechanisms for a single buyer and a single item. Their work, unlike ours, places much stronger assumptions on the weighting function: namely, they assume convexity (which in turn implies $w(x) \leq x$). In this paper we consider general weighting functions, but restrict the mechanism design space. Further afield, [11] study contract design in a crowd-sourcing setting with a prospect-theoretic model of workers. [15] demonstrate that equilibria may not exist in two-player games when players have prospect-theoretic preferences. [10] and [16] study mechanism design with risk-averse agents in a setting where risk-averse behavior is represented by a concave utility function, while more recently, in a similar setting, [22] study optimal mechanisms for risk-loving agents.

Our main result is that the better of selling separately and selling the grand bundle is a risk robust approximation to the optimal revenue. The approximation ratio of this mechanism has been studied extensively for risk-neutral buyers having a large class of valuations [3–5, 5, 8, 25]. Our result relies on this work, but our techniques are very different.

Roadmap. Section 2 poses our model and some preliminaries. We discuss the limits of our model in Sect. 3, and show that if the seller is allowed to use an arbitrary mechanism, then he can extract arbitrarily large revenue. In Sect. 3.1 we formally define two of the four mechanism classes considered in this paper. We proceed to analyze deterministic price mechanisms in Sect. 4. We study deterministic allocation mechanisms and binary lottery mechanisms in the full version of this paper.

2   Preliminaries

A risk-neutral seller, whose aim is to maximize revenue, is auctioning off m items to a single buyer with cumulative prospect theory preferences. The value of the buyer for item i is v_i, and is distributed according to a known distribution $\mathcal{D}_i$. We assume that the item distributions are independent, and denote the joint distribution by $\mathcal{D}$. We first go over the buyer's preference model in detail, and then formulate our mechanism design problem.

Weighted Expectation. In this paper we focus on a special case of cumulative prospect theory, *rank dependent utility theory*. In rank dependent utility theory a weighting function w distorts cumulative probabilities ([23]). The weighting function w satisfies the following properties: (1) $w : [0,1] \to [0,1]$, (2) w is non-decreasing, (3) $w(0) = 0$ and $w(1) = 1$. We use the notation $\mathcal{I}$ to indicate the risk-neutral weighting function; that is $\mathcal{I}(x) = x$. For a random variable Z over k outcomes, where the i-th outcome occurs with probability p_i and gives utility u_i, and $u_i \le u_{i+1}$, an agent with weighting function w has expected utility

$$\mathbb{E}_w[Z] = \sum_{i=1}^{k-1} u_i \left(w \left(\sum_{j=i}^k p_j \right) - w \left(\sum_{j=i+1}^k p_j \right) \right) + u_k w(p_k)$$

$$= u_1 + \sum_{i=2}^k (u_i - u_{i-1}) \cdot w \left(\sum_{j=i}^k p_j \right).$$

The intuitive interpretation (for the last expression) is that the agent always gets utility u_1. Then, the event that the agent gets an additional utility of at least $u_2 - u_1$ occurs with probability $1 - p_1 = \sum_{j=2}^k p_j$ (which is weighted by the function w). The agent gets an additional utility of at least $u_3 - u_2$ with probability $\sum_{j=3}^k p_j$, and so on. Note that this definition makes no assumption about the sign of u_i; that is, the u_is can be positive (corresponding to gains) or negative (corresponding to losses).

Mechanism Design. Back to mechanism design, any mechanism can be described by the allocation it makes and the payment it charges as a function of the buyer's report. For a report $v = (v_1, \ldots, v_m)$, we denote by $X(v)$ the random variable for the allocation, giving a probability to each possible allocation of the items in $\{0,1\}^m$. Similarly, $P(v)$ is the random variable for the payment when the report is v. $X(v)$ and $P(v)$ may be correlated. Importantly, common practices from mechanism design in the risk-neutral setting, like treating the allocation as a vector in $[0,1]^m$ or the payment as a real number (i.e. replacing the random variable of the payment with its expectation), are *with* loss of generality here.

We assume that the buyer has additive utility for the items and is quasilinear with respect to payments: if she receives a set of items S for a payment p, her total value for this outcome is $\sum_{i \in S} v_i - p$. The buyer's weighted expected utility from the mechanism's outcome is $\mathbb{E}_w[v \cdot X(v) - P(v)]$; we say that a mechanism is incentive compatible (IC) for a buyer with weighting function w if for all possible values v, v' of the buyer, it holds that $\mathbb{E}_w[v \cdot X(v) - P(v)] \ge \mathbb{E}_w[v \cdot X(v') - P(v')]$. It is without loss of generality to express an incentive compatible mechanism in the form of a menu $\mathcal{M}$, with each menu item corresponding to a particular (allocation, payment) pair of correlated random variables (X, P). Then, the allocation and payment of a buyer with value v and weighting function w is given by the utility-maximizing menu item[3] $(X_w(v), P_w(v)) = \arg\max_{(X,P)\in\mathcal{M}} \mathbb{E}_w[v \cdot X - P]$. The revenue of the mechanism is given by $\text{REV}_\mathcal{M}(w, \mathcal{D}) = \mathbb{E}[P(v)]$, where the expectation is with respect to the random valuation v (drawn from $\mathcal{D}$), as well as the random outcome of the payment random variable $P(v)$. A mechanism is individually rational (IR) if the buyer has non-negative expected utility when participating. Throughout the paper we focus on IC and IR mechanisms.

We slightly overload notation: let $\text{REV}(w, \mathcal{D})$ denote the optimal revenue achievable by an incentive compatible mechanism from selling m items to a buyer with weighting function w and values drawn from $\mathcal{D}$. We will frequently drop w to indicate the risk-neutral optimal revenue, i.e. we use $\text{REV}(\mathcal{D})$ to mean $\text{REV}(\mathcal{I}, \mathcal{D})$ (recall that $\mathcal{I}$ is the risk-neutral weighting function, $\mathcal{I}(x) = x$), and $\text{DREV}(\mathcal{D})$ for the optimal revenue from a deterministic mechanism. Note that $\text{DREV}(w, \mathcal{D}) = \text{DREV}(w', \mathcal{D})$, for all w, w'.

In this paper we show that the best of $\text{SREV}(\mathcal{D})$ (or just SREV), the auction that sells each item separately at its optimal posted price, and $\text{BREV}(\mathcal{D})$ (or just BREV), the auction that sells the grand bundle as a single item, is a risk-robust approximation for a prospect theoretic buyer. For a risk-neutral buyer, the following result is known.

Theorem 1 ([3, 4]). *For a single, risk-neutral, additive bidder and any independent item distribution $\mathcal{D}$ it holds that*

$$\text{REV}(\mathcal{I}, \mathcal{D}) \leq 2\text{BREV}(\mathcal{D}) + 4\text{SREV}(\mathcal{D}) \leq 6\max\{\text{SREV}(\mathcal{D}), \text{BREV}(\mathcal{D})\}.$$

3   Limits of the Model and Mechanism Classes

In this section, we demonstrate how our model, absent any additional assumptions on the mechanism or the weighting function, can lead to absurd results. Such results were known before our work. [2] show that under assumptions on the weighting functions a principal can extract unbounded revenue from a CPT agent, simply by offering a bet on a single coin-flip. Furthermore, [12] show that CPT behavior gives rise to time inconsistency, allowing a seller to extract the buyer's entire wealth over multiple rounds of interaction. We reproduce similar results in our context for completeness and to illustrate the variety of behaviors possible in this model. In later sections, we develop restrictions on the mechanism which preclude this sort of unreasonable behavior. First, the following simple lemma is instructive.

Lemma 1. *For every distribution $\mathcal{D}$, constant $R \in \mathbb{R}_{\geq 0}$, and weighting function w such that there exists $x^* < 1$ with $w(x^*) = 1$, there exists a mechanism $\mathcal{M}$ such that $\text{REV}_\mathcal{M}(w, \mathcal{D}) = R$.*

---

[3] We assume that any ties are broken in favor of menu items with a higher expected price.

Proof. Consider the following lottery, where (positive) Z represents a transfer *to* the agent.

$$Z = \begin{cases} 0 & \text{with probability } x^* \\ \frac{-R}{1-x^*} & \text{with probability } 1 - x^*. \end{cases} \tag{1}$$

The agent's utility is $\mathbb{E}_w[Z] = \frac{-R}{1-x^*}(1 - w(x^*)) = 0$, while the seller's revenue is $\mathbb{E}[-Z] = \frac{R}{1-x^*}(1-x^*) = R$. This lottery can be transformed into a mechanism for selling any number of items, by giving everything for free to the buyer, requiring only that she participates in the lottery. □

Lemma 1 relies on the dubious assumption that the buyer would assign no weight at all to an extremely negative—albeit potentially highly unlikely— outcome. However, even seemingly reasonable weighting functions can be exploited, as our next result shows.

Lemma 2. *For every distribution $\mathcal{D}$, constant $R \in \mathbb{R}_{\geq 0}$, and weighting function w such that there exists x^* with $1 > w(x^*) > x^*$, there exists a mechanism $\mathcal{M}$ such that $\mathrm{REV}_{\mathcal{M}}(w, \mathcal{D}) = R$.*

Proof. Consider the following lottery, where (positive) Z represents a transfer to the agent.

$$Z = \begin{cases} a & \text{with probability } x^* \\ -\rho a & \text{with probability } 1 - x^*, \end{cases} \tag{2}$$

where $a > 0$. The expected value of an agent with weighting function w is $\mathbb{E}_w[Z] = aw(x^*) - \rho a(1 - w(x^*))$. Pick $\rho = \frac{w(x^*)}{1-w(x^*)}$; then, for all a, $\mathbb{E}_w[Z] = 0$. That is, the buyer has utility exactly zero for this lottery.

On the other hand, the expected revenue of the seller, who pays a with probability x^* and gets paid ρa with probability $1 - x^*$, is equal to

$$\mathbb{E}[-Z] = \rho a(1 - x^*) - ax^* = a \cdot \left(\frac{w(x^*)(1 - x^*)}{1 - w(x^*)} - x^* \right) = a \cdot \frac{w(x^*) - x^*}{1 - w(x^*)}.$$

The lemma follows by setting $a = R\frac{1-w(x^*)}{w(x^*)-x^*}$; similarly to Lemma 1, this lottery can be turned into an auction by giving all the items for free to the agent after participating in the lottery. □

We note that the conditions of Lemma 2 are satisfied for nearly all weighting functions implied by experiments in the literature; we refer the reader to [27,28] for concrete examples. Furthermore, the issue exhibited by Lemma 2 persists even if one enforces ex-post individual rationality, so long as the seller is allowed to utilize a multi-round protocol.

Lemma 3. *For every distribution $\mathcal{D}$, constant $\epsilon > 0$ and weighting function w such that there exists x^* with $1 > w(x^*) > x^* + \frac{\epsilon}{1+\epsilon}$, there exists a multi-round, ex-post individually rational mechanism $\mathcal{M}$ such that $\mathrm{REV}_{\mathcal{M}}(w, \mathcal{D}) = \mathbb{E}[\mathcal{D}]$.*

Proof. For simplicity we only prove the $m = 1$ item case; the general case is identical. Consider again the random transfer defined in (2). Picking $\rho = \frac{w(x^*)}{1-w(x^*)} - \epsilon$ provides the buyer strictly positive utility. The seller's revenue is equal to $\mathbb{E}[-Z] = a \cdot (\frac{w(x^*)(1-x^*)}{1-w(x^*)} - \epsilon(1 - x^*) - x^*)$, which is again strictly positive for every $a > 0$. By picking a and x^* appropriately the seller can thus make both $\mathbb{E}_w[Z]$ and $\mathbb{E}[-Z]$ very small positive numbers.

This suffices to extract full buyer welfare as follows. The buyer and seller will interact over T rounds. In the first round, the buyer reports a bid b. In rounds $t > 1$, the seller will offer lottery Z (and the buyer has the option to not participate), unless the seller has already extracted an amount larger than the bid b. After T rounds have passed, the item will be awarded to the buyer for free. Of course, since $\mathbb{E}_w[Z] > 0$, the buyer always chooses to participate in round t, and (in expectation) loses a little bit of money. By picking T large enough, the buyer eventually goes bankrupt at some intermediate round, but since she eventually gets the item this mechanism is in fact ex-post IR. Notice that this mechanism is also truthful! Precisely because when the buyer is calculating (in the first round) her expected utility from reporting b she thinks that she will "come out on top", and therefore is indifferent between all bids b (and thus reports her true value v). □

As the previous lemmas exhibit, practical mechanisms cannot hope to compete against the theoretically optimal revenue maximizing mechanism in this model, and thus this theory does not give accurate predictions for the simple mechanisms that we observe in practice. There are multiple ways to proceed. A natural one is to put restrictions on the weighting functions considered. Indeed, this is the approach taken by [7] for the single item case, where the weighting function is restricted to be convex (therefore the buyer is always risk-averse). Another is to put restrictions on the mechanisms considered. In this paper we restrict our attention to specific mechanism classes; for some of our results this does not suffice and some mild restrictions on w are necessary as well.

3.1   Mechanism Classes

Here we define two classes of mechanisms; see the full version of this paper for the other mechanisms we consider. Recall that $\text{REV}_\mathcal{M}(w, \mathcal{D})$ denotes the seller's expected revenue from a mechanism $\mathcal{M}$, given that the buyer has weighting function w and her values are distributed according to $\mathcal{D}$. We denote the expected revenue of the optimal mechanism in a class $\mathcal{C}$ by $\text{REV}(w, \mathcal{D}, \mathcal{C})$. That is, $\text{REV}(w, \mathcal{D}, \mathcal{C}) = \max_{\mathcal{M} \in \mathcal{C}} \text{REV}_\mathcal{M}(w, \mathcal{D})$.

The Class $\mathcal{C}_{dp}$ of Deterministic Price Allocations. First, we consider mechanisms which use randomness only in the allocation. That is, the seller offers a menu of distributions over the items, each at a fixed price. The buyer may pay the price for a distribution over the items, after which she receives a draw from the distribution. We call this class *deterministic price (DP)* mechanisms, and denote it by $\mathcal{C}_{dp}$. It will be convenient to think of a mechanism $\mathcal{M}$ in this class as a menu, where the buyer selects her favorite menu item, of the form (p, X), where p is the payment and X is a (possibly correlated) distribution over items. This class remains completely general for risk-neutral buyers.

Unfortunately, general deterministic price mechanisms are technically difficult to work with. The arbitrary correlation allowed between items (in the allocation) makes arguing about the buyer's expected utility problematic. Specifically, different buyer types order outcomes of X differently, and therefore could have wildly different expected weighted utility for the same distribution X (since arbitrary correlation allows us to assign arbitrary probabilities to outcomes); this property can be used to tailor to each type v an allocation $X(v)$ that is attractive only to this type. Our understanding of general $\mathcal{C}_{dp}$ mechanisms is therefore partial. We show that $\max\{\text{SREV}, \text{BREV}\}$ gives a doubly exponential (in the number of items) approximation to the optimal deterministic price mechanism. This trivially implies a constant approximation for a constant number of items; we leave it as an open problem whether a constant approximation can be achieved for an arbitrary number of items.

To mitigate the problems caused by arbitrary correlation, we also consider a special case of deterministic price mechanisms, which imposes a specific form of correlation on the distribution over allocations: we ask that the allocations in the support of the allocation distribution form a nested set. We term this class *nested deterministic price (NDP)* mechanisms and denote it by $\mathcal{C}_{ndp}$. We say a random variable X supported in $2^{[m]}$ is a *monotone lottery* if X is supported on a chain of subsets $S_1, \cdots, S_k$, $k \leq m$, such that $S_i \subset S_{i+1}$ for all $i \in [k-1]$. We use $\Delta_N(2^{[m]})$ to denote the set of such correlated distributions over the set of m items. For a mechanism $\mathcal{M} \in \mathcal{C}_{ndp}$ the allocation distributions for each menu item are restricted to be in $\Delta_N(2^{[m]})$. Observe that nested deterministic price mechanisms are again completely general for risk-neutral buyers. This is so because the optimal mechanism for a risk-neutral buyer can be specified in terms of the marginal probabilities of allocation for each item. For any marginal probabilities, we can find a monotone lottery having the same marginal probabilities.

Observation 1. *For any distribution $\mathcal{D}$, the class $\mathcal{C}_{ndp}$ of nested deterministic price mechanisms contains an optimal mechanism for a risk-neutral buyer. That is, $\text{REV}(\mathcal{I}, \mathcal{D}) = \text{REV}(\mathcal{I}, \mathcal{D}, \mathcal{C}_{ndp})$.*

4   Deterministic Price Mechanisms

We first investigate general deterministic price mechanisms. We show that the optimal revenue of a deterministic price mechanism on independent items for a RDUT buyer can be upper bounded by doubly exponential times the optimal *risk-neutral* revenue of some items and the welfare on the distribution of the remaining items. Missing proofs can be found in the full version of this paper.

Theorem 2. *Let w be a weighting function, $\mathcal{D}_1$ be the product distribution of m_1 independent items, and $\mathcal{D}_2$ be the product distribution of m_2 independent items. $\mathcal{D} = \mathcal{D}_1 \times \mathcal{D}_2$ and $m = m_1 + m_2$. Then*

$$\text{REV}(w, \mathcal{D}, \mathcal{C}_{dp}) \leq 2^{2^{m_1}(m - \frac{1}{2}\log m_1)}\text{REV}(\mathcal{I}, \mathcal{D}_1) + \text{VAL}(\mathcal{D}_2).$$

Using standard techniques we get the following corollary.

Corollary 1. $\text{REV}(w, \mathcal{D}, \mathcal{C}_{dp}) \in O(2^{m2^m}) \max\{\text{SREV}, \text{BREV}\}.$

Though this approximation is doubly exponential in the number of items, we do get a constant approximation when the number of items is a constant. Notably, for the case of two items, we get $\text{REV}(w, \mathcal{D}, \mathcal{C}_{dp}) \leq 17\text{SREV}$; an improved analysis can reduce this to a factor of 2. We leave it as an open problem whether a constant approximation is possible for an arbitrary number of items.

4.1   Nested Deterministic Price Mechanisms

Our main result is that the class of nested deterministic price mechanisms does not offer the seller any means of exploiting the buyer's risk attitude: the optimal revenue within the class is equivalent to the optimal revenue obtainable from a risk-neutral mechanism.

Theorem 3. *Let w be an invertible weighting function and $\mathcal{D}$ be any distribution supported in $\mathbb{R}_{\geq 0}^m$. Then $\text{REV}(w, \mathcal{D}, \mathcal{C}_{ndp}) = \text{REV}(\mathcal{I}, \mathcal{D})$.*

Combining with Theorem 1 of [3] we get the following corollary.

Corollary 2. *Let $\mathcal{D}$ and w satisfy the conditions of Theorems 1 and 3. Then, it holds that $\text{REV}(w, \mathcal{D}, \mathcal{C}_{ndp}) \leq 6 \max\{\text{SREV}(\mathcal{D}), \text{BREV}(\mathcal{D})\}$.*

We prove Theorem 3 in two lemmas. We start by showing that for any invertible weighting function, there exists an NDP mechanism which recovers the optimal risk-neutral revenue. Next, we show the converse: that we can construct a mechanism for a risk-neutral buyer which obtains the same revenue as any DP mechanism for a buyer with weighting function w.

Lemma 4. *Let w be an invertible weighting function and $\mathcal{D}$ be any distribution supported in $\mathbb{R}_{\geq 0}^m$. Then $\text{REV}(w, \mathcal{D}, \mathcal{C}_{ndp}) \geq \text{REV}(\mathcal{I}, \mathcal{D})$.*

Lemma 5. *Let w be any weighting function and $\mathcal{D}$ any distribution supported in $\mathbb{R}_{\geq 0}^m$. Then $\text{REV}(w, \mathcal{D}, \mathcal{C}_{ndp}) \leq \text{REV}(\mathcal{I}, \mathcal{D})$.*

Proof. Consider a mechanism $\mathcal{M} \in \mathcal{C}_{ndp}$. Let $X(v)$ and $p(v)$ be the allocation and payment rule, respectively, of $\mathcal{M}$, where $X(v)$ is a random variable in $\Delta_N(2^{[m]})$ and $p(v) \in \mathbb{R}_{\geq 0}$. We construct a mechanism $\tilde{\mathcal{M}} = (\tilde{X}(v), \tilde{p}(v))$ for a risk-neutral buyer such that $\text{REV}_{\tilde{\mathcal{M}}}(\mathcal{I}, \mathcal{D}) = \text{REV}_{\mathcal{M}}(w, \mathcal{D})$.

Fix v. $X(v)$ is a monotone lottery by definition of $\mathcal{C}_{ndp}$, so let $S_1, \cdots, S_k$ be the support of $X(v)$, where $S_i \subset S_{i+1}$ for $i \in [k]$, and let $1 - F_i = \Pr[S_i \subseteq X(v)]$. Then the utility of an RDUT buyer is $u_w(v, X(v), p(v)) = \sum_{i=1}^{k} (v(S_i) - v(S_{i-1})) w(1 - F_i)$, where we take $S_0 = \emptyset$. Let $1 - \tilde{F}_i = w(1 - F_i)$, and define $\tilde{X}(v)$ such that $\Pr\left[\tilde{X}(v) = S_i\right] = \tilde{F}_{i+1} - \tilde{F}_i$. Lastly, let $\tilde{p}(v) = p(v)$. A risk-neutral buyer with *any* valuation v' has expected utility for the lottery $(\tilde{X}(v), \tilde{p}(v))$ equal to

$$u(v', \tilde{X}(v), \tilde{p}(v)) = \sum_{i=1}^{k} (v'(S_i) - v'(S_{i-1}))(1 - \tilde{F}_i) - \tilde{p}(v)$$
$$= \sum_{i=1}^{k} (v'(S_i) - v'(S_{i-1}))w(1 - F_i) - p(v),$$

which is just $u_w(v', X(v), p(v))$. Because this equality holds for every valuation v', $(\tilde{X}(v), \tilde{p}(v))$ is an IC, IR mechanism for a buyer with weighting function w, and furthermore obtains the same revenue from a buyer with weighting function w as $\mathcal{M}$ obtains from a risk-neutral buyer. □

Observe that the assumption of monotone lotteries was critical to the proof of Lemma 5. If $X(v)$ were an arbitrary distribution over subsets $S \in 2^{[m]}$, a buyer with valuation v' would order the outcomes differently from v. This would make it impossible to define the unweighted probability of allocation in the mechanism $\tilde{\mathcal{M}}$ in a way that would be simultaneously consistent with the weighted probability assigned to the outcome by all valuations v'.

Indeed a general deterministic-price mechanism (without the restriction to monotone lotteries) could exploit this discrepancy to obtain more revenue than a risk-neutral mechanism. That is, Lemma 5 does not hold for the class $\mathcal{C}_{dp}$, as the next claim shows.

Claim. There exists a distribution $\mathcal{D}$ over two items, and a weighting function w, such that $\textsc{Rev}(w, \mathcal{D}, \mathcal{C}_{dp}) > \textsc{Rev}(\mathcal{I}, \mathcal{D})$.

Proof. Let $\mathcal{D}_1, \mathcal{D}_2$ be independent and identical uniform distributions on $\{1, 3\}$. The revenue optimal auction that sells the two items to a risk-neutral buyer is the deterministic auction that sells the bundle of two items at the price 4. So $\textsc{Rev}(\mathcal{I}, \mathcal{D}_1 \times \mathcal{D}_2) = 4 \times \frac{3}{4} = 3$. Consider the weighting function

$$w(p) = \begin{cases} 0, & p \leq \frac{1}{2} \\ 4p - 2, & \frac{1}{2} < p < \frac{3}{4} \\ 1, & \frac{3}{4} \leq p \end{cases}.$$

Consider the auction $\mathcal{M}$ selling the two items in the following way: if the buyer reports type $(1, 1)$, the buyer gets the first item with probability $\frac{1}{2}$ and independently, get the second item with probability $\frac{1}{2}$, and the buyer pays 1 to the seller. Otherwise, the buyer gets both items and pays 4. It is easy to see that $\mathcal{M}$ is incentive compatible for a buyer with weighting function w. Furthermore, $\textsc{Rev}_{\mathcal{M}}(w, \mathcal{D}_1 \times \mathcal{D}_2) = 1 \times \frac{1}{4} + 4 \times \frac{3}{4} = \frac{13}{4} > 3$. □

References

1. Allais, M.: Fondements d'une théorie positive des choix comportant un risque et critique des postulats et axiomes de l'école américaine, vol. 144. Imprimerie nationale (1955)
2. Azevedo, E.M., Gottlieb, D.: Risk-neutral firms can extract unbounded profits from consumers with prospect theory preferences. J. Econ. Theory **147**(3), 1291–1299 (2012). https://doi.org/10.1016/j.jet.2012.01.002
3. Babaioff, M., Immorlica, N., Lucier, B., Weinberg, S.M.: A simple and approximately optimal mechanism for an additive buyer. SIGecom Exch. **13**(2), 31–35 (2015). https://doi.org/10.1145/2728732.2728736
4. Cai, Y., Devanur, N.R., Weinberg, S.M.: A duality-based unified approach to bayesian mechanism design. ACM SIGecom Exch. **15**(1), 71–77 (2016)
5. Cai, Y., Zhao, M.: Simple mechanisms for subadditive buyers via duality. In: Proceedings of the 49th Annual ACM SIGACT Symposium on Theory of Computing, pp. 170–183. ACM (2017)
6. Camerer, C.: Bounded rationality in individual decision making. Exp. Econ. **1**(2), 163–183 (1998). https://doi.org/10.1023/A:1009944326196
7. Chawla, S., Goldner, K., Miller, J.B., Pountourakis, E.: Revenue maximization with an uncertainty-averse buyer. In: Proceedings of the Twenty-Ninth Annual ACM-SIAM Symposium on Discrete Algorithms, pp. 2050–2068. SIAM (2018)
8. Chawla, S., Miller, J.B.: Mechanism design for subadditive agents via an ex ante relaxation. In: Proceedings of the 2016 ACM Conference on Economics and Computation, pp. 579–596. ACM (2016)
9. Devanur, N.R., Weinberg, S.M.: The optimal mechanism for selling to a budget constrained buyer: the general case. In: Proceedings of the 2017 ACM Conference on Economics and Computation, pp. 39–40. ACM (2017)
10. Dughmi, S., Peres, Y.: Mechanisms for risk averse agents, without loss. arXiv preprint arXiv:1206.2957 (2012)
11. Easley, D., Ghosh, A.: Behavioral mechanism design: optimal crowdsourcing contracts and prospect theory. In: Proceedings of the Sixteenth ACM Conference on Economics and Computation, pp. 679–696. ACM (2015)
12. Ebert, S., Strack, P.: Until the bitter end: on prospect theory in a dynamic context. Am. Econ. Rev. **105**(4), 1618–33 (2015). https://doi.org/10.1257/aer.20130896
13. Eden, A., Feldman, M., Friedler, O., Talgam-Cohen, I., Weinberg, S.M.: The competition complexity of auctions: a bulow-klemperer result for multi-dimensional bidders. In: Proceedings of the 2017 ACM Conference on Economics and Computation, pp. 343–343. ACM (2017)
14. Eden, A., Feldman, M., Friedler, O., Talgam-Cohen, I., Weinberg, S.M.: A simple and approximately optimal mechanism for a buyer with complements. In: Proceedings of the 2017 ACM Conference on Economics and Computation, pp. 323–323. ACM (2017)
15. Fiat, A., Papadimitriou, C.: When the players are not expectation maximizers. In: Kontogiannis, S., Koutsoupias, E., Spirakis, P.G. (eds.) SAGT 2010. LNCS, vol. 6386, pp. 1–14. Springer, Heidelberg (2010). https://doi.org/10.1007/978-3-642-16170-4_1
16. Fu, H., Hartline, J., Hoy, D.: Prior-independent auctions for risk-averse agents. In: Proceedings of the Fourteenth ACM Conference on Electronic Commerce, pp. 471–488. ACM (2013)

17. Hart, S., Nisan, N.: Approximate revenue maximization with multiple items. J. Econ. Theory **172**, 313–347 (2017)
18. Kahneman, D.: Prospect theory: an analysis of decisions under risk. Econometrica **47**, 278 (1979)
19. Li, X., Yao, A.C.C.: On revenue maximization for selling multiple independently distributed items. Proc. Nat. Acad. Sci. **110**(28), 11232–11237 (2013)
20. Liu, S., Psomas, C.A.: On the competition complexity of dynamic mechanism design. In: Proceedings of the Twenty-Ninth Annual ACM-SIAM Symposium on Discrete Algorithms, SODA 2018, pp. 2008–2025. Society for Industrial and Applied Mathematics, Philadelphia (2018). http://dl.acm.org/citation.cfm?id=3174304.3175436
21. Machina, M.J.: Choice under uncertainty: problems solved and unsolved. J. Econ. Perspect. **1**(1), 121–154 (1987)
22. Nikolova, E., Pountourakis, E., Yang, G.: Optimal mechanism design with risk-loving agents. In: Christodoulou, G., Harks, T. (eds.) WINE 2018. LNCS, vol. 11316, pp. 375–392. Springer, Cham (2018). https://doi.org/10.1007/978-3-030-04612-5_25
23. Quiggin, J.: A theory of anticipated utility. J. Econ. Behav. Organ. **3**(4), 323–343 (1982). https://doi.org/10.1016/0167-2681(82)90008-7
24. Quiggin, J.: Generalized Expected Utility Theory: The Rank-dependent Model. Springer Science & Business Media (2012)
25. Rubinstein, A., Weinberg, S.M.: Simple mechanisms for a subadditive buyer and applications to revenue monotonicity. In: Proceedings of the Sixteenth ACM Conference on Economics and Computation, pp. 377–394. ACM (2015)
26. Schmeidler, D.: Subjective probability and expected utility without additivity. Econom.: J. Econom. Soc. **57**, 571–587 (1989)
27. Tversky, A., Kahneman, D.: Advances in prospect theory: cumulative representation of uncertainty. J. Risk Uncertain. **5**(4), 297–323 (1992). https://doi.org/10.1007/BF00122574
28. Wakker, P.P.: Prospect Theory: For Risk and Ambiguity. Cambridge University Press (2010). https://doi.org/10.1017/CBO9780511779329
29. Weymark, J.A.: Generalized gini inequality indices. Math. Soc. Sci. **1**(4), 409–430 (1981)
30. Yaari, M.E.: The dual theory of choice under risk. Econom.: J. Econom. Soci. **55**, 95–115 (1987)
31. Yao, A.C.C.: An n-to-1 bidder reduction for multi-item auctions and its applications. In: Proceedings of the Twenty-Sixth Annual ACM-SIAM Symposium on Discrete Algorithms, pp. 92–109. Society for Industrial and Applied Mathematics (2015)

The Declining Price Anomaly Is Not Universal in Multi-buyer Sequential Auctions (But Almost Is)

Vishnu V. Narayan[1](✉), Enguerrand Prebet[2], and Adrian Vetta[1]

[1] McGill University, Montreal, Canada
vishnu.narayan@mail.mcgill.ca, adrian.vetta@mcgill.ca
[2] École normale supérieure de Lyon, Lyon, France
enguerrand.prebet@ens-lyon.fr

Abstract. The declining price anomaly states that the price weakly decreases when multiple copies of an item are sold sequentially over time. The anomaly has been observed in a plethora of practical applications. On the theoretical side, Gale and Stegeman [10] proved that the anomaly is guaranteed to hold in full information sequential auctions with exactly two buyers. We prove that the declining price anomaly is *not* guaranteed in full information sequential auctions with three or more buyers. This result applies to both first-price and second-price sequential auctions. Moreover, it applies regardless of the tie-breaking rule used to generate equilibria in these sequential auctions. To prove this result we provide a refined treatment of subgame perfect equilibria that survive the iterative deletion of weakly dominated strategies and use this framework to experimentally generate a very large number of random sequential auction instances. In particular, our experiments produce an instance with three bidders and eight items that, for a specific tie-breaking rule, induces a non-monotonic price trajectory. Theoretical analyses are then applied to show that this instance can be used to prove that for every possible tie-breaking rule there is a sequential auction on which it induces a non-monotonic price trajectory. On the other hand, our experiments show that non-monotonic price trajectories are extremely rare. In over six million experiments only a 0.000183 proportion of the instances violated the declining price anomaly.

1 Introduction

In a sequential auction identical copies of an item are sold over time. In a private values model with *unit-demand*, risk neutral buyers, Milgrom and Weber [19,26] showed that the sequence of prices forms a martingale. In particular, expected prices are constant over time.[1] In contrast, on attending a wine auction, Ashenfelter [1] made the surprising observation that prices for identical lots declined over time: "The law of the one price was repealed and no one even seemed to notice!" This *declining price anomaly* was also noted in sequential auctions for

---

[1] If the values are affiliated then prices can have an upwards drift.

© Springer Nature Switzerland AG 2019
D. Fotakis and E. Markakis (Eds.): SAGT 2019, LNCS 11801, pp. 109–122, 2019.
https://doi.org/10.1007/978-3-030-30473-7_8

the disparate examples of livestock (Buccola [7]), Picasso prints (Pesando and Shum [21]) and satellite transponder leases (Milgrom and Weber [19]). Indeed, the possibility of decreasing prices in a sequential auction was raised by Sosnick [23] nearly sixty years ago. In the case of wine auctions, proposed causes include absentee buyers utilizing non-optimal bidding strategies (Ginsburgh [11]) and the *buyer's option rule* where the auctioneer may allow the buyer of the first lot to make additional purchases at the same price (Black and de Meza [6]). Minor non-homogeneities amongst the items can also lead to falling prices. For example, in the case of art prints the items may suffer slight imperfections or wear-and-tear, and the auctioneer may sell the prints in decreasing order of quality (Pesando and Shum [21]). More generally, a decreasing price trajectory may arise due to risk-aversion, such as non-decreasing, absolute risk-aversion (McAfee and Vincent [17]) or aversion to price-risk (Mezzetti [18]); see also Hu and Zou [13]. Further potential economic and behavioural explanations have been provided in [2,11,25]. Of course, most of these explanations are context-specific. However, in practice the anomaly is ubiquitous: it has now been observed in sequential auctions for, among several other things, antiques (Ginsburgh and van Ours [12]), commercial real estate (Lusht [16]), flowers (van den Berg et al. [5]), fur (Lambson and Thurston [15]), jewellery (Chanel et al. [8]), paintings (Beggs and Graddy [4]) and stamps (Thiel and Petry [24]).

Given the plethora of examples, the question arises as whether this property is actually an anomaly. In groundbreaking work, Gale and Stegeman [10] proved that it is *not* in sequential auctions with *two* bidders. Specifically, in *second-price* sequential auctions with two multiunit-demand buyers, prices are weakly decreasing over time at the unique subgame perfect equilibrium that survives the iterative deletion of weakly dominated strategies. This result applies regardless of the valuation functions of the buyers, and also extends to the corresponding equilibrium in *first-price* sequential auctions. It is worth highlighting that Gale and Stegeman consider multiunit-demand buyers whereas prior theoretical work had focused on the simpler setting of unit-demand buyers. As well as being of more practical relevance (see the many examples above), multiunit-demand buyers can implement more sophisticated bidding strategies. Therefore, it is not unreasonable that equilibria in multiunit-demand setting may possess more interesting properties than equilibria in the unit-demand setting. The restriction to full information in [10] is extremely useful here as it separates away informational aspects and allows one to focus on the strategic properties caused purely by the sequential sales of items and not by a lack of information.

1.1   Results and Overview of the Paper

The result of Gale and Stegeman [10] prompts the question of whether or not the declining price anomaly is guaranteed to hold in general, that is, in sequential auctions with more than two buyers. We answer this question in the negative by exhibiting a sequential auction with three buyers and eight items where prices initially rise and then fall. In order to run our experiments that find this counterexample (to the conjecture that prices are weakly decreasing for multi-buyer sequential auctions) we study in detail the form of equilibria in sequential auctions.

First, it is important to note that there is a fundamental distinction between sequential auctions with two buyers and sequential auctions with three or more buyers. In the former case, each subgame reduces to a standard *auction with independent valuations*. In contrast, in a multi-buyer sequential auction each subgame reduces to an *auction with interdependent valuations*. We present these models in Sects. 2.1 and 2.2. Consequently to study multi-buyer sequential auctions we must study the equilibria of auctions with interdependent valuations. A theory of such equilibria was recently developed by Paes Leme et al. [20] via a correspondence with an ascending price mechanism. In particular, as we discuss in Sect. 2.3, this ascending price mechanism outputs a unique bid value, called the *dropout bid* β_i, for each buyer i. For first-price auctions it is known [20] that these dropout bids form a subgame perfect equilibrium and, moreover, the interval $[0, \beta_i]$ is the exact set of bids that survives *all* processes consisting of the iterative deletion of strategies that are weakly dominated. In contrast, we show that for second-price auctions it may be the case that no bids survive the iterative deletion of weakly dominated strategies; however, we prove in Sect. 2.3 that the interval $[0, \beta_i]$ is the exact set of bids for any losing buyer that survives *all* processes consisting of the iterative deletion of strategies that are weakly dominated *by a lower bid*.

In Sect. 3 we describe the counter-example. We emphasize that the form of the valuation functions used for the buyers are standard, namely, weakly decreasing marginal valuations. Furthermore, the non-monotonic price trajectory does not arise because of the use of an artificial tie-breaking rule; the three most natural tie-breaking rules, see Sect. 2.4, all induce the same non-monotonic price trajectory. Indeed, we present an even stronger result in Sect. 4: for *any* tie-breaking rule, there is a sequential auction on which it induces a non-monotonic price trajectory. This lack of weakly decreasing prices provides an explanation for why multi-buyer sequential auctions have been hard to analyze quantitatively. We provide a second explanation in the full paper, where we present a three-buyer sequential auction that does satisfy weakly decreasing prices but which has subgames where some agent has a negative value from winning against one of the two other agents. Again, this contrasts with the two-buyer case where every agent always has a non-negative value from winning against the other agent in every subgame.

Finally in Sect. 5, we describe the results obtained via our large scale experimentations. These results show that whilst the declining price anomaly is not universal, exceptions are extremely rare. Specifically, from a randomly generated dataset of over six million sequential auctions only a 0.000183 proportion of the instances produced non-monotonic price trajectories. Consequently, these experiments are consistent with the practical examples discussed in the introduction. Of course, it is perhaps unreasonable to assume that subgame equilibria arise in practice; we remark, though, that the use of simple bidding algorithms by bidders may also lead to weakly decreasing prices in a multi-buyer sequential auction. For example, Rodriguez [22] presents a method called the *residual monopsonist procedure* inducing this property in restricted settings.

2   The Sequential Auction Model

Here we present the full information sequential auction model. There are T identical items and n buyers. Exactly one item is sold in each time period over T time periods. Buyer i has a value $V_i(k)$ for winning exactly k items. Thus $V_i(k) = \sum_{\ell=1}^{k} v_i(\ell)$, where $v_i(\ell)$ is the marginal value buyer i has for an ℓth item. This induces an extensive form game. To analyze this game it is informative to begin by considering the 2-buyer case studied by Gale and Stegeman [10].

2.1   The Two-Buyer Case

During the auction, the relevant history is the number of items each buyer has currently won. Thus we may compactly represent the extensive form ("tree") of the auction using a directed graph with a node (x_1, x_2) for any pair of non-negative integers that satisfies $x_1 + x_2 \leq T$. The node (x_1, x_2) induces a subgame with $T - x_1 - x_2$ items for sale and where each buyer i already possesses x_i items. Note there is a *source node*, $(0,0)$, corresponding to the whole game, and *sink nodes* (x_1, x_2), where $x_1 + x_2 = T$. The values Buyer 1 and Buyer 2 have for a sink node (x_1, x_2) are $\Pi_1(x_1, x_2) = V_1(x_1)$ and $\Pi_2(x_1, x_2) = V_2(x_2)$, respectively. Take a node (x_1, x_2), where $x_1 + x_2 = T - 1$. This node corresponds to the final round of the auction, where the last item is sold, and has directed arcs to the sink nodes (x_1+1, x_2) and (x_1, x_2+1). For the case of second-price auctions, it is then a weakly dominant strategy for Buyer 1 to bid its marginal value $v_1(x_1 + 1) = V_1(x_1 + 1) - V_1(x_1)$; similarly for Buyer 2. Of course, this marginal value is just $v_1(x_1 + 1) = \Pi_1(x_1+1, x_2) - \Pi_1(x_1, x_2+1)$, the difference in value between winning and losing the final item. If Buyer 1 is the highest bidder at (x_1, x_2), that is, $\Pi_1(x_1+1, x_2) - \Pi_1(x_1, x_2 + 1) \geq \Pi_2(x_1, x_2 + 1) - \Pi_2(x_1 + 1, x_2)$, then we have that

$$\Pi_1(x_1, x_2) = \Pi_1(x_1 + 1, x_2) - \big(\Pi_2(x_1, x_2 + 1) - \Pi_2(x_1 + 1, x_2) \big)$$
$$\Pi_2(x_1, x_2) = \Pi_2(x_1 + 1, x_2)$$

Symmetric formulas apply if Buyer 2 is the highest bidder. Hence we may recursively define a value for each buyer for each node. The iterative elimination of weakly dominated strategies leads to a subgame perfect equilibrium [3, 10].

Example: Consider a two-buyer sequential auction with two items, where the marginal valuations are $\{v_1(1), v_1(2)\} = \{10, 8\}$ and $\{v_2(1), v_2(2)\} = \{6, 3\}$. This game is illustrated in Fig. 1. The base case with the values of the sink nodes is shown in Fig. 1(a). The first row in each node refers to Buyer 1 and shows the number of items won (in plain text) and the corresponding value (in bold); the second row refers to Buyer 2. The outcome of the second-price sequential auction, solved recursively, is then shown in Fig. 1(b). Arcs are labelled by the bid value; here arcs for Buyer 1 point left and arcs for Buyer 2 point right. Solid arcs represent winning bids and dotted arcs are losing bids. The equilibrium path is shown in bold. Figure 1(c) shows the corresponding first-price auction, where we make the standard assumption of a fixed small bidding increment, and the notation p^+ and p

are respectively used to denote a winning bid of value p and a losing bid equal to the maximum value smaller than p. For simplicity, all the figures we present in the rest of the paper will be for first-price auctions; equivalent figures can be drawn for the case of second-price auctions. Observe that this example exhibits the *declining price anomaly*: in the equilibrium, the first item has price 5 and the second item has price 3. As stated, Gale and Stegeman [10] showed that this example is not an exception.

Theorem 1 *[10]. In a 2-buyer second-price sequential auction there is a unique equilibrium that survives the iterative deletion of weakly dominated strategies. Moreover, at this equilibrium prices are weakly declining.*   □

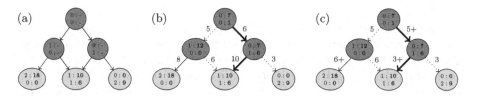

Fig. 1. Sequential auction examples

We remark that the subgame perfect equilibrium that survives iterative elimination is unique in terms of the values at the nodes. Moreover, given a fixed tie-breaking rule, the subgame perfect equilibrium also has a unique equilibrium path in each subgame. In addition, Theorem 1 also applies to first-price sequential auctions. The question of whether or not it applies to sequential auctions with more than two buyers remained open. We resolve this question in the rest of this paper. To do this, let's first study equilibria in the full information sequential auction model when there are more than two buyers.

2.2   The Multi-buyer Case

The underlying model of [10] extends simply to sequential auctions with $n \geq 3$ buyers. There is a node $(x_1, x_2, \ldots, x_n)$ for each set of non-negative integers satisfying $\sum_{i=1}^{n} x_i \leq T$. There is a directed arc from $(x_1, x_2, \ldots, x_n)$ to $(x_1, x_2, \ldots, x_{j-1}, x_j + 1, x_{j+1}, \ldots x_n)$ for each $1 \leq j \leq n$. Thus each non-sink node has n out-going arcs. This is problematic: whilst in the final time period each buyer has a value for winning and a value for losing, this is no longer the case recursively in earlier time periods. Specifically, buyer i has a value for winning, but $n - 1$ (different) values for losing depending upon the identity of the buyer $j \neq i$ who wins. Thus each node in the multi-buyer case corresponds to an *auction with interdependent valuations*. Formally, this is a single-item auction where each buyer i has a value $v_{i,i}$ for winning the item and a value $v_{i,j}$ if buyer j wins the item, for each $j \neq i$. These auctions, also called *auctions with externalities*, were introduced by Funk [9] and by Jehiel and Moldovanu [14]. Their motivations were applications where losing participants were not indifferent to the identity of the winner; examples include firms seeking to purchase a patented innovation, take-over acquisitions of a smaller company in an oligopolistic market, and sports teams competing to sign a star athlete. Therefore to understand multi-buyer sequential auctions we must first understand equilibria in auctions with interdependent valuations. This is not a simple task; indeed, such an understanding was only recently provided by Paes Leme et al. [20].

2.3   Equilibria in Auctions with Interdependent Valuations

We can explain the result of [20] via an ascending price auction. Imagine a two-buyer ascending price auction where the valuations of the buyers are $v_1 > v_2$. The requested price p starts at zero and continues to rise until the point where the second buyer drops out. Of course, this happens when the price reaches v_2, and so Buyer 1 wins for a payment $p^+ = v_2$, which is exactly the outcome expected from a first-price auction. To generalize this to multi-buyer settings we can view this process as follows. At a price p, buyer i remains in the auction as long as there is at least one buyer j *still in the auction* who buyer i is willing to pay a price p to beat; that is, $v_{i,i} - p > v_{i,j}$. The last buyer to drop out wins at the corresponding price. Even in this setting, this procedure produces a unique *dropout bid* β_i for each buyer i, as illustrated in Fig. 2. In these diagrams the label of an arc from buyer i to buyer j is $w_{i,j} = v_{i,i} - v_{i,j}$. That is, buyer i is willing to pay up to $w_{i,j}$ to win *if the alternative is that buyer j wins the item*. Now consider running our ascending price procedure for these auctions. In Fig. 2(a), Buyer 1 drops out when the price reaches 18. Since Buyer 1 is no longer active, Buyer 4 drops out at 23. Buyer 3 wins when Buyer 2 drops out at 31. Thus the drop-out bid of Buyer 3 is 31^+. Observe that Buyer 2 loses despite having very high values for winning against Buyer 1 and Buyer 4. The example of Fig. 2(b) is more subtle. Here Buyer 2 drops out at price 24. But Buyer 3 only wanted to beat Buyer 2 at this price so it then immediately drops out at the same price. Now Buyer 1 only wanted to beat Buyer 2 and Buyer 3 at this price, so it then immediately drops out at the same price. This leaves Buyer 4 the winner at price 24^+.

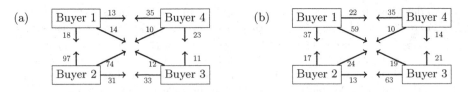

Fig. 2. DROP-OUT BID EXAMPLES. *In these two examples the dropout bid vectors* $(\beta_1, \beta_2, \beta_3, \beta_4)$ *are* $(18, 31, 31^+, 23)$ *and* $(24, 24, 24, 24^+)$, *respectively.*

As well as being solutions to the ascending price auction, the dropout bids have a much stronger property that makes them the natural and robust prediction for auctions with interdependent valuations. Specifically, Paes Leme et al. [20] proved that, for each buyer i, the interval $[0, \beta_i]$ is the set of strategies that survive *any* *sequence* consisting of the iterative deletion of weakly dominated strategies. This is formalized as follows. Take an n-buyer game with strategy sets $S_1, S_2, \ldots, S_n$ and utility functions $u_i : S_1 \times S_2 \times \cdots \times S_n \to \mathbb{R}$. Then $\{S_i^\tau\}_{i,\tau}$ is a *valid sequence* for the iterative deletion of weakly dominated strategies if for each τ there is a buyer i such that (i) $S_j^\tau = S_j^{\tau-1}$ for each buyer $j \neq i$ and (ii) $S_i^\tau \subset S_i^{\tau-1}$ where for each strategy $s_i \in S_i^{\tau-1} \setminus S_i^\tau$ there is an $\hat{s}_i \in S_i^\tau$ such that $u_i(\hat{s}_i, s_{-i}) \geq u_i(s_i, s_{-i})$ for all $s_{-i} \in \prod_{j:j \neq i} S_j^\tau$, and with strict inequality for at least one s_{-i}. We say that a strategy s_i for buyer i *survives* the iterative deletion if for any valid sequence $\{S_i^\tau\}_{i,\tau}$ we have $s_i \in \bigcap_\tau S_i^\tau$.

Theorem 2 *[20]. Given a first-price auction with interdependent valuations, for each buyer i, the set of bids that survive the iterative deletion of weakly dominated strategies is exactly $[0, \beta_i]$.* □

An exact analogue of Theorem 2 does *not* hold for second-price auctions with interdependent valuations. Indeed, there exist examples in which the set of strategies that survive iterative deletion is empty. However, consideration of these example shows that the problem occurs when a strategy is deleted because it is weakly dominated by a *higher* value bid. Observe that this can never happen for a potentially winning bid. Thus Theorem 2 still holds in first-price auctions when we restrict attention to sequences consisting of the iterative deletion of strategies that are weakly dominated by a *lower* bid. We can also show that the corresponding theorem holds for second-price auctions. The full technical details of the proof are deferred to the full paper.

Theorem 3. *Given a second-price auction with interdependent valuations, for each losing buyer i, the set of bids that survive the iterative deletion of strategies that are weakly dominated by a lower bid is exactly $[0, \beta_i]$.* □

We are now almost ready to be able to find equilibria in the sequential auction experiments we will conduct. This, in turn, will allow us to present a sequential auction with non-monotonic prices. Before doing so, one final factor remains to be discussed regarding the transition from equilibria in auctions with interdependent valuations to equilibria in sequential auctions.

2.4 Equilibria in Sequential Auctions: Tie-Breaking Rules

As stated, the dropout bid of each buyer is uniquely defined. However, our description of the ascending auction may leave some flexibility in the choice of winner. Specifically, it may be the case that simultaneously more than one buyer wishes to drop out of the auction. If this happens at the end of the ascending price procedure then any of these buyers could be selected as the winner. To fully define the ascending auction we must incorporate a tie-breaking rule to order the buyers when more

than one wish to drop out simultaneously. In an auction with interdependent valuations the tie-breaking rule only affects the choice of winner, but otherwise has no
structural significance. However, in a sequential auction, the choice of winner at
one node may affect the valuations at nodes higher in the tree. In particular, the
equilibrium path may vary with different tie-breaking rules, leading to different
prices, winners, and utilities.

As we will show in Sect. 4 there are a massive number of tie-breaking rules
even in small sequential auctions. We emphasize, however, that our main result
holds regardless of the tie-breaking rule: for *any* tie-breaking rule there is a
sequential auction on which it induces a non-monotonic price trajectory. First,
though, we will show that non-monotonicity occurs for perhaps the three most
natural choices, namely `preferential-ordering`, `first-in-first-out` and
`last-in-first-out`. Interestingly, these rules correspond to the fundamental
data structures of priority queues, queues, and stacks in computer science.

Preferential Ordering (Priority Queue): In `preferential-ordering` each
buyer is given a distinct rank. In case of a tie the buyer with the worst rank is
eliminated. Without loss of generality, we may assume that the ranks correspond
to a lexicographic ordering of the buyers. That is, the rank of a buyer is its index
label and given a tie amongst all the buyers that wish to dropout of the auction we
remove the buyer with the highest index. The preferential ordering tie-breaking
rule corresponds to the data structure known as a *priority queue*.

First-In-First-Out (Queue): The `first-in-first-out` tie-breaking rule corresponds to the data structure known as a *queue*. The queue consists of those buyers in the auction that wish to dropout. Amongst these, the buyer at the front of
the queue is removed. If multiple buyers request to be added to the queue simultaneously, they will be added lexicographically. Note though that this is different
from preferential ordering as the entire queue will not, in general, be ordered lexicographically. For example, when at a fixed price p we remove the buyer i at the
front of the queue this may cause new buyers to wish to dropout at price p, who
will be placed behind the other buyers already in the queue.

Last-In-First-Out (Stack): The `last-in-first-out` tie-breaking rule corresponds to the data structure known as a *stack*. Again the stack consists of those
buyers in the auction that wish to dropout. Amongst these, the buyer at the top
of the stack (i.e. the back of the queue) is removed. If multiple buyers request to
be added to the stack simultaneously, they will be added lexicographically. At first
glance, this `last-in-first-out` rule appears more unusual than the previous two,
but it still has a natural interpretation: it corresponds to settings where the buyer
whose situation has changed most recently reacts the quickest.

In order to understand these tie-breaking rules it is useful to see how they
apply on an example. In Fig. 3 the dropout vector is $(\beta_1, \beta_2, \beta_3, \beta_4, \beta_5) =$
$(40, 40, 40, 40, 40)$, but the three tie-breaking rules select three different winners.

On running the ascending price procedure, both Buyer 3 and Buyer 4 wish
to drop out when the price reaches 40. In `preferential-ordering`, our choice
set is then $\{3, 4\}$ and we remove the highest index buyer, namely Buyer 4.
With the removal of Buyer 4, neither Buyer 1 nor Buyer 5 have an incentive to

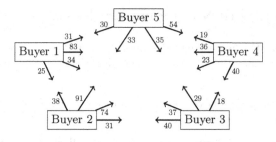

Fig. 3. An example to illustrate the three tie-breaking rules.

continue bidding so they both decide to dropout. Thus our choice set is now $\{1, 3, 5\}$ and `preferential-ordering` removes Buyer 5. With the removal of Buyer 5, now Buyer 2 no longer has an active participant it wishes to beat so the choice set is updated to $\{1, 2, 3\}$. The `preferential-ordering` rule now removes the buyers in the order Buyer 3, then Buyer 2 and lastly Buyer 1. Thus Buyer 1 wins under the `preferential-ordering` rule.

Now consider `first-in-first-out`. To allow for a consistent comparison between the three methods, we assume that when multiple buyers are simultaneously added to the queue they are added in decreasing lexicographical order. Thus our initial queue is 4 : 3 and `first-in-first-out` removes Buyer 4 from the front of the queue. With the removal of Buyer 4, neither Buyer 1 nor Buyer 5 have an incentive to continue bidding so they are added to the back of the queue. Thus the queue is now 3 : 5 : 1 and `first-in-first-out` removes Buyer 3 from the front of the queue. It then removes Buyer 5 from the front of the queue. With the removal of Buyer 5, we again have that Buyer 2 now wishes to dropout. Hence the queue is 1 : 2 and `first-in-first-out` then removes Buyer 1 from the front of the queue. Thus Buyer 2 wins under the `first-in-first-out` rule.

Finally, consider the `last-in-first-out` rule. Again, to allow for a consistent comparison we assume that when multiple buyers are simultaneously added to the stack they are added in increasing lexicographical order. Thus our initial stack is $\frac{4}{3}$ and `last-in-first-out` removes Buyer 4 from the top of the stack. Again, Buyer 1 and Buyer 5 both now wish to drop out so our stack becomes $\frac{5}{\frac{1}{3}}$. Therefore Buyer 5 is next removed from the the top of the stack. At this point, Buyer 2 wishes to dropout so the stack becomes $\frac{2}{\frac{1}{3}}$. The `last-in-first-out` rule now removes the buyers in the order Buyer 2, then Buyer 1 and lastly Buyer 3. Thus Buyer 3 wins under the `last-in-first-out` rule.

We have now developed all the tools required to implement our sequential auction experiments. We describe these experiments and their results in Sect. 5. Before doing so, we present in Sect. 3 one sequential auction obtained via these experiments and verify that it leads to a non-monotonic price trajectory with each of the three tie-breaking rules discussed above. We then explain in Sect. 4 how to generalize this conclusion to apply to every tie-breaking rule.

3   An Auction with Non-monotonic Prices

Here we prove that the decreasing price anomaly is *not* guaranteed for sequential auctions with more than two buyers. Specifically, in Sect. 4 we prove the following result:

Theorem 5. *For any tie-breaking rule τ, there is a sequential auction on which it produces non-monotonic prices.*

In the rest of this section, we show that for all three of the tie-breaking rules discussed (namely, `preferential-ordering`, `first-in-first-out` and `last-in-first-out`) there is a sequential auction with with non-monotonic prices. Specifically, we exhibit a sequential auction with three buyers and eight items that exhibits non-monotonic prices.

Theorem 4. *There is a sequential auction which exhibits a non-monotonic price trajectory for the* `preferential-ordering`, *the* `first-in-first-out` *and the* `last-in-first-out` *rules.*

Proof. Our counter-example to the conjecture is a sequential auction with three buyers and eight identical items for sale. We present the first-price version where at equilibrium the buyers bid their dropout values in each time period; as discussed, the same example extends to second-price auctions. In our example, Buyer 1 has marginal valuations $\{55, 55, 55, 55, 30, 20, 0, 0\}$, Buyer 2 has marginal valuations $\{32, 20, 0, 0, 0, 0, 0, 0\}$, and Buyer 3 has marginal valuations $\{44, 44, 44, 44, 0, 0, 0, 0\}$. Let's now compute the extensive forms of the auction under the three tie-breaking rules. We begin with the `preferential-ordering` rule. To compute its extensive form, observe that Buyer 1 is guaranteed to win at least two items in the auction because Buyer 2 and Buyer 3 together have positive value for six items. Therefore, the feasible set of sink nodes in the extensive form representation are shown in Fig. 4.

Fig. 4. Sink nodes of the extensive form game.

Given the valuations at the sink nodes we can work our way upwards recursively calculating the values at the other nodes in the extensive form representation. For example, consider the node $(x_1, x_2, x_3) = (4, 1, 2)$. This node has three children, namely $(5, 1, 2)$, $(4, 2, 2)$ and $(4, 1, 3)$; see Fig. 5(a). These induce a three-buyer auction as shown in Fig. 5(b). This can be solved using the ascending price procedure to find the dropout bids for each buyer. Thus we obtain that the value for the node $(x_1, x_2, x_3) = (4, 1, 2)$ is as shown in Fig. 5(c). Of course this node is particularly simple as, for the final round of the sequential auction, the corresponding auction with interdependent valuations is just a standard auction. That is, when the final

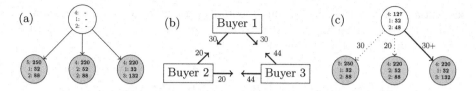

Fig. 5. Solving a subgame above the sinks.

item is sold, for any buyer i the value $v_{i,j}$ is independent of the buyer $j \neq i$. Nodes higher up the game tree correspond to more complex auctions with interdependent valuations. For example, the case of the source node $(x_1, x_2, x_3) = (0,0,0)$ is shown in Fig. 6. In this case, on applying the ascending price procedure, Buyer 1 is the first to dropout at price 15. At this point, both Buyer 2 and Buyer 3 no longer have a competitor that they wish to beat at this price, so they both want to dropout. With the `preferential-ordering` tie-breaking rule, Buyer 2 wins the item.

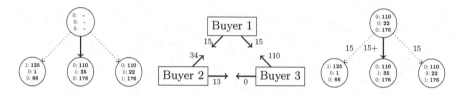

Fig. 6. Solving the subgame at the root.

Using similar arguments at each node verifies the concise extensive form representation of this example under the `preferential-ordering` tie-breaking rule. A figure showing the full extensive form tree is present in the full paper. The resultant price trajectory on the equilibrium path is $\{15, 17, 0, 0, 0, 0, 0, 0\}$. That is, the price rises and then falls to zero – a non-monotonic price trajectory.

Exactly the same example works with the other two tie-breaking rules. The node values under `preferential-ordering` and `first-in-first-out` are the same, but these two rules do produce different winners at some nodes, for example the node $(3, 0, 2)$. In contrast, the `last-in-first-out` rule gives an extensive for where some nodes have different valuations than those produced by the other two rules. For example, for the node $(2, 0, 0)$ and its subgame the equilibrium paths and their prices differ. However, for all three rules the equilibrium path and price trajectory *for the whole game* is exactly the same. We remark that these observations will play a role when we prove that, for any tie-breaking rule, there is a sequential auction with non-monotonic prices. □

Again, we emphasize that there is nothing inherently perverse about this example. The form of the valuation functions, namely decreasing marginal valuations, is standard. As explained, the equilibrium concept studied is the appropriate one for sequential auctions. Finally, the non-monotonic price trajectory is not the artifact of an aberrant tie-breaking rule; we will now prove that non-monotonic prices are exhibited under any tie-breaking rule.

4   General Tie-Breaking Rules: Non-monotonic Prices

Next we prove that for any tie-breaking rule there is a sequential auction on which it produces a non-monotonic price trajectory. To do this, we must first formally define the set of all tie-breaking rules. Our definition will utilize the concept of an *overbidding graph*, introduced by Paes Leme et al. [20]. For any price p and any set of bidders S, the overbidding graph $G(S, p)$ contains a labelled vertex for each buyer in S and an arc (i, j) if and only if $v_{i,i} - p > v_{i,j}$. For example, recall the auction with interdependent valuations seen in Fig. 3. This is reproduced in Fig. 7 along with its overbidding graph $G(\{1, 2, 3, 4, 5\}, 40)$.

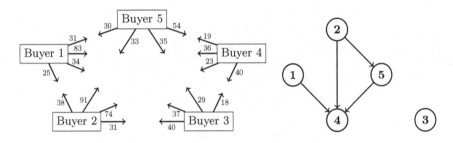

Fig. 7. The overbidding graph $G(\{1, 2, 3, 4, 5\}, 40)$.

But what does the overbidding graph have to do with tie-breaking rules? First, recall that the drop-out bid β_i is unique for any buyer i, regardless of the tie-breaking rule. Consequently, whilst the tie-breaking rule will also be used to order buyers that are eliminated at prices below the final price p^*, such choices are irrelevant with regards to the final winner. Thus, the only relevant factor is how a decision rule selects a winner from amongst those buyers S^* whose drop-out bids are p^*. Second, recall that a buyer *cannot* be eliminated if there remains another buyer still in the auction that it wishes to beat at price p^*. That is, buyer i must be eliminated after buyer j if there is an arc (i, j) in the overbidding graph. Thus, the order of eliminations given by the tie-breaking rule must be consistent with the overbidding graph. In particular, the winner can only be selected from amongst the *source vertices*[2] in the overbidding graph $G(S^*, p^*)$. For example, in Fig. 7 the source vertices are $\{1, 2, 3\}$. Note that this explains why the tie-breaking rules preferential-ordering, first-in-first-out and last-in-first-out chose Buyer 1, Buyer 2 and Buyer 3 as winners but none of them selected Buyer 4 or Buyer 5. Observe that the overbidding graph $G(S^*, p^*)$ is *acyclic*; if it contained

---

[2] A *source* is a vertex v with in-degree zero; that is, there no arcs pointing into v.

a directed cycle then the price in the ascending auction would be forced to rise further. Because every directed acyclic graph contains at least one source vertex, any tie-breaking rule does have at least one choice for winner. Thus a tie-breaking rule is simply a function $\tau : H \rightarrow \sigma(H)$, where the domain is the set of labelled, directed acyclic graphs and $\sigma(H)$ is the set of source nodes in H. Consequently, two tie-breaking rules are equivalent if they correspond to the same function τ. We are now ready to present our main result.

Theorem 5. *For any tie-breaking rule, there is a sequential auction with non-monotonic prices.*

We present here a sketch of our proof of this theorem; due to length restrictions the full proof is deferred. We consider the same example as in Theorem 4, and analyze the set of all possible tie-breaking rules in three-buyer auctions. We show that each tie-breaking rule produces an outcome from a set of exactly ten possible distinct extensive forms for this example. Of these ten classes, exactly five classes result in non-monotonicity. We then show that for any given tie-breaking rule from the other five classes it is possible to relabel the buyers in a way that the resulting equilibrium has a non-monotonic price trajectory.

5   Experiments

Our experiments were based on a dataset of over six million multi-buyer sequential auctions with non-increasing valuation functions randomly generated from different natural discrete probability distributions. Our goal was to observe the proportion of non-monotonic price trajectories and see how this varied with (i) the number of buyers, (ii) the number of items, (iii) the distribution of valuation functions, and (iv) the tie-breaking rule. For each auction we computed the subgame perfect equilibrium corresponding to the dropout bids and evaluated the prices on the equilibrium path to test for non-monotonicity. We repeated this test for each of the three tie breaking rules described in Sect. 2.4. The main conclusion to be drawn from these experiments is that non-monotonic prices are extremely rare. Of the 6,240,000 auctions, the `preferential-ordering`, `first-in-first-out` and `last-in-first-out` rules gave just 1,100, 986, and 1,334 violations of the declining price anomaly respectively. The overall observed rate of non-monotonicity over these 18 million tests was 0.000183. A detailed description of our dataset generation process and results are in the full paper.

References

1. Ashenfelter, O.: How auctions work for wine and art. J. Econ. Perspect. **3**(3), 23–36 (1989)
2. Ashta, A.: Wine auctions: more explanations for the declining price anomaly. J. Wine Res. **17**(1), 53–62 (2006)
3. Bae, J., Beigman, E., Berry, R., Honig, M., Vohra, R.: Sequential bandwidth and power auctions for distributed spectrum sharing. J. Sel. Areas Commun. **26**(7), 1193–1203 (2008)

4. Beggs, A., Graddy, K.: Declining values and the afternoon effect: evidence from art auctions. Rand J. Econ. **28**, 544–565 (1997)
5. van den Berg, G., van Ours, J., Pradhan, M.: The declining price anomaly in Dutch Dutch rose auctions. Am. Econ. Rev. **91**, 1055–1062 (2001)
6. Black, J., de Meza, D.: Systematic price differences between successive auctions are no anomaly. J. Econ. Manag. Strategy **1**(4), 607–628 (1992)
7. Buccola, S.: Price trends at livestock auctions. Am. J. Agric. Econ. **64**, 63–69 (1982)
8. Chanel, O., Gérard-Varet, L., Vincent, S.: Auction theory and practice: evidence from the market for jewellery. In: Ginsburgh, V., Menger, P. (eds.) Economics of the Arts: Selected Essays, pp. 135–149. North-Holland (1996)
9. Funk, P.: Auctions with interdependent valuations. Int. J. Game Theory **25**, 51–64 (1996)
10. Gale, I., Stegeman, M.: Sequential auctions of endogenously valued objects. Games Econ. Behav. **36**(1), 74–103 (2001)
11. Ginsburgh, V.: Absentee bidders and the declining price anomaly in wine auctions. J. Polit. Econ. **106**(6), 319–335 (1998)
12. Ginsburgh, V., van Ours, J.: On organizing a sequential auction: results from a natural experiment by Christie's. Oxford Econ. Pap. **59**(1), 1–15 (2007)
13. Hu, A., Zou, L.: Sequential auctions, price trends, and risk preferences. J. Econ. Theory **158**, 319–335 (2015)
14. Jehiel, P., Moldovanu, B.: Strategic nonparticipation. Rand J. Econ. **27**(1), 84–98 (1996)
15. Lambson, V., Thurston, N.: Sequential auctions: theory and evidence from the seattle fur exchange. Games Econ. Behav. **37**(1), 70–80 (2006)
16. Lusht, K.: Order and price in a sequential auction. J. Real Estate Finance Econ. **8**, 259–266 (1994)
17. McAfee, P., Vincent, D.: The declining price anomaly. J. Econ. Theory **60**, 191–212 (1993)
18. Mezzetti, C.: Sequential auctions with informational externalities and aversion to price risk: decreasing and increasing price sequences. Econ. J. **121**(555), 990–1016 (2011)
19. Milgrom, P., Weber, R.: A theory of auctions and competitive bidding II. In: Klemperer, P. (ed.) The Economic Theory of Auctions. Edward Elgar (2000)
20. Paes Leme, R., Syrgkanis, V., Tardos, E.: Sequential auctions and externalities. In: Proceedings of 23rd Symposium on Discrete Algorithms, pp. 869–886. Society for Industrial and Applied Mathematics (2012)
21. Pesando, J., Shum, P.: Price anomalies at auction: evidence from the market for modern prints. In: Ginsburgh, V., Menger, P. (eds.) Economics of the Arts: Selected Essays, pp. 113–134. North-Holland (1996)
22. Rodriguez, G.: Sequential auctions with multi-unit demands. B.E. J. Theor. Econ. **9**(1), 1–35 (2009)
23. Sosnick, S.: Bidding strategy at ordinary auctions. J. Farm Econ. **45**(1), 8–37 (1961)
24. Thiel, S., Petry, G.: Bidding behaviour in second-price auctions: rare stamp sales, 1923–1937. Appl. Econ. **27**(1), 11–16 (1995)
25. Tu, Z.: A resale explanation for the declining price anomaly in sequential auctions. Rev. Appl. Econ. **6**(1–2), 113–127 (2010)
26. Weber, R.: Multiple object auctions. In: Engelbrecht-Wiggans, R., Shubik, M., Stark, R. (eds.) Auctions, Bidding and Contracting: Use and Theory, pp. 165–191. New York University Press (1983)

Risk-Free Bidding in Complement-Free Combinatorial Auctions

Vishnu V. Narayan$^{(\boxtimes)}$, Gautam Rayaprolu, and Adrian Vetta

McGill University, Montreal, Canada
{vishnu.narayan,gautam.rayaprolu}@mail.mcgill.ca, adrian.vetta@mcgill.ca

Abstract. We study risk-free bidding strategies in combinatorial auctions with incomplete information. Specifically, what is the maximum profit a complement-free (subadditive) bidder can guarantee in an auction against individually rational bidders? Suppose there are n bidders and B_i is the value bidder i has for the entire set of items. We study the above problem from the perspective of the first bidder, Bidder 1. In this setting, the worst case profit guarantees arise in a duopsony, that is when $n = 2$, so this problem then corresponds to playing an auction against an individually rational, budgeted adversary with budget B_2. We present worst-case guarantees for two simple combinatorial auctions; namely, the sequential and simultaneous auctions, for both the first-price and second-price case. In the general case of distinct items, our main results are for the class of *fractionally subadditive* (XOS) bidders, where we show that for both first-price and second-price sequential auctions Bidder 1 has a strategy that guarantees a profit of at least $(\sqrt{B_1} - \sqrt{B_2})^2$ when $B_2 \leq B_1$, and this bound is tight. More profitable guarantees can be obtained for simultaneous auctions, where in the first-price case, Bidder 1 has a strategy that guarantees a profit of at least $\frac{(B_1 - B_2)^2}{2B_1}$, and in the second-price case, a bound of $B_1 - B_2$ is achievable. We also consider the special case of sequential auctions with identical items. In that setting, we provide tight guarantees for bidders with subadditive valuations.

1 Introduction

What strategy should a bidder use in a combinatorial auction for a collection I of items? This paper studies this question for sequential and simultaneous auctions. To motivate this question and to formalize the resultant problem, let's begin with sequential auctions which are perhaps the simplest and most natural method by which to sell multiple items. These auctions, where the items are ordered and sold one after another, are commonplace in auction house and online sale environments. The inherent simplicity of a sequential auction arises because a standard single-item mechanism, such as an *ascending-price, first-price,* or *second-price auction*, can then be used for each item in the collection. But there is a catch! Whilst a single-item auction is very well understood from both a theoretical perspective – see, for example, the seminal works of Vickrey [22]

© Springer Nature Switzerland AG 2019
D. Fotakis and E. Markakis (Eds.): SAGT 2019, LNCS 11801, pp. 123–136, 2019.
https://doi.org/10.1007/978-3-030-30473-7_9

and Myerson [18] – and a practical perspective, the concatenation of single-item auctions is not.

From a bidder's viewpoint, sequential auctions are perplexing for a variety of reasons. To understand this, observe that a sequential auction can be modelled as an *extensive form game*. In such games the basic notion of equilibrium is a *subgame perfect equilibrium* (SGPE). Unfortunately, these equilibria are, in general, hard to compute; see, for example, [7,8,20]. Intriguing structural properties can be derived for the equilibria of sequential auctions; see [12,19,21], but the recursive nature of this structure makes reasoning about equilibria complex. These equilibria suffer from an additional drawback in that they can change significantly with small changes to the payoff values. It follows that prescriptions derived from the complete information setting are unlikely to extend to more practical settings with incomplete information. Given that SGPE are very complex and informationally sensitive, it is extremely unlikely that the other bidders will be able to play their equilibrium strategies. In which case, why would you wish to play yours? But then what bidding strategy should you use instead? Similar computational and informational motivations also arise for the case of simultaneous auctions. In this paper, we consider the above question for both types of auctions.

Evidently, the answer to this question will depend upon the objective of the bidders, their computational resources, the informational structure inherent in the auction, etc. We study this problem from the perspective of Bidder 1 in the following very general incomplete information setting. What is the maximum *risk-free profit* that Bidder 1 can make assuming the other bidders are *rational*? Here the bidder knows her own entire valuation function but does not know the valuation function of Bidder 2 (we will see that the critical case to analyze is when there are just two bidders). Assume that the only information Bidder 1 has on the other bidder is an estimate that his value for the entire collection of items is at most B_2; beyond this trivial upper bound, she has no specific information on the values the other bidder has for any subset of the items. We will show that, in the worst case, to maximize her guaranteed profit, we can model this problem as Bidder 1 competing in the auction against an *individually rational*[1] adversary with a budget B_2. This type of approach is analogous to that of a *safety strategy* in bimatrix games. In this paper, we will then quantify the maximum risk-free profitability when the valuation function of Bidder 1 belongs to the class of subadditive (complement-free) functions and its subclasses. Interestingly, given the valuation class, tight bounds can be obtained that depend only on B_1 (the value Bidder 1 has for the entire set of items) and B_2. For example, the risk-free profitability of the class of fractionally subadditive (XOS) valuation functions is $(\sqrt{B_1} - \sqrt{B_2})^2$, for $B_2 \leq B_1$, and this bound is tight. For simultaneous auctions the risk-free profitability of the XOS class is at least $\frac{(B_1 - B_2)^2}{2B_1}$ and $(B_1 - B_2)$ for first-price and second-price auctions, respectively. Similarly, we present tight

---

[1] Recall that the only constraint on an individually rational agent is that it play a strategy that is guaranteed to provide non-negative utility; thus, an individually rational agent need not be utility maximizing (rational).

(but more complex) bounds for the class of subadditive valuation functions when the items are identical.

1.1 Related Literature

There is an extensive literature on sequential auctions. The study of incomplete information games was initiated by Milgrom and Weber [17,23]. Theoretical studies on equilibria in complete information games include [12,19,21]. Given the abundance of sequential auctions in practice, there is also a very large empirical literature covering an assortment of applications ranging from antiques [14] to wine [1] and from fish [13] to jewellery [5].

Recently there has been a strong focus in the computer science community on the design of simple mechanisms. For combinatorial auctions, *simultaneous auctions* are a notable example. These auctions are simple in that, as with a sequential auction, a standard single-item auction mechanism is used to sell each item. But in contrast, as the nomenclature suggests, these auctions are now held simultaneously rather than sequentially. Two important streams of research in this area concern the price of anarchy in simultaneous auctions (see, for example, [3,6,10,15]) and the hardness of computing an equilibrium (see [4]).

There has also been a range of papers examining the welfare of equilibria in sequential auctions. Bae et al. [2] consider the case of identical items and show that equilibria provide a factor $1 - \frac{1}{e}$ approximation guarantee if there are two bidders with non-decreasing marginal valuations. Paes Leme et al. [21] study the case of multi-bidder auctions. For sequential first-price auctions, they prove a factor 2 approximation guarantee for unit-demand bidders. In contrast, they show that equilibria can have arbitrarily poor welfare guarantees for bidders with submodular valuations. Feldman et al. [11] extend this result to the case where each bidder has either a unit-demand or additive valuation function.

Partly because of these negative results, a common assumption is that sequential auctions may not be a good mechanism by which to sell a collection of items. However, there are reasons to believe that, in practice, sequential auctions have the potential to proffer high welfare. For example, consider the influential paper of Lehmann et al. [16]. There, they present a simple greedy allocation mechanism with a factor 2 welfare guarantee for allocating items to agents with submodular valuation functions. One interesting implication of this result is that if the items are sold via a second-price sequential auction *and* every agent (assuming submodular valuations) truthfully bids their marginal value in each round then the outcome will have at least half the optimal social welfare.

1.2 Overview and Results

In Sect. 2 we explain the sequential auction model and give necessary definitions. We present our measure, the risk-free profitability of a bidder in incomplete information multi-bidder auctions, and explain how to quantify it via a two-bidder adversarial sequential auction. In Sect. 3 we present a simple sequential auction example (*uniform additive auctions*) to motivate the problem and to illustrate the difficulties that arise in designing risk-free bidding strategies, even in very small sequential auctions with at most three items.

Sections 4 and 5 contain our main results. In Sect. 4 we begin by presenting tight upper and lower bounds on the risk-free profitability of a fractionally subadditive (XOS) bidder. For the lower bound, in Sect. 4.1 we exhibit a bidding strategy that guarantees Bidder 1 a profit of at least $(\sqrt{B_1} - \sqrt{B_2})^2$.

In Sect. 4.2 we describe a sequence of sequential auctions that provide an upper bound that is asymptotically equal to the aforementioned lower bound as the number of items increases. We prove these bounds for *first-price* sequential auctions, but nearly identical proofs show the bounds also apply for *second-price* sequential auctions. Next we prove that the risk-free profitability of an XOS bidder is lower in sequential auctions than in simultaneous auctions. Equivalently, an individually rational adversary is stronger in a sequential auction than in the corresponding simultaneous auction. Specifically, in Sect. 4.3, we prove that an XOS bidder has a risk-free profitability of at least $\frac{(B_1-B_2)^2}{2B_1}$ in a first-price simultaneous auction and of at least $B_1 - B_2$ in a second-price simultaneous auction. Several other interesting observations arise from these results. First, unlike for sequential auctions, the power of the adversary differs in a simultaneous auction depending on whether a first-price or second-price mechanism is used: the adversary is stronger in a first-price auction. Second, the risk-free strategies we present for simultaneous auctions require *no* information about the adversary at all. The performance of the strategy (its risk-free profitability) is a function of B_2, but the strategy itself does not require that Bidder 1 have knowledge of B_2 (nor an estimate of it). Third, for the case of first-price simultaneous auctions, it is necessary that Bidder 1 use randomization in its risk-free strategy.

Finally, in Sect. 5 we study the risk-free profitability of a bidder with a subadditive valuation function. We give a possible explanation for why simple strategies fail to perform well in the general case. We then examine the special case where the items are identical. We derive tight lower and upper bounds for this setting. Due to space restrictions, most proofs are deferred to the full paper.

2   The Model

2.1   Sequential Auctions and Valuation Functions

There are n bidders and a collection $I = \{a_1, \ldots, a_m\}$ of m items to be sold using a sequential auction. In the ℓth round of the auction item a_ℓ is sold via a first-price (or second-price) auction. We view the auction from the perspective of Bidder 1 who has a publicly-known valuation function $v_1 : 2^I \to \mathbb{R}_{\geq 0}$ that assigns a non-negative value to every subset of items. We denote v_1 by v where no confusion arises. This valuation function is assumed to satisfy $v(\emptyset) = 0$ and to be *monotone*, that is, $v(S) \leq v(T)$, for all $S \subseteq T$. When all the items have been auctioned, the *utility* or *profit* π_1 of Bidder 1 is her value for the set of items she was allocated minus the sum of prices of these items.

The sequential auction setting is captured by extensive form games. A *strategy* for player i is a function that assigns a bid b_i^t for the item a_t, depending on the previous bids $\{b_i^\tau\}_{i,\tau<t}$ of all players (and the allocation of the first $t-1$ items). The utility (profit) of a strategy profile **b** for Bidder 1 is the profit Bidder 1 obtains when all bidders bid according to **b**.

The question we then study is how much profit Bidder 1 can guarantee itself. We examine the case where v is in the class of *subadditive* or *complement-free* valuation functions. Belonging to this class, of particular interest in this paper are *additive* functions, *submodular* functions, and *fractionally subadditive* or *XOS* functions. These functions are defined as follows.

- **Subadditive (Complement-Free)**. A function v is subadditive if $v(S \cup T) \leq v(S) + v(T)$ for all $S, T \subseteq I$.
- **Additive (Linear)**. A function v is additive if $v(S) = \sum_{a \in S} v(a)$ for each $S \subseteq I$.
- **Submodular (Decreasing Marginal Valuations)**. A function v is submodular if $v(S \cup T) + v(S \cap T) \leq v(S) + v(T)$ for all $S, T \subseteq I$.
- **Fractionally Subadditive (XOS)**. A function v is fractionally subadditive if there exists a non-empty collection of additive functions $\{\gamma_1, \gamma_2 \ldots, \gamma_\ell\}$ on I such that for every $S \subseteq I$, $v(S) = \max_{j \in [\ell]} \gamma_j(S)$.[2]

Lehmann et al. [16] showed that these valuation classes form the following hierarchy:

ADDITIVE $\subseteq$ SUBMODULAR $\subseteq$ FRACTIONALLY SUBADDITIVE $\subseteq$ SUBADDITIVE

Other important classes in this hierarchy include unit-demand and gross substitutes valuation functions, but they will not be needed here.

2.2 Bidding Against an Adversary

To quantify the maximum profit that Bidder 1 can obtain, without loss of generality, we may *normalize* the valuation function (and corresponding auction) by scaling the values so that $v(I) = v_1(I) = 1$. Now the maximum guaranteed profit will depend on the strength of the other bidders. We quantify this by a parameter B: in the setting where each player $j \geq 2$ has valuation function v_j, B is the sum of the total values of the other bidders, i.e., $B = \sum_{j=2}^{n} v_j(I)$. This corresponds to an incomplete information auction where the only common knowledge are upper bounds on the value each agent has for the entire set of items. From the perspective of Bidder 1, it will be apparent that the worst case arises when $n = 2$, and so $B = B_2 = v_2(I)$. Thus we may assume that $n = 2$, and can view Bidder 1 as playing against an *adversary* with a budget B. To see this, observe that for a fixed $B = \sum_{j=2}^{n} v_j(I)$ if there are $n >= 3$ bidders then the worst case for Bidder 1 arises when the other bidders coordinate to act as a single adversary: however, when the budget is split between two or more other bidders then their ability to buy a single item of high value decreases.

---

[2] This is the standard definition of XOS functions. Fractionally subadditive functions are defined in terms of fractional set covers; the equivalence between fractionally subadditive and XOS functions was shown by Feige [9].

Here the adversary is individually rational in that the budget constraint is tight: in time step t, if Bidder 2 paid p_2^{t-1} for the items that have already sold, then his next bid b_2^t is at most $B - p_2^{t-1}$. Bidder 1's profit π_1 in this game is her value for her allocated set minus the sum of the prices of the items. Viewing Bidder 2 as an adversary lets us take $\pi_2 = -\pi_1$, making this auction a special case of a zero-sum game. We call this the *risk-free sequential auction game* $\mathcal{R}(v, B)$. The *guaranteed profit* for Bidder 1 is the minimum profit obtainable by playing a safety strategy in this game (i.e. the *value* of this game). For any normalized valuation v, we denote this profit by $\pi_1^*(v, B)$ or simply π_1^* where there is no ambiguity. For any class of set functions $\mathcal{C}$ and any budget $B \in (0, 1)$, we want to find the maximum profit Bidder 1 can guarantee in *any* instance $\mathcal{R}(v, B)$ where $v \in \mathcal{C}$, which is precisely $\inf_{v \in \mathcal{C}} \pi_1^*(v, B)$. We call this the *risk-free profitability* $\mathcal{P}(\mathcal{C}, B)$ of the class $\mathcal{C}$ (and define risk-free profitability analogously for simultaneous auctions). The focus of this paper is to quantify the risk-free profitability of the aforementioned classes of valuation functions.

3   Example: Uniform Additive Auctions

We now present a simple example of a sequential auction with an agent (Bidder 1) that strategizes against an adversary (Bidder 2), which will be helpful for two reasons. First, it illustrates some of the strategic issues facing the agent and, implicitly, the adversary in a sequential auction. Second, these examples form base cases in our proof in Sect. 4.2.

The auction is defined as follows. Bidder 1 has an additive valuation function where each item has exactly the same value. That is, for an auction with m items, we have that $v(a_t) = \frac{1}{m}$. The adversary Bidder 2 has a budget B. We call this the *uniform additive auction* on m items and denote it by $\mathcal{A}_m$. For our example, we are interested in uniform additive auctions where $m \leq 3$. We denote by b_i^j Bidder i's bid on item j.

One Item. First, consider the case $\mathcal{A}_1$. We have a single item a_1 with $v(\{a_1\}) = 1$ for Bidder 1. Clearly if $b_1 < B$ then the adversary's best response is to bid $b_2 = b_1^+$ and win, so $\pi_1 = \pi_2 = 0$. If $b_1 \geq B$, then the adversary is constrained by his budget, thus Bidder 1 wins and obtains a profit of $\pi_1 = 1 - b_1$. So we have

$$\pi_1^* = \begin{cases} 1 - B \text{ if } 0 \leq B < 1 \\ 0 \quad\;\; \text{if } 1 \leq B \end{cases} \tag{1}$$

Two Items. Now consider the case $\mathcal{A}_2$. So there are two items a_1 and a_2 and Bidder 1 has an additive valuation function with $v(\{a_1\}) = v(\{a_2\}) = \frac{1}{2}$ and $v(\{a_1, a_2\}) = 1$. We divide our analysis into three cases.

- $B < \frac{1}{4}$: If $B < \frac{1}{4}$, then Bidder 1 can bid B on each item and win both items at price B each, so her guaranteed profit is at least $1 - 2B > \frac{1}{2}$. If Bidder 1 bids less than B on either item, then Bidder 2 can win that item, ensuring that Bidder 1's profit is less than her value of the other item, that is $\frac{1}{2}$. Bidder 1's risk-free strategy is thus to bid B on both items for a profit $\pi_1^* = 1 - 2B$.

– $\frac{1}{4} \leq B < \frac{1}{2}$: If Bidder 1 bids $b_1^1 = x$ on a_1, with $0 \leq x \leq \frac{1}{2}$, then Bidder 2 can either win by bidding $b_2^1 > x$ or lose by bidding $b_2^1 < x$ (for now, we assume $x < B$). In the former case, the adversary's budget in the second auction is $B - b_2^1$, and there is only one item remaining. It is easy to see that Bidder 1's profit from the second item is $\pi_1 = \frac{1}{2} - (B - b_2^1) = \frac{1}{2} - B + b_2^1$. This is minimized (with value $\frac{1}{2} - B + x$) when Bidder 2 bids an amount negligibly larger than x. In the latter case, the adversary loses the first item, so he has budget B in the second auction. Bidder 1's combined profit (on both items) is then $\pi_1 = (\frac{1}{2} - x) + (\frac{1}{2} - B) = 1 - B - x$. For $x = 0$ we have $\frac{1}{2} - B + x < 1 - B - x$ and for $x = B$ we have $\frac{1}{2} - B + x \geq 1 - B - x$, since $B \geq \frac{1}{4}$. But $\frac{1}{2} - B + x$ is increasing in x and $1 - B - x$ is decreasing in x. Therefore, assuming Bidder 2 plays a best response, we see that π_1 is maximized when the minimum of these values is maximized. That is $\pi_1^* = \max_{0 \leq x < B} \min \left[\frac{1}{2} - B + x, 1 - B - x \right]$. The optimal choice is $x = \frac{1}{4}$ giving $\pi_1^* = \frac{3}{4} - B$. Note that our assumption that $x < B$ is validated: if Bidder 1 bids an amount x that is greater than or equal to B on the first item the she will win both items for a total profit $(\frac{1}{2} - x) + (\frac{1}{2} - B) = 1 - x - B \leq 1 - 2B \leq \frac{3}{4} - B$.

– $\frac{1}{2} \leq B < 1$: Reasoning as we did for the previous case, we see that $\pi_1^* = \frac{1}{2} - \frac{B}{2}$ when $\frac{1}{2} \leq B < 1$.

Putting this all together we have that

Budget	$0 \leq B < \frac{1}{4}$	$\frac{1}{4} \leq B < \frac{1}{2}$	$\frac{1}{2} \leq B < 1$	$1 \leq B$
Profit π_1^*	$1 - 2B$	$\frac{3}{4} - B$	$\frac{1}{2} - \frac{B}{2}$	0

(2)

Before proceeding to the three-item case, we emphasize that even the very simple case $\mathcal{A}_2$ illustrates many of the strategic considerations that arise in more complex sequential auctions. To wit, in the first time period Bidder 1 faces the standard conundrum that bidding high increases her chances of winning but at the expense of receiving a smaller profit if she does win. Interestingly, in this adversarial setting, Bidder 1 has an additional incentive for bidding high: if she bids high *and* loses then adversary's budget is significantly reduced in the auction for the second item. Counterintuitively, therefore, in adversarial sequential actions, Bidder 1 has an incentive to lose some of the items! More interestingly, the adversary has perhaps even stronger incentives to lose than Bidder 1. Whilst winning the first item does hurt Bidder 1, this also reduces the strength of the adversary in the subsequent round. Thus, the optimal outcome for adversary is that he lose the first item at a high price; this keeps the profit of Bidder 1 low and increases the relative strength of the adversary in the second auction. This is in stark contrast with the simultaneous case, where both Bidder 1 and the adversary have an incentive to win every item.

Three Items. Now there are three items a_1, a_2 and a_3 and Bidder 1 has an additive valuation function with $v(\{a_1\}) = v(\{a_2\}) = v(\{a_3\}) = \frac{1}{3}$. Applying a similar case analysis, her maximum guaranteed profits are then:

B	$B < \frac{1}{9}$	$\frac{1}{9} \leq B < \frac{1}{6}$	$\frac{1}{6} \leq B < \frac{1}{3}$	$\frac{1}{3} \leq B < \frac{5}{9}$	$\frac{5}{9} \leq B < \frac{2}{3}$	$\frac{2}{3} \leq B < 1$	$1 \leq B$
π_1^*	$1 - 3B$	$\frac{8}{9} - 2B$	$\frac{7}{9} - \frac{4B}{3}$	$\frac{7}{12} - \frac{3B}{4}$	$\frac{4}{9} - \frac{B}{2}$	$\frac{1}{3} - \frac{B}{3}$	0

(3)

We remark that this profit function is still piecewise linear and is so for $\mathcal{A}_m$ in general. However the complexity of the profit function grows rapidly as the number of items increases.

4   Tight Bounds for XOS Valuation Functions

In this section we prove tight bounds on the risk-free profitability of Bidder 1 with a fractionally subadditive (XOS) valuation function. In Sect. 4.1, we show that the agent has a strategy in the normalized auction that gives a guaranteed profit of $(1 - \sqrt{B})^2$, equivalent to a profit of $(\sqrt{B_1} - \sqrt{B_2})^2$ in the unnormalized auction. In Sect. 4.2, we prove that no strategy can guarantee a profit that is greater than this by an (asymptotically zero) additive amount.

4.1   The XOS Lower Bound

It is quite straightforward to obtain a lower bound: for each item, Bidder 1 computes her marginal value under the assumption that she wins every other item and bids a fixed fraction of this value. This guarantees a profit of at least $(1 - \sqrt{B})^2$ against any strategy utilized by an adversary with budget $B \in (0, 1)$.

Theorem 4.1. $\mathcal{P}(XOS, B) \geq (1 - \sqrt{B})^2$.

Let $I = \{a_1, \ldots, a_m\}$ be the set of auctioned items, and v be Bidder 1's valuation function. Since v is XOS, there is a set $\{\gamma_1, \gamma_2, \ldots, \gamma_\ell\}$ of (normalized) additive set functions on I, such that for any $S \subseteq I$ we have $v(S) = \max_{i \in [\ell]} \gamma_i(S)$. Let $\gamma^* = \arg\max_{i \in [\ell]} \gamma_i(I)$ be an additive function that induces the value of v on the entire set of items I. Thus $v(I) = \gamma^*(I)$. Moreover, by definition of v, we have that

$$v(S) \geq \gamma^*(S) \qquad \forall S \subseteq I \qquad (4)$$

Bidder 1's strategy is then to bid $b_1^t = \sqrt{B} \cdot \gamma^*(a_t)$ on item $a_t \in I$, for all $t \in [m]$. It can be seen that with this strategy, Bidder 1 wins a bundle of items of value at least $(1 - \sqrt{B})$ and makes a profit of at least a $(1 - \sqrt{B})$-fraction of the value, giving $\pi_1 \geq (1 - \sqrt{B})^2$ as required. Full details are provided in the full paper.

Next we will show this bound is tight by providing instances where the adversary has a strategy limiting the profitability of Bidder 1 to this amount. This is surprising because the bidding strategy described above is *non-adaptive* – it does not adapt to the history of the auction. Given the extra flexibility afforded by adaptive strategies, one would expect a priori the optimal risk-free strategy to be adaptive. However, as we will see in the next section, the simple bidding strategy presented above is optimal for Bidder 1.

4.2   The XOS Upper Bound

In this section, we present a sequential auction with an XOS valuation function where the game value is at most $(1 - \sqrt{B})^2 + \frac{1}{\sqrt{m}}$. Specifically, we use the uniform additive auction $\mathcal{A}_m$. Consequently, rather surprisingly, the upper bound

applies to every class of valuation functions that contains the additive functions! Together with the lower bound, this resolves the profitability of several
well-studied classes, including the additive, submodular and gross substitutes
valuation classes. We will see later on, in Sect. 5, that the situation is not as
simple for subadditive valuations (that are not contained in XOS). Denote by
XOS_m the class of XOS functions on m items. The following theorem, together
with the lower bound, gives our main result: $\mathcal{P}(XOS, B)$ is asymptotically equal
to $(1 - \sqrt{B})^2$ when $B \in (0, 1)$.

Theorem 4.2. $\mathcal{P}(XOS_m, B) \leq (1 - \sqrt{B})^2 + \frac{1}{\sqrt{m}}.$

Proof. We prove this result by induction on m. We start with a simple observation: after the first item has been sold in the uniform additive auction $\mathcal{A}_m$ then
the sequential auction on items $\{a_2, \ldots, a_m\}$ is simply the auction $\mathcal{A}_{m-1}$ but
with the additive values scaled by a multiplicative factor $\frac{m-1}{m}$. Consequently, by
appropriately scaling the values *and* the budget of the adversary we will be able
to analyze the auction $\mathcal{A}_m$ by studying the first round of that auction and then
applying induction on the remaining rounds.

Formally, for any positive integer m let $f_m : \mathbb{R}_{\geq 0} \to [0, 1]$ be a function
giving the highest guaranteed profit $f_m(x)$ of a risk-free strategy in $\mathcal{A}_m$ given
the adversary has a budget $B = x$. Clearly, for all m, we have that $f_m(0) = 1$
and that $f_m(x) = 0$ for any $x \geq 1$. Set $f(x) = (1 - \sqrt{x})^2$. Then we want to prove
by induction that

$$f_m(x) \leq f(x) + \frac{1}{\sqrt{m}} \qquad\qquad \forall m \geq 1, \forall x \in (0, 1) \qquad\qquad (5)$$

Base Cases: For the base cases, consider $m \in \{1, 2, 3\}$. Note that we have
already studied the auctions $\mathcal{A}_1, \mathcal{A}_2$ and $\mathcal{A}_3$ in Sect. 3. Specifically, we found
that $f_1(x) = (1 - x)$, and that $f_2(x)$ is given by (2) and $f_3(x)$ is given by (3).
It can be easily verified that each of the above functions $f_m(x)$, $m \in \{1, 2, 3\}$, is
at most $f(x) + \frac{1}{\sqrt{m}}$, for any $x \in [0, 1]$.

Induction Hypothesis: Assume that $f_k(x) \leq f(x) + \frac{1}{\sqrt{k}}$ for all $k < m$.

Induction Step: We will now prove that $f_m(x) \leq f(x) + \frac{1}{\sqrt{m}}$. We will present a
strategy for the adversary and prove that this strategy guarantees that Bidder 1
cannot make a profit greater than $f(x) + \frac{1}{\sqrt{m}}$ in the uniform additive auction
$\mathcal{A}_m$. Specifically, we consider the auction for the first item a_1 in $\mathcal{A}_m$, and we let
$b_2^1 = \alpha \cdot \frac{1}{m}$ be the adversary's bid on this item. Since Bidder 1 has an additive
value $\frac{1}{m}$ for this item, the adversary will never make a bid $b_2^1 > \frac{1}{m}$. Thus we may
assume that the adversary makes a bid $b_2^1 = \alpha \cdot \frac{1}{m}$ for some $0 \leq \alpha \leq 1$. We then
show that for some particular choice of α, even with an optimal response Bidder 1
does not make a profit greater than $f(x) + \frac{1}{\sqrt{m}}$. In determining Bidder 1's optimal
response, we have two possibilities:

- *Bidder 1 wins item a_1.*
  In this case it is easy to see that Bidder 1 will bid $b_1^1 = b_2^{1+}$ (which is $b_2^1 + \epsilon$

for any negligibly small ϵ) as any higher bid will lead to a strictly smaller profit as this is a first price auction. Thus, Bidder 1 makes an immediate profit of $\frac{1}{m} - \alpha \cdot \frac{1}{m} = \frac{1-\alpha}{m}$ on the first item. The rest of the sequential auction is an instance of $\mathcal{A}_{m-1}$, where the additive valuations of Bidder 1 and the budget of the adversary are both scaled. As the adversary lost the first item his budget remains x, which corresponds to a budget of $B = \frac{m}{m-1} \cdot x$ in the scaled auction $\mathcal{A}_{m-1}$. Then, assuming that the bidders play optimal strategies in the remaining rounds, the maximum profit Bidder 1 can make is:

$$g_m(x, \alpha) = \frac{1-\alpha}{m} + \frac{m-1}{m} \cdot f_{m-1}\left(\frac{mx}{m-1}\right) \tag{6}$$

- *Bidder 1 loses item a_1.*
  If Bidder 1 loses the first item, then Bidder 1 makes no profit on a_1. Since this is a first-price auction the adversary will pay b_2^1 if he wins regardless of the bid of Bidder 1. Thus Bidder 1 is indifferent between any bids less than b_2^1. After the first round we again have a scaled version of $\mathcal{A}_{m-1}$. As the adversary won the first item his scaled budget is now $B = \frac{m}{m-1} \cdot \left(x - \frac{\alpha}{m}\right) = \frac{mx-\alpha}{m-1}$. Then, assuming that the bidders play optimal strategies in the remaining rounds the maximum profit Bidder 1 can make is:

$$h_m(x, \alpha) = \frac{m-1}{m} \cdot f_{m-1}\left(\frac{mx-\alpha}{m-1}\right) \tag{7}$$

Evidently, the best response of Bidder 1 to a bid $b_2^1 = \alpha \cdot \frac{1}{m}$ is given by the maximum of $g_m(x, \alpha)$ and $h_m(x, \alpha)$. Thus, the adversary should select α to minimize this maximum. Specifically,

$$f_m(x) = \min_{0 \le \alpha \le 1} \max\left(g_m(x, \alpha), h_m(x, \alpha)\right).$$

Thus, our goal is to prove that there exists a bid $b_2^1 = \tilde{\alpha} \cdot \frac{1}{m}$ by the adversary such that both $g_m(x, \tilde{\alpha})$ and $h_m(x, \tilde{\alpha})$ are at most $f(x) + \frac{1}{\sqrt{m}}$. This will ensure that the maximum guaranteed profit of Bidder 1 is $f_m(x) \le f(x) + \frac{1}{\sqrt{m}}$ as required. Our proof of this fact requires examination of three cases depending upon the magnitude of the budget of the adversary. The low budget case (where $0 \le x < \frac{1}{m^2}$) and high budget case (where $\frac{m-1}{m} < x \le 1$) do not require the induction hypothesis (nor consideration of the functions $g_m(x, \tilde{\alpha})$ and $h_m(x, \tilde{\alpha})$) but constitute a part of our inductive step. The proofs of these cases are quite straightforward and are in the full paper. For the remainder of this section, we assume that we are in the third case, where $\frac{1}{m^2} \le x \le \frac{m-1}{m}$. This case is more difficult and represents one of the main technical contributions of this paper.

Recall that by the induction hypothesis $f_{m-1}(x) \le f(x) + \frac{1}{\sqrt{m-1}}$. Rather than calculate $f_m(x)$ exactly, our approach is to find a feasible choice $\tilde{\alpha}$ for the adversary that ensures that both $g_m(x, \tilde{\alpha})$ and $h_m(x, \tilde{\alpha})$ are at most $f(x) + \frac{1}{\sqrt{m}}$. To do this, we begin by investigating the properties of the functions $g_m(x, \alpha)$

and $h_m(x, \alpha)$. Using these properties, we find a candidate choice $\tilde{\alpha}$ which we first prove is feasible and second prove gives the desired upper bound.

Let's start by showing that $g_m(x, \alpha)$ and $h_m(x, \alpha)$ are both monotonic functions. For any fixed m, since the valuation is additive and the space of strategies for the adversary is constrained only by his budget, any strategy available with budget $\bar{x} < x$ is also available with budget x. Hence f_m is non-increasing in x. So for fixed x $g_m(x, \alpha)$ is non-increasing and $h_m(x, \alpha)$ is non-decreasing in α. Now the minimum choice the adversary can make for α is zero. Since $g_m(x, 0) = \frac{1}{m} + \frac{m-1}{m} f_{m-1} \left(\frac{mx}{m-1} \right)$ and $h_m(x, 0) = \frac{m-1}{m} f_{m-1} \left(\frac{mx}{m-1} \right)$, we have $g_m(x, 0) \geq h_m(x, 0)$.

Now consider the maximum choice the adversary can make for α. We denote this value by α_{max}. We have two cases. Suppose $x \geq \frac{1}{m}$. Then the adversary may set $\alpha = 1$ and bid $\frac{1}{m}$ on the first item. In this case, both $g_m(x, 1)$ and $h_m(x, 1)$ are well defined, and we have $g_m(x, 1) = \frac{m-1}{m} f_{m-1} \left(\frac{mx}{m-1} \right)$ and $h_m(x, 1) = \frac{m-1}{m} f_{m-1} \left(\frac{mx-1}{m-1} \right)$. Because f_{m-1} is non-increasing, we have that $g_m(x, 1) \leq h_m(x, 1)$. Now suppose $x < \frac{1}{m}$. Because of the budget constraint, the maximum possible value of α is mx. Suppose the adversary bids x on the first item (corresponding to the choice $\alpha = mx$) and loses. Bidder 1 then makes a profit of $\frac{1}{m} - x$ on the first item. The adversary can subsequently play the following strategy: bid x on every item until he wins an item. It can be seen that Bidder 1's best response to this strategy is to give up the first item at price x and wins the remaining $m - 1$ items for free. Thus $g_m(x, mx) \leq h_m(x, mx)$.

So we have shown that $\alpha_{max} = \min(1, mx)$, and that $g_m(x, 0) \geq h_m(x, 0)$ and $g_m(x, \alpha_{max}) \leq h_m(x, \alpha_{max})$. Then, because $g_m(x, \alpha)$ is non-increasing in α and $h_m(x, \alpha)$ is non-decreasing in α when x is fixed, our upper bound of $\max(g_m(x, \alpha), h_m(x, \alpha))$ is minimized at any bid $\bar{\alpha} \cdot \frac{1}{m}$ such that $0 \leq \bar{\alpha} \leq \alpha_{max}$ and $g_m(x, \bar{\alpha}) = h_m(x, \bar{\alpha})$. This is also precisely equal to a risk-free bid $\alpha^* \cdot \frac{1}{m}$ placed by Bidder 1 on the first item, since from her perspective, if the adversary plays a best response then she gets the *minimum* of $g_m(x, \alpha^*)$ and $h_m(x, \alpha^*)$, and this minimum is maximized when they are equal.

We now use the above observations to establish an upper bound on the highest guaranteed profit of a risk-free strategy. We choose $\tilde{\alpha} = 1 - 2m(1 - \sqrt{x}) + 2\sqrt{m(m-1)}(1 - \sqrt{x})$, where the adversary bids $\tilde{\alpha}\frac{1}{m}$. We then prove that both $g_m(x, \tilde{\alpha})$ and $h_m(x, \tilde{\alpha})$ are well-defined for all $x \in [\frac{1}{m^2}, \frac{m-1}{m}]$, and are both at most $f(x) + \frac{1}{\sqrt{m}}$. We rely on the three technical claims below.

Claim 4.3. *For any* $x \in [\frac{1}{m^2}, \frac{m-1}{m}]$, $0 \leq \tilde{\alpha} \leq \alpha_{max}$.

Claim 4.4. $g_m(x, \tilde{\alpha}) \leq f(x) + \frac{1}{\sqrt{m}}$.

Claim 4.5. $h_m(x, \tilde{\alpha}) \leq f(x) + \frac{1}{\sqrt{m}}$.

Since $f_m(x) \leq \max(g_m(x, \tilde{\alpha}), h_m(x, \tilde{\alpha}))$, we have $f_m(x) \leq f(x) + \frac{1}{\sqrt{m}}$ when $\frac{1}{m^2} \leq x \leq \frac{m-1}{m}$. With this third case (intermediate budget) completed so is the proof of Theorem 4.2. □

4.3 Risk-Free Bidding in Simultaneous Auctions

In this section we consider risk-free bidding in a simultaneous auction. Here, for an individually rational adversary, the analogue of budget-constrained bidding is that the *sum* of the adversary's bids is at most B. Intuitively, an individually rational adversary is weaker in a simultaneous auction than in a sequential auction, since in the sequential case he has the option to "overbid" on an item but suffers no consequence *if he loses the item*. The issue then is whether or not the resultant broader range of strategies available to an adversary in a sequential auction makes it provably more powerful than the corresponding adversary in a simultaneous auction. We show that this is indeed the case in the following theorems.

Theorem 4.6. *The two-player simultaneous first-price auction with a normalized XOS valuation function and an adversary with normalized budget $B \in (0,1)$ has a (randomized) risk-free strategy for Bidder 1 that guarantees a profit of at least $\frac{(1-B)^2}{2}$ in expectation.*

Theorem 4.7. *The two-player simultaneous second-price auction with a normalized XOS valuation function and an adversary with normalized budget $B \in (0,1)$ has a risk-free strategy for Bidder 1 that guarantees a profit of at least $(1 - B)$.*

The proof of Theorem 4.7 is quite simple; the proof of Theorem 4.6 is more intricate and relies on consideration of the Lagrangian dual of an appropriate quadratic program. We also show that an analogue of Theorem 4.7 does *not* hold for first-price auctions. We remark that the strategies used in proving these theorems require no knowledge of the adversary's budget. Bidder 1 can implement them based solely on her own valuation function so these profit guarantees are extremely robust. In addition, unlike for sequential auctions, the power of the adversary differs in a simultaneous auction depending on whether a first-price or second-price mechanism is used: the adversary is stronger in a first-price auction.

5 Bounds for Subadditive Valuation Functions

In this section we return to sequential auctions, and study the risk-free profitability of Bidder 1 when her valuation function is subadditive. The relationship between XOS and subadditive functions was explored by Bhawalkar and Roughgarden [3], via the class of β-*fractionally subadditive* valuation functions. The following proposition is tight.

Prop. 5.1. *[3] Every subadditive valuation is $\ln m$-fractionally subadditive.*

Since there exist subadditive functions that are not XOS, the simple strategy from Sect. 4.1 is no longer guaranteed to work. Indeed, we show in the full paper that an analogous strategy fails to guarantee non-zero profit for Bidder 1 when her valuation is subadditive but not XOS. However, we make progress on an important special case, namely subadditive valuations on identical items.

5.1   The Subadditive Lower Bound with Identical Items

We obtain our lower bound on the profitability of Bidder 1 with a simple strategy: Bidder 1 chooses a constant price $\tilde{p}$ and a target allocation $\tilde{q}$ in advance, and bids $\tilde{p}$ on every item, stopping when she wins $\tilde{q}$ items.

Claim 5.2. *For any set $S \subseteq I$, where $|S| = q$, $v(S) \geq \frac{v(I)}{\lceil \frac{m}{q} \rceil}$.*

Now, for an appropriate choice of $\tilde{p}$ and $\tilde{q}$, we show that Bidder 1 can guarantee a profit of at least $t^*(B) - O(\frac{1}{m})$, where

$$t^*(B) = \max_{k \in \mathbb{Z}_{\geq 1}} t_k(B).$$

Interestingly, $t_k(B) = \frac{1}{k+1} - \frac{B}{k}$ is the tangent to our earlier lower bound of $f(B) = (1 - \sqrt{B})^2$ at $B = (\frac{k}{k+1})^2$. Denote by SI_m the subadditive valuation functions on m identical items.

Theorem 5.3. $\mathcal{P}(SI_m, B) \geq t^*(B) - O(\frac{1}{m})$.

5.2   The Subadditive Upper Bound with Identical Items

Interestingly, we can show a matching upper bound for the range $0 < B < \frac{1}{4}$, so the lower bound is fully tight when the budget B is in $(0, \frac{1}{4})$ and at every B of the form $(\frac{k}{k+1})^2$ for any positive integer k. We conjecture that this tightness extends to all $B \in (0, 1)$.

Theorem 5.4. $\mathcal{P}(SI_m, B) \leq t^*(B) + O(\frac{1}{\sqrt{m}})$ when $B \in (0, \frac{1}{4})$ and m is larger than some constant m_0 that depends only on B.

An important consequence of the above result is that the lower bound for XOS valuations does not hold for subadditive valuations. This differentiates the class of subadditive valuations from the additive, submodular and XOS classes in that Bidder 1 can no longer guarantee a profit of $(1 - \sqrt{B})^2$ when his valuation function is subadditive.

References

1. Ashenfelter, O.: How auctions work for wine and art. J. Econ. Perspect. **3**(3), 23–36 (1989)
2. Bae, J., Beigman, E., Berry, R., Honig, M., Vohra, R.: On the efficiency of sequential auctions for spectrum sharing. In: 2009 International Conference on Game Theory for Networks, pp. 199–205 (2009)
3. Bhawalkar, K., Roughgarden, T.: Welfare guarantees for combinatorial auctions with item bidding. In: Proceedings of the 22nd Symposium on Discrete Algorithms (SODA), pp. 700–709 (2011)
4. Cai, Y., Papadimitriou, C.: Simultaneous Bayesian auctions and computational complexity. In: Proceedings of 15th ACM Conference on Economics and Computation (EC), pp. 895–910 (2014)

5. Chanel, O., Gérard-Varet, L., Vincent, S.: Auction theory and practice: evidence from the market for jewellery. In: Ginsburgh, V., Menger, P. (eds.) Economics of the Arts: Selected Essays, pp. 135–149. North-Holland, Amsterdam (1996)
6. Christodoulou, G., Kovács, A., Schapira, M.: Bayesian combinatorial auctions. J. ACM **63**(2), 11:1–11:19 (2016)
7. Daskalakis, C., Fabrikant, A., Papadimitriou, C.H.: The game world is flat: the complexity of Nash equilibria in succinct games. In: Bugliesi, M., Preneel, B., Sassone, V., Wegener, I. (eds.) ICALP 2006. LNCS, vol. 4051, pp. 513–524. Springer, Heidelberg (2006). https://doi.org/10.1007/11786986_45
8. Etessami, K.: The complexity of computing a (quasi-)perfect equilibrium for an n-player extensive form game of perfect recall. Games Econ. Behav. (2019, to appear)
9. Feige, U.: On maximizing welfare when utility functions are subadditive. SIAM J. Comput. **39**(1), 122–142 (2009)
10. Feldman, M., Fu, H., Gravin, N., Lucier, B.: Simultaneous auctions are (almost) efficient. In: Proceedings of 49th Symposium on Theory of Computing (STOC), pp. 201–210 (2013)
11. Feldman, M., Lucier, B., Syrgkanis, V.: Limits of efficiency in sequential auctions. In: Chen, Y., Immorlica, N. (eds.) WINE 2013. LNCS, vol. 8289, pp. 160–173. Springer, Heidelberg (2013). https://doi.org/10.1007/978-3-642-45046-4_14
12. Gale, I., Stegeman, M.: Sequential auctions of endogenously valued objects. Games Econ. Behav. **36**(1), 74–103 (2001)
13. Gallegati, M., Giulioni, G., Kirman, A., Palestrini, A.: What's that got to do with the price of fish?: Buyer's behavior on the Ancona fish market. J. Econ. Behav. Organ. **80**(1), 20–33 (2011)
14. Ginsburgh, V., van Ours, J.: On organizing a sequential auction: results from a natural experiment by Christie's. Oxf. Econ. Pap. **59**(1), 1–15 (2007)
15. Hassidim, A., Kaplan, H., Mansour, Y., Nisan, N.: Non-price equilibria in markets of discrete goods. In: Proceedings of 12th Conference on Electronic Commerce (EC), pp. 295–296 (2011)
16. Lehmann, B., Lehmann, D., Nisan, N.: Combinatorial auctions with decreasing marginal utilities. Games Econ. Behav. **55**(2), 270–296 (2006)
17. Milgrom, P., Weber, R.: A theory of auctions and competitive bidding. Econometrica **50**, 1089–1122 (1982)
18. Myerson, R.: Optimal auction design. Math. Oper. Res. **6**(1), 58–73 (1981)
19. Narayan, V.V., Prebet, E., Vetta, A.: The declining price anomaly is not universal in multi-buyer sequential auctions (but almost is). In: Fotakis, D., Markakis, E. (eds.) SAGT 2019. LNCS, vol. 11801, pp. 109–122. Springer, Cham (2019)
20. Paes Leme, R., Syrgkanis, V., Tardos, E.: The curse of simultaneity. In: Proceedings of the 3rd Innovations in Theoretical Computer Science Conference (ITCS), pp. 60–67 (2012)
21. Paes Leme, R., Syrgkanis, V., Tardos, E.: Sequential auctions and externalities. In: Proceedings of the 23rd Symposium on Discrete Algorithms (SODA), pp. 869–886 (2012)
22. Vickrey, W.: Counterspeculation, auctions, and competitive sealed tenders. J. Financ. **16**(1), 8–37 (1961)
23. Weber, R.: Multiple object auctions. In: Engelbrecht-Wiggans, R., Shubik, M., Stark, R. (eds.) Auctions, Bidding and Contracting: Use and Theory, pp. 165–191. New York University Press, New York (1983)

Computational Aspects of Games

On the Existence of Nash Equilibrium in Games with Resource-Bounded Players

Joseph Y. Halpern[1], Rafael Pass[1], and Daniel Reichman[2(✉)]

[1] Cornell University, Ithaca, USA
{halpern,pass}@cs.cornell.edu
[2] Princeton University, Princeton, USA
daniel.reichman@gmail.com

Abstract. We consider *computational games*, sequences of games $\mathcal{G} = (G_1, G_2, \ldots)$ where, for all n, G_n has the same set of players. Computational games arise in electronic money systems such as Bitcoin, in cryptographic protocols, and in the study of generative adversarial networks in machine learning. Assuming that one-way functions exist, we prove that there is 2-player zero-sum computational game $\mathcal{G}$ such that, for all n, the size of the action space in G_n is polynomial in n and the utility function in G_n is computable in time polynomial in n, and yet there is no ϵ-Nash equilibrium if players are restricted to using strategies computable by polynomial-time Turing machines, where we use a notion of Nash equilibrium that is tailored to computational games. We also show that an ϵ-Nash equilibrium may not exist if players are constrained to perform at most T computational steps in each of the games in the sequence. On the other hand, we show that if players can use arbitrary Turing machines to compute their strategies, then every computational game has an ϵ-Nash equilibrium. These results may shed light on competitive settings where the availability of more running time or faster algorithms can lead to a "computational arms race", precluding the existence of equilibrium. They also point to inherent limitations of concepts such as "best response" and Nash equilibrium in games with resource-bounded players.

Keywords: Nash equilibrium · Bounded rationality · Turing machines

1 Introduction

One of the most widely used solution concepts in game theory is Nash equilibrium (NE). In a Nash equilibrium, no player can improve his utility by deviating unilaterally from his strategy. A key property of NE is that it exists in every normal-form game, making it a potential candidate for an equilibrium rational players may end up in. However, the proof of existence of NE is silent with respect to the computational resources players may or may not have. But if a Nash equilibrium is hard to compute, it is hard to imagine how computationally

© Springer Nature Switzerland AG 2019
D. Fotakis and E. Markakis (Eds.): SAGT 2019, LNCS 11801, pp. 139–152, 2019.
https://doi.org/10.1007/978-3-030-30473-7_10

bounded players could play it.[1] The importance of taking computational concerns into account in game theory has been recognized since at least the work of Simon [26]. Our goal here is to examine how considering computationally bounded players influences notions such as best response and NE.

We will be mainly interested in players that are polynomially bounded, continuing a long line of work in game theory on resource-bounded players (e.g., [16,19,21,23]). To make sense of polynomial-time players, we need to have a set of inputs that grow as a function of n. But game theorists typically study individual games, which have a fixed size. To deal with this, we consider not single games, but *computational games* [12], which have the form $(G_1, G_2, \ldots)$, where for all n, G_n is a finite game. We assume that each player chooses a Turing machine (TM) that, given n, computes a strategy for the player in G_n. If a player is polynomial-time bounded, then the player's action in the nth game can be computed in time polynomial in n.

Computational games arise in a number of settings of interest. One example is "crypto-currencies" such as Bitcoin. An essential ingredient of Bitcoin [17] is miners who solve challenging cryptographic problems, whose solution is later used in verifying transactions in the system. Bitcoin keeps the average time at which puzzles are solved a constant, despite technological advances, by making the cryptographic problem needed to be solved harder and harder over time, forcing miners to examine a larger number of possible solutions. This can be modeled by viewing Bitcoin as a sequence of games, where in the nth game the miner is required to solve a cryptographic puzzle P_n such that the number of candidate solutions that need to be examined in order to solve P_n is a function of n.

Cryptographic protocols such as *commitment schemes* [2] provide another example of computational games. A commitment scheme consists of two parties; a sender and a receiver. In the first step of this protocol, the sender chooses a bit b and sends an encryption of b to the receiver, committing the sender to b without revealing b to the receiver. Next, the receiver chooses a bit. Finally the sender reveals the bit to the receiver. This protocol can be viewed as a game where the receiver wins if the bit he chooses matches the bit revealed; the sender wins if they do not match. Clearly, if the receiver can break the scheme and deduce the sender's bit, the receiver wins; if the sender can cheat ("reveal" a bit that does not necessarily match what he committed to), the sender wins. The encryption at the first step involves a security parameter k, where larger security parameters provide more security (i.e., more running time is required to break the scheme). This can be modeled as a sequence of games, where in the kth game the sender encrypts the bit using a security parameter k [12]. Many cryptographic protocols, including secret sharing and multiparty computation, can be viewed as computational games in this way.

---

[1] The celebrated PPAD-completeness results [3,5] indicate that finding a NE in a fixed game is intractable. Our setting is very different from the setting that is considered in these PPAD-hardness results. For more details see the discussion of related work in the end of this section.

Yet one more example of computational games arises in the study of GANs *generative advsersarial networks* in machine learning; as argued by Oliehook et al. [22], GANs can be viewed as computational games that end up converging to a "local resource-bounded NE".

The computational games that we consider are actually sequences of *Bayesian* games, where the action of a player may depend on his *type*, which encodes some private information that the player may have. In a computational game, the action spaces and types spaces all have to be finite, and the utility functions and probability distribution over types have to be computable. We focus here on a subclass of computational games that we call *polynomial games*; these are sequences of games where the action space and type space in the nth game have size polynomial in n, and the utility function and probability distribution over types in the nth game can be computed in time polynomial in n. These restrictions all apply to the games that we are interested in, such as Bitcoin.[2]

An analogue of NE can be defined in computational games $(G_1, G_2, \ldots)$ [12]. We assume that every game G_j is a k-player game and that for $1 \le i \le k$, player i uses a TM M_i that computes his actions in G_j given j. Roughly speaking, a machine profile $(M_1, \ldots, M_k)$ consisting of TMs is a NE if, for every player, replacing his TM by a different TM gives him at most a negligible improvement to his utility. We can get a notion of *polynomial-time* NE by replacing "TM" with "polynomial-time TM" everywhere in the definition. (There are certain subtleties in this definition; see Definition 5 and the discussion thereafter for more detail.)

In contrast to fixed games, where NE always exists, we show that in computational games, NE may not exist. Specifically, we show (Theorem 1) that, assuming the existence of one-way functions, there are polynomial 2-player zero-sum games for which no polynomial-time Nash equilibrium exists. This is done by simulating the "largest integer game" in this setting, the game where players simultaneously output an integer, and the player who chooses the largest integer wins. Clearly this game has no Nash equilibrium [15]. We can effectively simulate this game by presenting players with multiple one-way function puzzles, requiring players to invert as many puzzles as possible. We can ensure that a player with sufficiently more (but only polynomially more) running time can invert more puzzles. Thus, we get an "arms race" with no equilibrium. This example points to an inherent difficulty in analyzing games with polynomially-bounded players. Namely, in such games, there is often no best response; players can use longer and longer running times to improve their payoffs. Interestingly, a similar phenomenon has been observed in Bitcoin, where miners use increasingly more sophisticated computational devices for the mining operation (see [4] and the reference therein).

---

[2] The games used to model protocols such as Bitcoin are actually *extensive-form* games, which are played over time. Our impossibility results show that there are computational Bayesian games where there is no NE when we restrict to polynomial-time players. Since Bayesian games are a special case of extensive-form games, our non-existence results carry over to extensive-form games.

We then demonstrate (Theorem 2) that Nash equilibrium may fail to exist even if players are constrained to run for at most T steps for a fixed integer T, without asymptotics kicking in. The idea is to let players first play a game (matching pennies) that requires randomization to achieve equilibrium, and then effectively give the player with greater remaining running time an additional bonus. Assuming that the generation of a random bit requires computational effort, this game cannot have a Nash equilibrium. Our impossibility results hold even if we replace "Nash equilibrium" by "ϵ-Nash equilibrium". By way of contrast, we show (Theorem 3) that if players are *not* computationally bounded (i.e., can use arbitrary Turing machines), then there is *always* an ϵ-NE in a computational game. The key idea behind Theorem 3 is that an algorithm similar to that of Lipton and Markakis [14] for finding an ϵ-NE in a fixed game can be used by the players to find ϵ-NE in computational games.

It is worthwhile at this point to examine our result in the context of the literature on bounded rationality in game theory. Two high-level approaches to incorporating complexity-theoretic considerations into game theory have been considered:

- Rubinstein [24] did not limit the complexity, but charged for it.
- Neyman [20] limited the players (e.g., to being finite automata).

Halpern and Pass [10] extended Rubinstein's approach to TMs: players choose a TM, and then they are charged for the running time/space used/amount of randomization used by the TM on a given input. The approach of charging for complexity of Turing machines was also considered by Fortnow and Santhanam [9], who discount the payoffs of players by the amount of time they use to compute their response. The effect of charging players for the strategies they use on the convergence of learning dynamics to Nash Equilibrium was considered by Ben-Sasson, Tauman-Kalai, and Kalai [1].

In this work we follow the approach of Neyman [20]: we limit players to using polynomial-time TMs, but don't charge for computation. Thus, unlike Halpern and Pass [10] and Fortnow and Santhanam [9], we limit computation, rather than charging for it. Just as we do, Halpern and Pass [10] prove both the existence and non-existence of NE, depending on assumptions. However, the *reasons* for these results are very much framework-dependent. For example, Halpern and Pass [10] show that NE may not exist if we charge players for randomness and it does exist in their framework if we do not charge for randomness. By way of contrast, our main result concerning the non-existence of NE (Theorem 1) holds even if we do not charge for the time taken to generate a random bit. Fortnow and Santhanam's result on the existence of ϵ-NE in their version of computational games [9] depends heavily on their assumption that utilities are discounted; we have no analogue of this assumption, and thus must use quite different techniques in our proof of the existence of ϵ-NE.

Despite all the work on resource-bounded players, to the best of our knowledge, very little work has been done on games where players are limited to using polynomial-time Turing Machines. One exception is the work of Megiddo and

Wigderson [16], who consider playing repeated prisoner dilemma (for finitely many rounds) with TMs. Their main interest is whether, in finitely repeated prisoners dilemma, there exist "almost cooperative" equilibria (where "defect" is played $o(n)$ times). They restrict attention to deterministic TM. With this restriction it is not difficult to give examples of games (with polynomially-bounded players) for which an ϵ-NE does not exist.

Polynomial games bear some similarities to *succinct games*. In succinct games, there exists a circuit C that calculates the utility $C(x_1, x_2, \ldots, x_k)$ of the players once they choose the actions $x_1, x_2, \ldots, x_k \in \{0,1\}^m$. It is known that, given a 2-player zero-sum succinct game, it is EXP-hard to find a NE [7,8] (see also [25]). Our results regarding the non-existence of NE in polynomial games are incomparable to these results. We are concerned with *polynomial-time computable* strategies. Considering polynomially-bounded players (as opposed to unbounded players) may drastically change the set of Nash equilibria in succinct games. Indeed, a NE for a computational game $(G_1, G_2, \ldots)$ with polynomially-bounded players may fail to be a Nash equilibrium for G_n for all $n \geq 1$: for an example, see the end of Sect. 3. Moreover, for any fixed game G, a computational NE for the computational game $(G, G, G, \ldots)$ can always be found in polynomial time. Thus, the PPAD-hardness results of finding a Nash equilibrium in a fixed game [3,5] cannot be applied in our setting either.

2   Preliminaries

We begin by defining Bayesian games.

Definition 1. *A k-player normal-form Bayesian game is described by a tuple (J, B, T, P, v), where*

- *J is a set of k players (we identify J with $[k] = \{1, \ldots, k\}$);*
- *$B = \prod_{i=1}^{k} B_i$, where B_i is a finite set for all $i \in [k]$ consisting of the available actions of player i;*
- *$T = \prod_{i=1}^{k} T_i$, where T_i is a finite set called the type space of player i;*
- *P is a probability distribution over T;*
- *$v = (v_1, \ldots, v_k)$, where for all i, v_i is a function from $B \times T$ to the real numbers.*

In our settings, it will often be the case that all types are perfectly correlated: all players have the same type and all players know the type of every other player. Observe that normal-form games can be viewed as a special case of Bayesian games (where the type space is a singleton). Finally, since we are concerned here mainly with Bayesian games, when we write "game" we mean "Bayesian game", unless explicitly stated otherwise.

A *pure strategy* s_i for player i is a map $s_i : T_i \to B_i$; a strategy s_i maps the type $t_i \in T_i$ of player i to an action $s_i(t_i) \in B_i$. We denote by $\Delta(B_i)$ the set of all probability distribution over B_i; let $\Delta = \prod_{i=1}^{k} \Delta(B_i)$. A mixed-strategy s_i for player i is a function mapping type $t_i \in T_i$ to an element of $\Delta(B_i)$. We denote by $s_i(t_i, b_i)$ the probability assigned by a mixed strategy $s_i(t_i)$ to $b_i \in B_i$. The expected utility of player i with the mixed strategy profile $s = (s_1, \ldots, s_k)$ (where $t = (t_1, \ldots, t_k) \in T$, $b = (b_1, \ldots, b_k) \in B$, and $(s_1(t_1), \ldots, s_k(t_k)) \in \Delta$) is given by

$$V_i(s) = \sum_{t \in T} P(t) \sum_{b \in B} \left(\prod_{j=1}^{k} s_j(t_j, b_j) \right) v_i(t, b). \tag{1}$$

Note that there are two sources of uncertainty in the utility of a player choosing a mixed action: the probability distribution over other players actions and the distribution P over the type space.

Definition 2. *Let $G = (J, B, T, P, v)$ be a k-player Bayesian game and suppose that $\epsilon \geq 0$. A mixed-strategy profile $s = (s_1, \ldots, s_k)$ is an ϵ-Nash equilibrium (ϵ-NE for short) if, for all players i and all mixed strategies s_i', we have that*

$$V_i(s) \geq V_i(s_i', s_{-i}) - \epsilon.$$

(As usual, if $s = (s_1, \ldots, s_k)$ then $s_{-i} = (s_1, \ldots, s_{i-1}, s_{i+1}, \ldots, s_k)$ is the tuple excluding s_i.) When $\epsilon = 0$, we have a Nash equilibrium.

To reason about resource-bounded players in games, we consider a sequence $(G_1, G_2, \ldots)$ of games where, for all n, $G_n = (J, B^n, T^n, P^n, v^n)$ is a k-player game (k is fixed and does not depend on n). We adapt the definition of [12], which in turn is based on earlier definitions by Dodis, Halevi and Rabin [6] and is applied to Bayesian games. For an integer s, recall that $\{0, 1\}^{\leq s}$ is the set of all bit strings of length at most s.

Definition 3. *A computational game $\mathcal{G} = (G_1, G_2, \ldots)$ is a sequence of normal-form Bayesian games, where $G_n = ([k], B^n, T^n, P^n, v^n)$, such that*

- *The set of players in G_n, $[k]$, is the same for all n.[3]*
- *For all n and all i, $B_i^n \subseteq \{0, 1\}^{\leq m}$ for some finite m (that may depend on n).*
- *For all n and all i, $T_i^n \subseteq \{0, 1\}^{\leq r}$ for some finite r (that may depend on n).*
- *For all $i \in [k]$ and n, there is a TM M such that, given $b \in B^n$, $t \in T^n$, and 1^n, computes $v_i^n(b, t)$.*
- *For all $i \in [k]$ and n, there is a TM M' such that given $t \in T^n$ and 1^n, computes $P^n(t)$.[4]*

$\mathcal{G}$ *is* bounded *if there exist constants $0 < c < C$ such that for all $n, b \in B^n$, and $t \in T^n$ we have that $v_i^n(b, t) \neq 0 \Rightarrow |v_i^n(b, t)| \in [c, C]$.*

When dealing with games with polynomial-time players, we require slightly stronger properties summarized in the definition below. Following the definition of polynomial games for extensive-form games [12], we define polynomial games for a sequence of Bayesian games.

---

[3] It is also possible to allow k to depend on n, but we focus on the case where k is a constant for concreteness.

[4] We restrict our attention to utilities and probabilities that are rational numbers.

Definition 4. *A computational game* $\mathcal{G} = (G_1, G_2, \ldots)$ *is a polynomial game if the following conditions hold:*

- *There exist a polynomial p such that, for all n and all i, $B_i^n = \{0,1\}^{\leq p(n)}$.*
- *There exist a polynomial q such that, for all n and all i, $T_i^n = \{0,1\}^{\leq q(n)}$.*
- *For all $i \in [k]$ and n, there is a TM M such that, given $b = (b_1, \ldots, b_k) \in B^n$, $t \in T^n$, and 1^n, computes $v_i^n(b, t)$ and runs in time polynomial in n.*
- *For all $i \in [k]$ and n, there is a TM M' such that given $t \in T^n$ and 1^n, computes $P^n(t)$ in time polynomial in n.*

Throughout, we take the *size* of the action set (or type set) to be the maximal number of bits needed to encode an action (or type). Observe that while we require that size in polynomial games is polynomial in n for every n, the *cardinality* of the action or type set can be exponential.

A *strategy* for player j in a computational game $\mathcal{G}$ is a TM M_j that, given 1^n and the type $t_j \in T_j^n$, outputs a distribution $M_j(1^n, t_j)$ over actions in B_j^n in the game G_n (so that, given some additional random bits, it outputs an action in B_j^n).[5] $M_j(1^n)$ is the strategy defined by taking $M_j(1^n)(t_j) = M_j(1^n, t_j)$. Observe that there are two sources of randomness in $M_j(1^n)$: the distribution of the type t_j and the randomness of M_j once t_j has been determined. We stress that randomized strategies in our setting are obtained by using probabilistic TMs rather than by mixing over TMs. That is, the randomization is part of the computation, not external to it. The utility of player i in G_n given a machine profile $(M_1 \ldots M_k)$ is $V_i^n(M_1(1^n), \ldots, M_k(1^n))$ (as defined in (1)).

To analyze computational games $\mathcal{G} = (G_1, G_2, \ldots)$, we would like to be able to apply classical game-theoretic notions, such as best response and Nash equilibrium, to sequences of games. However, there are certain difficulties in generalizing these notions to computational games. A first obstacle is that sequences of infinite games may allow resource-bounded players to improve over any strategy by doing additional polynomial-time computations. For example, consider a player who gets a payoff of 1 by breaking an encrypted massage $E(s)$ with $s \in \{0,1\}^n$ and a payoff of 0 if he does not break it, where the player's running time is polynomial in n. Assuming that there is no polynomial-time algorithm (in n) for finding s given $E(s)$, there is no best response in this game, as a player can always make polynomially many additional "guesses" on top of his current action, increasing his expected utility. As pointed out by Dodis, Halevi, and Rabin [6], this observation applies to many problems of interest, such as those arising from cryptographic protocols.

One way around this problem, suggested by Dodis, Haley and Rabin [6] and Halpern, Pass, and Seeman [12], is to ignore *negligible* additive changes in the

---

[5] One question is how to deal with players who use Turing machines that fail to halt or return an action that does not belong to the action space. We deal with this issue by assigning to each player i a special action a_0^i that we take to be the action played if i's TM does not halt or if i's output is not an action in the action space. Any profile that includes a_0^i gives utility $-\infty$ to all players, thus discouraging players from using TMs that fail to halt or return inappropriate actions.

utility of players, where a sequence $\delta(n)$ is negligible if for every polynomial $p, p(n) = o(\delta(n)^{-1})$. That is, deviations that result in a negligible increase in utility are not considered to be improvements. Ignoring negligible terms suffices to ensure the existence of equilibrium in a number of games of interest for which there would not be an equilibrium otherwise [6].

If we ignore negligible change, then given a machine profile $\overline{M}$, changing the behavior of a TM M in finitely many games will not be a deviation breaking an alleged equilibrium, as altering a sequence $\delta(n)$ on finitely many n's does not change the fact that $\delta(n)$ is negligible. On the other hand, a deviation that improves a given player utility on infinitely many n's by a constant $\delta > 0$ implies that the machine profile is not a NE. Finally, it is worth noting that if the utilities of players are exponentially small (say, on the order of $1/2^n$ in the game G_n), a negligible additive term can have a noticeable effect on the utility of players; on the other hand, if utilities are exponentially large, even a (non-negligible) constant change in utilities would be viewed as negligible. In order to avoid such scaling issues, we deal exclusively with bounded games when considering solution concepts for computational games.[6]

Definition 5. *Let $\mathcal{M}$ be a set of TMs and let $\epsilon \geq 0$ be a constant independent of n. A profile $\overline{M} = (M_1, \ldots, M_k)$ of TMs is an ϵ-$\mathcal{M}$-NE for a bounded computational game $\mathcal{G}$ with respect to $\mathcal{M}$, if (a) for all i, $M_i \in \mathcal{M}$, and (b) there exists a negligible sequence $\delta(n)$ such that, for all $M_i' \in \mathcal{M}$ and all $n > 0$ and all $i \in [k]$ we have that*

$$V_i(M_i(1^n), M_{-i}(1^n)) \geq V_i(M_i'(1^n), M_{-i}(1^n)) - \epsilon - \delta(n). \tag{2}$$

When $\epsilon = 0$, we say that $\overline{M}$ is a $\mathcal{M}$-Nash equilibrium. If $\mathcal{M}$ is the set of all probabilistic polynomial-time TMs, we say $\overline{M}$ is a polynomial ϵ-NE.

We can consider polynomial-time players, best response, and equilibrium even if the action space of every player is of super-polynomial size. However, in this case, there are trivial examples showing that a NE may not exist. For example, one can take G_n to be the 2-player zero-sum game where each player outputs an integer of length at most 2^{2^n} (written in binary) and the player outputting the larger integer receives payoff 1, with both players getting 0 in case of equality. Clearly this sequence of games does not have a polynomial equilibrium.

In contrast to previous work [9], we require the utilities of players to be computable. Without this requirement, it is not difficult to give examples of polynomial games that do not have a NE. Indeed, let $x_1, x_2 \ldots$ be an enumeration of $\{0, 1\}^*$ and let L be an arbitrary non-recursive language. Furthermore, suppose that for all i, j, $i < j$ implies that $|x_i| \leq |x_j|$ (ensuring that for every n the type x_n can be represented by at most $poly(n)$ bits). Consider the sequence $G = (G_1, G_2, \ldots)$ of two-player games such that the type of each player in G_n is x_n and a player gets a payoff of 1 if it correctly determines whether x_n belongs to L and 0 otherwise. Clearly, G does not have a polynomial-time NE (and the utility function in G is not computable).

---

[6] Our results also hold in a more general setting where the absolute value of a (nonzero) utility is at most polynomial and at least inversely polynomial in n.

3   Polynomial Games with No Polynomial Equilibrium

As we now show, there is a polynomial game for which there is no polynomial NE, assuming one-way functions exist. We find it convenient to use the definition of one-way function given in [13].

Definition 6. *Given* $s : I\!N \to I\!N, t : I\!N \to I\!N$, *a one-way function with security parameter* s *against a* t*-bounded inverter is a family of functions* $f_k : \{0,1\}^k \to \{0,1\}^m$, $k = 1, 2, 3, \ldots$, *satisfying the following properties:*

- $m = k^b$ *for some positive constant* b;
- *there is a TM* M *such that, given* x *with* $|x| = k$ *computes* $f_k(x)$ *in time polynomial in* k;
- *for all but finitely many* k*'s and all probabilistic TM* M', *running in time at most* $t(k)$ *for a given input* $f_k(x)$,

$$\Pr[f_k(M'(f_k(x))) = f_k(x)] < \frac{1}{s(k)},$$

where the probability $\Pr$ *is taken over* x *sampled uniformly from* $\{0,1\}^k$ *and the randomness of* M'.

We assume that exponential one-way functions exist. Specifically, we assume that there exists a one-way function that is $2^{k/10}$-secure against a $2^{k/30}$-bounded inverter. The existence of a one-way function with these parameters follows from an assumption made by Wee [27] regarding the existence of exponential non-uniform one-way functions. Given $f_k(x)$, we say an algorithm *inverts* $f_k(x)$ if it finds some z such that $f_k(x) = f_k(z)$.

We can now demonstrate the non-existence of polynomial-time computable equilibrium in a polynomial game.

Theorem 1. *If there exists a one-way function that is* $2^{k/10}$-*secure against a* $2^{k/30}$-*inverter, then, for all* $\epsilon > 0$, *there exists a 2-player zero-sum polynomial game* $\mathcal{G}$ *that has no polynomial* ϵ-*NE.*

Proof. Let $\mathcal{G} = (G_1, G_2, \ldots)$ be the following polynomial game, which we call the *one-way function game*. For all n, we define G_n as follows. There are two players, 1 and 2. Fix a one-way function $\{f_k\}_{k \geq 1}$ that is $2^{k/10}$-secure against a $2^{k/30}$-bounded inverter. The type space is the same for each player, and consists of tuples of $l = \lceil \log n \rceil$ bitstrings of the form $(f_{\lceil \log n \rceil}(x_1), \ldots, f_{\lceil \log n \rceil^2}(x_l))$. The distribution on types is generated by choosing $x_i \in \{0,1\}^{i \lceil \log n \rceil}$ uniformly at random, and choosing the x_i's independently. Given his type t_n, player j outputs $y_1^j, \ldots, y_l^j$. A *hit* for player j is an index i such that $f_{i \lceil \log n \rceil}(y_i^j) = f_{i \lceil \log n \rceil}(x_i)$. Let a_j denote how many hits player j gets. The payoff of player j is 1 if $a_j - a_{3-j} > 0$. If $a_j - a_{3-j} = 0$, both players receive a payoff of 0. Observe that the utility function of each player is polynomial-time computable in n. Clearly the length of every action of G_n is polynomial in n and so is the length of the type t_n. Hence the one-way function game is a polynomial game. In the full paper [11], we prove that there cannot be a polynomial-time ϵ-NE for $\mathcal{G}$. ∎

Similar ideas can be applied to show there is a 2-player *extensive-form* polynomial game that has no polynomial ϵ-NE, where we no longer need to use a type space. (See [12] for the definition of extensive-form polynomial game and polynomial ϵ-NE in extensive-form polynomial games; we hope that our discussion suffices to give the reader an intuitive sense.) In the game G_n, instead of the tuple $(f_{\lceil \log n \rceil}(x_1^j), \ldots, f_{l \lceil \log n \rceil}(x_l^j))$ being player j's type, player j chooses $x_1^j, \ldots, x_l^j$ at random and sends this tuple to player $3-j$. Again, player j attempts to invert as many of $f_{\lceil \log n \rceil}(x_1^{3-j}), \ldots, f_{l \lceil \log n \rceil}(x_l^{3-j})$ as it can; their payoffs are just as in the Bayesian game above. A proof similar to that of Theorem 1 shows that this game does not have a polynomial NE.

The one-way function game also shows the effect of restricting strategies to be polynomial-time computable. Clearly, without this restriction, the game has a trivial NE: all players correctly invert every element of their tuple. On the other hand, consider a modification of the game where in G_n, a player's type consists of a single element $f_n(x_n)$, with x_n a bitstring of length n chosen uniformly at random. If both players simultaneously invert or fail to invert $f_n(x_n)$, then both get zero. Otherwise, the player who correctly inverts gets 1 and the other player gets -1. Again, it is easy to see that if we take $\mathcal{M}$ to be the family of all TMs, the only Nash equilibrium is to find y_n, z_n such that $f_n(y_n) = f_n(z_n) = f(x_n)$. But if $\mathcal{M}$ consists of only polynomial-time TMs, then it is a polynomial-time NE for both players to simply output a random string, as neither player can invert f with non-negligible probability, and we ignore negligible additive increase to the utilities of players.

4   Equilibrium with Respect to Concrete Time Bounds

The previous example may lead one to speculate that lack of Nash equilibrium in computational games hinges on asymptotic issues, namely, our ability to consider larger and larger action and type spaces. This raises the question of what happens if we restrict our attention to games where players are constrained to execute at most T computational steps, where $T > 0$ is a fixed integer. It turns out that if the use of randomness is counted as a computational action, then there may not be Nash equilibria, as the following example shows. We assume from now on that $T > 2$.

In our computational game, the family of admissible TMs, which we denote by $\mathcal{M}_T$, is the set of all probabilistic TMs whose running time is upper-bounded by T. The operation of printing a character takes one computational step, and so does the movement of the cursor to a different location on the tape. The generation of a random bit (or alternatively querying a bit in a designated tape that contains random bits) requires at least one computational step (we allow arbitrary bias of a bit, as it does not affect the proof).

Consider the following 2-player zero-sum normal-form computational game $\mathcal{F}$ between Alice (A) and Bob (B). For every n, F_n is the same game F. The action space of each player is $\{0, 1\}^T$. By our choice of $\mathcal{M}_T$, both players are constrained to perform at most T computational steps. The game proceeds as follows. A and B use TMs $M_A, M_B \in \mathcal{M}_T$ respectively, to compute their strategies. M_A outputs a single bit a_1. M_B outputs $b_1 \in \{0, 1\}$. Based on a_1 and b_1, a game of *matching pennies* is played. Namely, if $a_1 = b_1$, A gets 1, otherwise B gets 1. In the second phase of the game, the TM of each player prints as many characters as possible without violating the constraint of performing at most T steps. If the final number of characters is the same for both players, then both get a payoff of 0 for the second phase. Otherwise the player with a larger number of printed characters gets an additional bonus of 1, and the player with fewer printed characters incurs a loss of 1.

Theorem 2. *The computational game $\mathcal{F}$ does not have an ϵ-$\mathcal{M}_T$-NE, for all $\epsilon < 1$.*

Proof. Assume, by way of contradiction, that (M_A, M_B) is a Nash equilibrium for $\mathcal{F}$. Since TMs in $\mathcal{M}_T$ are constrained to query at most T bits, it follows that the strategy computed by M_A (or M_B) given 1^n, will be the same for all $n > T$. As the outcomes of the games F_m, $m \leq T$, do not effect, by our definition of NE in computational games, whether (M_A, M_B) is an equilibrium, we can assume w.l.o.g that both M_A and M_B compute the same strategy (whether mixed or pure) in all games $F_n, n \geq 1$.

Suppose that one of the players uses randomization. Assume this is player A. Namely, M_A generates a random bit before outputting a_1. Then A can guarantee a payoff for the first phase (the matching pennies game) that is no smaller than his current payoff by choosing a TM M'_A that outputs a *deterministic* best response a_1 against the strategy of B in the matching penny game. Observe that we can assume that a_1 is "hardwired" to M'_A. In particular outputting a_1 can be done in a single computational step. Then A can print strictly more 1's in the second phase of the game by configuring M'_A to print $T - 1$ 1's (which can be done in $T - 1$ steps). If B prints $T - 1$ in the second phase of the game, we have that A can increase its payoff in F_n for all n by switching to M'_A. If, on the other hand, M_B prints less than $T - 1$ characters in the second step, an analogous argument shows that B can strictly increase its payoff in F_n for all n, by using a TM that runs in at most T steps. In any event, we get a contradiction to the assumption that (M_A, M_B) is a NE.

Suppose now that A does not use randomization. In this case, it follows immediately by the definition of matching-pennies that either A or B can strictly improve their payoff in the first phase of F_n for all n, by outputting the (deterministic) best response to their opponent and printing $T - 1$ characters afterwards. As before, we can assume this response is hardwired to the appropriate TM, such that outputting it consumes one computational step, allowing players to print $T - 1$ characters in the second phase of the game.

Finally, it is not difficult to verify that the argument above establishes that $\mathcal{F}$ does not have an ϵ-NE for ϵ-NE for all $\epsilon \in (0, 1)$. This concludes the proof. ∎

One might wonder whether the non-existence of NE in computational games follows from the fact that we are dealing with an infinite sequence of games with infinitely many possible TMs (e.g., $|\mathcal{M}| = \infty$). Nash Theorem regarding the existence of NE requires that the action space of every player is finite; without this requirement a NE may fail to exist. Hence it is natural to ask whether limiting $|\mathcal{M}|$ to be finite (for example, taking $\mathcal{M}$ to be the family of all TMs over a fixed alphabet with at most S states for some bound S) may force the existence of NE in computational games. Theorem 2 illustrates that this is not the case: $\mathcal{F}$ will not have a NE even if we take $\mathcal{M}$ to consist only of TMs whose number of states is upper bounded by a large enough positive number S (S should allow for using the TM that is hardwired to output the appropriate best response in the matching pennies game and print $T - 1$ characters in the second phase). The reason why NE does not exist despite the finiteness of $\mathcal{M}$, is that in contrast to ordinary games, where a mixed actions of best responses is a best response, in our setting this is not necessarily true: mixing over actions may consume computational resources, forcing players to choose actions that are suboptimal when using randomized strategies.

5   The Existence of ϵ-NE in Computational Games

Our previous results show that if we restrict players to be computationally bounded, then there are polynomial games with no ϵ-NE. Here we demonstrate that the restriction to computationally bounded players is critical. If we allow players to choose arbitrary TMs (or TMs that are guaranteed to halt on every input), we show that for all $\epsilon > 0$, there is an ϵ-NE in every computational game (and thus, a fortiori, in every polynomial game). The reason that we need ϵ-NE rather than NE (although ϵ can be arbitrarily small) is that there are 3-player games in which, in every NE, some actions are chosen with irrational probabilities (even if all utilities are rational and nature's moves are made with rational probabilities) [18]. By considering ϵ-NE, we can avoid representational issues involving irrational numbers.

Let $\epsilon > 0$ be a fixed constant. Suppose that $\mathcal{G} = (G_1, G_2, \ldots)$ is a computational game. Let $\mathcal{M}$ be any set of TMs that includes all TMs that are guaranteed to halt on every input (thus, $\mathcal{M}$ could consist of all TMs). At a high level, the argument for the existence of ϵ-NE is a straightforward application of ideas of Lipton and Markakis [14]. As they observe, given a game G, we can represent the conditions required for a strategy to be a NE using a single algebraic equation (in several variables), where a NE must be a root of the equation. We can compute a strategy profile that is arbitrarily close to a root of this algebraic equation; it can be shown that a strategy vector that is sufficiently close to a root is an ϵ-NE. We can now obtain an ϵ-NE for the computational game $\mathcal{G} = (G_1, G_2, \ldots)$ as follows: Given ϵ, the nth game G_n, and type $t \in T_i$, player i computes a profile $(s_1^n, s_2^n \ldots s_k^n)$ of distributions over actions that is an ϵ-NE of G_n (conditional on t) and plays according to $s_i^n(t)$. (If there are several ϵ-NEs, one is chosen in a consistent way, so that all players are playing a component of the same profile.) Using these ideas we can prove the following result, whose proof can be found in the full paper [11].

Theorem 3. *If $\mathcal{G} = (G_1, G_2 \ldots)$ is a computational game, $\epsilon > 0$, and $\mathcal{M}$ includes all TMs that halt on all inputs, then G has an ϵ-$\mathcal{M}$-NE.*

6 Conclusion

We have considered computational games, where TMs compute strategies of players. We showed that a NE for polynomial-time players may not exist. This suggests that classic notions in game theory, such as best response, must be treated carefully when considering computational games with resource-bounded players.

As we showed, for unbounded players, an ϵ-NE always exists in a computational game. Even for bounded players, there may exist circumstances under which an ϵ-NE exists. For example, it may be that there exists an equilibrium if we bound the number of states in TMs used by players. Studying properties of games or TMs used by players that ensure the existence of (ϵ)-NE in computational games is an interesting direction for future research. It might also prove worthwhile to study the effect of limiting resources other than time, such as space or the amount of randomness used by players. Finally, our paper also leaves open the question as to whether there exists a NE in our model when we restrict players to TM whose running time is at most n^r for a fixed integer r.

Acknowledgments. Halpern was supported in part by NSF grants IIS-178108 and IIS-1703846, a grant from the Open Philanthropy Foundation, ARO grant W911NF-17-1-0592, and MURI grant W911NF-19-1-0217. Pass was supported in part by NSF grant IIS-1703846. Most of the work was done while Reichman was a postdoc at Cornell University.

References

1. Ben-Sasson, E., Tauman-Kalai, A., Kalai, E.: An approach to bounded rationality. In: Proceedings of the 19th Neural Information Processing Systems Conference, pp. 145–152 (2007)
2. Brassard, G., Chaum, D., Crépeau, C.: Minimum disclosure proofs of knowledge. J. Comput. Syst. Sci. **37**, 156–189 (1988)
3. Chen, X., Deng, X., Teng, S.H.: Settling the complexity of two-player Nash equilibrium. J. ACM **53**(3) (2009)
4. Courtois, N.T., Bahack, L.: On subversive miner strategies and block with holding attack in bitcoin digital currency (2014). arXiv preprint: http://arxiv.org/abs/1402.1718
5. Daskalakis, C., Goldberg, P.W., Papadimitriou, C.H.: The complexity of computing a Nash equilibrium. In: Proceedings of the 38th ACM Symposium on Theory of Computing, pp. 71–78 (2006)
6. Dodis, Y., Halevi, S., Rabin, T.: A cryptographic solution to a game theoretic problem. In: Bellare, M. (ed.) CRYPTO 2000. LNCS, vol. 1880, pp. 112–130. Springer, Heidelberg (2000). https://doi.org/10.1007/3-540-44598-6_7

7. Feigenbaum, J., Koller, D., Shor, P.W.: A game-theoretic classification of inter-active complexity classes. In: Proceedings of the Structure in Complexity Theory Conference, pp. 227–237 (1995)
8. Fortnow, L., Impagliazzo, R., Kabanets, V., Umans, C.: On the complexity of succinct zero-sum games. Comput. Complex. **17**(3), 353–376 (2008)
9. Fortnow, L., Santhanam, R.: Bounding rationality by discounting time. In: Proceedings of Innovations in Computer Science Conference, pp. 143–155 (2010)
10. Halpern, J.Y., Pass, R.: Algorithmic rationality: game theory with costly computation. J. Econ. Theory **156**, 246–268 (2015). https://doi.org/10.1016/j.jet.2014.04.007
11. Halpern, J.Y., Pass, R., Reichman, D.: On the nonexistence of equilibrium in computational games (2019)
12. Halpern, J.Y., Pass, R., Seeman, L.: Computational extensive-form games. In: Proceedings of 17th ACM Conference on Electronic Commerce (EC 2016), pp. 681–698 (2016)
13. Holenstein, T.: Pseudorandom generators from one-way functions: a simple construction for any hardness. In: Halevi, S., Rabin, T. (eds.) TCC 2006. LNCS, vol. 3876, pp. 443–461. Springer, Heidelberg (2006). https://doi.org/10.1007/11681878_23
14. Lipton, R.J., Markakis, E.: Nash equilibria via polynomial equations. In: Farach-Colton, M. (ed.) LATIN 2004. LNCS, vol. 2976, pp. 413–422. Springer, Heidelberg (2004). https://doi.org/10.1007/978-3-540-24698-5_45
15. Maschler, M., Solan, E., Zamir, S.: Game Theory. Cambridge University Press, Cambridge (2013)
16. Megiddo, N., Wigderson, A.: On play by means of computing machines. In: Theoretical Aspects of Reasoning About Knowledge: Proceedings of 1986 Conference, pp. 259–274 (1986)
17. Nakamoto, S.: Bitcoin: a peer-to-peer electronic cash system (2008). http://www.bitcoin.org/bitcoin.pdf
18. Nash, J.: Non-cooperative games. Ann. Math. **54**, 286–295 (1951)
19. Neyman, A.: Bounded complexity justifies cooperation in finitely repeated prisoner's dilemma. Econ. Lett. **19**, 227–229 (1985)
20. Neyman, A.: The positive value of information. Games Econ. Behav. **3**, 350–355 (1991)
21. Neyman, A.: Finitely repeated games with finite automata. Math. Oper. Res. **23**, 513–552 (1998)
22. Oliehook, F., Savani, R., Gallego, J., van der Poel, E., Gross, R.: Beyond local Nash equilibria for adversarial networks (2018). http://arxiv.org/abs/1806.07268
23. Papadimitriou, C.H., Yannakakis, M.: On complexity as bounded rationality. In: Proceedings of 26th ACM Symposium on Theory of Computing, pp. 726–733 (1994)
24. Rubinstein, A.: Finite automata play the repeated prisoner's dilemma. J. Econ. Theory **39**, 83–96 (1986)
25. Schoenebeck, G., Vadhan, S.: The computational complexity of nash equilibria in concisely represented games. Theory Comput. **4**, 270–279 (2006)
26. Simon, H.A.: A behavioral model of rational choice. Quart. J. Econ. **49**, 99–118 (1955). https://doi.org/10.2307/1884852
27. Wee, H.: On obfuscating point functions. In: Proceedings of the 37th ACM Annual Symposium on Theory of Computing, pp. 523–532 (2005)

On the Computational Complexity of Decision Problems About Multi-player Nash Equilibria

Marie Louisa Tølbøll Berthelsen and Kristoffer Arnsfelt Hansen[✉] [iD]

Aarhus University, Aarhus, Denmark
marielouisaberthelsen@gmail.com, arnsfelt@cs.au.dk

Abstract. We study the computational complexity of decision problems about Nash equilibria in m-player games. Several such problems have recently been shown to be computationally equivalent to the decision problem for the existential theory of the reals, or stated in terms of complexity classes, $\exists\mathbb{R}$-complete, when $m \geq 3$. We show that, unless they turn into trivial problems, they are $\exists\mathbb{R}$-hard even for 3-player *zero-sum* games.

We also obtain new results about several other decision problems. We show that when $m \geq 3$ the problems of deciding if a game has a Pareto optimal Nash equilibrium or deciding if a game has a strong Nash equilibrium are $\exists\mathbb{R}$-complete. The latter result rectifies a previous claim of NP-completeness in the literature. We show that deciding if a game has an irrational valued Nash equilibrium is $\exists\mathbb{R}$-hard, answering a question of Biló and Mavronicolas, and address also the computational complexity of deciding if a game has a rational valued Nash equilibrium. These results also hold for 3-player zero-sum games.

Our proof methodology applies to corresponding decision problems about symmetric Nash equilibria in symmetric games as well, and in particular our new results carry over to the symmetric setting. Finally we show that deciding whether a symmetric m-player games has a *non-symmetric* Nash equilibrium is $\exists\mathbb{R}$-complete when $m \geq 3$, answering a question of Garg, Mehta, Vazirani, and Yazdanbod.

1   Introduction

Given a finite strategic form m-player game the most basic algorithmic problem is to compute a Nash equilibrium, shown always to exist by Nash [20]. The computational complexity of this problem was characterized in seminal work by Daskalakis, Goldberg, and Papadimitriou [12] and Chen and Deng [10] as PPAD-complete for 2-player games and by Etessami and Yannakakis [13] as FIXP-complete for m-player games, when $m \geq 3$. Any 2-player game may be viewed

---

This paper forms an extension of parts of the master's thesis of the first author. The second author is supported by the Independent Research Fund Denmark under grant no. 9040-00433B.

D. Fotakis and E. Markakis (Eds.): SAGT 2019, LNCS 11801, pp. 153–167, 2019.
https://doi.org/10.1007/978-3-030-30473-7_11

as a 3-player *zero-sum* game by adding a dummy player, thereby making the class of 3-player zero-sum games a natural class of games intermediate between 2-player and 3-player games. The problem of computing a Nash equilibrium for a 3-player zero-sum game is clearly PPAD-hard and belongs to FIXP, but its precise complexity appears to be unknown.

Rather than settling for *any* Nash equilibrium, one might be interested in a Nash equilibrium that satisfies a given property, e.g. giving each player at least a certain payoff. Such a Nash equilibrium might of course not exist and therefore results in the basic computational problem of deciding existence. In the setting of 2-player games, the computational complexity of several such problems was proved to be NP-complete by Gilboa and Zemel [16]. Conitzer and Sandholm [11] revisited these problems and showed them, together with additional problems, to be NP-complete even for symmetric games.

Only recently was the computational complexity of analogous problems in m-player games determined, for $m \geq 3$. Schaefer and Štefankovič [22] obtained the first such result by proving $\exists\mathbb{R}$-completeness of deciding existence of a Nash equilibrium in which no action is played with probability larger than $\frac{1}{2}$ by any player. Garg, Mehta, Vazirani, and Yazdanbod [14] used this to also show $\exists\mathbb{R}$-completeness for deciding if a game has more than one Nash equilibrium, whether each player can ensure a given payoff in a Nash equilibrium, and for the two problems of deciding whether the support sets of the mixed strategies of a Nash equilibrium can belong to given sets or contain given sets. In addition, by a symmetrization construction, they show that the analogue to the latter two problems for symmetric Nash equilibria are $\exists\mathbb{R}$-complete as well. Biló and Mavronicolas [4,5] subsequently extended the results of Garg et al. to further problems both about Nash equilibria and about symmetric Nash equilibria. They show $\exists\mathbb{R}$-completeness of deciding existence of a Nash equilibrium where all players receive at most a given payoff, where the total payoff of the players is at least or at most a given amount, whether the size of the supports of the mixed strategies all have a certain minimum or maximum size, and finally whether a Nash equilibrium exists that is *not* Pareto optimal or that is *not* a strong Nash equilibrium. All the analogous problems about symmetric Nash equilibria are shown to be $\exists\mathbb{R}$-complete as well.

1.1   Our Results

We revisit the problems about existence of Nash equilibria in m-player games, with $m \geq 3$, considered by Garg et al. and Biló and Mavronicolas. In a zero-sum game the total payoff of the players in any Nash equilibrium is of course 0, and any Nash equilibrium is Pareto optimal. This renders the corresponding decision problems trivial in the case of zero-sum games. We show that except for these, all the problems considered by Garg et al. and Biló and Mavronicolas remain $\exists\mathbb{R}$-hard for 3-player zero-sum games. We obtain our results building on a recent more direct and simple proof of $\exists\mathbb{R}$-hardness of the initial $\exists\mathbb{R}$-complete problem of Schaefer and Štefankovič due to Hansen [17]. We can also give comparably simpler proofs of $\exists\mathbb{R}$-hardness for the problems about total payoff and existence of a non Pareto optimal Nash equilibrium.

We next show that deciding existence of a strong Nash equilibrium in an m-player game with $m \geq 3$ is $\exists\mathbb{R}$-complete, and likewise for the similar problem of deciding existence of a Pareto optimal Nash equilibrium. Gatti, Rocco, and Sandholm [15] proved earlier that deciding if a given (rational valued) strategy profile x is a strong Nash equilibrium can be done in polynomial time. They then erroneously concluded that the problem of deciding existence of a strong Nash equilibrium is, as a consequence NP-complete. A problem with this reasoning is that if a strong Nash equilibrium exists, there is no guarantee that a rational valued strong Nash equilibrium exists. Even if one disregards a concern about irrational valued strong Nash equilibria, it is possible that even when a rational valued strong Nash equilibrium exists, any rational valued strong Nash equilibrium would require *exponentially* many bits to describe in the standard binary encoding of the numerators and denominators of the probabilities of the equilibrium strategy profile. Nevertheless, our proof of $\exists\mathbb{R}$-membership builds on the idea behind the polynomial time algorithm of Gatti et al.

In another work, Biló and Mavronicolas [3] considered the problems of deciding whether an irrational valued Nash equilibrium exists and whether a rational valued Nash equilibrium exists, proving both problems to be NP-hard. Biló and Mavronicolas asked if the problem about existence of irrational valued Nash equilibria is hard for the so-called square-root-sum problem. We confirm this, showing the problem to be $\exists\mathbb{R}$-hard. We relate the problem about existence of rational valued Nash equilibria to the existential theory of the rationals.

We next use a symmetrization construction similar to Garg et al. to translate all problems considered to the analogous setting of decision problems about symmetric Nash equilibria. Here we do not obtain qualitative improvements on existing results. A final problem we consider is of deciding existence of a *nonsymmetric* Nash equilibrium in a given symmetric game. Mehta, Vazirani, and Yazdanbod [19] proved that this problem is NP-complete for 2-player games, and Garg et al. [14] raised the question of the complexity for m-player games with $m \geq 3$. We show this problem to be $\exists\mathbb{R}$-complete.

Our results about irrational valued and rational valued Nash equilibrium, all results about symmetric games, as well as several other proofs are omitted in this version of the paper due to lack of space.

2 Preliminaries

2.1 Existential Theory of the Reals and Rationals

The existential theory $\mathrm{Th}_\exists(\mathbb{R})$ of the reals is the set of all true sentences over $\mathbb{R}$ of the form $\exists x_1, \ldots, x_n \in \mathbb{R} : \phi(x_1, \ldots, x_n)$, where ϕ is a quantifier free Boolean formula of equalities and inequalities of polynomials with integer coefficients. The complexity class $\exists\mathbb{R}$ is defined [22] as the closure of $\mathrm{Th}_\exists(\mathbb{R})$ under polynomial time many-one reductions. Equivalently, $\exists\mathbb{R}$ is the constant-free Boolean part of the class $\mathrm{NP}_\mathbb{R}$ [7], which is the analogue class to NP in the Blum-Shub-Smale model of computation [6]. It is straightforward to see that $\text{Th}_\exists(\mathbb{R})$ is NP-hard (cf. [8]) and the decision procedure by Canny [9] shows that $\text{Th}_\exists(\mathbb{R})$ belongs to PSPACE. Thus it follows that $\text{NP} \subseteq \exists\mathbb{R} \subseteq \text{PSPACE}$.

The basic complete problem for $\exists\mathbb{R}$ is the problem QUAD of deciding whether a system of quadratic equations with integer coefficients has a solution over $\mathbb{R}$ [6].

2.2 Strategic Form Games and Nash Equilibrium

A finite strategic form game $\mathcal{G}$ with m players is given by sets $S_1, \ldots, S_m$ of actions (*pure strategies*) together with *utility functions* $u_1, \ldots, u_m : S_1 \times \cdots \times S_m \to \mathbb{R}$. A choice of an action $a_i \in S_i$ for each player together form a pure strategy profile $a = (a_1, \ldots, a_m)$.

The game $\mathcal{G}$ is *symmetric* if $S_1 = \cdots = S_m$ and for every permutation π on $[m]$, every $i \in [m]$ and every $(a_1, \ldots, a_m) \in S_1 \times \cdots \times S_m$ it holds that $u_i(a_1, \ldots, a_m) = u_{\pi(i)}(a_{\pi(1)}, \ldots, a_{\pi(m)})$. In other words, a game is symmetric if the players share the same set of actions and the utility function of a player depends only on the action of the player together with the *multiset* of actions of the other players.

Let $\Delta(S_i)$ denote the set of probability distributions on S_i. A *(mixed) strategy* for Player i is an element $x_i \in \Delta(S_i)$. The *support* $\text{Supp}(x_i)$ is the set of actions given strictly positive probability by x_i. We say that x_i is *fully mixed* if $\text{Supp}(x_i) = S_i$. A strategy x_i for each player i together form a strategy profile $x = (x_1, \ldots, x_m)$. The utility functions naturally extend to strategy profiles by letting $u_i(x) = \mathbb{E}_{a \sim x} u_i(a_1, \ldots, a_m)$. We shall also refer to $u_i(x)$ as the *payoff* of Player i.

Given a strategy profile x we let $x_{-i} = (x_1, \ldots, x_{i-1}, x_{i+1}, \ldots, x_m)$ denote the strategies of all players except Player i. Given a strategy $y \in S_i$ for Player i, we let $(x_{-i}; y)$ denote the strategy profile $(x_1, \ldots, x_{i-1}, y, x_{i+1}, \ldots, x_m)$ formed by x_{-i} and y. We may also denote $(x_{-i}; y)$ by $x \setminus y$. We say that y is a *best reply* for Player i to x (or to x_{-i}) if $u_i(x \setminus y) \geq u_i(x \setminus y')$ for all $y' \in \Delta(S_i)$.

A *Nash equilibrium* (NE) is a strategy profile x where each individual strategy x_i is a best reply to x. As shown by Nash [20], every finite strategic form game $\mathcal{G}$ has a Nash equilibrium. In a symmetric game $\mathcal{G}$, a *symmetric Nash equilibrium* (SNE) is a Nash equilibrium where the strategies of all players are identical. Nash also proved that every symmetric game has a symmetric Nash equilibrium.

A strategy profile x is *Pareto optimal* if there is no strategy profile x' such that $u_i(x) \leq u_i(x')$ for all i, and $u_j(x) < u_j(x')$ for some j. A Nash equilibrium strategy profile need not be Pareto optimal and a Pareto optimal strategy profile need not be a Nash equilibrium. A strategy profile that is both a Nash equilibrium and is Pareto optimal is called a Pareto optimal Nash equilibrium. The existence of a Pareto optimal Nash equilibrium is not guaranteed.

A *strong Nash equilibrium* [1] (strong NE) is a strategy profile x for which there is no non-empty set $B \subseteq [m]$ for which *all* players $i \in B$ can increase their payoff by different strategies assuming players $j \in [m] \setminus B$ play according to x. Equivalently, x is a strong Nash equilibrium if for every strategy profile $x' \neq x$

there exist i such that $x_i \neq x_i'$ and $u_i(x') \leq u_i(x)$. The existence of a strong Nash equilibrium is not guaranteed.

3 Decision Problems About Nash Equilibria

Below we define the decision problems under consideration with names generally following Biló and Mavronicolas [4]. The given input is a finite strategic form game $\mathcal{G}$, together with auxiliary input depending on the particular problem. We let u denote a rational number, k an integer, and $T_i \subseteq S_i$ a set of actions of Player i, for every i. We describe the decision problem by stating the property a Nash equilibrium x whose existence is to be determined should satisfy. The problems are grouped together in four groups which we cover separately.

Problem	Condition
∃NEWithLargePayoffs	$u_i(x) \geq u$ for all i
∃NEWithSmallPayoffs	$u_i(x) \leq u$ for all i
∃NEWithLargeTotalPayoff	$\sum_i u_i(x) \geq u$
∃NEWithSmallTotalPayoff	$\sum_i u_i(x) \leq u$
∃NEInABall	$x_i(a_i) \leq u$ for all i and $a_i \in S_i$
∃SecondNE	x is not the only NE
∃NEWithLargeSupports	$|\mathrm{Supp}(x_i)| \geq k$ for all i
∃NEWithSmallSupports	$|\mathrm{Supp}(x_i)| \leq k$ for all i
∃NEWithRestrictingSupports	$T_i \subseteq \mathrm{Supp}(x_i)$ for all i
∃NEWithRestrictedSupports	$\mathrm{Supp}(x_i) \subseteq T_i$ for all i
∃NonParetoOptimalNE	x is not Pareto optimal
∃NonStrongNE	x is not a strong NE
∃ParetoOptimalNE	x is Pareto optimal
∃StrongNE	x is a strong NE
∃IrrationalNE	$x_i(a_i) \notin \mathbb{Q}$ for some i and $a_i \in S_i$
∃RationalNE	$x_i(a_i) \in \mathbb{Q}$ for all i and $a_i \in S_i$

Except for the last four problems above, it is straightforward to prove membership in $\exists\mathbb{R}$ by an explicit existentially quantified first-order formula. We prove $\exists\mathbb{R}$ membership of ∃ParetoOptimalNE and ∃StrongNE in Subsect. 3.3.

A key step (implicitly present) in the proof of the first $\exists\mathbb{R}$-hardness result about Nash equilibrium in 3-player games by Schaefer and Štefankovič is a result due to Schaefer [21] that QUAD remains $\exists\mathbb{R}$-hard under the *promise* that either the given quadratic system has no solutions or a solution exists in the unit ball $B(\mathbf{0}, 1)$. For our purposes the following variation [17, Proposition 2] will be more directly applicable (and may easily be proved from the former). Here we denote by Δ_c^n the standard corner n-simplex $\{x \in \mathbb{R}^n \mid x \geq 0 \wedge \sum_{i=1}^n x_i \leq 1\}$.

Proposition 1. *It is $\exists\mathbb{R}$-hard to decide if a given system of quadratic equations in n variables and with integer coefficients has a solution under the promise that either the system has no solutions or a solution z exists that is in the interior of Δ_c^n and also satisfies $z_i \leq \frac{1}{2}$ for all i and that $\sum_{i=1}^n z_i \geq \frac{1}{2}$.*

Schaefer and Štefankovič showed that $\exists\text{NEINABALL}$ is $\exists\mathbb{R}$-hard for 3-player games by first proving that the following problem is $\exists\mathbb{R}$-hard: Given a continuous function $f : B(\mathbf{0}, 1) \to B(\mathbf{0}, 1)$ mapping the unit ball to itself, where each coordinate function f_i is given as a polynomial, and given a rational number r, is there a fixed point of f in the ball $B(\mathbf{0}, r)$? The proof was then concluded by a transformation of Brouwer functions into 3-player games by Etessami and Yannakakis [13]. This latter reduction is rather involved and goes though an intermediate construction of 10-player games. More recently, Hansen [17] gave a simple and direct reduction from the above promise version of QUAD to $\exists\text{NEINABALL}$.

The first step of this as well as our reductions is to transform the given quadratic system over the corner simplex Δ_c^n into a homogeneous bilinear system over the standard n-simplex $\{x \in \mathbb{R}^{n+1} \mid x \geq 0 \wedge \sum_{i=1}^{n+1} x_i = 1\}$ which we denote by Δ^n. We can obtain the following statement (cf. [17, Proposition 3]).

Proposition 2. *It is $\exists\mathbb{R}$-complete to decide if a system of homogeneous bilinear equations $q_k(x, y) = 0$, $k = 1, \ldots, \ell$ with integer coefficients has a solution $x, y \in \Delta^n$. It remains $\exists\mathbb{R}$-hard under the promise that either the system has no such solution or a solution (x, x) exists where x belongs to the relative interior of Δ^n and further satisfies $x_i \leq \frac{1}{2}$ for all i.*

3.1 Payoff Restricted Nash Equilibria

For proving the $\exists\mathbb{R}$-hardness results we start by showing that it is $\exists\mathbb{R}$-hard to decide if a given zero-sum game has a Nash equilibrium in which each player receives payoff 0. This is in contrast to the earlier work of Garg et al. [14] and Biló and Mavronicolas [4,5] that reduce from the $\exists\text{NEINABALL}$ problem. On the other hand we do show $\exists\mathbb{R}$-hardness even under the promise that the Nash equilibrium also satisfies the condition of $\exists\text{NEINABALL}$. The construction and proof below are modification of proofs by Hansen [17, Theorem 1 and Theorem 2].

Definition 3 (The 3-player zero-sum game $\mathcal{G}_0$). *Let $\mathcal{S}$ be a system of homogeneous bilinear polynomials $q_1(x, y), \ldots, q_\ell(x, y)$ with integer coefficients in variables $x = (x_1, \ldots, x_{n+1})$ and $y = (y_1, \ldots, y_{n+1})$,*

$$q_k(x, y) = \sum_{i=1}^{n+1}\sum_{j=1}^{n+1} a_{ij}^{(k)} x_i y_j.$$

We define the 3-player game $\mathcal{G}_0(\mathcal{S})$ as follows. The strategy set of Player 1 is the set $S_1 = \{1, -1\} \times \{1, 2, \ldots, \ell\}$. The strategy sets of Player 2 and Player 3 are $S_2 = S_3 = \{1, 2, \ldots, n+1\}$. The (integer) utility functions of the players are defined by $\frac{1}{2}u_1((s, k), i, j) = -u_2((s, k), i, j) = -u_3((s, k), i, j) = sa_{ij}^{(k)}$.

When the system $\mathcal{S}$ is understood by the context, we simply write $\mathcal{G}_0 = \mathcal{G}_0(\mathcal{S})$. We think of the strategy (s, k) of Player 1 as corresponding to the polynomial q_k together with a sign s, the strategy i of Player 2 as corresponding to x_i and the strategy j of Player 3 as corresponding to y_j. We may thus identify mixed strategies of Player 2 and Player 3 as assignments to variables $x, y \in \Delta^n \subseteq \mathbb{R}^{n+1}$. The following observation is immediate from the definition of $\mathcal{G}_0$.

Lemma 4. *Any strategy profile (x, y) of Player 2 and Player 3 satisfies for every $(s, k) \in S_1$ the equation*

$$\tfrac{1}{2}u_1((s, k), x, y) = -u_2((s, k), x, y) = -u_3((s, k), x, y) = sq_k(x, y). \qquad (1)$$

Hence $u_1(z, x, y) = u_2(z, x, y) = u_3(z, x, y) = 0$ when z is the uniform distribution on S_1. Consequentially, any Nash equilibrium payoff profile is of the form $(2u, -u, -u)$, where $u \geq 0$.

Next we relate solutions of the system $\mathcal{S}$ to Nash equilibria in $\mathcal{G}_0$.

Proposition 5. *Let $\mathcal{S}$ be a system of homogeneous bilinear polynomials $q_k(x, y)$. If $\mathcal{S}$ has a solution $(x, y) \in \Delta^n \times \Delta^n$, then letting z be the uniform distribution on S_1, the strategy profile $\sigma = (z, x, y)$ is a Nash equilibrium of $\mathcal{G}_0$ in which every player receives payoff 0. If in addition (x, y) satisfies the promise of Proposition 2, then σ is fully mixed, Player 2 and Player 3 use identical strategies, and no action is chosen with probability more than $\tfrac{1}{2}$ by any player. Conversely, if (z, x, y) is a Nash equilibrium of $\mathcal{G}_0$ in which every player receives payoff 0, then (x, y) is a solution to $\mathcal{S}$.*

Proof. Suppose first that $(x, y) \in \Delta^n \times \Delta^n$ is a solution to $\mathcal{S}$ and let z be the uniform distribution on S_1. By Eq. (1) the strategy profile (x, y) of Player 2 and Player 3 ensures that all players receive payoff 0 regardless of which strategy is played by Player 1, and likewise the strategy z of Player 1 ensures that all players receive payoff 0 regardless of the strategies of Player 2 and Player 3. This shows that σ is a Nash equilibrium of $\mathcal{G}_0$, in which by Lemma 4 every player receives payoff 0. If (x, y) in addition satisfies the promise of Proposition 2 we have $0 < x_i = y_i \leq \tfrac{1}{2}$. From this and our choice of z, we have that σ is a fully mixed and that no action is chosen by a strategy of σ with probability more than $\tfrac{1}{2}$.

Suppose on the other hand that $\sigma = (z, x, y)$ is a Nash equilibrium of $\mathcal{G}_0$ with payoff 0 for every player and suppose that $q_k(x, y) \neq 0$ for some k. Then by Eq. (1) we get that $u_1((\text{sgn}(q_k(x, y)), k), x, y) = |2q_k(x, y)| > 0$, contradicting that σ is a Nash equilibrium. Thus (x, y) is a solution to $\mathcal{S}$.

Theorem 6. $\exists$NEWITHLARGEPAYOFFS *and* $\exists$NEWITHSMALLPAYOFFS *are $\exists\mathbb{R}$-complete, even for 3-player zero-sum games.*

Proof. For a strategy profile x in a zero-sum game $\mathcal{G}$ we have that $u_i(x) = 0$, for all i, if and only if $u_i(x) \geq 0$, for all i, if and only if $u_i(x) \leq 0$, for all i. Thus Proposition 5 gives a reduction from the promise problem of Proposition 2, thereby establishing $\exists\mathbb{R}$-hardness of the problems $\exists$NEWITHLARGEPAYOFFS and $\exists$NEWITHSMALLPAYOFFS.

A simple change to the game $\mathcal{G}_0$ gives $\exists\mathbb{R}$-hardness for the two problems $\exists$NEWITHLARGETOTALPAYOFF and $\exists$NEWITHSMALLTOTALPAYOFF. Naturally we must give up the zero-sum property of the game. We omit the proof.

Theorem 7 (Bilò and Mavronicolas [4]). $\exists$NEWITHLARGETOTALPAYOFF *and* $\exists$NEWITHSMALLTOTALPAYOFF *are* $\exists\mathbb{R}$-*complete, even for 3-player games.*

3.2   Probability Restricted Nash Equilibria

A key property of the game $\mathcal{G}_0$ is that Player 1 may ensure all players receive payoff 0. We now give *all* players this choice by playing a new additional action $\bot$. We then design the utility functions involving $\bot$ in such a way that the pure strategy profile $(\bot, \bot, \bot)$ is always a Nash equilibrium, and every other Nash equilibrium is a Nash equilibrium in $\mathcal{G}_0$ in which all players receive payoff 0.

Definition 8. *For $u \geq 0$, let $\mathcal{H}_1 = \mathcal{H}_1(u)$ be the 3-player zero-sum game where each player has the action set $\{G, \bot\}$ and the payoff vectors are given by the entries of the following two matrices, where Player 1 selects the matrix, Player 2 selects the row, Player 3 selects the column.*

G	G	$\bot$
G	$(2u, -u, -u)$	$(1, -1, 0)$
$\bot$	$(1, \quad 0, -1)$	$(-4, \quad 2, 2)$

$\bot$	G	$\bot$
G	$(0, 0, \quad 0)$	$(2, -3, 1)$
$\bot$	$(2, 1, -3)$	$(-2, \quad 1, 1)$

It is straightforward to determine the Nash equilibria of $\mathcal{H}_1$.

Lemma 9. *When $u > 0$, the only Nash equilibrium of $\mathcal{H}_1(u)$ is the pure strategy profile $(\bot, \bot, \bot)$. When $u = 0$ the only Nash equilibria of $\mathcal{H}_1(u)$ are the pure strategy profiles (G, G, G) and $(\bot, \bot, \bot)$.*

We use the game $\mathcal{H}_1(u)$ to extend the game $\mathcal{G}_0$. The action G of $\mathcal{H}_1$ represents selecting an action from $\mathcal{G}_0$, and the payoff vector $(2u, -u, -u)$ that is the result of all players playing the action G is precisely of the form of the Nash equilibrium payoff profile of $\mathcal{G}_0$.

Definition 10 (The 3-player zero-sum game $\mathcal{G}_1$). *Let $\mathcal{G}_1 = \mathcal{G}_1(\mathcal{S})$ be the game obtained from $\mathcal{G}_0(\mathcal{S})$ as follows. Each player is given an additional action $\bot$. When no player plays the action $\bot$, the payoffs are the same as in $\mathcal{G}_0$. When at least one player is playing the action $\bot$ the payoff are the same as in $\mathcal{H}_1$, where each action different from $\bot$ is translated to action G.*

We next characterize the Nash equilibria in $\mathcal{G}_1$.

Proposition 11. *The pure strategy profile $(\bot, \bot, \bot)$ is a Nash equilibrium of $\mathcal{G}_1$. Any other Nash equilibrium x in $\mathcal{G}_1$ is also a Nash equilibrium of $\mathcal{G}_0$ and is such that every player receives payoff 0.*

Proof. By Lemma 4 any Nash equilibrium of $\mathcal{G}_1$ induces a Nash equilibrium of $\mathcal{H}_1(u)$, where $(2u, -u, -u)$ is a Nash equilibrium payoff profile of $\mathcal{G}_0$, by letting each player play the action G with the total probability of which the actions of $\mathcal{G}_0$ are played. By Lemma 9, any Nash equilibrium in $\mathcal{G}_1$ different from $(\bot, \bot, \bot)$ must then be a Nash equilibrium of $\mathcal{G}_0$ with Nash equilibrium payoff profile $(0, 0, 0)$ as claimed.

Theorem 12. *The following problems are $\exists\mathbb{R}$-complete, even for 3-player zero-sum games:* $\exists$NEInABall, $\exists$SecondNE, $\exists$NEWithLargeSupports, $\exists$NEWithRestrictingSupports, *and* $\exists$NEWithRestrictedSupports.

Proof. Propositions 5 and 11 together gives a reduction from the promise problem of Proposition 2 to all of the problems under consideration when setting the additional parameters as follows. For $\exists$NEInABall we let $u = \frac{1}{2}$, for $\exists$NEWithLargeSupports we let $k = 2$, and lastly for both of $\exists$NEWithRestrictingSupports and $\exists$NEWithRestrictedSupports we let T_i be the set of all actions of Player i except $\bot$.

To adapt the reduction of Theorem 12 to $\exists$NEWithSmallSupports we need to replace the trivial Nash equilibrium $(\bot, \bot, \bot)$ by a Nash equilibrium with large support. We define a game $\mathcal{H}_2(k)$ for this purpose and omit its easy analysis.

Definition 13. *Define the 2-player zero-sum game $\mathcal{H}_2(k)$ as follows. The two players, which we denote Player 2 and Player 3, have the same set of pure strategies $S_2 = S_3 = \{0, 1, \ldots, k-1\}$. The utility functions $u_2(a_2, a_3) = -u_3(a_2, a_3)$ are defined by $u_2(a_2, a_3) = 1$ if $a_2 = a_3$, $u_2(a_2, a_3) = -1$ if $a_2 \equiv a_3 + 1 \pmod{k}$, and $u_2(a_2, a_3) = 0$ otherwise.*

Lemma 14. *For any $k \geq 2$, in the game $\mathcal{H}_2(k)$ the strategy profile in which each action is played with probability $\frac{1}{k}$ is the unique Nash equilibrium and yields payoff 0 to both players.*

Definition 15 (The 3-player zero-sum game $\mathcal{G}_2$). *Let $\mathcal{G}_2 = \mathcal{G}_2(\mathcal{S})$ be the game obtained from $\mathcal{G}_1$ as follows. The action $\bot$ of Player 2 and Player 3 are replaced by the set of actions $(\bot, i)$, $i \in \{0, 1, \ldots, k-1\}$, where k is the maximum number of actions of a player in $\mathcal{G}_1$. The payoff vector of the pure strategy profile $(\bot, (\bot, a_2), (\bot, a_3))$ is $(-2, 1 + u_2(a_2, a_3), 1 + u_3(a_2, a_3))$, where u_2 and u_3 are the utility functions of the game $\mathcal{H}_2(k)$. Otherwise, when at least one player plays the action G, the payoff is as in $\mathcal{H}_1$, where actions of the form $(\bot, i)$ are translated to the action $\bot$.*

Theorem 16. $\exists$NEWithSmallSupports *is $\exists\mathbb{R}$-complete, even for 3-player zero-sum games.*

Proof. In $\mathcal{G}_2$, the strategy profile where Player 1 plays $\bot$ and Player 2 and Player 3 play $(\bot, i)$, with i chosen uniformly at random, is a Nash equilibrium that takes the role of the Nash equilibrium $(\bot, \bot, \bot)$ in $\mathcal{G}_1$. Consider now an arbitrary Nash equilibrium in $\mathcal{G}_2$. In case all players play the action G with probability less than 1, Player 2 and Player 3 must choose each action of the form $(\perp, i)$ with the same probability, since $\mathcal{H}_2$ has a unique Nash equilibrium. The Nash equilibrium induces a strategy profile in $\mathcal{G}_1$, letting Player 2 and Player 3 play the action $\perp$ with the total probability each player placed on the actions $(\perp, i)$. By definition of $\mathcal{H}_2(k)$ the payoff vector of $(\perp, \perp, \perp)$ in $\mathcal{G}_1$ differs by at most 1 in each entry from the payoff vectors of $(\perp, (\perp, a_2), (\perp, a_3))$. The proof of Lemma 9 and Proposition 11 still holds when changing the payoff vector of $(\perp, \perp, \perp)$ by at most 1 in each coordinate. The strategy profile induced in $\mathcal{G}_1$ must therefore be a Nash equilibrium in $\mathcal{G}_1$. We conclude that in a Nash equilibrium x of $\mathcal{G}_2$, either Player 2 and Player 3 use strategies with support of size k or x is a Nash equilibrium of $\mathcal{G}_0$, where every player uses a strategy of support size strictly less than k and where every player receives payoff 0. Proposition 5 thus give a reduction showing $\exists\mathbb{R}$-hardness.

3.3   Pareto Optimal and Strong Nash Equilibria

For showing $\exists\mathbb{R}$-hardness for $\exists$NonStrongNE we first analyze the strong Nash equilibria in the game $\mathcal{H}_1$.

Lemma 17. *For $u \geq 0$, the Nash equilibrium $(\perp, \perp, \perp)$ of $\mathcal{H}_1(u)$ is a strong Nash equilibrium. For $u = 0$, the Nash equilibrium (G, G, G) of $\mathcal{H}_1(u)$ is not a strong Nash equilibrium.*

Proof. Consider first $u = 0$ and the Nash equilibrium (G, G, G). This is not a strong Nash equilibrium, since for instance Player 1 and Player 2 could both increase their payoff by playing the strategy profile $(\perp, \perp, G)$. Consider next $u \geq 0$ and the Nash equilibrium $(\perp, \perp, \perp)$. Since $\mathcal{H}_1$ is a zero-sum game it is sufficient to consider possible coalitions of two players. Player 2 and Player 3 are already receiving the largest possible payoff given that Player 1 is playing the strategy $\perp$, and hence they do not have a profitable deviation. Consider then, by symmetry, the coalition formed by Player 1 and Player 2, and let them play G with probabilities p_1 and p_2. A simple calculation shows that to increase the payoff of Player 1 requires $p_1 p_2 + 4p_2 - 2p_1 > 0$ and to increase the payoff of Player 2 requires $p_1 p_2 - 4p_2 + p_1 > 0$. Adding these gives $p_1(2p_2 - 1) > 0$ which implies $p_2 > \frac{1}{2}$. But then $p_1 p_2 - 4p_2 + p_1 < 0$. Thus $(\perp, \perp, \perp)$ is a strong Nash equilibrium.

Theorem 18. $\exists$NonStrongNE *is $\exists\mathbb{R}$-complete, even for 3-player zero-sum games.*

Proof. Propositions 5 and 11 together gives a reduction showing $\exists\mathbb{R}$-hardness, since by Lemma 17 the Nash equilibrium $(\perp, \perp, \perp)$ is a strong Nash equilibrium, and a Nash equilibrium of $\mathcal{G}_0$ where every player receives payoff 0 is not a strong Nash equilibrium.

In a zero-sum game, every strategy profile is Pareto optimal. Thus for showing $\exists\mathbb{R}$-hardness of $\exists$NonParetoOptimalNE we need to consider non-zero-sum games, thus leading to the result of Bilò and Mavronicolas.

We next consider the problems $\exists$StrongNE and $\exists$ParetoOptimalNE. We first outline a proof of membership in $\exists\mathbb{R}$, building on ideas of Gatti et al. [15] and Hansen, Hansen, Miltersen, and Sørensen [18]. Gatti et al. proved that deciding whether a given strategy profile x of an m-player game $\mathcal{G}$ is a strong Nash equilibrium can be done in polynomial time. The crucial insight behind this result is that the question of whether a coalition of $k \leq m$ players may all improve their payoff by together changing their strategies can be recast into a question in a derived game about the minmax value of an additional *fictitious* player that has only k strategies. Hansen et al. proved that in such a game, the minmax value may be achieved by strategies of the other players that have support size at most k.

Lemma 19 (Hansen et al. [18]). *Let $\mathcal{G}$ be a $m + 1$ player game and let $k = |S_{m+1}|$. If there exists a strategy profile x of the first m players such that $u_{m+1}(x; a) \leq 0$ for all $a \in S_{m+1}$ then there also exists a strategy profile x' of the first m players in which each strategy has support size at most k and $u_{m+1}(x'; a_{m+1}) \leq 0$ for all $a \in S_{m+1}$.*

We next give a generalization of the auxiliary game construction of Gatti et al. that also allows us to treat Pareto optimal Nash equilibria at the same time.

Definition 20 (cf. Gatti et al. [15]). *Let $\mathcal{G}$ be an m-player game with strategy sets S_i and utility functions u_i. Let x be a strategy profile of $\mathcal{G}$ and let $B_1 \dot\cup B_2 \dot\cup B_3 = [m]$ be a partition of the players, let $k_i = |B_i|$ and $k = k_1 + k_2$. For $\varepsilon > 0$ consider the $(m+1)$-player auxiliary game $\mathcal{G}' = \mathcal{G}'_{x,\varepsilon,(B_1,B_2,B_3)}$ defined as follows. For $i \in B_1 \cup B_2$ the strategy set of Player i is $S'_i = S_i$. For $i \in B_3$ the strategy set of Player i is $S'_i = \{\bot\}$. Finally, the strategy set of Player $m+1$ is $B_1 \cup B_2$. The utility function of Player $m + 1$ is defined as follows. Let $a = (a'_1, \ldots, a'_m, j)$ be a pure strategy profile of $\mathcal{G}'$. Define the strategy profile x^a of $\mathcal{G}$ letting $x^a_i = a_i$ for $i \in B_1 \cup B_2$ and $x^a_i = x_i$ for $i \in B_3$. We then let $u'_{m+1}(a) = u_j(x) - u_j(x^a) + \varepsilon$ for $j \in B_1$ and $u'_{m+1}(a) = u_j(x) - u_j(x^a)$ for $j \in B_2$.*

The following is immediate from the definition of $\mathcal{G}'$.

Lemma 21. *There exist a strategy profile x' in $\mathcal{G}$ that satisfies $u_i(x') > u_i(x)$ when $i \in B_1$, $u_i(x') \geq u_i(x)$ when $i \in B_2$, and $x'_i = x_i$ when $i \in B_3$ if and only if there exist $\varepsilon > 0$ and a strategy x' in $\mathcal{G}'_{x,\varepsilon,(B_1,B_2,B_3)}$ of the first m players such that $u'_{m+1}(x', j) \leq 0$ for all $j \in B_1 \cup B_2$.*

The task of deciding if a strategy x is Pareto optimal amounts to checking the condition of Lemma 21 for $B_1 = \{i\}$ and $B_2 = [m] \setminus \{i\}$ for all i and to decide whether x is a strong Nash equilibrium amounts to checking the condition for all nonempty $B_1 \subseteq [m]$ while letting $B_2 = \emptyset$.

According to Lemma 19 we may restrict our attention to strategies x' in $\mathcal{G}'$ of supports of size at most m. Fixing such a set of supports $T_i \subseteq S_i$ for $i \in B_1 \cup B_2$, we may formulate the question of existence of a strategy x', with $\mathrm{Supp}(x'_i) \subseteq T_i$ for $i \in B_1 \cup B_2$ that satisfies the conditions of Lemma 21 as an existentially quantified first-order formula over the reals. For a fixed x we need only $1 + m^2$ existentially quantified variables to describe ε and the strategy x'. Since this is a constant number of variables, when as in our case m is a constant, the general decision procedure of Basu, Pollack, and Roy [2] runs in polynomial time in the bitsize of coefficients, number of polynomials, and their degrees, resulting in an overall polynomial time algorithm. Now, adding a step of simply enumerating over all nonempty $B_1 \subseteq [m]$ and all support sets of size m we obtain the result of Gatti et al. that deciding whether a given strategy profile x is a strong Nash equilibrium can be done in polynomial time. The same holds in a similar way for checking that a strategy profile is a Pareto optimal Nash equilibrium.

In our case, when proving $\exists \mathbb{R}$ membership the only input is the game $\mathcal{G}$, whereas the strategy profile x will be given by a block of existentially quantified variables. We then need to show how to express that x is a Pareto optimal or a strong Nash equilibrium by a quantifier free formula over the reals with free variables x. This will be possible by the fact that quantifier elimination, rather than just decision, is possible for the first order theory of the reals. The quantifier elimination procedure of Basu et al. [2] runs in time exponential in the number of free variables, so we cannot apply it directly with x being the set of free variables.

Instead we express the condition of Lemma 21 for a strategy profile x' that is constrained by $\mathrm{Supp}(x'_i) \subseteq T_i$ for $i \in B_1 \cup B_2$ in terms of auxiliary free variables $\widetilde{u}'$ that take the place of the values of the utility function u' of $\mathcal{G}'$. Since the supports of x' are restricted to size m, we need just m^{m+1} variables to represent the utility to Player $m + 1$ on every such pure strategy profile. Since this is a constant number of variables, the quantifier elimination procedure of Basu et al. runs in polynomial time and outputs a quantifier free formula over the reals with free variables $\widetilde{u}'$ that expresses the condition of Lemma 21 when the utilities u' are given by $\widetilde{u}'$. After this we substitute expressions for the utilities u' in terms of the variables x for the variables $\widetilde{u}'$. The final formula is obtained, in an analogous way to the decision question, by enumerating over the appropriate sets B_1 and B_2 as well as all possible supports T_i, obtaining a formula for each such choice and combining them to a single formula with free variables x expressing either that x is Pareto optimal or that x is a strong Nash equilibrium. To the former we add the simple conditions of x being a Nash equilibrium. Finally we existentially quantify over x and obtain a formula expressing either that $\mathcal{G}$ has a Pareto optimal Nash equilibrium or that $\mathcal{G}$ has a strong Nash equilibrium. Since this formula was computed in polynomial time given $\mathcal{G}$ we obtain the following result.

Proposition 22. $\exists$STRONGNE *and* $\exists$PARETOOPTIMALNE *both belong to* $\exists \mathbb{R}$.

For showing $\exists \mathbb{R}$-hardness we construct a new extension of $\mathcal{G}_0$.

Definition 23. *For $u \geq 0$, let $\mathcal{H}_4 = \mathcal{H}_4(u)$ be the 3-player game given by the following matrices, where Player 1 selects the matrix, Player 2 selects the row, Player 3 selects the column.*

	G	$\perp$
G	$(2u, -u, -u)$	$(-3, -3, \; 0)$
$\perp$	$(-3, \; 0, -3)$	$(-2, -2, -2)$

$$G$$

	G	$\perp$
G	$(\; 0, -3, -3)$	$(-2, -2, -2)$
$\perp$	$(-2, -2, -2)$	$(-1, -1, -1)$

$$\perp$$

Lemma 24. *When $u > 0$, the only Nash equilibrium of $\mathcal{H}_4(u)$ is the pure strategy profile $(\perp, \perp, \perp)$. When $u = 0$, the only Nash equilibria of $\mathcal{H}_4(u)$ are the pure strategy profiles (G, G, G) and $(\perp, \perp, \perp)$. Furthermore, when $u = 0$, the Nash equilibrium (G, G, G) is both Pareto optimal and a strong Nash equilibrium.*

Proof. When $u = 0$, clearly (G, G, G) is a Nash equilibrium, which is both Pareto optimal and a strong Nash equilibrium. Likewise, clearly $(\perp, \perp, \perp)$ is always a Nash equilibrium. When $u > 0$, the action G is strictly dominated by the action $\perp$ for Player 2 and Player 3, and hence they play $\perp$ with probability 1 in a Nash equilibrium. The only best reply of Player 1 is to play $\perp$ with probability 1 as well.

Analogously to Definition 10 we define the game $\mathcal{G}_4 = \mathcal{G}_4(\mathcal{S})$ to be the game extending $\mathcal{G}_0$ with $\mathcal{H}_4$ replacing the role of $\mathcal{H}_1$. We next establish $\exists\mathbb{R}$-hardness

Theorem 25. $\exists\text{PARETOOPTIMALNE}$ *and* $\exists\text{STRONGNE}$ *are* $\exists\mathbb{R}$-*complete, even for 3-player games.*

Proof. In $\mathcal{G}_4$, the strategy profile $(\perp, \perp, \perp)$, with payoff profile $(-1, -1, -1)$, is a Nash equilibrium that is neither Pareto optimal or a strong Nash equilibrium, since by Lemma 4 a strategy profile in $\mathcal{G}_0$ in which Player 1 plays an action according to the uniform distribution has payoff profile $(0, 0, 0)$.

Similarly to the proof of Theorem 12, any Nash equilibrium x in $\mathcal{G}_4$ different from $(\perp, \perp, \perp)$ must by Lemma 24 be a Nash equilibrium of $\mathcal{G}_0$ with payoff profile $(0, 0, 0)$. Since $\mathcal{G}_0$ is a zero-sum game, any strategy that is Pareto dominating x must involve the strategy $\perp$ and is thus ruled out by Lemma 24. Therefore x is Pareto-optimal. Now, x is not necessarily a strong Nash equilibrium, but by Lemma 4, letting Player 1 instead play an action of $\mathcal{G}_0$ according to the uniform distribution is also a Nash equilibrium of $\mathcal{G}_0$ with payoff profile $(0, 0, 0)$, that furthermore ensures that any strategy profile of Player 2 and Player 3 in $\mathcal{G}_0$ does not improve their payoffs. Also, by Lemma 4, no coalition involving Player 1 can improve their payoff without playing the action $\perp$. No coalition can however improve their payoff by a strategy profile involving the action $\perp$, since all such payoff profiles result in a player receiving negative payoff. Thus x' is a strong Nash equilibrium.

We conclude that Proposition 5 gives a reduction showing $\exists\mathbb{R}$-hardness of both $\exists\text{PARETOOPTIMALNE}$ and $\exists\text{STRONGNE}$, thereby together with Proposition 22 completing the proof.

References

1. Aumann, R.J.: Acceptable points in games of perfect information. Pacific J. Math. **10**(2), 381–417 (1960). https://doi.org/10.2140/pjm.1960.10.381
2. Basu, S., Pollack, R., Roy, M.-F.: Algorithms in Real Algebraic Geometry, 2nd edn. Springer, Heidelberg (2008). https://doi.org/10.1007/3-540-33099-2
3. Bilò, V., Mavronicolas, M.: Complexity of rational and irrational Nash equilibria. Theory Comput. Syst. **54**(3), 491–527 (2014). https://doi.org/10.1007/s00224-013-9523-7
4. Bilò, V., Mavronicolas, M.: A catalog of ∃ℝ-complete decision problems about Nash equilibria in multi-player games. In: Ollinger, N., Vollmer, H. (eds.) STACS 2016. LIPIcs, vol. 47, pp. 17:1–17:13. Schloss Dagstuhl - Leibniz-Zentrum für Informatik, Wadern (2016). https://doi.org/10.4230/LIPIcs.STACS.2016.17
5. Bilò, V., Mavronicolas, M.: ∃ℝ-complete decision problems about symmetric Nash equilibria in symmetric multi-player games. In: Vollmer, H., Vallé, B. (eds.) STACS 2017. LIPIcs, vol. 66, pp. 13:1–13:14. Schloss Dagstuhl-Leibniz-Zentrum für Informatik, Wadern (2017). https://doi.org/10.4230/LIPIcs.STACS.2017.13
6. Blum, L., Shub, M., Smale, S.: On a theory of computation and complexity over the real numbers: NP-completeness, recursive functions and universal machines. Bull. Amer. Math. Soc. **21**(1), 1–46 (1989). https://doi.org/10.1090/S0273-0979-1989-15750-9
7. Bürgisser, P., Cucker, F.: Exotic quantifiers, complexity classes, and complete problems. Found. Comput. Math. **9**(2), 135–170 (2009). https://doi.org/10.1007/s10208-007-9006-9
8. Buss, J.F., Frandsen, G.S., Shallit, J.O.: The computational complexity of some problems of linear algebra. J. Comput. Syst. Sci. **58**(3), 572–596 (1999). https://doi.org/10.1006/jcss.1998.1608
9. Canny, J.F.: Some algebraic and geometric computations in PSPACE. In: Simon, J. (ed.) Proceedings of the 20th Annual ACM Symposium on Theory of Computing (STOC 1988), pp. 460–467. ACM (1988). https://doi.org/10.1145/62212.62257
10. Chen, X., Deng, X.: Settling the complexity of two-player Nash equilibrium. In: 47th Annual IEEE Symposium on Foundations of Computer Science (FOCS 2006), pp. 261–272. IEEE Computer Society Press (2006). https://doi.org/10.1109/FOCS.2006.69
11. Conitzer, V., Sandholm, T.: New complexity results about Nash equilibria. Games Econ. Behav. **63**(2), 621–641 (2008). https://doi.org/10.1016/j.geb.2008.02.015
12. Daskalakis, C., Goldberg, P.W., Papadimitriou, C.H.: The complexity of computing a Nash equilibrium. SIAM J. Comput. **39**(1), 195–259 (2009). https://doi.org/10.1137/070699652
13. Etessami, K., Yannakakis, M.: On the complexity of Nash equilibria and other fixed points. SIAM J. Comput. **39**(6), 2531–2597 (2010). https://doi.org/10.1137/080720826
14. Garg, J., Mehta, R., Vazirani, V.V., Yazdanbod, S.: ∃ℝ-completeness for decision versions of multi-player (symmetric) Nash equilibria. ACM Trans. Econ. Comput. **6**(1), 1:1–1:23 (2018). https://doi.org/10.1145/3175494
15. Gatti, N., Rocco, M., Sandholm, T.: On the verification and computation of strong Nash equilibrium. In: Gini, M.L., Shehory, O., Ito, T., Jonker, C.M. (eds.) AAMAS 2013, pp. 723–730. IFAAMAS, Richland (2013)
16. Gilboa, I., Zemel, E.: Nash and correlated equilibria: some complexity considerations. Games Econ. Behav. **1**(1), 80–93 (1989). https://doi.org/10.1016/0899-8256(89)90006-7

17. Hansen, K.A.: The real computational complexity of minmax value and equilibrium refinements in multi-player games. Theory Comput. Syst. (2018). https://doi.org/10.1007/s00224-018-9887-9

18. Hansen, K.A., Hansen, T.D., Miltersen, P.B., Sørensen, T.B.: Approximability and parameterized complexity of minmax values. In: Papadimitriou, C., Zhang, S. (eds.) WINE 2008. LNCS, vol. 5385, pp. 684–695. Springer, Heidelberg (2008). https://doi.org/10.1007/978-3-540-92185-1_74

19. Mehta, R., Vazirani, V.V., Yazdanbod, S.: Settling some open problems on 2-player symmetric Nash equilibria. In: Hoefer, M. (ed.) SAGT 2015. LNCS, vol. 9347, pp. 272–284. Springer, Heidelberg (2015). https://doi.org/10.1007/978-3-662-48433-3_21

20. Nash, J.: Non-cooperative games. Ann. Math. **2**(54), 286–295 (1951). https://doi.org/10.2307/1969529

21. Schaefer, M.: Complexity of some geometric and topological problems. In: Eppstein, D., Gansner, E.R. (eds.) GD 2009. LNCS, vol. 5849, pp. 334–344. Springer, Heidelberg (2010). https://doi.org/10.1007/978-3-642-11805-0_32

22. Schaefer, M., Štefankovič, D.: Fixed points, Nash equilibria, and the existential theory of the reals. Theory Comput. Syst. **60**, 172–193 (2017). https://doi.org/10.1007/s00224-015-9662-0

Computing Stackelberg Equilibria
of Large General-Sum Games

Avrim Blum[1], Nika Haghtalab[2], MohammadTaghi Hajiaghayi[3],
and Saeed Seddighin[4(✉)]

[1] Toyota Technological Institute, Chicago, IL 60637, USA
avrim@ttic.edu
[2] Microsoft Research, Cambridge, MA 02142, USA
nika.haghtalab@microsoft.com
[3] University of Maryland, College Park, MD 20742, USA
hajiagha@cs.umd.edu
[4] Harvard University, Cambridge, MA 02138, USA
saeedreza.seddighin@gmail.com

Abstract. We study the computational complexity of finding Stackelberg Equilibria in general-sum games, where the set of pure strategies of the leader and the followers are exponentially large in a natural representation of the problem.

In *zero-sum* games, the notion of a Stackelberg equilibrium coincides with the notion of a *Nash Equilibrium* (Korzhyk et al. 2011b). Finding these equilibrium concepts in zero-sum games can be efficiently done when the players have polynomially many pure strategies or when (in additional to some structural properties) a best-response oracle is available (Ahmadinejad et al. 2016; Dudík et al. 2017; Kalai and Vempala 2005). Despite such advancements in the case of zero-sum games, little is known for general-sum games.

In light of the above, we examine the computational complexity of computing a Stackelberg equilibrium in large general-sum games. We show that while there are natural large general-sum games where the Stackelberg Equilibria can be computed efficiently if the Nash equilibrium in its zero-sum form could be computed efficiently, in general, structural properties that allow for efficient computation of Nash equilibrium in zero-sum games are not sufficient for computing Stackelberg equilibria in general-sum games.

1 Introduction

Recent years have witnessed significant interest in Stackelberg games and their equilibria. A *Stackelberg game* models an interaction between two players, a *leader* and a *follower*, where the leader's goal is to *commit* to a randomized strategy that yields the highest utility, given that the follower responds by choosing an action that is best for itself. Such a pair of strategies is called a *Stackelberg equilibrium (SE)*. The interest in these games is driven, in part, by their applications to security (Tambe 2011) and their adoption by major security agencies

© Springer Nature Switzerland AG 2019
D. Fotakis and E. Markakis (Eds.): SAGT 2019, LNCS 11801, pp. 168–182, 2019.
https://doi.org/10.1007/978-3-030-30473-7_12

such as the US Coast Guard, the Federal Air Marshals Service, and the Los Angeles Airport Police.

Standard approaches for finding a Stackelberg equilibrium, such as the Multiple LPs approach of Conitzer and Sandholm (2006), run in time polynomial in the number of pure strategies of the leader and follower. As Stackelberg games and their applications have become more prevalent, they are increasingly used to model complex scenarios where one or both players' strategy sets are *exponentially large in a natural representation of the problem*, in which case existing approaches are not computationally feasible. In this work, we consider such "large" games and ask whether there are computationally efficient algorithms for finding their Stackelberg equilibria.

Of course, such algorithms cannot exist without some assumptions on the problem structure. Here, we review the common assumptions and approaches for computing minimax-optimal solutions in large *zero-sum* games, where minimax strategies, Nash equilibria, and Stackelberg equilibria all coincide. Computing these equilibrium concepts in 2-player zero-sum games has received significant attention (Ahmadinejad et al. 2016; Behnezhad et al. 2018a, 2019, 2017a, 2018b; Conitzer and Sandholm 2006; Dudík et al. 2017; Freund and Schapire 1995; Garg et al. 2011; Immorlica et al. 2011; Von Neumann and Morgenstern 1945; Wang and Shroff 2017; Xu 2016). For large zero-sum games, two structural assumptions that have proven useful in computing a Nash equilibrium are the ability to efficiently optimize a linear function over the strategy space of each player (Ahmadinejad et al. 2016) and the ability to compute the best-response of each player against a mixed strategy of the other combined with a decomposibility property of the action set (Dudík et al. 2017).

In general-sum games, however, Stackelberg, Nash, and Minimax equilibria diverge. In general-sum games the leader can benefit from committing to a mixed strategy and obtain a more favorable Stackelberg equilibrium than any Nash equilibrium. From the algorithmic perspective, a Stackelberg equilibrium in a general-sum game can be computed efficiently when the game is small. That is, there are algorithms, such as the Multiple LPs approach of Conitzer and Sandholm (2006), that run in time poly($|S_L|, |S_F|$) where S_L and S_F are the set of pure strategies of the leader and follower, respectively. While this method is an efficient approach for computing a Stackelberg equilibrium in small games, it become computationally inefficient in many natural scenarios where the set of actions of the leader or follower is exponential in a natural representation of the game. Examples of such settings include games inspired by applications to security, where either the actions of the leader or the follower represent sets of edges in a graph. As opposed to the zero-sum case for which existence of certain structural properties are known to lead to efficient computation of the equilibrium concepts, computation of Stackelberg equilibrium in large general-sum games has remained mostly unexplored.

1.1   Our Results and Contributions

In light of the above, we examine the computational complexity of computing Stackelberg equilibria in large general-sum games. Specifically, we consider two classes of general-sum games, both of which demonstrate structural properties that under the zero-sum assumption would lead to efficient algorithms for computing the minimax optimal strategies. For the first class of games, we give an efficient algorithm for computing a Stackelberg Equilibrium. In the second class of games, we show that even approximating the Stackelberg equilibrium is NP-Hard. This drives home the main message of this work, that is *while there are natural large general-sum games where the Stackelberg Equilibria can be computed efficiently if the Nash equilibrium in its zero-sum form could be computed efficiently, in general, structural properties that allow for efficient computation of Nash equilibrium in zero-sum games are not sufficient for computing Stackelberg equilibria in general-sum games.*

In more details, the two classes of games we work with are as follow.

INCENTIVE GAMES. In Sect. 3, we introduce a class of games called INCENTIVE GAMES. In these games, the actions of the leader can be described as two-part actions, the first part of the action is an element of a set and the second part of the action is a set of incentives to the follower for playing certain actions.

As a motivating example consider a taxation scenario. In this setting, a government agency (e.g., IRS) takes the role of the leader and a taxpayer is the follower. A number of investments, indicated by the set E, are available to the taxpayer. Each investment e has a return of c_e to the taxpayer. The taxpayer invests in a package of investments $S \subseteq E$ that has the highest net payoff. The government agency is interested in taxing these investments in order to maximize the tax revenue. To do so, the agency allocates 1 unit of taxes between these investments. There are two types of taxation mechanisms. First is taxing an individual investment e by some amount x_e. The second is to provide tax relief v_S for a package of options the taxpayer has invested in. Examples of the second type of taxation mechanism include United States federal residential renewable energy tax credit that offers a tax break to individuals who have invested in home electric power storage, e.g., batteries, and home-generated renewable energy, e.g., solar panels, but no tax break to those who have invested in the former without the latter (EnergySage 2019; U.S. Department of Energy 2019). The tax revenue and the net payoff the taxpayer respectively receive from individual taxes x_e and combinatorial tax reliefs v_S when the taxpayer invests in investments S are $\sum_{e \in S} x_e - v_S$ and $\sum_{e \in S} (-x_e + c_e) + v_S$. It is not hard to see that these tax breaks play an essential role in the design of tax systems. Not only they increase the total tax revenue obtainable by a tax system (See Example 1 for an illustration) but they can also be used to incentivize the taxpayers to take actions that are more beneficial to the government.

More generally, we consider Stackelberg games and we consider a family of sets $\mathcal{S} \subseteq \{0,1\}^E$ and leader and follower element payoffs, C_e and c_e, respectively, for all $e \in E$. A pure strategy of the leader is to choose $e \in E$ and a vector of *incentives* $\mathbf{v} \in [0,1]^{|\mathcal{S}|}$, such that $\|\mathbf{v}\|_0 \in \text{poly}(|E|)$.[1] A pure strategy of the follower is to choose one set $S \in \mathcal{S}$. The payoff of the leader and follower are defined, respectively, by

$$\mathsf{U}_\mathsf{L}((e,\mathbf{v}),S) = 1_{e \in S} - v_S + C_e,$$

$$\mathsf{U}_\mathsf{F}((e,\mathbf{v}),S) = -1_{e \in S} + v_S + \sum_{e' \in S} c_{e'},$$

that is, the players receive non-zero-sum utilities from their individual choices, i.e., C_e and $\sum_{e \in S} c_{e'}$, and zero-sum utilities from choosing actions that intersect, i.e., $\pm 1_{e \in S}$, and from the incentives provided on the followers actions sets, i.e., $\pm v_S$.

We first note that when c_e and C_e are set to 0 for all $e \in E$, this game is zero-sum and can be efficiently solved when each player can compute its best-response to any choice of mixed strategy of the other player, i.e., optimize a linear function over the strategy space of the other player using existing results (Ahmadinejad et al. 2016; Dudík et al. 2017; Kalai and Vempala 2005).

When c_e and C_e are non-zero, we show that the leader can obtain a higher payoff equilibrium if it could make additional commitments in the form of incentives for the follower, i.e., can play non-zero $\mathbf{v}$. An interesting aspect of this game is that it is derived from a simple Stackelberg game model (where $\mathbf{v} = \mathbf{0}$) by adding zero-sum payoffs that *only benefit the follower*. Yet, the leader's payoff in the Stackelberg equilibrium of the new game is much higher than its payoff in the original game. Moreover, as we show in Theorem 1 there is a polynomial time algorithm for finding the Stackelberg equilibrium of such games when the leader can optimize a linear function over the actions of the follower, which is a similar condition to the ones used for computing Stackelberg equilibria in large zero-sum games (Ahmadinejad et al. 2016; Dudík et al. 2017; Kalai and Vempala 2005).

PERMUTED MATCHING GAME. In Sect. 4, we introduce a non-zero-sum game called PERMUTED MATCHING. In this game, there is a graph $G = (V, E)$ and a permutation $\pi : E \to E$. The set of pure strategies of the leader and follower is the set of all matchings in G. The goal of the leader is to maximize the intersection of its matching with the π-transformation of the matching of the follower. On the other hand, the goal of the follower is to maximize the intersection of the two matchings, with no regards to π. More formally, for $S \subseteq E$ we define $\pi(S) = \{e \in S | \pi(e)\}$. Then the utility of the leader and follower are defined, respectively, by

$$\mathsf{U}_\mathsf{L}(M_1, M_2) = |M_1 \cap \pi(M_2)| \qquad \mathsf{U}_\mathsf{F}(M_1, M_2) = |M_1 \cap M_2|.$$

It is not hard to see that, in this game, the problem of finding a best response for a player reduces to computing maximum weighted matching of G and can

---

[1] The sparsity requirement is such that the leader can communicate its strategy to the follower efficiently.

be solved in polynomial time. This would have been sufficient for getting a polynomial time algorithm for finding a Stackelberg equilibrium had the game been a zero-sum (Ahmadinejad et al. 2016; Dudík et al. 2017; Kalai and Vempala 2005). In a sharp contrast, however, we show that computing a Stackelberg equilibrium of this general-sum game is APX-hard, even though, we can compute player's best-response efficiently.

We obtain this hardness result via two reductions. First, we define the following computational problem:

π-TRANSFORMATION-IDENTICAL-MATCHING: Given a graph G and a mapping $\pi : E(G) \to E(G)$ over the edges of G, find a matching M of G that maximizes $|M \cap \pi(M)|$.

We next show that computing an approximate Stackelberg equilibrium of the PERMUTED MATCHING game is at least as hard as computing an approximate solution for the π-TRANSFORMATION-IDENTICAL-MATCHING problem. The crux of the argument is that if in an instance of the π-TRANSFORMATION-IDENTICAL-MATCHING problem there exists a matching which is almost identical to its π-transformation, then a Stackelberg equilibrium of the PERMUTED MATCHING game is closely related to that matching. Thus, any solution for the PERMUTED MATCHING game can be turned into a solution for π-TRANSFORMATION-IDENTICAL-MATCHING with almost the same quality. In the second step, we reduce the π-TRANSFORMATION-IDENTICAL-MATCHING problem to the MAXIMUM 3D MATCHING problem, which is known to be APX-hard (Petrank 1994).

We note that our results strengthen the existing hardness results of Letchford and Conitzer (2010), Li et al. (2016) that showed that computing Stackelberg equilibrium is *NP-hard*[2]. Our APX-hardness result shows that one cannot even approximate the Stackelberg equilibria of large games within *an arbitrary constant factor*, even when best-response can be efficiently computed.

1.2   Related Work

There is an extensive body of work investigating the complexity of solving Security games, which is a special case of computing Stackelberg equilibria (see *e.g.* Basilico et al. (2009), Behnezhad et al. (2017b), Letchford and Vorobeychik (2011), Tambe (2011), Xu (2016), Xu et al. (2014)).

Zero-Sum Games. Several algorithms have been proposed for finding the Stackelberg equilibria of a special case of security games called the spatio-temporal security games (Behnezhad et al. 2017b; Xu et al. 2014). These games are zero-sum by definition, where Stackelberg equilibria, Nash equilibria, and Minimax equilibria all coincide. In comparison, our work focuses on general-sum games.

---

[2] Interestingly, it is not hard to show that player best-response can also be computed efficiently in the games used by Letchford and Conitzer (2010), Li et al. (2016), although this was not central to their results.

Smaller General-Sum Games. Several works have introduced polynomial time algorithms for computing Stackelberg equilibria in games where *only one player's* strategy set is exponentially large (Kiekintveld et al. 2009; Xu 2016). A common approach used in this case is the Multiple LPs approach of Conitzer and Sandholm (2006) that runs in $\text{poly}(|S_L|, |S_F|)$. In this approach one creates a separate Linear Program for every action $y \in S_F$ of the follower, where the variables represent the probability assigned to the actions of the leader, the objective maximizes the expected payoff of the leader, and the constraints assure that action y is the best-response of the follower. This method can be implemented efficiently even when the leader's strategy set is exponentially large, e.g., when a separation oracle can be implemented efficiently. In comparison, our main computational result focus on settings where both the leader and follower have exponentially many strategies.

Existing Hardness Results. Letchford and Conitzer (2010) studied the computational complexity of extensive form games and proved a closely related hardness result.

They showed that computing Stackelberg equilibrium of a game is weakly NP-hard using a reduction from Knapsack. Interestingly, one can efficiently compute player best-response in their setting. In comparison, our hardness result improves over these results by showing that Stackelberg equilibria are hard to approximate within arbitrary constant factor even when player best-response can be computed efficiently.

A number of works have investigated the relationship between the Stackelberg equilibria and Nash equilibria of security games and have shown that computing a Stackelberg equilibrium is at least as hard as computing a Nash equilibrium of general-sum games. Korzhyk et al. (2011a) studied a special class of general-sum Stackelberg Security games where any Stackelberg Equilibrium is also a Nash equilibrium. This shows that computing Stackelberg equilibria is harder than computing Nash equilibria. Li et al. (2016) studied Bayesian Stackelberg Games, where there is additional uncertainty about the attacker and show that computing the Stackelberg equilibrium is hard, and introduce an exponential time algorithm for computing the Nash equilibria. In comparison, our work shows that Stackelberg equilibria are hard to approximate even when players best-response is easy to compute. That is, we show a gap between the computational complexity of approximating Stackelberg equilibrium of a general-sum game and that of its corresponding zero-sum variant.

2 Preliminaries

Throughout this paper, we study Stackelberg equilibria of large games. Our emphasis is on two player games and therefore we denote the players by L (leader) and F (follower). Let S_L and S_F be the set of actions (pure strategies) of players L and F. For a pair of pure strategies $x \in S_L$ and $y \in S_F$, we denote the payoffs of players L and F by $U_L(x, y)$ and $U_F(x, y)$, respectively. Similarly, for a pair of mixed strategies X and Y we denote the payoffs by

$$U_L(X, Y) = \mathbb{E}_{x \sim X, y \sim Y}[U_L(x, y)]$$
$$U_F(X, Y) = \mathbb{E}_{x \sim X, y \sim Y}[U_F(x, y)].$$

In Stackelberg games, the leader commits to a (possibly mixed) strategy X and plays this strategy. The follower then plays a best response against X, $b(X)$, according to her payoff function. Since the follower goes second its best-response is a deterministic action $b(X) = \max_y U_F(X, y)$. In case there is more than one best response for the follower, we assume she plays the one that maximizes the payoff of the leader. A pair of strategies X and y are in Stackelberg equilibrium if y is a best response of the follower against X and X maximizes the payoff of the leader, subject to the follower playing a best response.

3   INCENTIVE GAMES

In this section, we discuss a class of Stackelberg games where the leader has the ability to make additional commitments in the form of additional incentives to the follower. Recall that a natural scenario that can be addressed by this Stackelberg model is *taxation*. In this case the leader can set taxes on individual investments but can also provide tax breaks on bundles of investments that the tax payer has invested in. We first show how these additional combinatorial incentives can improve the leader's payoff significantly and then show polynomial time algorithms for computing a Stackelberg equilibrium in this model.

Let us first recall the definition of INCENTIVE GAMES. In this model, we consider a set of elements E, a family of its subsets $\mathcal{S} \subseteq \{0, 1\}^E$ and rewards C_e and c_e for all $e \in E$. The set of pure strategies of the leader is $S_L = E \times [0, 1]^{|\mathcal{S}|}$. That is, each action of the leader has two parts, the first part is an element $e \in E$ and the second part is a vector of *incentives* $\mathbf{v} \in [0, 1]^{|\mathcal{S}|}$. We assume that the leader is restricted to playing incentive vectors $\|\mathbf{v}\|_0 \in \text{poly}(|E|).^3$ The follower's strategy set is $S_F = \mathcal{S}$. The leader and follower payoffs are as follows.

$$U_L((e, \mathbf{v}), S) = 1_{e \in S} - v_S + C_e, \text{ and } U_F((e, \mathbf{v}), S) = -1_{e \in S} + v_S + \sum_{e' \in S} c_{e'}, \quad (1)$$

that is, the players receive non-zero-sum utilities from their individual choices, i.e., C_e and $\sum_{e \in S} c_{e'}$, and zero-sum utilities from choosing actions that intersect, i.e., $\pm 1_{e \in S}$, and from the incentives provided on the followers actions sets, i.e., $\pm v_S$. For ease of exposition and by the linearity of the payoffs, we denote a mixed strategy of the leader by $(\mathbf{x}, \mathbf{V})$, where x_e is the probability with which the first part of the leader's action is e and V_S is the expected incentive provided

---

[3] The sparsity requirement is such that the leader can communicate its strategy to the follower efficiently.

on action S in the second part of the leader's action. Note that in this case, the expected utilities of the leader and follower are

$$U_L((\mathbf{x}, \mathbf{V}), S) = \sum_{e \in S} x_e - V_S + \sum_{e \in E} x_e C_e \tag{2}$$

$$U_F((\mathbf{x}, \mathbf{V}), S) = \sum_{e \in S} (-x_e + c_e) + V_S. \tag{3}$$

Let us first consider a variation of INCENTIVE GAMES where the leader cannot provide additional incentives to the follower, i.e, $S_L = E \times \mathbf{0}$. The only difference between these games is that INCENTIVE GAMES are amended by allowing zero-sum non-negative payments $\mathbf{v}$ that benefit the follower solely. One might wonder if the commitment to make additional payments $\mathbf{v}$ to the follower can ever be beneficial to the leader. This is exactly what we demonstrate in the next example. That is, by allowing the leader to make additional zero-sum payoffs that only benefit the follower, we can obtain Stackelberg equilibria that have much higher payoff to the leader.

Example 1. Consider a graph instance in Fig. 1, E is the set of all edges, S is the set of all *s-t* paths, there are no edge payoff to the leader, i.e., $C_e = 0$ for all $e \in E$, and the edge payoff to the follower, c_e, are the *negative* of the edges costs that are denoted below each edge. That is, this is an instance where the follower is responding by choosing a shortest path with respect to the edge weights that correspond to the probability with which the leader plays them. Note that, since there are many parallel edges sb and at, the leaders optimal strategy (with or without additional commitment) is only supported on edges, sa, ab, and bt. It is not hard to see that without any additional commitment, the Stackelberg equilibrium involves the leader playing edge sa with probability $x_{sa} = 0.4$ and edge bt with probability $x_{bt} = 0.6$, and all other edges with probability $x_e = 0$. Note that in such a mixed strategy the follower chooses path a, b, t and the leader's payoff is 0.6. On the other hand, when the leader commits to providing additional incentive (or discount in the cost of a path) of $v_{sabt} = 0.2$ on path s, a, b, t, the follower best responds by choosing path s, a, b, t and the leader's payoff is 0.8.

Our main theorem in this section show that there is a polynomial time algorithm for finding the Stackelberg equilibrium of this modified game.

Theorem 1. *There is a polynomial time algorithm for finding a Stackelberg equilibrium if one can solve the following problem in polynomial time: Given* $\mathbf{x}$ *and value* W, *return* $S \in S$, *such that* $\sum_{e \in S} (-x_e + c_e) \leq -W$, *or return "None" if no such* $S \in S$ *exists.*[4]

At a high level, we show that a Stackelberg equilibrium, $(\mathbf{x}^*, \mathbf{V}^*)$, can be found by finding the optimal solution $(\mathbf{x}, \mathbf{0})$ (with no additional incentives) that

---

[4] An example of a game where this linear program can be solved efficiently is the shortest path game in Example 1.

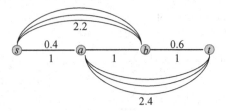

Fig. 1. An example where additional commitment increases the leader's payoff in the Stackelberg equilibrium. The follower's cost for each edge is denoted *below* the edge, i.e., $-c_{sa} = -c_{ab} = -c_{bt} = 1$ and $-c_{ab} = 2.2$ and $-c_{at} = 2.4$ for all the edges between s and b, and between a and t. The mixed strategy of the leader is denoted in gray *above* the edges.

involves maximizing the followers payoff of the best response, and then providing enough incentive on one of the follower's actions. In particular, we choose to provide incentive on the specific $S \in \mathcal{S}$ that constitutes the best response of the follower to the mixed strategy $(\mathbf{0}, \mathbf{0})$.

For the first step of this proof, we consider the following LP, which can be efficiently solved by the separation oracle given in Theorem 1,

$$
\begin{aligned}
\max_{\mathbf{x}, W} \ & W + \sum_{e \in E} x_e C_e \\
\forall S \in \mathcal{S}, \ & \sum_{e \in S}(-x_e + c_e) \leq -W.
\end{aligned}
\tag{4}
$$

Let $\mathbf{x}^*, W^*$ be the solution to the above LP. Furthermore, let $S^* = \arg\max_{S \in \mathcal{S}} \sum_{e \in S} c_e$, and consider the incentive commitments $V_{S^*}^* = -W^* - \sum_{e \in S^*}(-x_e + c_e)$, and $V_0^* = 0$ for all $S \neq S^*$. That is, we provide enough incentive on set S^* such that it becomes the best response for the follower.

To prove Theorem 1, we first prove a lower bound on the incentive needed to make an action the best response of the follower.

Lemma 1. *Let $(\mathbf{x}', \mathbf{V}')$ be any mixed strategy of the leader and let $S' = b(\mathbf{x}', \mathbf{V}')$ be the corresponding best response of the follower. Let $W' = -\max_{S \in \mathcal{S}} \sum_{e \in S}(-x'_e + c_e)$. We have,*

$$
V_{S'}' \geq -W' - \sum_{e \in S'}(-x'_e + c_e).
$$

In the interest of space, we skip the proof of Lemma 1 here. The reader may find a complete proof in the full-version of the paper.

Proof (of Theorem 1). Let $(\mathbf{x}^*, W^*)$ be the solution to Eq. 4. Let $S^* = \arg\max_{S \in \mathcal{S}} \sum_{e \in S} c_e$, and let $V_{S^*}^* = -W^* - \sum_{e \in S^*}(-x_e^* + c_e)$, and $V_S^* = 0$ for all $S \neq S^*$. It is clear that $b(\mathbf{x}^*, \mathbf{V}^*) = S^*$. Here, we show that $(\mathbf{x}^*, \mathbf{V}^*)$ is indeed the optimal leader strategy.

For any leader strategy $(\mathbf{x}', \mathbf{V}')$, let $S' = b(\mathbf{x}', \mathbf{V}')$ be the follower's best response. Moreover, let $W' = -\max_{S \in \mathcal{S}} \sum_{e \in S}(-x'_e + c_e)$. We have

$$U_L((\mathbf{x}^*, \mathbf{V}^*), S^*) = \sum_{e \in S^*} x_e^* + \sum_{e \in E} x_e^* C_e^* - V_{S^*}^* \tag{5}$$

$$= \sum_{e \in S^*} x_e^* + \sum_{e \in E} x_e^* C_e^* + W^* + \sum_{e \in S^*}(-x_e^* + c_e) \tag{6}$$

$$= \sum_{e \in E} x_e^* C_e^* + W^* + \sum_{e \in S^*} c_e \tag{7}$$

$$\geq \sum_{e \in E} x_e' C_e' + W' + \sum_{e \in S'} c_e, \tag{8}$$

where the second equation is by the definition of $V_{S^*}^*$) and the last inequality follows by the fact that $(\mathbf{x}', W')$ form a valid solution for the LP in Eq. 4, for which $(\mathbf{x}^*, W^*)$ is the optimal solution and the fact that S^* is chosen to maximize $\sum_{e \in S^*} c_e$.

Using Lemma 1 on the value of $V_{S'}'$, we have

$$U_L((\mathbf{x}', \mathbf{V}'), S') = \sum_{e \in S'} x_e' + \sum_{e \in E} x_e' C_e' - V_{S'}' \tag{9}$$

$$\leq \sum_{e \in S'} x_e' + \sum_{e \in E} x_e' C_e' + W' + \sum_{e \in S'}(-x_e' + c_e) \tag{10}$$

$$= \sum_{e \in E} x_e' C_e' + W' + \sum_{e \in S'} c_e. \tag{11}$$

Equations 8 and 11 complete the proof.

4   The PERMUTED MATCHING Game

In this section, we introduce a large but structured general-sum Stackelberg game, called PERMUTED MATCHING, and examine the computational complexity of computing its Stackelberg equilibrium. We show two sets of results for this game. In Sect. 4.1, we show that this problem is APX-hard. This implies that unlike zero-sum games, finding a Stackelberg equilibrium is computationally hard even if best-response oracles are provided.

The PERMUTED MATCHING game is defined as follows. Consider the leader and follower, L and F. Consider a multigraph $G = (V, E)$ and a one-to-one mapping (permutation) $\pi : E \to E$. Note that π may take different values on parallel edges of a multi-graph. In the remainder of this section, we refer to a multi-graph G as a *graph*. In PERMUTED MATCHING, the set of pure strategies of both players is the set of all matchings in G. Given matchings M_L and M_F played by the leader and follower, respectively, we define

$$U_L(M_L, M_F) = |M_L \cap \pi(M_F)|, \text{ and } \quad U_F(M_L, M_F) = |M_L \cap M_F|,$$

where for a set $S \subseteq E$, we define $\pi(S) = \{e \in S | \pi(e)\}$. Note that G and π are fixed and known to both players.

Let us highlight an important aspect of our hardness result in advance. As the next observation shows, the strategy space of the players in PERMUTED MATCHING, though large, is very structured. At a high level, the reward of each player is a linear function of the action of the other and each player can efficiently optimize a linear function over the strategy space of the other player, for example, each player can compute a best-response to a mixed strategy of the other.

Observation 1. *There is a polynomial time algorithm such that for every vector* $\mathbf{w} \in [0,1]^{|E|}$ *finds a strategy of the players whose corresponding representation vector* $\mathbf{v}$ *maximizes* $\mathbf{v} \cdot \mathbf{w}$.

Proof (Sketch). This problem reduces to computing a maximum weighted matching of a graph with edge weights w_e for all $e \in E$, which can be performed efficiently Cormen (2009).

In a zero-sum game, existence of such a structure leads to efficient algorithms for computing the Nash or Stackelberg equilibria Ahmadinejad et al. (2016), Dudík et al. (2017), Kalai and Vempala (2005). On the other hand, our APX-hardness result for the PERMUTED MATCHING game shows that existence of this structure does not necessarily lead to efficient algorithms for computing Stackelberg equilibria in general-sum games. With this in mind, we present our hardness results next.

4.1   Hardness of Approximation

In this section, we show that it is impossible to approximate a Stackelberg equilibrium of the PERMUTED MATCHING game in polynomial time within an arbitrarily small constant factor unless P=NP.

Before we proceed to the proof, we define an auxiliary problem and show a hardness result for this problem. Then, we take advantage of this hardness result and show that computing a Stackelberg equilibrium of PERMUTED MATCHING is APX-hard. We call the intermediary problem π-TRANSFORMATION-IDENTICAL-MATCHING and define it as follows: In the input comes an unweighted undirected graph $G = (V, E)$ and a permutation $\pi : E \to E$. The goal is to find A matching M of G that maximizes $|M \cap \pi(M)|$. For an instance I of π-TRANSFORMATION-IDENTICAL-MATCHING, we denote by Opt(I) the optimal solution to I and refer to the value of this solution by Val(I).

We show that π-TRANSFORMATION-IDENTICAL-MATCHING has a *hard gap* at *gap location* 1. That is, it is NP-hard to decide whether for a given graph G with n vertices and a function π, the solution of the π-TRANSFORMATION-IDENTICAL-MATCHING problem is exactly equal to $n/2$ or at most $(1-\varepsilon)n/2$ for some $\varepsilon > 0$.

Lemma 2. *There exists an* $\varepsilon > 0$ *such that it is NP-hard to decide whether the solution of the* π-TRANSFORMATION-IDENTICAL-MATCHING *problem is exactly* equal to $n/2$ or less than $(1 - \varepsilon)n/2$ where n is the number of the vertices of the input graph.

In the interest of space, we defer the proof of Lemma 2 to the full-version of the paper and just show how this lemma can be used to prove the main result of this section.

Theorem 2. *Computing a Stackelberg equilibrium of* PERMUTED MATCHING *is APX-hard.*

Proof. More generally, we show that approximating a Stackelberg equilibrium of the PERMUTED MATCHING game has a hard gap at gap location 1. This immediately implies a hardness of approximation. We show this by a reduction from the π-TRANSFORMATION-IDENTICAL-MATCHING problem. Suppose we are given an instance $\mathsf{I} = \langle G, \pi \rangle$ of the π-TRANSFORMATION-IDENTICAL-MATCHING problem and wish to decide for some $\varepsilon' > 0$, whether the solution of this problem achieves a value that is exactly $n/2$ or is bounded above by $(1 - \varepsilon')n/2$ where n is the size of G. Based on I, we construct an instance $\mathsf{Cor}(\mathsf{I})$ of the PERMUTED MATCHING game with the same graph G and permutation π and seek to find a Stackelberg equilibrium in this game. Note that by definition, $\mathsf{Val}(\mathsf{I})$ is equal to $n/2$ if and only if G contains a perfect matching that is identical to its π-transformation. Otherwise, $\mathsf{Val}(\mathsf{I})$ is at most $(1 - \varepsilon')n/2$ and thus any matching of G shares no more than $(1 - \varepsilon')n/2$ edges with its π-transformation.

Since for small enough ε', it is NP-hard to distinguish the two cases (Lemma 2), we show that it is NP-hard to approximate a Stackelberg equilibrium of the leader in $\mathsf{Cor}(\mathsf{I})$. If $\mathsf{Val}(\mathsf{I}) = n/2$, then there exists a perfect matching in G that is identical to its π-transformation. Thus, if both players play this matching in $\mathsf{Cor}(\mathsf{I})$, they both get a payoff of $n/2$. Notice that $n/2$ is the maximum possible payoff for any player in this game, therefore, such a strategy pair is a Stackelberg equilibrium. Hence, in case $\mathsf{Val}(\mathsf{I}) = n/2$, the leader achieves a payoff of $n/2$ in a Stackelberg equilibrium of the corresponding PERMUTED MATCHING game.

Now, suppose for $\varepsilon < \varepsilon'/13$ we have a $1 - \varepsilon$ approximation solution for $\mathsf{Cor}(\mathsf{I})$. If $\mathsf{Val}(\mathsf{I}) = n/2$, then the payoff of the leader in an exact solution of $\mathsf{Cor}(\mathsf{I})$ is $n/2$ and therefore a $1 - \varepsilon$ approximation solution guarantees a payoff of at least $n(1 - \varepsilon)/2$ for the leader. Let the strategies of the leader and follower be X and y in such a solution. Therefore, $\mathsf{U_L}(X, y) \geq n(1 - \varepsilon)/2$. Notice that X may be a mixed strategy, but we can assume w.l.g that y is a pure strategy since there always exists a best response for the follower which is pure. Also, let y^* be the π-transformation of strategy y. Let for two matchings x and y, $\mathsf{common}(x, y)$ denote the number of edges that x and y have in common and define $\mathsf{dist}(x, y) = |x| + |y| - 2\mathsf{common}(x, y)$. Recall that the payoff of the leader in this game can be formulated as $\mathbb{E}_{x \sim X}[\mathsf{common}(x, y^*)]$. Since this value is at least $n(1 - \varepsilon)/2$ we have:

$$\mathbb{E}_{x \sim X}[\mathsf{common}(x, y^*)] = \mathsf{U_L}(X, y) \geq n(1 - \varepsilon)/2$$

and thus

$$\begin{aligned}
\mathbb{E}_{x \sim X}[\mathsf{dist}(x, y^*)] &= \mathbb{E}_{x \sim X}[|x| + |y^*| - 2\mathsf{common}(x, y^*)] \\
&\leq \mathbb{E}_{x \sim X}[n - 2\mathsf{common}(x, y^*)] \\
&= n - 2\mathbb{E}_{x \sim X}[\mathsf{common}(x, y^*)] \\
&\leq n - 2n(1 - \varepsilon)/2 \\
&= n\varepsilon.
\end{aligned} \tag{12}$$

Inequality (12) shows that y^* is very similar (in expectation) to a random matching drawn from strategy X. This intuitively implies that pure strategies of X should have a considerable amount of edges in common. It follows from the definition that for three matchings x, y, and z we have $\mathsf{dist}(x, z) \leq \mathsf{dist}(x, y) + \mathsf{dist}(y, z)$. Therefore, we have

$$\begin{aligned}
\mathbb{E}_{x \sim X, x' \sim X}[\mathsf{dist}(x, x')] &\leq \mathbb{E}_{x \sim X, x' \sim X}[\mathsf{dist}(x, y^*) + \mathsf{dist}(y^*, x')] \\
&= \mathbb{E}_{x \sim X, x' \sim X}[\mathsf{dist}(x, y^*) + \mathsf{dist}(x', y^*)] \\
&= \mathbb{E}_{x \sim X}[\mathsf{dist}(x, y^*)] + \mathbb{E}_{x' \sim X}[\mathsf{dist}(x', y^*)] \\
&= 2\mathbb{E}_{x \sim X}[\mathsf{dist}(x, y^*)] \\
&\leq 2\varepsilon n
\end{aligned} \tag{13}$$

Recall that the payoff of the follower is determined by the number of edges his matching shares with that of the leader. Moreover, since $\mathsf{U_L}(X, y) \geq n(1 - \varepsilon)/2$, this implies that $\mathbb{E}_{x \sim X}[|x|] \geq n(1 - \varepsilon)/2$. What Inequality (13) implies is that if the follower plays X instead of y, he gets a payoff of at least $\mathbb{E}_{x \sim X}|x| - 2\varepsilon n \geq (1 - 5\varepsilon)n/2$ against X. In other words $\mathsf{U_F}(X, X) \geq (1 - 5\varepsilon)n/2$. Since y is a best response of the follower against the leader's strategy, we have $\mathsf{U_F}(X, y) \geq \mathsf{U_F}(X, X) \geq (1 - 5\varepsilon)n/2$ and thus

$$\begin{aligned}
\mathbb{E}_{x \sim X}[\mathsf{common}(x, y)] &= \mathsf{U_F}(X, y) \\
&\geq \mathsf{U_F}(X, X) \\
&= \mathbb{E}_{x \sim X, x' \sim X}[\mathsf{common}(x, x')] \\
&\geq (1 - 5\varepsilon)n/2.
\end{aligned}$$

Hence

$$\begin{aligned}
\mathbb{E}_{x \sim X}[\mathsf{dist}(x, y)] &= \mathbb{E}_{x \sim X}[|x| + |y| - 2\mathsf{common}(x, y)] \\
&\leq \mathbb{E}_{x \sim X}[n - 2\mathsf{common}(x, y)] \\
&= n - 2\mathbb{E}_{x \sim X}[\mathsf{common}(x, y)] \\
&\leq n - 2(1 - 5\varepsilon)n/2 \\
&\leq 5\varepsilon n.
\end{aligned} \tag{14}$$

Combining Inequalities (12) and (14) yields

$$\mathsf{dist}(y, y^*) \leq \mathbb{E}_{x \sim X}[\mathsf{dist}(x, y)] + \mathbb{E}_{x \sim X}[\mathsf{dist}(x, y^*)] \leq 6\varepsilon n.$$

Therefore, we have $\mathsf{common}(y, y^*) \geq |y^*| - 6\varepsilon n$ and since $|y^*| \geq (1 - \varepsilon)n/2$ we have $\mathsf{common}(y, y^*) \geq (1 - 13\varepsilon)n/2 > (1 - \varepsilon')n/2$. If $\mathsf{Val}(\mathsf{I}) \neq n/2$, then $\mathsf{Val}(\mathsf{I})$

is bounded by $(1 - \varepsilon')n/2$. Therefore, $\mathsf{common}(y, y^*) > (1 - \varepsilon')n/2$ holds if and only if $\mathsf{Val}(I) = n/2$. Thus, an approximation solution for $\mathsf{Cor}(I)$ within a factor $(1 - \varepsilon) > (1 - \varepsilon'/13)$ can be used to decide if the solution of I is $n/2$ or bounded by $(1 - \varepsilon')n/2$. This implies a hard gap for the π-TRANSFORMATION-IDENTICAL-MATCHING problem at gap location 1.

References

Ahmadinejad, A., Dehghani, S., Hajiaghay, M., Lucier, B., Mahini, H., Seddighin, S.: From duels to battlefields: computing equilibria of Blotto and other games. In: Thirtieth AAAI Conference on Artificial Intelligence (2016)

Basilico, N., Gatti, N., Amigoni, F.: Leader-follower strategies for robotic patrolling in environments with arbitrary topologies. In: Proceedings of the 8th International Conference on Autonomous Agents and Multiagent Systems, vol. 1, pp. 57–64. International Foundation for Autonomous Agents and Multiagent Systems (2009)

Behnezhad, S., et al.: From battlefields to elections: winning strategies of Blotto and auditing games. In: Proceedings of the Twenty-Ninth Annual ACM-SIAM Symposium on Discrete Algorithms, pp. 2291–2310. SIAM (2018a)

Behnezhad, S., Blum, A., Derakhshan, M., Hajiaghayi, M., Papadimitriou, C.H., Seddighin, S.: Optimal strategies of Blotto games: beyond convexity. In: Proceedings of the 2019 ACM Conference on Economics and Computation. ACM (2019)

Behnezhad, S., Dehghani, S., Derakhshan, M., Hajiaghayi, M., Seddighin, S.: Faster and simpler algorithm for optimal strategies of Blotto game. In: Thirty-First AAAI Conference on Artificial Intelligence (2017a)

Behnezhad, S., Derakhshan, M., Hajiaghayi, M., Seddighin, S.: Spatio-temporal games beyond one dimension. In: Proceedings of the 2018 ACM Conference on Economics and Computation. ACM (2018b)

Behnezhad, S., Derakhshan, M., Hajiaghayi, M., Slivkins, A.: A polynomial time algorithm for spatio-temporal security games. In: Proceedings of the 2017 ACM Conference on Economics and Computation. ACM (2017b)

Conitzer, V., Sandholm, T.: Computing the optimal strategy to commit to. In: Proceedings of the 7th ACM Conference on Economics and Computation (EC), pp. 82–90. ACM (2006)

Cormen, T.H.: Introduction to Algorithms. MIT Press, Cambridge (2009)

Dudík, M., Haghtalab, N., Luo, H., Schapire, R.E., Syrgkanis, V., Vaughan, J.W.: Oracle-efficient online learning and auction design. In: Proceedings of the 58th Symposium on Foundations of Computer Science (FOCS) (2017)

EnergySage: Using the solar investment tax credit for energy storage (2019). https://www.energysage.com/solar/solar-energy-storage/energy-storage-tax-credits-incentives/

Freund, Y., Schapire, R.E.: A desicion-theoretic generalization of on-line learning and an application to boosting. In: Vitányi, P. (ed.) EuroCOLT 1995. LNCS, vol. 904, pp. 23–37. Springer, Heidelberg (1995). https://doi.org/10.1007/3-540-59119-2_166

Garg, J., Jiang, A.X., Mehta, R.: Bilinear games: polynomial time algorithms for rank based subclasses. In: Chen, N., Elkind, E., Koutsoupias, E. (eds.) WINE 2011. LNCS, vol. 7090, pp. 399–407. Springer, Heidelberg (2011). https://doi.org/10.1007/978-3-642-25510-6_35

Immorlica, N., Kalai, A.T., Lucier, B., Moitra, A., Postlewaite, A., Tennenholtz, M.: Dueling algorithms. In: Proceedings of the Forty-Third Annual ACM Symposium on Theory of Computing, pp. 215–224. ACM (2011)

Kalai, A., Vempala, S.: Efficient algorithms for online decision problems. J. Comput. Syst. Sci. **71**(3), 291–307 (2005)

Kiekintveld, C., Jain, M., Tsai, J., Pita, J., Ordóñez, F., Tambe, M.: Computing optimal randomized resource allocations for massive security games. In: Proceedings of the 8th International Conference on Autonomous Agents and Multiagent Systems, vol. 1, pp. 689–696. International Foundation for Autonomous Agents and Multiagent Systems (2009)

Korzhyk, D., Conitzer, V., Parr, R.: Security games with multiple attacker resources. In: IJCAI Proceedings of the International Joint Conference on Artificial Intelligence, vol. 22, p. 273 (2011a)

Korzhyk, D., Yin, Z., Kiekintveld, C., Conitzer, V., Tambe, M.: Stackelberg vs. nash in security games: an extended investigation of interchangeability, equivalence, and uniqueness. J. Artif. Intell. Res. (JAIR) **41**, 297–327 (2011b)

Letchford, J., Conitzer, V.: Computing optimal strategies to commit to in extensive-form games. In: Proceedings of the 11th ACM Conference on Electronic Commerce, pp. 83–92. ACM (2010)

Letchford, J., Vorobeychik, Y.: Computing randomized security strategies in networked domains. In: Workshops at the Twenty-Fifth AAAI Conference on Artificial Intelligence (2011)

Li, Y., Conitzer, V., Korzhyk, D.: Catcher-evader games. In: Proceedings of the Twenty-Fifth International Joint Conference on Artificial Intelligence, IJCAI 2016, 9–15 July 2016, New York, NY, USA, pp. 329–337 (2016). http://www.ijcai.org/Abstract/16/054

Petrank, E.: The hardness of approximation: gap location. Comput. Complex. **4**(2), 133–157 (1994)

Tambe, M.: Security and Game Theory: Algorithms, Deployed Systems, Lessons Learned. Cambridge University Press, Cambridge (2011)

U.S. Department of Energy: Residential Renewable Energy Tax Credit (2019). https://www.energy.gov/savings/residential-renewable-energy-tax-credit

Von Neumann, J., Morgenstern, O.: Theory of Games and Economic Behavior. Princeton University Press, Princeton (1945)

Wang, S., Shroff, N.: Security game with non-additive utilities and multiple attacker resources (2017). arXiv preprint: arXiv:1701.08644

Xu, H.: The mysteries of security games: equilibrium computation becomes combinatorial algorithm design. In: Proceedings of the 2016 ACM Conference on Economics and Computation, pp. 497–514. ACM (2016)

Xu, H., Fang, F., Jiang, A.X., Conitzer, V., Dughmi, S., Tambe, M.: Solving zero-sum security games in discretized spatio-temporal domains. In: Twenty-Eighth AAAI Conference on Artificial Intelligence (2014)

Network Games and Congestion Games

The Impact of Tribalism
on Social Welfare

Seunghee Han, Matvey Soloviev$^{(\boxtimes)}$, and Yuwen Wang

Cornell University, Ithaca, NY, USA
msoloviev@cs.cornell.edu, ywang@math.cornell.edu

Abstract. We explore the impact of mutual altruism among the players belonging to the same set – their *tribe* – in a partition of all players in arbitrary strategic games upon the quality of equilibria attained. To this end, we introduce the notion of a τ-*tribal extension* of an arbitrary strategic game, in which players' subjective cost functions are updated to reflect this, and the associated *Price of Tribalism*, which is the ratio of the social welfare of the worst Nash equilibrium of the tribal extension to that of the optimum of social welfare. We show that in a well-known game of friendship cliques, network contribution games as well as atomic linear congestion games, the Price of Tribalism is higher than the Price of Anarchy of either the purely selfish players or fully altruistic players (i.e. ones who seek to maximise the social welfare). This phenomenon is observed under a variety of equilibrium concepts. In each instance, we present upper bounds on the Price of Tribalism that match the lower bounds established by our example.

1 Introduction

According to the standard narrative around the concept of Nash equilibrium, one of its great contributions is that it shed light on why multi-agent systems in the real world often "race to the bottom", or otherwise fail to exhibit behaviour anywhere near the social optimum. Perhaps fortunately, the picture suggested by the theory is not always reflected in real-world systems, which often appear to stabilise in states that are better than self-interested Nash equilibria. On the other hand, many a carefully designed political and economic system fails to deliver on its theoretical promises in reality.

Beside the "spherical cow" class of model-reality disagreements such as computational limitations and insufficient rationality of the agents, the good half of this discrepancy is often rationalised by saying that players exhibit a degree of

We would like to thank Jerry Anunrojwong, Ioannis Caragiannis, Artur Gorokh, Bart de Keijzer, Bobby Kleinberg, Guido Schaefer and Éva Tardos, as well as the anonymous reviewers, for helpful feedback and discussions regarding the paper. M.S. was supported by NSF grants IIS-1703846 and IIS-1718108, ARO grant W911NF-17-1-0592, and a grant from the Open Philanthropy project. Y.W.'s research was partially supported by NSF grant DMS-1645643. For a full version of this paper, please see https://arxiv.org/abs/1907.06862.

D. Fotakis and E. Markakis (Eds.): SAGT 2019, LNCS 11801, pp. 185–199, 2019.
https://doi.org/10.1007/978-3-030-30473-7_13

altruism – that is, they seek to optimise not just their own welfare but some weighted combination of it with the sum of the welfare of all players. This approach may partially explain the more favourable dynamics we observe. However, recent results [CKK+10, CdKKS14] demonstrate that in some cases, altruism can give rise to equilibria that are even worse than those that exist if all players are purely self-interested.

Once identified, this might not seem that unrealistic: for instance, all historical industrial revolutions, implemented by arguably selfish agents seeking to maximise profit, were accompanied by a temporary dip in social welfare [Szr04]. A society whose members are altruistic but do not coordinate may therefore never have implemented these changes, remaining stuck in the pre-industrial local optimum without electricity, mass production or modern medicine. Looking at the workings of the lower bounds for "altruistic anarchy", we find that the ways in which altruistic players get stuck in local optima appear quite different from those enabling bad selfish equilibria. Could it then be that there are realistic settings in which both mechanisms occur together?

Tribalism and political polarisation are often argued to be a feature of human interactions that predates those interactions even being, strictly speaking, human at all, but according to news media and sociological analyses alike, their impact on public life in Western societies has been steadily increasing [Eco17, CRF+11, FA08, BB07]. A tribalistic agent, broadly speaking, is concerned with the welfare of other agents belonging to the same tribe, rather than the overall social welfare of everyone participating in the system. A game or system in which the agents are tribalistic, then, could be said to exhibit both altruism (within a tribe) and selfishness (in how the tribes interact with each other). Inspired by the failure of real-world systems, we set out to investigate if this mixture of altruism and selfishness could in fact lead to even worse outcomes than either altruism or selfishness alone.

We find that this is indeed the case. In the following sections, we will show that tribalism leads to a greater Price of Anarchy than either altruism or selfishness in a folklore model of friend cliques, network contribution games [AH12] and atomic linear routing games [Ros73b]. In each case, we also present upper bounds for the tribal Price of Anarchy that match the lower bounds demonstrated by our examples.

2   Main Results

2.1   Definition of Tribalism

Our definition of games in which the players exhibit tribalism is designed to resemble the definition of α-*altruistic extensions* that [CdKKS14] make for their analysis of universal altruism. An analogous definition can be made for utility-maximisation games. We will represent cost-minimisation games G as the triple $(N, (\Sigma_i)_{i \in N}, (c_i)_{i \in N})$, where N is the set of players, Σ_i are the strategies available to player i and $c_i(\mathbf{s})$ is the cost for player i when the vector of strategies chosen by all players is $\mathbf{s} \in \prod_{i \in N} \Sigma_i$. We will use $(t; \mathbf{s}_{-i})$ to denote the vector $\mathbf{s}$ with player i's entry replaced with t.

Definition 1. *Suppose $G = (N, (\Sigma_i)_{i \in N}, (c_i)_{i \in N})$ is a finite cost-minimisation game. Let $\tau : N \to \mathbb{N}$ be a function that assigns each player a unique tribe, identified by a natural number. The τ-tribal extension of G is the cost-minimisation game $G^\tau = (N, (\Sigma_i)_{i \in N}, (c_i^\tau)_{i \in N})$, where the cost experienced by every player is the sum of costs of all players in the same tribe in the original game: for every $i \in N$ and $s \in \Sigma = \Sigma_1 \times \cdots \times \Sigma_n$,*

$$c_i^\tau(s) = \sum_{j \in N : \tau(i) = \tau(j)} c_j(s).$$

When the partition function τ is constant, our definition agrees with the one in [CdKKS14] with $\alpha = 1$, and we say the players are *(fully) altruistic*. When $\tau(i) \neq \tau(j)$ for all $i \neq j$, $G^\tau = G$ and we say the players are *selfish*.

We then define the Price of Tribalism for a class of games $\mathcal{G}$ and class of partition functions $\{\mathcal{T}_G\}_{G \in \mathcal{G}}$ as the supremum of ratios between the social welfare $C_G(s) = \sum_{i \in N} c_i(s)$ of any Nash equilibrium *of any τ-tribal extension* for $\tau \in \mathcal{T}_G$ and the social optimum, i.e. the lowest attainable social cost. In other words, the PoT captures how bad a "tribal equilibrium" can get for any game in $\mathcal{G}$ and any pattern of tribal allegiance of the players therein. The definition for utility-maximisation games is again analogous.

Definition 2. *The pure (resp. correlated, strong, mixed...) tribal Price of Anarchy (Price of Tribalism, PoT) of a class of games $\mathcal{G}$ and class of partition functions for each game $\mathcal{T} = \{\mathcal{T}_G\}_{G \in \mathcal{G}}$ is*

$$\text{PoT}(\mathcal{T}, \mathcal{G}) = \sup_{G \in \mathcal{G}, \tau \in \mathcal{T}_G} \frac{\sup_{s \in S_{G^\tau}} C_G(s)}{\inf_{s \in \Sigma} C_G(s)},$$

where S_{G^τ} is the set of pure (correlated, strong, mixed...) Nash equilibria of G^τ.

We can control the tribal structures that we want to consider by choosing an appropriate class $\mathcal{T}$ of partition functions. We will denote the class of all functions which sort the players into exactly k tribes by $\mathcal{T}_G^{(k)} = \{\tau : |\tau(N)| = k\}$, and that of all possible functions as $\mathcal{T}_G^{(*)} = \bigcup_{i=1}^{\infty} \mathcal{T}_G^{(k)}$. The Price of Anarchy given full altruism (i.e. in the game G^1 of [CdKKS14]) then equals $\text{PoT}(\mathcal{T}^{(1)}, \mathcal{G})$.

2.2   Impact of Tribalism on Known Games

We first consider a folklore game which is often invoked as a simple model of friendship. This game can be seen as a special case of the *party affiliation game* of [FPT04], with positive payoffs only. Players choose to associate with one of two cliques A and B. Each pair of players has an associated utility of being friends u_{ij}, which they can only enjoy if they choose to associate with the same clique. For this game, a folklore bound we revisit shows that the pure Price of Anarchy is 2.

Game	PoA	Altruistic PoA	PoT
Social grouping with 2 cliques	2 (folklore)	2 (Theorem 5)	3 (Theorem 6)
Social grouping with k cliques [HSW19]	k (Theorem B.1)	k (Theorem B.1)	$2k - 1$ (Theorem B.2)
Network contribution with additive rewards	1 [AH12]	1 ([AH12], Cor 1)	2 (Theorem 7)
Network contribution with convex rewards	2 [AH12]	2 (Theorem B.3 [HSW19])	4 (Theorem 8)
Atomic linear routing	5/2 [CK05, AAE13]	3 [CKK+10]	4 (Theorem 9)

Theorem 1 *(Theorems 5; 6; Theorem B.2 in [HSW19]). The pure Price of Anarchy for the social 2-grouping game $\mathcal{F}_2$ under full altruism satisfies*

$$\mathrm{PoT}(\mathcal{T}^{(1)}, \mathcal{F}_2) = 2$$

as well. However, the pure Price of Tribalism is

$$\mathrm{PoT}(\mathcal{T}^{(*)}, \mathcal{F}_2) = 3.$$

More generally, in the social k-grouping game (i.e. with k cliques) with at least k tribes $\mathcal{F}_k$,

$$\mathrm{PoT}(\mathcal{T}^{(1)}, \mathcal{F}_k) = k \quad and \quad \mathrm{PoT}(\mathcal{T}^{(*)}, \mathcal{F}_k) = 2k - 1,$$

while the pure Price of Anarchy is k.

A more involved model of friendship networks was first described by Anshelevich and Hoefer [AH12]. In the *network contribution game* $\mathcal{N}_{\mathcal{F}}$, we are given a social graph of vertices representing players and edges representing potential relationships between them. Each player has a fixed budget b_i, which they seek to allocate among the edges adjacent to them. They then receive a payoff from each edge based on a symmetric function $f_e(x, y) = f_e(y, x) \in \mathcal{F}$ of their own and the other player's contribution to that edge.

In the original paper, the authors show different bounds on the Price of Anarchy for this game depending on the form the functions f_e can take: among others, for $f_e(x, y) = c_e(x + y)$ (we call this class of games $\mathcal{N}_+$), they show a PoA of 1, and when $f_e(x, y)$ satisfies $f_e(x, 0) = 0$ and each f_e is convex in each coordinate (denoted by $\mathcal{N}_C$), the PoA is 2. Moreover, instead of pure Nash equilibria, they invoke *pairwise* ones, which are resilient against any pair of associated players deviating together. In the presence of tribalism, we demonstrate that both of these bounds deteriorate.

Theorem 2 *(Theorems 7; 8; Theorem B.3 in [HSW19]). The pure and pairwise Price of Tribalism for the network contribution game with additive rewards is*

$$\mathrm{PoT}(\mathcal{T}^{(*)}, \mathcal{N}_+) = 2.$$

The pure and pairwise Price of Tribalism for each of the network contribution games with coordinate-convex reward functions is

$$\mathrm{PoT}(\mathcal{T}^{(*)}, \mathcal{N}_C) = 4.$$

Meanwhile, the altruistic Price of Anarchy is still 1 for $\mathcal{N}_+$ and 2 for $\mathcal{N}_C$.

Finally, we will turn our attention to atomic linear routing games $\mathcal{R}$ [Ros73a], a popular class of games that model a set of players seeking to each establish a point-to-point connection over a shared network (such as the internet or roads) represented by a graph, where the cost to all players using an edge increases linearly with the number of players using it. In the case of selfish behaviour, these games are well-known to exhibit a pure Price of Anarchy of 5/2 [CK05, AAE13]. Caragiannis et al. [CKK+10] show that in the case of universal altruism, this deteriorates to $\text{PoT}(\mathcal{T}^{(1)}, \mathcal{R}) = 3$. We show matching lower and upper bounds that demonstrate that in the face of tribalism, significantly worse equilibria can arise. It should be noted that in fact, our result applies to the more general class of atomic linear congestion games.

Theorem 3 *Theorem (9). The Price of Tribalism for the atomic linear routing game is*

$$\text{PoT}(\mathcal{T}^{(2)}, \mathcal{R}) = \text{PoT}(\mathcal{T}^{(*)}, \mathcal{R}) = 4.$$

3 Related Work

A number of papers [CKK+10, CK08, CdKKS14] concern themselves with how the Price of Anarchy is affected if players exhibit some concern for social welfare. In [CKK+10], the players of an atomic linear congestion game are taken to optimise for a linear combination of their own utility and social welfare, and this is shown to result in a greater price of anarchy. We draw particular inspiration from the later treatment by [CdKKS14]. The authors of this paper define a generalisation of this construction, called the *α-altruistic extension* of a game, where α is a parameter determining the level of altruism.

A separate line of work (e.g. [KLS01, AKT08]) considers altruism as a more localised phenomenon, where the degree of mutual concern between players is captured by a weighted graph called the *social context graph*. The model obtained by augmenting a given game with arbitrary social context graphs defines a strictly larger class of games than the tribal extensions of this paper, which accordingly may enable still greater Prices of Anarchy – e.g. 17/3 for atomic linear congestion games in [BCFG13]. It is in this context that [RS13] defines a generalisation of Roughgarden's *smoothness* [Rou09], which we use a close analogue of. A more comprehensive discussion of related work can be found in the extended version of this paper [HSW19].

4 Examples

4.1 Social Grouping Games

A simple folklore example of a game with nontrivial and clearly suboptimal Nash equilibria is the *social (k-)grouping game*. In this game, the players are nodes of a directed graph, with edge weights $u_{ij} \geq 0$ to be thought of as the benefit a friendship between players i and j would give player j. We can assume the graph to be complete, with previously absent edges having weight 0. In general, we do not assume $u_{ij} = u_{ji}$, but both our lower and upper bounds satisfy this assumption. Each player must declare their membership in one of two, or more generally k "cliques" or "friend groups", and receives utility $u_i(s) = \sum_{j:s(i)=s(j)} u_{ji}$, that is, the sum of benefits from all other players in the same clique.

Here, we will focus on the lower bounds 2-clique case; for the fully general version of these theorems and proofs of the upper bounds, see [HSW19]. It is not hard to see that in this game, it is optimal for everyone to declare membership in the same clique, therefore being able to reap the benefits of all possible friendships. However, there exist locally optimal pure Nash equilibria that fall short of the optimum by up to a factor of two. The following theorem is known to us from private communication with Éva Tardos.

Theorem 4. *The pure Price of Anarchy for the social 2-grouping game is 2.*

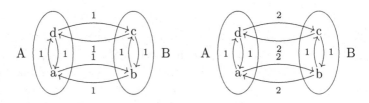

Fig. 1. Left: An OPT/2 selfish equilibrium. Right: An OPT/3 tribal equilibrium. (Color figure online)

This circumstance does not change when players are fully altruistic, and in fact, the lower bound is established by the following Nash equilibrium in the same example as seen for the preceding theorem.

In Fig. 1 on the left, we see that by switching to the other clique, each player would gain (from the unique player that benefits them in the other group) 1 unit of utility, and lose (from the unique player that benefits them in the current group) 1 as well. Also, the net loss to the rest of the community (as the person sharing the clique with the switching player would no longer benefit from their friendship) and the net gain (as one person in the clique they are switching to would now benefit) each are 1 as well, and so switching would indeed make no difference to the player's subjective utility. On the other end, we can establish a matching upper bound with little effort.

Theorem 5. *The Price of Tribalism for the social 2-grouping game and constant partition functions $\tau \in \mathcal{T}^{(1)}$ is 2.*

What happens once different tribes enter the picture? In the above example, the player's own loss due to lost friends was neatly cancelled by gained benefit due to newly gained friends, and likewise the loss to the community was cancelled by the gain experienced by the members of the defector's new peer group.

With multiple tribes, we no longer have to make a putative defector value the gains and losses of all other players equally. How much worse could we make the equilibrium by making the player care about the friends he is currently in a clique with, but wouldn't care about the benefit he could bring to the members of the other clique? The defector would have to value the benefit *to himself* of the other group's friendship higher than the sum of the benefit to himself of his current group's *and* the benefit his current group derives from him. Therefore, the foregone friendship of those who are in the other clique could be worth up to twice as much to the player before he is incentivised to switch: if we mark the members of one tribe red and the members of the other tribe blue, the choice of cliques shown in Fig. 1 on the right is pure Nash. Here, each player would gain 2 by defecting, and their tribe would gain 0; at the same time, they would lose 1, and their community would also lose 1, and so defecting is zero-sum. This example turns out to be tight for 2-grouping, regardless of how many tribes we allow the players to belong to.

Theorem 6. *The Price of Tribalism for the social 2-grouping game and arbitrary partition functions $\tau \in \mathcal{T}^{(*)}$ is 3.*

The above argument can be extended to show that k tribes are in fact always strictly worse than $k-1$ whenever there are at least k distinct friendship cliques to be formed; see Theorem B.2 in [HSW19].

To the extent to which the social grouping game is a useful model of real-world social networks, we see this result as confirmation of an intuitively relatable phenomenon: when individuals "fall in with the wrong crowd", they can get stuck in local minima that are quite bad for everyone.

4.2 Network Contribution Games

A more involved model of social relationships are the *network contribution games* $\mathcal{N}$ described by Anshelevich and Hoefer [AH12]. We model a social graph in which the vertices represent players who can divide up a personal *budget* of effort $B_i \geq 0$ among their potential relationships, represented by edges. The benefit each player derives from a relationship $e = \{i, j\}$ is given by a non-negative, non-decreasing and symmetric *reward function* $f_e : \mathbb{R}^2_{\geq 0} \to \mathbb{R}_{\geq 0}$ in terms of the amount of effort each of them invests, and each player's total utility is just the sum of benefits from all relationships.

In the original paper, players are both allowed to deviate individually if it benefits themselves and to coordinate a joint deviation with a neighbouring player if it benefits them both; this is both seen as more realistic for pairwise relationships, and necessary to enable interesting defection patterns for some classes of payoff functions to exist at all. We will follow this approach in the tribal extension, calling deviations by single players and connected pairs *unilateral* and *bilateral* respectively. We will denote an equilibrium stable against bilateral deviations as a *pairwise equilibrium*, giving rise to a *pairwise Price of Tribalism*. Furthermore, we note that all lower-bound examples in this section work even when entire tribes are allowed to coordinate their deviation; we will call the corresponding PoT *coordinated*.

Definition 3. *Some notation for the rest of this section.*

(i) *Denote by $s_i(e)$ the amount that player i contributes into edge e in strategy s.*

(ii) *For each edge $e = \{i,j\}$, let $w_e(s) = f_e(s_i(e), s_j(e))$ be the reward that the edge pays to both i and j.*

(iii) *An edge $e = \{i,j\}$ is called* tight *in strategy s if $s_i(e) \in \{0, B_i\}$ and $s_j(e) \in \{0, B_j\}$, and a strategy s is tight if all edges are tight.*

When the reward is just a weighted sum of investments, [AH12] shows that the PoA is 1. It is easy to establish that this is also the case when players are altruistic. The proof of the following corollary, as well as the remaining missing proofs in this section, can be found in the extended version [HSW19].

Corollary 1 (of Theorem 2.8 in [AH12]). $\mathrm{PoT}(\mathcal{T}^{(1)}, \mathcal{N}_+) = 1$. *However:*

Theorem 7. *Suppose all reward functions are of the form $c_e(x + y)$, where $c_e > 0$. Then the pure, pairwise and coordinated PoT each is $\mathrm{PoT}(\mathcal{T}^{(*)}, \mathcal{N}_+) = 2$.*

Proof (upper bound). As noted in the proof of Theorem 2.8 in [AH12], the social optimum s^* is attained when all players invest their entire budget in the respective adjacent edge e^* with maximum c_e. For a configuration s to be a Nash equilibrium, no player may want to deviate to investing their budget like this. If a player i invests $s_i(e)$ units into an adjacent edge $e \sim i$, their tribe earns $2s_i(e)c_e$ units of utility if the player on the other end is also in the same tribe, and $s_i(e)c_e$ units otherwise. So $\sum_{e \sim i} 2s_i(e)c_e \geq B_i c_{e^*}$. Summing over all players, we find that

$$U(s) = \sum_i \sum_{e \sim i} s_i(e)c_e \geq \frac{1}{2} \sum_i B_i c_{e^*(i)} = U(s^*),$$

and so the PoT is bounded above by 2.   □

Proof (lower bound). A tight lower bound is given by the following graph:

Here, the two players on the right are in the same tribe, but only the middle player has any budget. It would be socially optimal for them to invest this in the edge on the left, attaining a social welfare of 4; however, the configuration where they instead invest in the edge on the right – yielding a social welfare of 2 – is stable, since the blue tribe's total utility is 2 regardless of how the middle player's budget is allocated.   □

When all reward functions are convex in each coordinate, [AH12] shows a PoA of 2. In Theorem B.3 of [HSW19], we show that this is also the case given altruism. Again, though, tribalism leads to deterioration.

Theorem 8. *Suppose all reward functions are convex in each coordinate. Then the pure, pairwise and coordinated Price of Tribalism each is* $\mathrm{PoT}(\mathcal{T}^{(*)}, \mathcal{N}_C) = 4$.

Proof (upper bound). By Claim 2.10 in [AH12], since all reward functions are coordinate convex, we can assume that the optimum s^* is tight. Fix a pairwise tribal Nash equilibrium s. Note that we can normalise the f_e's so that $f_e(0,0) = 0$. Since the reward functions are non-decreasing, the normalised functions will still be valid reward functions, and subtracting a constant from utility at both OPT and Nash can only increase their ratio.

In a tight strategy, each player can invest their budget in at most one edge. When a player i invests in an edge e in the optimum solution s^*, we will say that i is a *witness* to e. Let $e = \{i,j\}$ be an edge where $s_i^*(e) = B_i$ and $s_j^*(e) = B_j$, so i and j are both witnesses to e. By the Nash condition, if i and j were to bilaterally deviate to their strategies in s^*, then it must not be beneficial for at least one of the two players' tribes. Suppose WLOG that this is i. In the worst case, i and j were in the same tribe *and* benefitting other members of their tribe, and so the tribe loses $2(u_i(s) + u_j(s))$. On the other hand, the worst-case gain occurs when i and j are in different tribes, and so the switch only benefits i's tribe one lot of $w_e(s^*)$. So by the Nash condition, we can derive

$$u_i^\mathcal{T}(s) \geq u_i^\mathcal{T}(s_i^*; s_j^*; s_{-i,j}) \geq u_i^\mathcal{T}(s) - 2(u_i(s) + u_j(s)) + w_e(s^*).$$

Rearranging the inequality, we have $2(u_i(s) + u_j(s)) \geq w_e(s^*)$. So $w_e(s^*)$ is less than two times the sum of the utilities of its witnesses in s. So suppose instead $e = \{i,j\}$ is an edge where $s_i^*(e) = B_i$ and $s_j^*(e) = 0$, so only i is a witness to e. By the same reasoning as above, we have

$$u_i^\mathcal{T}(s) \geq u_i^\mathcal{T}(s_i^*; s_{-i}) \geq u_i^\mathcal{T}(s) - 2u_i(s) + w_e(s^*).$$

So again, $w_e(s^*)$ is less than two times the sum of the utilities of its witnesses in s.

Since each player is marked as a witness to exactly one edge, we can sum the above inequalities, treating one side as a sum over all edges and the other as a sum over all players. We thus conclude

$$U(s^*) = 2 \sum_e w_e(s^*) \leq 4 \sum_{i \in V} u_i(s) = 4U(s). \qquad \square$$

Proof (lower bound). The following example in fact provides a matching lower bound for any function class that contains a coordinate convex function f satisfying $f(x,0) = 0$ and is closed under scalar multiplication.

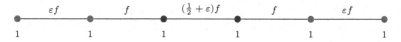

The social optimum, with welfare $4f(1,1)$, is attained when the four players in the middle invest their budgets in the respective adjacent edge with payoff f. However, we can show that the configuration in which all budget is invested in the first, third and fifth edge, for a total payoff of $(2\varepsilon + 1 + 2\varepsilon + 2\varepsilon)f(1,1)$, is stable against unilateral, bilateral and whole-tribe deviations: No set of players who are in the same tribe will want to deviate, as this would involve diverting budget from an edge that has investments on both ends (thus losing utility) to one that has no investment on the other end (thus not gaining any). Also, the two (distinct-tribe) players at the second and fourth edge in the graph will not want to deviate together, because this will not benefit the blue tribe player closer to the center: supposing they divert b units and the red player diverts a units to their shared edge, we have

$$f(a,b)+2(1/2+\varepsilon)f(1,1-b) < 2(1/2+\varepsilon)(f(1,b)+f(1,1-b)) < 2(1/2+\varepsilon)f(1,1)$$

by non-decreasingness and coordinate convexity.   □

4.3   Atomic Linear Routing Games

Atomic linear routing games were first defined in [Ros73a], and their prices of anarchy were first studied in [STZ07] in the context of asymmetric scheduling games; an exposition of this is given in [Rou16]. In these games, each player i is associated with a pair of vertices (s_i, t_i) of a directed graph, called its *source* and *sink* respectively. We think of the game as modelling multiple players traversing a road network, incurring some delays along the way depending on the total congestion on each road segment traversed. In the linear case, these delay functions are assumed to be linear, so each edge e is associated with a positive factor α_e such that when k players are on the edge, each of them incurs a delay of $\alpha_e k$ (and hence the sum of their delays is $\alpha_e k^2$). Formally, the strategies available to player i are the set of paths from s_i to t_i in the graph, and the cost incurred by the player is $c_i(\boldsymbol{s}) = \sum_{e \in \boldsymbol{s}_i} \alpha_e \#\{j : e \in \boldsymbol{s}_j\}$.

By [CK05,AAE13], the pure Price of Anarchy for atomic linear routing games is exactly $\frac{5}{2}$. In [CKK+10], a weaker upper bound of 3 is shown to hold if players are at least partially altruistic (optimising some convex combination of their own utility and social welfare), and this bound is tight when players are fully altruistic. We will demonstrate that this bound does not hold when players show tribal altruism towards two or more tribes. The example that gives rise to our lower bound relies on tribal behaviour that is quite intuitive: at certain interior nodes (case 1 below), a tribally altruistic player prefers to continue paying a greater cost (while also causing a great cost to an "outgroup" member) over switching to a configuration that would benefit both the player and the commons, but result in a greater cost being paid by the player's tribe. The matching upper bound uses smoothness.

Theorem 9. *The Price of Tribalism for atomic linear routing games $\mathcal{R}$ with 2-tribe partition functions $\tau \in \mathcal{T}^{(2)}$, as well as arbitrary partition functions in $\mathcal{T}^{(*)}$, is*

$$\mathrm{PoT}(\mathcal{T}^{(2)}, \mathcal{R}) = \mathrm{PoT}(\mathcal{T}^{(*)}, \mathcal{R}) = 4.$$

Proof (lower bound). Our construction is inspired by the construction in [CKK+10]. As in that paper, our example will be formulated not as a routing game, but as a specific load-balancing game in which each player (represented as an edge) can choose between one of exactly two "servers" or congestible elements with linear cost functions (represented as the endpoints of the edge). This representation can be converted back into a routing game by the following scheme:

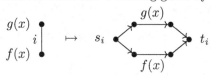

For every k, we will now construct a game G_k and describe a tribal Nash $\mathbf{s}^k$. The game is played on a binary tree with $k+1$ layers of nodes (and hence k layers of edges). Unlike the construction of [CKK+10] (Theorem 2), we do not require to introduce additional edges below the tree, since the costs in the layers of our construction decay fast enough that the total weight of the final layer is dominated by the rest of the tree.

We set the delay function of the nodes at depths (distances from the root) $i = 0, 1, \ldots, k-1$ to be $f_i(x) = (1/2)^i x$. The cost of the nodes in the final layer shall instead be *twice* that of the preceding layer: $f_k(x) = (1/2)^{k-1} \cdot 2 \cdot x$. Each of the two players (edges) under a node shall belong to different tribes, say the left edge to tribe 1 and the right edge to tribe 2.

The overall construction will then look like this:

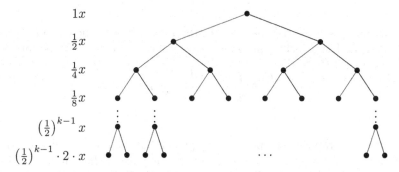

We claim that the strategy profile $\mathbf{s}^k$ in which every player-edge chooses to occupy the "upper" (closer to the root) vertex is Nash.

Indeed, by analysing the environment of each edge depending on the layer it is situated in, we can verify the Nash condition for all players.

1. Intermediate layers, up to exchange of tribes:

In this case, at Nash, the top red player incurs a cost of $1c \cdot 2 + \frac{1}{2}c2 = 3c$ (for himself on the two-player node above, and his tribesman on the two-player node below). If he were to switch down, his cost would be $0 + \frac{1}{2}c3 \cdot 2$, which is also $3c$.

2. Final layer, up to exchange of tribes:

In this case, at Nash, the red player incurs a cost of $2c$, as nobody is using the bottom node and he is sharing the top node with a player from the other tribe. If he were to switch down, he would be using the node alone, but his cost would still be $2c$.

Summing by layer, the total cost of this assignment then is

$$C_{G_k}(\mathbf{s}^k) = \sum_{i=0}^{k-1} 4 \cdot 2^i \cdot \left(\frac{1}{2}\right)^i = 4k.$$

Here, the cost factor due to congestion on each vertex is red, and the number of vertices in each layer is blue. The cost factor on each vertex is black. On the other hand, the social optimum is at least as good as the strategy $(\mathbf{s}^k)^*$ where every player uses the node further "down" (away from the root). In this assignment, every vertex except for the root is occupied by exactly one player, so the cost of the optimum is bounded above by the total cost

$$C_{G_k}(\text{opt}) \leq C_{G_k}((\mathbf{s}^k)^*) = \sum_{i=0}^{k-1} 1 \cdot 2^i \cdot \left(\frac{1}{2}\right)^i \underbrace{-1}_{\text{root}} + \underbrace{1 \cdot 2^k \left(\frac{1}{2}\right)^{k-1} \cdot 2}_{\text{bottom-most row}}$$

$$= k - 1 + 4.$$

Hence we can conclude that as $k \to \infty$, the ratio between the cost of the Nash equilibrium and the social optimum goes to 4 from below: that is, for any ε, there is a k such that

$$\frac{C_{G_k}(\mathbf{s}^k)}{C_{G_k}(\text{opt})} \geq \frac{4k}{k+3} \geq 4 - \varepsilon$$

as claimed.   □

In order to establish the upper bound (which holds for any number of tribes), we will first need to introduce an appropriate instance of the common notion of *smoothness*, originally due to Roughgarden [Rou09]. Broadly speaking, a smooth game is one in which in expectation, a unilateral deviation towards a different strategy profile moves the deviating player's welfare towards some multiple of its welfare in the target profile. This property can be used to deduce a generic bound on the Price of Anarchy.

Definition 4. *Let G^τ be the tribal extension of a finite cost-minimisation game G. G is (λ, μ, τ)-smooth if for any strategy profiles $\mathbf{s}, \mathbf{s}' \in \Sigma$,*

$$\sum_{i \in N} (c_i^\tau(\mathbf{s}_i'; \mathbf{s}) - (c_i^\tau(\mathbf{s}) - c_i(\mathbf{s}))) \leq \lambda C(\mathbf{s}') + \mu C(\mathbf{s}).$$

Other work in the literature on altruism and social context uses generalisations of smoothness. Of particular note is the notion of (μ, λ, α)-altruistic smoothness in [CdKKS14] and Rahn and Schäfer's $\mathcal{SC}$-smoothness [RS13]. Our definition agrees with Roughgarden's when τ assigns each player to his own tribe, and with $(\mu, \lambda, 1)$-altruistic smoothness when τ assigns all players to the same tribe; it also turns out that $\mathcal{SC}$-smoothness is a straightforward generalisation.

Theorem 10. ([RS13]). *Let* $\mathcal{G}$ *be a class of games and* $\mathcal{T} = \{\mathcal{T}_G\}_{G \in \mathcal{G}}$ *be a class of partition functions for each game. If for every* $G \in \mathcal{G}$ *and* $\tau \in \mathcal{T}_G$, G^τ *is* (λ, μ, τ)-*smooth, then* $\mathrm{PoT}(\mathcal{T}, \mathcal{G}) \leq \lambda/(1 - \mu)$.

Proof. See [HSW19] or [RS13]. ◻

The following bound, which is in the spirit of several similar ones in the literature (e.g. [CdKKS14] Lemma 4.4), will be a key ingredient in the proof to follow.

Lemma 1. *For integers* $x, y \geq 0$, $x(y - x) + xy + x + y \leq \frac{8}{3}y^2 + \frac{1}{3}x^2$.

Proof. See [HSW19]. ◻

Lemma 2. *Let* G^τ *be a* τ-*tribal extension of an atomic linear routing game. Then* G^τ *is* $(8/3, 1/3, \tau)$-*smooth.*

Proof. Let $\mathbf{s}$ and $\mathbf{s}^*$ be two strategy profiles in any τ-extension of any atomic linear congestion game. We will use $n_e(\mathbf{s}) = \#\{i \mid e \in \mathbf{s}_i\}$ to denote the number of players using edge e in strategy $\mathbf{s}$, and $n_e^t(\mathbf{s}) = \#\{i \mid e \in \mathbf{s}_i, \tau(i) = t\}$ be the number of players on edge e that belong to tribe t. Then $c_i^\tau(\mathbf{s}) = \sum_e \alpha_e n_e^{\tau(i)}(\mathbf{s}) n_e(\mathbf{s})$. For each player i, we can compute the change in cost of i's tribe as she switches from $\mathbf{s}$ to $\mathbf{s}^*$,

$$c_i^\tau(\mathbf{s}_i^*; \mathbf{s}_{-i}) - c_i^\tau(\mathbf{s}) = \sum_{e \in \mathbf{s}_i^* \setminus \mathbf{s}_i} \alpha_e((n_e^{\tau(i)}(\mathbf{s}) + 1)(n_e(\mathbf{s}) + 1) - n_e^{\tau(i)}(\mathbf{s}) n_e(\mathbf{s}))$$
$$+ \sum_{e \in \mathbf{s}_i \setminus \mathbf{s}_i^*} \alpha_e((n_e^{\tau(i)}(\mathbf{s}) - 1)(n_e(\mathbf{s}) - 1) - n_e^{\tau(i)}(\mathbf{s}) n_e(\mathbf{s}))$$
$$\leq \sum_{e \in \mathbf{s}_i*} \alpha_e(n_e^{\tau(i)}(\mathbf{s}) + n_e(\mathbf{s}) + 1) + \sum_{e \in \mathbf{s}_i} \alpha_e(1 - n_e^{\tau(i)}(\mathbf{s}) - n_e(\mathbf{s})).$$

Here, the last inequality is because we can add the (always positive) contribution of edges $e \in \mathbf{s}_i \cap \mathbf{s}_i^*$. Then, substituting into the left hand side of Definition 4 and using that $c_i(\mathbf{s}) = \sum_{e \in \mathbf{s}_i} \alpha_e n_e(\mathbf{s})$, we find that

$$\sum_{i \in N}(c_i^\tau(\mathbf{s}_i^*; \mathbf{s}_{-i}) - c_i^\tau(\mathbf{s}) + c_i(\mathbf{s}))$$

$$\leq \sum_{\text{tribes } t} \sum_{i \in N : \tau(i) = t} \left(\sum_{e \in \mathbf{s}_i^*} \alpha_e(n_e^t(\mathbf{s}) + n_e(\mathbf{s}) + 1) + \sum_{e \in \mathbf{s}_i} \alpha_e(1 - n_e^t(\mathbf{s})) \right)$$

$$= \sum_{\text{tribes } t} \sum_{\text{edges } e} \alpha_e \left(n_e^t(\mathbf{s}^*)(n_e^t(\mathbf{s}) + n_e(\mathbf{s}) + 1) + n_e^t(\mathbf{s})(1 - n_e^t(\mathbf{s})) \right)$$

by changing the order of summation and combining the $n_e^t(\mathbf{s}^*)$ (resp. $n_e^t(\mathbf{s})$) identical summands on each edge; this is

$$= \sum_{\text{tribes } t} \sum_{\text{edges } e} \alpha_e \left(n_e^t(\mathbf{s})(n_e^t(\mathbf{s}^*) - n_e^t(\mathbf{s})) + n_e^t(\mathbf{s}^*)n_e(\mathbf{s}) + n_e^t(\mathbf{s}^*) + n_e^t(\mathbf{s}) \right)$$

$$\leq \sum_{\text{edges } e} \alpha_e \left(n_e(\mathbf{s})(n_e(\mathbf{s}^*) - n_e(\mathbf{s})) + n_e(\mathbf{s}^*)n_e(\mathbf{s}) + n_e(\mathbf{s}^*) + n_e(\mathbf{s}) \right)$$

by summing over tribes and using $n_e^t(\mathbf{s}) \leq n_e(\mathbf{s})$ (as the tribes are a partition of all players using edge). By Lemma 1, we conclude that this is

$$\leq \sum_{\text{edges } e} \alpha_e \left(\frac{8}{3}n_e(\mathbf{s}^*)^2 + \frac{1}{3}n_e(\mathbf{s})^2 \right) = \frac{8}{3}C(\mathbf{s}^*) + \frac{1}{3}C(\mathbf{s}). \qquad \square$$

Proof (upper bound of Theorem 9). Follows from Lemma 2 and Theorem,10. $\square$

References

[AAE13] Awerbuch, B., Azar, Y., Epstein, A.: The price of routing unsplittable flow. SIAM J. Comput. **42**(1), 160–177 (2013)

[AH12] Anshelevich, E., Hoefer, M.: Contribution games in networks. Algorithmica **63**(1), 51–90 (2012)

[AKT08] Ashlagi, I., Krysta, P., Tennenholtz, M.: Social context games. In: Papadimitriou, C., Zhang, S. (eds.) WINE 2008. LNCS, vol. 5385, pp. 675–683. Springer, Heidelberg (2008). https://doi.org/10.1007/978-3-540-92185-1_73

[BB07] Baldassarri, D., Bearman, P.: Dynamics of political polarization. Am. Sociol. Rev. **72**(5), 784–811 (2007)

[BCFG13] Bilò, V., Celi, A., Flammini, M., Gallotti, V.: Social context congestion games. Theor. Comput. Sci. **514**, 21–35 (2013)

[CdKKS14] Chen, P.-A., de Keijzer, B., Kempe, D., Schäfer, G.: Altruism and its impact on the price of anarchy. ACM Trans. Econ. Comput. **2**(4), 17:1–17:45 (2014)

[CK05] Christodoulou, G., Koutsoupias, E.: The price of anarchy of finite congestion games. In: Proceedings of the 37th Annual ACM Symposium on Theory of Computing, Baltimore, MD, USA, 22–24 May 2005, pp. 67–73 (2005)

[CK08] Chen, P.-A., Kempe, D.: Altruism, selfishness, and spite in traffic routing. In: Proceedings 9th ACM Conference on Electronic Commerce (EC-2008), Chicago, IL, USA, 8–12 June 2008, pp. 140–149 (2008)

[CKK+10] Caragiannis, I., Kaklamanis, C., Kanellopoulos, P., Kyropoulou, M., Papaioannou, E.: The impact of altruism on the efficiency of atomic congestion games. In: Wirsing, M., Hofmann, M., Rauschmayer, A. (eds.) TGC 2010. LNCS, vol. 6084, pp. 172–188. Springer, Heidelberg (2010). https://doi.org/10.1007/978-3-642-15640-3_12

[CRF+11] Conover, M., Ratkiewicz, J., Francisco, M.R., Gonçalves, B., Menczer, F., Flammini, A.: Political polarization on twitter. In: Proceedings of the Fifth International Conference on Weblogs and Social Media, Barcelona, Catalonia, Spain, 17–21 July 2011 (2011)

[Eco17] The Economist. Whither nationalism? - Vladimir's choice, pp. 53–58, December 2017

[FA08] Fiorina, M.P., Abrams, S.J.: Political polarization in the American public. Ann. Rev. Polit. Sci. 11(1), 563–588 (2008)

[FPT04] Fabrikant, A., Papadimitriou, C.H., Talwar, K.: The complexity of pure Nash equilibria. In: Proceedings of the 36th Annual ACM Symposium on Theory of Computing, Chicago, IL, USA, 13–16 June 2004, pp. 604–612 (2004)

[HSW19] Han, S., Soloviev, M., Wang, Y.: The impact of tribalism on social welfare (2019). arXiv:1907.06862 [cs.GT]

[KLS01] Kearns, M.J., Littman, M.L., Singh, S.P.: Graphical models for game theory. In: UAI 2001: Proceedings of the 17th Conference in Uncertainty in Artificial Intelligence, University of Washington, Seattle, Washington, USA, 2–5 August 2001, pp. 253–260 (2001)

[Ros73a] Rosenthal, R.W.: A class of games possessing pure-strategy Nash equilibria. Int. J. Game Theory 2, 65–67 (1973)

[Ros73b] Rosenthal, R.W.: The network equilibrium problem in integers. Networks 3(1), 53–59 (1973)

[Rou09] Roughgarden, T.: Intrinsic robustness of the price of anarchy. In: Proceedings of the 41st Annual ACM Symposium on Theory of Computing, STOC 2009, Bethesda, MD, USA, 31 May–2 June 2009, pp. 513–522 (2009)

[Rou16] Roughgarden, T.: Twenty Lectures on Algorithmic Game Theory. Cambridge University Press, Cambridge (2016)

[RS13] Rahn, M., Schäfer, G.: Bounding the inefficiency of altruism through social contribution games. In: Chen, Y., Immorlica, N. (eds.) WINE 2013. LNCS, vol. 8289, pp. 391–404. Springer, Heidelberg (2013). https://doi.org/10.1007/978-3-642-45046-4_32

[STZ07] Suri, S., Tóth, C.D., Zhou, Y.: Selfish load balancing and atomic congestion games. Algorithmica 47(1), 79–96 (2007)

[Szr04] Szreter, S.: Industrialization and health. Br. Med. Bull. 69(1), 75–86 (2004)

The Online Best Reply Algorithm for Resource Allocation Problems

Max Klimm[1] , Daniel Schmand[2]([✉]) , and Andreas Tönnis[3]

[1] Operations Research, HU Berlin, Berlin, Germany
max.klimm@hu-berlin.de
[2] Institute for Computer Science, Goethe University Frankfurt, Frankfurt, Germany
schmand@em.uni-frankfurt.de
[3] Department of Computer Science, University of Bonn, Bonn, Germany
atoennis@uni-bonn.de

Abstract. We study the performance of a best reply algorithm for online resource allocation problems with a diseconomy of scale. In an online resource allocation problem, we are given a set of resources and a set of requests that arrive in an online manner. Each request consists of a set of feasible allocations and an allocation is a set of resources. The total cost of an allocation vector is given by the sum of the resources' costs, where each resource's cost depends on the total load on the resource under the allocation vector. We analyze the natural online procedure where each request is allocated greedily to a feasible set of resources that minimizes the individual cost of that particular request. In the literature, this algorithm is also known as a one-round walk in congestion games starting from the empty state. For unweighted resource allocation problems with polynomial cost functions with maximum degree d, upper bounds on the competitive ratio of this greedy algorithm were known only for the special cases $d \in \{1, 2, 3\}$. In this paper, we show a general upper bound on the competitive ratio of $d(d/W(\frac{1.2d-1}{d+1}))^{d+1}$ for the unweighted case where W denotes the Lambert-W function on $\mathbb{R}_{\geq 0}$. For the weighted case, we show that the competitive ratio of the greedy algorithm is bounded from above by $(d/W(\frac{d}{d+1}))^{d+1}$.

Keywords: Online algorithms · Resource allocation problems · Congestion games

1 Introduction

We consider a greedy best reply algorithm for online resource allocation problems. The set of feasible allocations for each request is a set of subsets of the resources. Each resource is endowed with a cost function that is a polynomial with non-negative coefficients depending on the total load of that resource. In

A. Tönnis—Partially supported by CONICYT grant PCI PII 20150140 and ERC Starting Grant 306465 (BeyondWorstCase).

D. Fotakis and E. Markakis (Eds.): SAGT 2019, LNCS 11801, pp. 200–215, 2019.
https://doi.org/10.1007/978-3-030-30473-7_14

the online variant considered in this paper, the requests arrive one after another. Upon arrival of a request, we immediately and irrevocably choose a feasible allocation for that request without any knowledge about requests arriving in the future. After the sequence of requests terminates, we evaluate the solution quality of the best reply algorithm in terms of its *competitive ratio* defined as the worst-case over all instances of the ratio of the cost of an online solution and the cost of an offline optimal solution. Here, the cost of a solution is defined as the sum of the resources' cost under the allocation vector. The cost of a resource is the sum of the personal costs of each request on that resource. Specifically, we consider the natural greedy best reply algorithm that assigns each request to the allocation that minimizes the personal cost of the request. More formally, in an unweighted resource allocation problem all requests have a unit weight, the cost of each resource depends on the number of requests using it. In a weighted resource allocation problem, each request i has a weight w_i, the personal cost of each request on the resource depends on the total weight of requests using it.

A prominent application of this model is energy efficient algorithm design. Here, resources model machines or computing devices that can run at different speeds. A sequence of jobs is revealed in an online manner and has to be allocated to a set of machines such that all machines process the tasks allocated to them within a certain time limit. As a consequence, machines have to run at higher speed when more tasks are are allocated to the machine. As its speed is increased, the energy consumption of a machine increases superlinearly; a typical assumption in the literature is that the energy consumption is a polynomial with non-negative coefficients and maximal degree 3 as a function of the load [2]. The aim is to find an allocation with minimal energy consumption. Our results imply bounds on the competitive ratio of the natural online algorithm that assigns each task to a (set of machines) that greedily minimizes the energy consumption of that task.

Another application of the resource allocation problems considered in this paper arises in the context of congestion games. Here, requests correspond to players and feasible allocations correspond to feasible strategies of that player. In a network congestion game, e.g., the set of strategies of a player is equal to the set of paths from some designated start node to a designated destination node in a given graph. Congestion occurs on links that are chosen by multiple users and is measured in terms of a load-dependent cost function. Polynomial cost functions play a particularly important role in the field of traffic modeling; a typical class of cost functions used in traffic models are polynomials with non-negative coefficients and maximal degree 4 as proposed by the US Bureau of Public Roads [31]. The online variant of the resource allocation problem models the situation where users arrive one after another and choose a path that minimizes their private cost with respect to the users that are already present. This scenario is very natural, e.g., in situations where requests to a supplier of connected automotive navigation systems appear online and requests are routed such that the travel time for each request is minimized individually. Our results imply bounds on the competitive ratio of the natural greedy algorithm were each player chooses a strategy that minimizes their total cost given the set of players already present in the network.

1.1 Related Work

Already the approximation of optimal solutions to the offline version of resource allocation problems with polynomial cost functions is very challenging. Roughgarden [28] showed that there is a constant $\beta > 0$ such that the optimal solution cannot be approximated in polynomial time by a factor better than $(\beta d)^{d/2}$ when cost functions are polynomials of maximum degree d with non-negative coefficients. This holds even for the unweighted case. For arbitrary cost functions, the optimal solution cannot be approximated by a constant factor in polynomial time [25]. For polynomials of maximal degree d, currently the best known approximation algorithm is due to Makarychev and Srividenko [24] and uses a convex programming relaxation. They showed that randomly rounding an optimal fractional solution gives an $\mathcal{O}\left(\left(\frac{0.792d}{\ln(d+1)}\right)^d\right)$ approximate solution. This approach is highly centralized and relies on the fact that all requests are initially known, which both might be unrealistic assumptions for large-scale problems.

Online and decentralized algorithms that have been studied in the literature are local search algorithms and multi-round best-reply dynamics. The analysis of both algorithms is technically very similar to the now-called *smoothness* technique to establish bounds on the price of anarchy of Nash equilibria in congestion games [1,3,14,29]. The price of anarchy is equal to the worst-case ratio of the cost of a Nash equilibrium and the cost of an optimal solution. To obtain tight bounds, one solves an optimization problem of the form

$$\min_{\lambda>0,\mu\in[0,1)} \left\{ \frac{\lambda}{1-\mu} : c(x+y) \le \lambda x c(x) + \mu y c(y), \forall x,y \in \mathbb{N}, c \in \mathcal{C} \right\},$$

where $\mathcal{C}$ is the set of cost functions in the game. For the case that $\mathcal{C}$ is the set of polynomial functions with maximum degree d and positive coefficients, Aland et al. [1] used this approach to show that the price of anarchy is Φ_d^{d+1} where $\Phi_d \in \Theta(\frac{d}{\ln(d)})$ is the unique solution to $(x+1)^d = x^{d+1}$. The price of stability, defined as the worst case of the ratio of the cost of a best Nash equilibrium and that of a system optimum, was not as well understood until recently when Christodoulou et al. [13] showed that the price of stability is at least $(\Phi_d/2)^{d+1}$ for large d. For unweighted congestion games, Christodoulou and Gairing [12] showed a tight bound on the price of stability in the order of $\Theta(d)$. Unfortunately, a best-response walk towards a Nash equilibrium can take exponential time [16], even for unweighted congestion games, so that price of anarchy results do not give rise to polynomial approximation algorithms. For weighted games, best-response walks may even cycle [22]. On the other hand, random walks [19] or walks using approximate best-response steps [4] converge to approximate Nash equilibria in polynomial time. In contrast to this, one-round walks in congestion games, or equivalently, the best reply algorithm for online resource allocation problems touches every request only once. Fanelli et al. [17] have shown a linear lower bound even for linear cost functions if the requests are restricted to make one best-response starting from a bad initial configuration. This lower bound does not hold for one round walks starting in the empty state.

The best reply algorithm with respect to the personal cost of a request, which we analyze here, has also been studied for the online setting before. For weighted resource allocation problems with linear cost functions, it turns out that the algorithm admits the same competitive ratio as the best reply algorithm with respect to the total cost function. There is a tight bound on the competitive ratio of $3 + 2\sqrt{2} \approx 5.83$, where the lower bound is due to Caragiannis et al. [11] and the upper bound is due to Harks et al. [21]. For $d > 1$, a first lower bound of $\Omega((d/\ln 2)^{d+1})$ has been shown by Caragiannis et al. [10]. A first upper bound dates back to the mid 90s when Awerbuch et al. [5] gave an upper bound on the competitive ratio of personal cost best replies of Ψ_d^{d+1}, where Ψ_d is defined to be the unique solution to the equation $(d + 1)(x + 1)^d = x^{d+1}$. However, they only considered the setting of singleton requests where each allocation contains a single resource only and all cost functions are equal to the identity. Bilò and Vinci [8] show that the worst-case competitive ratio is in fact attained for singletons and mention that the tight competitive ratio is Ψ_d^{d+1}, but their paper does not contain a proof of the latter result. Prior to that, Harks et al. [21] noted that the competitive ratio in the order of $\mathcal{O}(1.77^d d^{d+1})$. We here slightly improve the bound to a closed form as we note that $\Psi_d \leq d/W(\frac{d}{d+1})$, where W is the Lambert-W function. This recovers the bound by Harks et al. in the limit since $1/W(\frac{d}{d+1}) \approx 1.77$ for d large enough.

For unweighted instances it turns out that the personal cost best reply algorithm admits a better competitive ratio than the total cost best reply algorithm, where requests are allocated greedily such that the total cost of the current solution is minimized. For $d = 1$, the tight bound is $\frac{(\phi+1)^2}{\phi} \approx 4.24$ where $\phi = \frac{1+\sqrt{5}}{2}$ is the golden ratio. The lower bound is due to Bilò et al. [7] and the upper bound is due to Christodoulou et al. [15]. For arbitrary d, the lower bound of $(d+1)^{d+1}$ by Farzad et al. [18] also holds in this setting. There was no general upper bound known.

Harks et al. [20,21] and Farzad et al. [18] both also studied a setting that is equivalent to ours. However, they measure cost slightly different as they define the cost of a resource as the integral of its cost function from 0 to the current load. This different cost measure leads also to a different notion of the private cost of a request. However, as remarked by Farzad et al. [18], the models are equivalent when only polynomial costs are considered. Additionally, they introduced a generalization of the model, where requests are only present during certain time windows. It is easy to see that all our results also hold in this slightly more general setting.

Close to our work is the analysis of best reply algorithms with different personal cost functions. Mirrokni and Vetta [26] were the first to study best-response dynamics with respect to social cost. Bjelde et al. [9] analyzed the solution quality of local minima of the total cost function both for weighted and unweighted resource allocation problems. By a result of Orlin et al. [27], this admits a PTAS in the sense that an $(1 + \epsilon)$-approximate local optimal solution can be computed in polynomial time via local improvement steps. Best reply algorithms with respect to the social cost function instead of the personal cost function have also been well studied for the online setting. For weighted resource allocation problems and $d = 1$, there is a tight bound of 5.83. The upper bound is due to Awerbuch et al. [5] and the lower bound due to Caragiannis et al. [11]. For larger d, there is a lower bound of $\Omega((d/\ln 2)^{d+1})$ by Caragiannis et al. [10] and an upper bound of $\mathcal{O}((d/\ln 2)^{d+1})$ by Bilò and Vinci [8]. For unweighted resource allocation problems there is a tight bound of 5.66 in the linear case. The upper bound is due to Suri et al. [30] and the lower bound due to Caragiannis et al. [11]. For larger d, there is a lower bound of $(d + 1)^{d+1}$ by Farzad et al. [18]. Up to our knowledge, there is no known upper bound that separates the unweighted case from the weighted case.

1.2   Our Contribution

We show upper bounds on the competitive ratio of the best reply algorithm in online resource allocation problems with cost functions that are polynomials of maximal degree d with non-negative coefficients. For unweighted instances, we provide the first bound that hold for any fixed value of d. Prior to our results, non-trivial upper bounds were only known for the cases $d = 1$ by Christodoulou et al. [15] and for the cases $d \in \{2, 3\}$ by Bilò [6]. To the best of our knowledge, despite the wealth of results for weighted problems, prior to this work, no competitive ratio for any $d > 3$ or the asymptotic behavior as $d \to \infty$ has been known that holds specifically for the unweighted case. We close this gap and show that the best reply algorithm is $d(\Xi_d d)^{d+1}$ competitive, where $\Xi_d \leq 1/W(\frac{1.2d-1}{d+1})$. Here W is the Lambert-W function on $\mathbb{R}_{\geq 0}$. Thus, we obtain a concrete factor that holds for any d. We further show that $\lim_{d \to \infty} \Xi_d \approx 1.523$ thus also giving the asymptotic behavior of the bound.

For weighted resource allocation problems, previous work [5,8] has established an upper bound of Ψ_d^{d+1}, where Ψ_d is defined to be the solution to the equation $(d+1)(x+1)^d = x^{d+1}$. In the full version [23], we also add a short proof for this result since the proof of Awerbuch et al. [5] does only consider singletons with identical resource cost functions and the paper by Bilò and Vinci [8] mentions the result without a proof. We also show that $\Psi_d \leq d/W(\frac{d}{d+1})$. This refines an upper bound of $\mathcal{O}(1.77^d d^{d+1})$ obtained by Harks et al. [21]. Note that in the limit, both results coincide as $\lim_{d \to \infty} \Psi_d \approx 1.77$.

Both our main result concerning unweighted games and our proofs for weighted games allow for the first time to separate the competitive ratios of greedy personal best replies for unweighted and weighted problems, respectively. While for weighted games the competitive ratio is about $(1.77d)^{d+1}$, for unweighted games it is bounded from above by $(1.523d)^{d+1}$ for d large enough.

Due to space constraints, some proofs and technical lemmas are deferred to the full version [23].

2 Preliminaries

We consider online algorithms for unsplittable resource allocation problems. Let R be a finite set of resources r each endowed with a non-negative cost function $c_r : \mathbb{R}_{\geq 0} \to \mathbb{R}_{\geq 0}$. There is a sequence of requests $\mathcal{R} = (w_1, \mathcal{S}_1), \ldots, (w_n, \mathcal{S}_n)$. At time step i, the existence of request $(w_i, \mathcal{S}_i)$ is revealed, where w_i is its weight, and $\mathcal{S}_i \subseteq 2^R$ is the set of feasible allocations. If $w_i = 1$ for all $i \in \{1, \ldots, n\}$, we call the instance *unweighted*. Upon arrival of request i, an allocation $S_i \in \mathcal{S}_i$ has to be fixed irrevocably by an online algorithm.

We use the notation $[n] = \{1, \ldots, n\}$. For $i \in [n]$, let $\mathcal{S}_{\leq i} = \mathcal{S}_1 \times \cdots \times \mathcal{S}_{i-1} \times \mathcal{S}_i$ be the set of all feasible allocation vectors up to request i. For a resource $r \in R$ and an allocation vector $\mathbf{S}_{\leq i} = (S_1, \ldots, S_{i-1}, S_i) \in \mathcal{S}_{\leq i}$ we denote the load of r under $\mathbf{S}_{\leq i}$ by $w_r(\mathbf{S}_{\leq i})$. In the following, we write $\mathcal{S}$ and $\mathbf{S}$ instead of $\mathcal{S}_{\leq n}$ and $\mathbf{S}_{\leq n}$. The total cost of an allocation vector $\mathbf{S}$ is defined as

$$C(\mathbf{S}) = \sum_{i \in [n]} \sum_{r \in S_i} w_i c_r(w_r(\mathbf{S})) = \sum_{r \in R} w_r(\mathbf{S}) c_r(w_r(\mathbf{S})) .$$

Given a sequence of requests $\mathcal{R}$, the offline optimal solution value is denoted by $\mathrm{OPT}(\mathcal{R}) = \min_{\mathbf{S} \in \mathcal{S}} C(\mathbf{S})$. As a convention, the allocations used in the optimal solution are denoted by $\mathbf{S}^*$. For a sequence of requests $\mathcal{R}$ and $i \in [n]$, denote by $\mathcal{R}_{\leq i} = (w_1, \mathcal{S}_1), \ldots, (w_{i-1}, \mathcal{S}_{i-1}), (w_i, \mathcal{S}_i)$ the subsequence of requests up to request i. An online algorithm ALG is a family of functions $f_i : \mathcal{R}_{\leq i} \to \mathcal{S}_i$ mapping partial requests up to request i to a feasible allocation for request i. For a sequence of requests $\mathcal{R}$, the cost of an online algorithm ALG with a family of functions $(f_i)_{i \in N}$ is the given by $\mathrm{ALG}(\mathcal{R}) = C(\mathbf{S})$ where $\mathbf{S} = S_1 \times \cdots \times S_n$ and $S_i = f_i(\mathcal{R}_{\leq i})$.

We measure the performance of an online algorithm by its competitive ratio which is $\rho = \sup_{\mathcal{R}} \mathrm{ALG}(\mathcal{R})/\mathrm{OPT}(\mathcal{R})$ where the supremum is taken over all finite sequences of requests for which $\mathrm{OPT}(\mathcal{R}) > 0$. When the sequence of requests $\mathcal{R}$ is clear from context, we write ALG and OPT instead of $\mathrm{ALG}(\mathcal{R})$ and $\mathrm{OPT}(\mathcal{R})$.

We analyze a very easy and natural online algorithm, which we call best reply algorithm. Let again denote $\mathbf{S}_{<i}$ the allocation vector of the algorithm before the i-th request is revealed. Then, $w_i c_r(w_r(\mathbf{S}_{<i}))$ is the per request cost at the arrival of request i on resource r. Upon arrival of request i, the best reply algorithm chooses an allocation $S_i \in \mathcal{S}_i$ that minimizes the cost of that request, i.e. we choose some allocation S_i such that,

$$\sum_{r \in S_i} w_i c_r(w_r(\mathbf{S}_{<i}) + w_i) \leq \sum_{r \in S_i'} w_i c_r(w_r(\mathbf{S}_{<i}) + w_i) , \tag{1}$$

for all other feasible allocations $S_i' \in \mathcal{S}_i$. This choice is motivated by best response moves in the corresponding congestion game. Note that the response steps used by the best reply algorithm are typically tractable and therefore the competitive ratio ρ is also the approximation factor for the corresponding approximation algorithm.

3   Unweighted Resource Allocation Problems

In this section, we consider unweighted resource allocation problems with polynomial cost functions and derive an upper bound on the competitive ratio for the best reply algorithm. We give a general analysis and show a bound of $d(\Xi_d d)^{d+1} \in \mathcal{O}(((\Xi_d + \epsilon)d)^{d+1})$ for cost functions in $\mathcal{C}_d$. Here, $\mathcal{C}_d$ denotes the set of polynomials with non-negative coefficients and maximum degree d. Ξ_d is the unique solution to the equation $d\left(2xe^{1/x} + x^2 - e^{2/x} - x^2e^{1/x}\right) = e^{\frac{2}{x}}$ and $\lim_{d \to \infty} \Xi_d \approx 1.523$. This implies an exponential gap between the weighted and the unweighted case since for the weighted case the competitive ratio is about $(1.77d)^{d+1}$.

Recall that the sequence of requests is given by $\mathcal{R} = (1, S_1), \ldots, (1, S_n)$ for unweighted resource allocation problems, i.e., $w_i = 1$ for all $i \in [n]$. This implies that the cost functions c_r directly define the per request cost on this resource, since $w_i c_r = c_r$ in this case. The cost functions c_r now only depend on the *number* of requests that have chosen some resource r.

We use the definition of the algorithm in (1) to derive an optimization problem for all resources such that the solution to the problem relates to the competitive ratio of the algorithm. This approach is also known as the (λ, μ)-smoothness framework. In this section, the constraints in the optimization problem only have to hold for integral $x = w_r(\mathbf{S}^*)$ and $y = w_r(\mathbf{S})$.

To retain generality, let $d_r \leq d$ denote the maximal degree of the cost function c_r. We show that for every resource r, the smoothness condition is fulfilled for $\lambda_d = (\Xi_d d)^{d+1}$ and $\mu_d = 1 - \frac{1}{d}$. Towards this end, we make a case distinction between (1) $x = 0$, (2) $x \neq 0, y \leq \frac{1}{W(1.27)}d_r$, (3) $x = 1, y > \frac{1}{W(1.27)}d_r$ and (4) $x \geq 2, y > \frac{1}{W(1.27)}d_r$.

Theorem 1. *For any $d \in \mathbb{N}$, the best reply algorithm is $d(\Xi_d d)^{d+1}$ competitive for unweighted resource allocation problems with cost functions in $\mathcal{C}_d$. Here, Ξ_d is the unique solution to*

$$d\left(2xe^{1/x} + x^2 - e^{2/x} - x^2e^{1/x}\right) = e^{\frac{2}{x}}$$

and $\Xi_d \leq \frac{1}{W\left(\frac{1.20d}{d+1}\right)}$.

Throughout this proof, we reference Lemmas 5–13 which are all included in the full version [23].

Proof. First, note that the best reply algorithm is 4.24 competitive for $d = 1$ due to the work of Christodoulou, Mirrokni and Sidiropoulos [15]. We show in Lemma 6 that $\Xi_1 \geq \frac{1}{W(1.27/2)} \approx 2.39$, i.e. we can assume $d \geq 2$.

For the proof, we will assume wlog. that resource r has cost equal to x^{d_r} with $d_r \leq d$. If this is not the case, we can achieve this setting by splitting up resources. Additionally, we assume that there are no resources r with $w_r(\mathbf{S}) = 0$ and $w_r(\mathbf{S}^*) > 0$. If this is not the case, ignoring any contribution of these resources to $C(S^*)$ does only increase the competitive ratio.

The algorithm minimizes the current request's cost in each step, that is, $\sum_{r \in S_i} (w_r(\mathbf{S}_{<i}) + 1)^{d_r} \leq \sum_{r \in S_i^*} (w_r(\mathbf{S}_{<i}) + 1)^{d_r}$. The total cost can be written as the sum of the marginal increases to the total cost functions, i.e. we can write

$$C(\mathbf{S}) = \sum_{i \in [n]} \sum_{r \in S_i} \left((w_r(\mathbf{S}_{<i}) + 1)^{d_r+1} - w_r(\mathbf{S}_{<i})^{d_r+1} \right)$$

$$= \sum_{i \in [n]} \sum_{r \in S_i} \left(\sum_{k=0}^{d_r} w_r(\mathbf{S}_{<i})^k \binom{d_r+1}{k} + (w_r(\mathbf{S}_{<i}) + 1)^{d_r} \cdot (d_r + 1) \right)$$
$$- \sum_{i \in [n]} \sum_{r \in S_i} \left((w_r(\mathbf{S}_{<i}) + 1)^{d_r} \cdot (d_r + 1) \right)$$

$$= \sum_{i \in [n]} \sum_{r \in S_i} \left((d_r + 1)(w_r(\mathbf{S}_{<i}) + 1)^{d_r} + \sum_{k=0}^{d_r} w_r(\mathbf{S}_{<i})^k \frac{(d_r+1)!}{k!(d_r+1-k)!} \right)$$
$$- \sum_{i \in [n]} \sum_{r \in S_i} \sum_{k=0}^{d_r} \left(w_r(\mathbf{S}_{<i})^k \frac{d_r!}{k!(d_r-k)!} \cdot (d_r + 1) \right),$$

where we used that $\sum_{k=0}^{d_r+1} a^k \binom{d_r+1}{k} = (a+1)^{d_r+1}$ in the first step. We get

$$C(\mathbf{S}) \leq \sum_{i \in [n]} \sum_{r \in S_i^*} (d_r + 1)(w_r(\mathbf{S}_{<i}) + 1)^{d_r}$$
$$- \sum_{i \in [n]} \sum_{r \in S_i} \sum_{k=0}^{d_r-1} w_r(\mathbf{S}_{<i})^k \binom{d_r+1}{k}(d_r - k),$$

by using the definition of the algorithm. In the following, we use $w_r(\mathbf{S}_{<i}) \leq w_r(\mathbf{S})$ and that $w_r(\mathbf{S}_{<i})^k$ can be written as $j - 1$ in the second sum, if i is the j-th request that has been allocated to r in $\mathbf{S}$. We obtain

$$C(\mathbf{S}) \leq \sum_{r \in R} (d_r+1)w_r(\mathbf{S}^*)(w_r(\mathbf{S})+1)^{d_r} - \sum_{r \in R} \sum_{j=1}^{w_r(\mathbf{S})} \sum_{k=0}^{d_r-1} (j-1)^k \binom{d_r+1}{k}(d_r-k)$$

Now we bound the inner sum of the triple sum with the integral, which yields

$$\sum_{r \in R} \sum_{j=1}^{w_r(\mathbf{S})} \sum_{k=0}^{d_r-1} (j-1)^k \binom{d_r+1}{k}(d_r-k)$$

$$\geq \sum_{r \in R} \sum_{k=0}^{d_r-1} \binom{d_r+1}{k}(d_r-k) \int_1^{w_r(\mathbf{S})} (j-1)^k dj$$

$$= \sum_{r \in R} \sum_{k=0}^{d_r-1} \binom{d_r+1}{k}(d_r-k)\frac{1}{k+1}(w_r(\mathbf{S})-1)^{k+1}$$

$$= \sum_{r \in R} \sum_{k=1}^{d_r} \binom{d_r+1}{k}\frac{d_r-k+1}{d_r-k+2}(w_r(\mathbf{S})-1)^k.$$

At this point, we split the sum and apply a variant of the binomial theorem used above as well as $\sum_{k=0}^{d+1}(a-1)^k\binom{d+2}{k} = a^{d+2} - (a-1)^{d+2}$. We get

$$\sum_{r \in R} \sum_{k=1}^{d_r} \binom{d_r + 1}{k} \frac{d_r - k + 1}{d_r - k + 2} (w_r(\mathbf{S}) - 1)^k$$

$$= \sum_{r \in R} \sum_{k=0}^{d_r+1} \left((w_r(\mathbf{S}) - 1)^k \binom{d_r + 1}{k} \frac{d_r - k + 1}{d_r - k + 2} \right) - \frac{d_r + 1}{d_r + 2}$$

$$= \sum_{r \in R} \sum_{k=0}^{d_r+1} \left((w_r(\mathbf{S}) - 1)^k \binom{d_r + 1}{k} \left(1 - \frac{1}{d - k + 2} \right) \right) - \frac{d_r + 1}{d_r + 2}$$

$$= \sum_{r \in R} w_r(\mathbf{S})^{d+1} - \sum_{k=0}^{d_r+1} \left((w_r(\mathbf{S}) - 1)^k \binom{d_r + 2}{k} \frac{1}{d_r + 2} \right) - \frac{d_r + 1}{d_r + 2}$$

$$= \sum_{r \in R} w_r(\mathbf{S})^{d+1} - \frac{(w_r(\mathbf{S})^{d_r+2} - (w_r(\mathbf{S}) - 1)^{d_r+2})}{d_r + 2} - \frac{d_r + 1}{d_r + 2}.$$

In Proposition 1, we will show that choosing $\lambda_d = (\Xi_d d)^{d+1}$ and $\mu_d = 1 - \frac{1}{d}$ fulfills the condition

$$(d_r+1)(y+1)^{d_r}x - y^{d_r+1} + \frac{y^{d_r+2} - (y-1)^{d_r+2}}{d_r + 2} + \frac{d_r+1}{d_r+2} \le \lambda_d x^{d_r+1} + \mu_d y^{d_r+1}$$

$$(2)$$

for all $x \in \mathbb{N}_{\ge 0}, y \in \mathbb{N}_{\ge 1}$ and $d_r \le d$. Note that we will only show the inequality for $y \ge 1$. However, omitting resources with $w_r(S) = 0$ can only increase the approximation bound. Applying this to all $r \in R$ yields

$$C(\mathbf{S}) \le \lambda_d C(\mathbf{S}^*) + \mu_d C(\mathbf{S}),$$

that is, a competitive ratio of $\frac{\lambda_d}{1-\mu_d}$. Thus, we seek to minimize $\frac{\lambda_d}{1-\mu_d}$ subject to the inequality (2). We will show in Proposition 1 that there are λ_d and μ_d such that $\frac{\lambda_d}{1-\mu_d}$ is upper bounded by $d(\Xi_d d)^{d+1}$. In Lemma 6, we show that $\frac{1}{W(\frac{1.27d-1}{d+1})} \le \Xi_d \le \frac{1}{W(\frac{1.20d-1}{d+1})}$. Note that a numerical analysis shows that $\lim_{d \to \infty} \sqrt[d+1]{d}\Xi_d \approx 1.523$, i.e. the value for Ξ_d seems to be quite close to our lower bound for large d, since $\lim_{d \to \infty} \frac{1}{W(\frac{1.27d-1}{d+1})} \approx 1.520$. $\square$

It remains to show that there are choices for λ_d and μ_d that give the claimed competitive ratio. The proof of the following proposition relies on several technical lemmas which are, due to space constraints, deferred to the full version [23].

Proposition 1. *For any $1 \le d_r \le d$, there are λ_d, μ_d with*

$$\frac{\lambda_d}{1 - \mu_d} \le d(\Xi_d d)^{d+1},$$

and $(d_r + 1)(y+1)^{d_r} x - y^{d_r+1} + \frac{y^{d_r+2}-(y-1)^{d_r+2}}{d_r+2} + \frac{d_r+1}{d_r+2} \le \lambda_d x^{d_r+1} + \mu_d y^{d_r+1}$
for all $x \in \mathbb{N}_{\ge 0}, y \in \mathbb{N}_{\ge 1}$ and $d_r \le d$, where Ξ_d is the solution to the equation
$d(2xe^{\frac{1}{x}} + x^2 - e^{\frac{2}{x}} - x^2 e^{\frac{1}{x}}) = e^{\frac{2}{x}}$.

Proof. Let $1 \le d_r \le d$. In this proof, we will distinguish 4 cases. First, we show that the inequality holds for $x = 0$. Then, we show the result for $x \ne 0$, $y \le \frac{d_r}{W(1.27)}$. Third, we consider the case $x = 1$, $y > \frac{d_r}{W(1.27)}$ and finally we finish the proof with the case $x \ge 2$, $y > \frac{d_r}{W(1.27)}$. For all 4 cases, we choose $\mu_d = 1 - \frac{1}{d}$ and $\lambda_d = (\Xi_d d)^{d+1}$.

Case 1: $x = 0$.
In this case, $(d_r+1)(y+1)^{d_r} x = 0$. In Lemma 7, we show $\frac{y^{d_r+2}-(y-1)^{d_r+2}+d_r+1}{d_r+2} \le y^{d_r+1}$ $\forall y \in \mathbb{N}_{\ge 1}$, i.e. we get

$$(d_r + 1)(y+1)^{d_r} x - y^{d_r+1} + \frac{y^{d_r+2} - (y-1)^{d_r+2}}{d_r+2} + \frac{d_r+1}{d_r+2} \le 0$$

for $x = 0$, all $d_r \in \mathbb{N}_{\ge 1}$ and all $y \in \mathbb{N}_{\ge 1}$. This finishes the proof of Case 1.

For Cases 2–4, we can assume $x \ge 1$. In order to show that the constraint is fulfilled for all $x, y \in \mathbb{N}_{\ge 1}$ for the choice $\mu_d = 1 - \frac{1}{d}$, $\lambda_d = (\Xi_d d)^{d+1}$, note that the constraint is equivalent to

$$\max_{x,y \in \mathbb{N}_{\ge 1}} \left\{ \frac{(d_r+1)(y+1)^{d_r} x - y^{d_r+1} + \frac{y^{d_r+2}-(y-1)^{d_r+2}+d_r+1}{d_r+2} - (1-\frac{1}{d})y^{d_r+1}}{x^{d_r+1}} \right\}$$
$$\le (\Xi_d d)^{d+1}.$$

Case 2: $x \ne 0, y \le \frac{d_r}{W(1.27)}$.
First, we will reconsider the inequality $\frac{y^{d_r+2}-(y-1)^{d_r+2}+d_r+1}{d_r+2} \le y^{d_r+1}$, which has been proven in Case 1. We get

$$\max_{\substack{x,y \in \mathbb{N}_{\ge 1}, \\ y \le \frac{d_r}{W(1.27)}}} \left\{ \frac{(d_r+1)(y+1)^{d_r} x - y^{d_r+1} + \frac{y^{d_r+2}-(y-1)^{d_r+2}+d_r+1}{d_r+2} - (1-\frac{1}{d})y^{d_r+1}}{x^{d_r+1}} \right\}$$

$$\le \max_{\substack{x,y \in \mathbb{N}_{\ge 1}, \\ y \le \frac{d_r}{W(1.27)}}} \left\{ \frac{(d_r+1)(y+1)^{d_r} x}{x^{d_r+1}} \right\} \le \max_{\substack{x,y \in \mathbb{N}_{\ge 1}, \\ y \le \frac{d_r}{W(1.27)}}} \left\{ (d_r+1)(y+1)^{d_r} \right\}$$

$$\le \left(\frac{d_r}{W(1.27)} + 1 \right)^{d_r+1} \le \left(\frac{d}{W(1.27)} + 1 \right)^{d+1} \le (\Xi_d d)^{d+1},$$

where the last inequality is shown in Lemma 8.

Case 3: $x = 1, y > \frac{d_r}{W(1.27)}$.

We start the proof of this case by plugging in $x = 1$. This yields

$$M := \max_y \left\{ (d_r + 1)(y + 1)^{d_r} - (2 - \frac{1}{d})y^{d_r+1} + \frac{y^{d_r+2} - (y-1)^{d_r+2} + d_r + 1}{d_r + 2} \right\}$$

$$\leq \max_y \left\{ y^{d_r} \left((d_r + 1)\left(1 + \frac{1}{y}\right)^{d_r} - \left(2 - \frac{1}{d_r}\right)y + y^2 \frac{1 - (1 - \frac{1}{y})^{d_r+2}}{d_r + 2} \right) \right\}$$

$$+ \frac{d_r + 1}{d_r + 2}.$$

We use that $(1 + \frac{1}{y})^y \leq e \leq (1 + \frac{1}{y})^{y+1}$ and get

$$M \leq \max_y \left\{ y^{d_r} \left((d_r + 1)e^{d_r/y} - \left(2 - \frac{1}{d}\right)y + y^2 \frac{1 - e^{-\frac{d_r+2}{y-1}}}{d_r + 2} \right) \right\} + \frac{d_r + 1}{d_r + 2}.$$

In the following, we define $c = \frac{y}{d_r}$, replace y with cd_r, and maximize over c instead of y. We then obtain

$$M < \max_c \left\{ (cd_r)^{d_r} \left((d_r + 1)e^{\frac{1}{c}} - \left(2 - \frac{1}{d}\right)cd_r + c^2 d_r \left(1 - e^{-\frac{1}{c}}\right) \right) \right\} + \frac{d_r + 1}{d_r + 2}.$$

Here we use $\frac{1 - e^{-\frac{1}{c}}}{d_r} > \frac{1 - e^{-\frac{d_r+2}{cd_r-1}}}{d_r+2} \Leftrightarrow 2 > (d_r + 2)e^{-\frac{1}{c}} - d_r e^{-\frac{d_r+2}{cd_r-1}}$. This is shown in Lemma 9.

In the following, we will not explicitly calculate the maximizing c, but will derive an upper bound Ξ_{d_r} on the maximum possible c, dependent on d_r. In order to do so, note that we show in Lemma 10 that the term $(d_r + 1)(e)^{\frac{1}{c}} - (2 - \frac{1}{d})cd_r + c^2 d_r(1 - e^{-\frac{1}{c}})$ is monotonically decreasing in c. Additionally, note that the whole expression is negative, if the term $(d_r + 1)e^{1/c} - (2 - \frac{1}{d})cd_r + c^2 d_r(1 - e^{-\frac{1}{c}})$ is negative. Thus, we conclude that the maximum is not attained for all c such that

$$d_r < -\frac{e^{\frac{1}{c}}}{e^{\frac{1}{c}} - (2 - \frac{1}{d})c + c^2 - c^2 e^{-\frac{1}{c}}} = \frac{e^{\frac{2}{c}}}{(2 - \frac{1}{d})ce^{\frac{1}{c}} + c^2 - e^{\frac{2}{c}} - c^2 e^{\frac{1}{c}}}.$$

We conclude that Ξ_{d_r} is defined such that

$$d_r = \frac{e^{\frac{2}{\Xi_{d_r}}}}{(2 - \frac{1}{d})\Xi_{d_r} e^{\frac{1}{\Xi_{d_r}}} + \Xi_{d_r}^2 - e^{\frac{2}{\Xi_{d_r}}} - \Xi_{d_r}^2 e^{\frac{1}{\Xi_{d_r}}}}$$

is an upper bound on the maximizing c. Lemma 5 shows that Ξ_{d_r} is unique. We conclude that $cd_r \leq \Xi_{d_r} d_r$, then we argue in Lemma 12 that $\Xi_{d_r} d_r$ is monotonically increasing in d_r and thus, can be upper bounded by $\Xi_d d$. We use that the second part of the product is decreasing in c to get

$$(\Xi_d d)^d \max_c \left\{ \left((d_r + 1)e^{\frac{1}{c}} - (2 - \frac{1}{d})cd_r + c^2 d_r \left(1 - e^{-\frac{1}{c}} \right) \right) \right\} + \frac{d+1}{d+2}$$

$$\leq (\Xi_d d)^d \left((d_r + 1)e^{W(1.27)} - \frac{2d_r - 1}{W(1.27)} + \frac{d_r \left(1 - e^{-W(1.27)} \right)}{W(1.27)^2} \right) + \frac{d+1}{d+2}$$

$$\leq (\Xi_d d)^d (0.0042d + 3.46) + \frac{d+1}{d+2} \leq (\Xi_d d)^{d+1},$$

where the last inequality can be checked separately for $d = 2$ and $d \geq 3$ with the help of the lower bound on Ξ_d shown in Lemma 6.

Case 4: $x \geq 2, y > \frac{d_r}{W(1.27)}$.
We use the inequality $\frac{y^{d_r+2} - (y-1)^{d_r+2} + d_r + 1}{d_r + 2} \leq y^{d_r+1}$ from Case 1 again and get

$$M := \max_{\substack{x \geq 2, \\ y > \frac{d_r}{W(1.27)}}} \left\{ \frac{(d_r + 1)(y + 1)^{d_r} x - y^{d_r+1} + \frac{y^{d_r+2} - (y-1)^{d_r+2}}{d_r + 2} - (1 - \frac{1}{d})y^{d_r+1}}{x^{d_r+1}} \right\}$$

$$\leq \max_{x \geq 2, y > \frac{d_r}{W(1.27)}} \left\{ \frac{(d_r + 1)(y + 1)^{d_r} x - (1 - \frac{1}{d})y^{d_r+1}}{x^{d_r+1}} \right\}.$$

Again, we write $y = c \cdot d_r$ and optimize over all $c > \frac{1}{W(1.27)}$ instead. We write

$$M = \max_{x \geq 2, c > \frac{1}{W(1.27)}} \left\{ \frac{(d_r + 1)(cd_r + 1)^{d_r} x - (1 - \frac{1}{d})(cd_r)^{d_r+1}}{x^{d_r+1}} \right\}$$

$$\leq \max_{x \geq 2, c > \frac{1}{W(1.27)}} \left\{ \left(\frac{c}{x} d_r \right)^{d_r} \left((d_r + 1) \left(1 + \frac{1}{cd_r} \right)^{d_r} - \left(1 - \frac{1}{d} \right) \frac{cd_r}{x} \right) \right\}$$

$$\leq \max_{x \geq 2, c > \frac{1}{W(1.27)}} \left\{ \left(\frac{c}{x} d_r \right)^{d_r} \left((d_r + 1)e^{\frac{1}{c}} - (d_r - 1)\frac{c}{x} \right) \right\}. \tag{3}$$

We will proceed by deriving an upper bound on $\frac{c}{x}$ for the maximizing c and x. In order to do so, set $z = \frac{c}{x}$ and consider the second part of the product.

$$(d_r + 1)e^{\frac{1}{c}} - (d_r - 1)\frac{c}{x} = (d_r + 1)e^{\frac{1}{xz}} - (d_r - 1)z \leq (d_r + 1)e^{\frac{1}{2z}} - (d_r - 1)z$$

This term is monotonically decreasing in z, thus we get an upper bound on z, maximizing term (3) by setting

$$(d_r + 1)e^{\frac{1}{2z}} - (d_r - 1)z = 0 \Leftrightarrow \frac{1}{2z}e^{\frac{1}{2z}} = \frac{d_r - 1}{2(d_r + 1)} \Leftrightarrow z = \frac{1}{2W\left(\frac{d_r - 1}{2(d_r + 1)} \right)},$$

i.e. we assume without loss of generality in term (3) that $z := \frac{c}{x} \leq \frac{1}{2W\left(\frac{d_r-1}{2(d_r+1)}\right)}$.
This leads to

$$M \leq \max_{z \leq \frac{1}{2W\left(\frac{d_r-1}{2(d_r+1)}\right)}} \left\{ (zd_r)^{d_r}\left((d_r+1)e^{W(1.27)} - (d_r-1)z\right) \right\} .$$

Lemma 13 shows that not only the maximum of (3) but also the maximum of the upper bound is attained at $z = \frac{1}{2W\left(\frac{d_r-1}{2(d_r+1)}\right)}$. We conclude that we can upper bound the term by

$$\left(\frac{d_r}{2W\left(\frac{d_r-1}{2(d_r+1)}\right)}\right)^{d_r} \left((d_r+1)e^{W(1.27)} - (d_r-1)\frac{1}{2W\left(\frac{d_r-1}{2(d_r+1)}\right)}\right) .$$

It remains to show that this is in fact upper bounded by $(\Xi_d d)^{d+1}$. This can be easily checked for $d = 2,3,4,5$ by using $\Xi_d \geq \frac{1}{W\left(\frac{1.27d-1}{d+2}\right)}$ from Lemma 6. For $d \geq 6$, the right side of the expression is bounded by

$$\left(d_r\left(e^{W(1.27)} - \frac{1}{2W\left(\frac{d_r-1}{2(d_r+1)}\right)}\right) + e^{W(1.27)} + \frac{1}{2W\left(\frac{d_r-1}{2(d_r+1)}\right)}\right)$$

$$\leq \left(d_r\left(e^{W(1.27)} - \frac{1}{2W\left(\frac{1}{2}\right)} + \frac{1}{6}\left(e^{W(1.27)} + \frac{1}{2W\left(\frac{1}{6}\right)}\right)\right)\right) \leq 1.41d_r .$$

This gives us

$$\left(\frac{d_r}{2W\left(\frac{d_r-1}{2(d_r+1)}\right)}\right)^{d_r} (1.41d_r) \leq (\Xi_{d_r} d_r)^{d_r+1} \leq (\Xi_d d)^{d+1} .$$

Here, we used that $\Xi_d \geq \frac{1}{W(1.27)} \approx 1.52$ (Lemma 6) and that $\Xi_{d_r} d_r$ is increasing in d_r, see Lemma 12. □

4   Weighted Resource Allocation Problems

In this section, we revisit some upper bound on the competitive ratio of the best reply algorithm for weighted resource allocation problems with polynomial cost functions in $\mathcal{C}_d$. Awerbuch et al. [5] have shown an upper bound of $(\Psi_d)^{d+1}$, where Ψ_d is the solution to the equation $(d+1)(x+1)^d = x^{d+1}$ for singleton instances where the cost of each resource is the identity. Bilò and Vinci [8] showed that the worst case for the competitive ratio is obtained for singletons but their proof crucially relies on non-identical cost functions. Bilò and Vinci also claimed that the Ψ_d is the correct competitive ratio for arbitrary games, but their paper does not contain a proof of this result. For completeness, a proof of Theorem 2 is contained in the full version.

Theorem 2 (Bilò and Vinci [8]). *For polynomial costs in $\mathcal{C}_d$, the competitive ratio of the best reply algorithm is at most Ψ_d^{d+1} where Ψ_d is the unique solution to the equation $(d+1)(x+1)^d = x^{d+1}$.*

The following theorem provides a closed expression that approximates Ψ_d with small error.

Theorem 3. *The equation $(d+1)(x+1)^d = x^{d+1}$ has a unique solution Ψ_d in $\mathbb{R}_{\geq 0}$ for all $d \in \mathbb{R}_{\geq 0}$. Moreover, $\Psi_d \in [d/W(\frac{d}{d+1}) - 1, d/W(\frac{d}{d+1})]$, where $W : \mathbb{R}_{\geq 0} \to \mathbb{R}_{\geq 0}$ is the Lambert-W function on $\mathbb{R}_{\geq 0}$.*

Proof. We first show that the equation $(d+1)(x+1)^d = x^{d+1}$ has a unique solution. Since this solution is not $x = 0$ we may assume $x \neq 0$, take logarithms and obtain the equivalent equation $\log(d+1) + d\log(x+1) = (d+1)\log x$. Rearranging terms yields

$$\log(x+1) - \log(x) = \frac{\log x - \log(d+1)}{d} . \tag{4}$$

The left hand side of this equation is decreasing in x and takes values in $(0, \infty)$. The right hand side is increasing in x and for $x \in [d+1, \infty)$, it takes values in $(0, \infty)$. Thus, the equation has a unique solution which we denote by Ψ_d.

To get an approximate closed form expression for Ψ_d, we use (4) and the mean value theorem to obtain $\frac{1}{\xi} = \frac{\log \Psi_d - \log(d+1)}{d}$ for some $\xi \in (\Psi_d, \Psi_d + 1)$. We obtain

$$\Psi_d \in \left\{ x \in [d+1, \infty) : \frac{1}{x+1} \leq \frac{\log x - \log(d+1)}{d} \leq \frac{1}{x} \right\}.$$

As $\log x$ is strictly increasing in x and both $\frac{1}{x+1}$ and $\frac{1}{x}$ are decreasing, we obtain $\Psi_d \in [a, b]$ where a is the unique solution to the equation $\frac{d}{x+1} = \log x - \log(d+1)$ and b is the unique solution to the equation $\frac{d}{x} = \log x - \log(d+1)$. The latter equation gives

$$\frac{d}{b} = \log \frac{b}{d+1} \qquad \Leftrightarrow \qquad e^{d/b}\frac{d}{b} = \frac{d}{d+1} .$$

Using that W is bijective on $\mathbb{R}_{\geq 0}$ and that $W(xe^x) = x$ for all $x \in \mathbb{R}_{\geq 0}$, this implies $\frac{d}{b} = W(\frac{d}{d+1})$ and, hence, $b = d/W(\frac{d}{d+1})$. To get a bound on a, note that $a \geq a'$ where a' solves $\frac{d}{a'+1} = \log(a'+1) - \log(d+1)$. Substituting $b = a' + 1$, we obtain $a' = d/W(\frac{d}{d+1}) - 1$ as before. □

References

1. Aland, S., Dumrauf, D., Gairing, M., Monien, B., Schoppmann, F.: Exact price of anarchy for polynomial congestion games. SIAM J. Comput. **40**, 1211–1233 (2011)
2. Albers, S.: Energy-efficient algorithms. Commun. ACM **53**(5), 86–96 (2010)

3. Awerbuch, B., Azar, Y., Epstein, A.: The price of routing unsplittable flow. SIAM J. Comput. **42**(1), 160–177 (2013)
4. Awerbuch, B., Azar, Y., Epstein, A., Mirrokni, V.S., Skopalik, A.: Fast convergence to nearly optimal solutions in potential games. In: Proceedings of the 9th ACM Conference on Electronic Commerce (EC), pp. 264–273 (2008)
5. Awerbuch, B., Azar, Y., Grove, E.F., Kao, M., Krishnan, P., Vitter, J.S.: Load balancing in the l_p norm. In: Proceedings of the 36th Annual IEEE Symposium on Foundations of Computer Science (FOCS), pp. 383–391 (1995)
6. Bilò, V.: A unifying tool for bounding the quality of non-cooperative solutions in weighted congestion games. Theory Comput. Syst. **62**(5), 1288–1317 (2018). https://doi.org/10.1007/s00224-017-9826-1
7. Bilò, V., Fanelli, A., Flammini, M., Moscardelli, L.: Performance of one-round walks in linear congestion games. Theory Comput. Syst. **49**(1), 24–45 (2011)
8. Bilò, V., Vinci, C.: On the impact of singleton strategies in congestion games. In: Proceedings of the 25th Annual European Symposium on Algorithms (ESA), pp. 17:1–17:14 (2017)
9. Bjelde, A., Klimm, M., Schmand, D.: Brief announcement: approximation algorithms for unsplittable resource allocation problems with diseconomies of scale. In: Proceedings of the 29th ACM Symposium on Parallelism in Algorithms and Architectures (SPAA), pp. 227–229 (2017)
10. Caragiannis, I.: Better bounds for online load balancing on unrelated machines. In: Proceedings of the 19th Annual ACM-SIAM Symposium on Discrete Algorithms (SODA), pp. 972–981 (2008)
11. Caragiannis, I., Flammini, M., Kaklamanis, C., Kanellopoulos, P., Moscardelli, L.: Tight bounds for selfish and greedy load balancing. Algorithmica **61**(3), 606–637 (2011)
12. Christodoulou, G., Gairing, M.: Price of stability in polynomial congestion games. ACM Trans. Econ. Comput. **4**(2), 10:1–10:17 (2016)
13. Christodoulou, G., Gairing, M., Giannakopoulos, Y., Spirakis, P.G.: The price of stability of weighted congestion games. In: Proceedings of the 45th International Colloquium on Automata, Languages, and Programming (ICALP), pp. 150:1–150:16 (2018)
14. Christodoulou, G., Koutsoupias, E.: The price of anarchy of finite congestion games. In: Proceedings of the 37th Annual ACM Symposium on Theory of Computing (STOC), pp. 67–73 (2005)
15. Christodoulou, G., Mirrokni, V.S., Sidiropoulos, A.: Convergence and approximation in potential games. Theor. Comput. Sci. **438**, 13–27 (2012)
16. Fabrikant, A., Papadimitriou, C.H., Talwar, K.: The complexity of pure nash equilibria. In: Proceedings of the 36th Annual ACM Symposium on Theory of Computing (STOC), pp. 604–612 (2004)
17. Fanelli, A., Flammini, M., Moscardelli, L.: The speed of convergence in congestion games under best-response dynamics. ACM Trans. Algorithms **8**(3), 25:1–25:15 (2012)
18. Farzad, B., Olver, N., Vetta, A.: A priority-based model of routing. Chicago J. Theor. Comput. Sci. **2008** (2008). Article 1
19. Goemans, M.X., Mirrokni, V.S., Vetta, A.: Sink equilibria and convergence. In: Proceedings of the 46th Annual IEEE Symposium on Foundations of Computer Science (FOCS), pp. 142–154 (2005)
20. Harks, T., Heinz, S., Pfetsch, M.E.: Competitive online multicommodity routing. Theory Comput. Syst. **45**(3), 533–554 (2009)

21. Harks, T., Heinz, S., Pfetsch, M.E., Vredeveld, T.: Online multicommodity routing with time windows. ZIB Report 07-22, Zuse Institute Berlin (2007)
22. Harks, T., Klimm, M.: On the existence of pure nash equilibria in weighted congestion games. Math. Oper. Res. **37**, 419–436 (2012)
23. Klimm, M., Schmand, D., Tönnis, A.: The online best reply algorithm for resource allocation problems. CoRR abs/1805.02526. https://arxiv.org/abs/1805.02526
24. Makarychev, K., Sviridenko, M.: Solving optimization problems with diseconomies of scale via decoupling. In: Proceedings of the 55th Annual IEEE Symposium on Foundations of Computer Science (FOCS), pp. 571–580 (2014)
25. Meyers, C.A., Schulz, A.S.: The complexity of welfare maximization in congestion games. Networks **59**(2), 252–260 (2012)
26. Mirrokni, V.S., Vetta, A.: Convergence issues in competitive games. In: Jansen, K., Khanna, S., Rolim, J.D.P., Ron, D. (eds.) APPROX/RANDOM 2004. LNCS, vol. 3122, pp. 183–194. Springer, Heidelberg (2004). https://doi.org/10.1007/978-3-540-27821-4_17
27. Orlin, J.B., Punnen, A.P., Schulz, A.S.: Approximate local search in combinatorial optimization. SIAM J. Comput. **33**(5), 1201–1214 (2004)
28. Roughgarden, T.: Barriers to near-optimal equilibria. In: Proceedings of the 55th Annual IEEE Symposium on Foundations of Computer Science (FOCS), pp. 71–80 (2014)
29. Roughgarden, T.: Intrinsic robustness of the price of anarchy. J. ACM **62**(5), 32:1–32:42 (2015)
30. Suri, S., Tóth, C.D., Zhou, Y.: Selfish load balancing and atomic congestion games. Algorithmica **47**(1), 79–96 (2007)
31. U.S. Bureau of Public Roads: Traffic assignment manual. U.S. Department of Commerce, Urban Planning Division, Washington, DC (1964)

Connected Subgraph Defense Games

Eleni C. Akrida[1], Argyrios Deligkas[1,2], Themistoklis Melissourgos[1(✉)],
and Paul G. Spirakis[1,3]

[1] Department of Computer Science, University of Liverpool, Liverpool, UK
{E.Akrida,Argyrios.Deligkas,T.Melissourgos,P.Spirakis}@liverpool.ac.uk
[2] Leverhulme Research Centre for Functional Materials Design, Liverpool, UK
[3] Computer Engineering and Informatics Department,
University of Patras, Patras, Greece

Abstract. We study a security game over a network played between a
defender and k *attackers*. Every attacker chooses, probabilistically, a node
of the network to damage. The defender chooses, probabilistically as well,
a connected induced subgraph of the network of λ nodes to scan and clean.
Each attacker wishes to maximize the probability of escaping her cleaning by the defender. On the other hand, the goal of the defender is to
maximize the expected number of attackers that she catches. This game
is a generalization of the model from the seminal paper of Mavronicolas et al. [11]. We are interested in Nash equilibria of this game, as well
as in characterizing *defense-optimal* networks which allow for the best
equilibrium defense ratio; this is the ratio of k over the expected number of attackers that the defender catches in equilibrium. We provide
characterizations of the Nash equilibria of this game and defense-optimal
networks. This allows us to show that the equilibria of the game coincide
independently from the coordination or not of the attackers. In addition, we
give an algorithm for computing Nash equilibria. Our algorithm requires
exponential time in the worst case, but it is polynomial-time for λ con-
stantly close to 1 or n. For the special case of tree-networks, we further
refine our characterization which allows us to derive a polynomial-time
algorithm for deciding whether a tree is defense-optimal and if this is the
case it computes a defense-optimal Nash equilibrium. On the other hand,
we prove that it is NP-hard to find a best-defense strategy if the tree is not
defense-optimal. We complement this negative result with a polynomial-
time constant-approximation algorithm that computes solutions that are
close to optimal ones for general graphs. Finally, we provide asymptotically
(almost) tight bounds for the *Price of Defense* for any λ; this is the worst
equilibrium defense ratio over all graphs.

Keywords: Defense games · Defense ratio · Defense-optimal

This work was supported by the NeST initiative of the EEE/CS School of the University
of Liverpool and by the EPSRC grant EP/P02002X/1.

© Springer Nature Switzerland AG 2019
D. Fotakis and E. Markakis (Eds.): SAGT 2019, LNCS 11801, pp. 216–236, 2019.
https://doi.org/10.1007/978-3-030-30473-7_15

1   Introduction

With technology becoming a ubiquitous and integral part of our lives, we find ourselves using several different types of "computer" networks. An important issue when dealing with such networks, which are often prone to security breaches [6], is to prevent and monitor unauthorized access and misuse of the network or its accessible resources. Therefore, the study of network security has attracted a lot of attention over the years [18]. Unfortunately, such breaches are often inevitable, since some parts of a large system are expected to have weaknesses that expose them to security attacks; history has indeed shown several successful and highly-publicized such incidents [17]. Therefore, the challenge for someone trying to keep those systems and networks of computers secure is to counteract these attacks as efficiently as possible, once they occur.

To that end, inventing and studying appropriate theoretical models that capture the essence of the problem is an important line of research, ongoing for a few years now [13,14]. Here, extending some known models for very simple cases of attacks and defenses [11,12], we introduce and analyze a more general model for a scenario of network attacks and defenses modeling it as a *defense game*.

The Network Security Game. We follow the terminology established by the seminal paper of Mavronicolas et al. [12]. We consider a network whose nodes are vulnerable to infection by threats called *attackers*; think of those as viruses, worms, Trojan horses or eavesdroppers [7] infecting the components of a computer network. Available to the network is a security software (or firewall), called the *defender*. The defender is only able to "clean" a limited part of the network from threats that occur; the reason for the limited cleaning capacity of the defender may be, for example, the cost of purchasing a global security software. The defender seeks to protect the network as much as possible, and on the other hand, every attacker seeks to increase the likelihood of not being caught. Both the attackers and the defender make individual decisions for their positioning in the network with the aim to maximize their own objectives.

Every attacker targets (and attacks) a node chosen via her own probability distribution over the nodes of the network. The defender cleans a connected induced subgraph of the network with size λ, chosen via her own probability distribution over all connected induced subgraphs of the graph with λ nodes. The attack of a particular attacker is successful unless the node chosen by the attacker is incident to an edge (link) being cleaned by the defender, i.e. to an edge belonging in the induced subgraph chosen by the defender. One could equivalently think of the defender selecting a set of λ connected nodes to defend, and an attacker is successful if and only if she attacks a node that is not being defended. Since attacks and defenses over a large computer network are self-interested procedures that seek to maximize damage and protection, respectively, it is natural to model this network security scenario as a non-cooperative *strategic game* on graphs with two kinds of players: $k \geq 1$ *attackers*, each playing a *vertex* of the graph, and a single *defender* playing a *connected induced subgraph* of the graph. The *(expected) payoff* of an attacker is the probability that she is not caught by the defender; the *(expected) payoff* of the defender is the (expected) number of attackers she catches. We are interested in the Nash equilibria [15,16] associated with this graph theoretic game, where no player can unilaterally improve her (expected) payoff by switching to another probability distribution. We are also interested in understanding and characterizing the networks that allow for a good *defense ratio*: given a strategy profile, i.e. a combination of strategies for the network entities (attackers and defender), the defense ratio of a network is the ratio of the total number of attackers over the defender's expected payoff in that strategy profile.

1.1   Our Results

In this paper we depart from and significantly extend the line of work of Mavronicolas et al. in their seminal paper [12] on defense games in graphs; we term the type of games we consider *CSD games*. In our model the defender is more powerful than in [12], since her power is parameterized by the size, λ, of the defended part of the network. We allow λ to take values from 1 to n, while in [12] only the case where $\lambda = 2$ was studied. We study many questions related to CSD games. We extend the notions of *defense ratio* and *defense-optimal graphs* for CSD games. In fact, the defense ratio of a given graph G and a given strategy profile S of the attackers and the defender is the ratio of the number of attackers, k, over the defender's expected payoff (the number of attackers she catches on expectation). We thoroughly investigate the notion of the defense ratio for Nash equilibria strategy profiles.

Firstly, we precisely characterize the Nash equilibria and defense-optimal graphs in CSD games. This allows us to show that, in equilibrium, the game version of k uncoordinated attackers and a single defender is equivalent to the version in which a single leader coordinates the k attackers, meaning that both versions of the game have the same defense ratio. We present an LP-based algorithm to compute an exact equilibrium of any given CSD game, whose running time is polynomial in $\binom{n}{\lambda}$. Then, we focus on tree-graphs. There, we further refine our equilirbium characterization which allows us to derive a polynomial-time algorithm for deciding whether a tree is defense-optimal and, if this is the case, it computes a defense-optimal Nash equilibrium. A tree is defense-optimal if and only if it can be partitioned into $\frac{n}{\lambda}$ disjoint sub-trees. On the other hand, we prove that it is NP-hard to find a best-defense strategy if the tree is not defense-optimal. We remark that a very crucial parameter for defense-optimality of a graph G is the "best" probability with which any vertex of G is defended in a NE; we call that probability *MaxMin probability* and denote it by $p^*(G)$. Then, for any graph G, the defense ratio in equilibrium is shown to be exactly $\frac{1}{p^*(G)}$. Although it is hard to exactly compute $p^*(G)$ even for trees, we complement this negative result with a polynomial-time constant-approximation algorithm that computes solutions that are close to the optimal ones for any λ, for any given general graph. In particular, we approximate the (best) defense ratio of any graph within a factor of $2 + \frac{\lambda-3}{n}$. Finally, we provide asymptotically tight bounds for the Price of Defense for any $\lambda \in \omega(1) \cap o(n)$, and almost tight bounds for any other value of λ.

For detailed proofs and auxiliary figures of the results presented, we refer the reader to the full version of the paper [1].

1.2 Related Work

Our graph-theoretic game is a direct generalization of the defense game considered by Mavronicolas et al. [11,12]. In the latter, the authors examined the case where the size of the defended part of the network is $\lambda = 2$, i.e. where the defender "cleans" an edge. This lead to a nice connection between equilibria and (fractional) matchings in the graph [13]. But when λ is greater than 2, one has to investigate (as we shall see here) how to sparsely cover the graph by as small a number as possible of connected induced subgraphs of size λ. This direction can be seen as an extension of fractional matchings to covers of the graph by equisized connected subgraphs. Sparse covering of graphs by connected induced subgraphs (clusters), not necessarily equisized, is a notion known to be useful also for distributed algorithms, since it affects message communication complexity [5].

In another line of work, Kearns and Ortiz [9] study *Interdependent Security games* in which a large number of players must make individual decisions regarding security. Each player's *safety* may depend on the actions of the entire population (in a complex way). The graph-theoretic game that we consider could be seen as a particular instance of such games with some sort of limited interdependence: the actions of the defender and an attacker are interdependent, while the actions of the attackers are not dependent on each other.

Aspnes et al. [4] consider a graph-theoretic game that models containment of the spread of viruses on a network; each node individually must choose to either install anti-virus software at some cost, or risk infection if a virus reaches it without being stopped by some intermediate node with installed anti-virus software. Aspnes et al. [4] prove several algorithmic properties for their graph-theoretic game and establish connections to a certain graph-theoretic problem called *Sum-of-Squares Partition*.

A game on a weighted graph with two players, the *tree player* and the *edge player*, was studied by Alon et al. [2]. At each play, the tree player chooses a spanning tree and the edge player chooses an edge of the graph, and the payoffs of the players depend on whether the chosen edge belongs in the spanning tree. Alon et al. investigate the theoretical aspects of the above game and its connections to the *k-server problem* and *network design*.

Finally, there is a long line of work on security games [3] where many scenarios are modelled using graph theoretic problems [8,10,19,20].

2 Preliminaries

The Game. A *Connected-Subgraph Defense (CSD) game* is defined by a graph $G = (V, E)$, a *defender*, $k \geq 1$ *attackers*, and a positive integer λ. Throughout the paper, λ is considered to be a *given* parameter of the game. A pure strategy for the defender is any induced connected subgraph H of G with λ vertices, which we term λ-*subgraph*. For any λ-subgraph H of G we denote $V(H)$ its set of vertices. Since $V(H)$ uniquely defines an induced subgraph of G, we will use the term λ-subgraph to denote either $V(H)$ or H. The *action set* of the defender is $D := \{V(H)|H$ is a λ-subgraph of $G\}$ and we will denote its cardinality by θ, i.e. $\theta := |D|$. For ease of presentation, we will also refer to D as $[\theta] := \{1, 2, \ldots, \theta\}$. A pure strategy for each of the attackers is any vertex of G. So, the action set of every attacker is V, the vertex set of G; we denote $n := |V|$ and we similarly refer to V also as $[n]$.

To play the game, the defender chooses a *defense (mixed) strategy*, i.e. a probability distribution over her action set, and each attacker chooses an *attack (mixed) strategy*, i.e. a probability distribution over the vertices of G. We denote a strategy by $s := (s_1, \ldots, s_d) \in \Delta_d$, i.e. by the probability distribution over d enumerated pure strategies, where $\Delta_d := \{x_1, \ldots, x_d \geq 0 | \sum_{i=1}^{d} x_i = 1\}$ is the $(d-1)$-unit simplex. In a defense strategy $q \in \Delta_\theta$ each pure strategy $j \in [\theta]$ is assigned a probability q_j.

We say that a pure strategy of the defender, i.e. a specific λ-subgraph H of G, *covers* a vertex $v \in V$ if $v \in V(H)$. A defense strategy covers a vertex $v \in V$ if it assigns strictly positive probability to at least one λ-subgraph H of G which contains v.

Definition 1 (Vertex Probability). *The* vertex probability p_i *of vertex* $i \in [n]$, *is the probability that i will be covered, formally* $p_i := \sum_{j \in [\theta]:\ i \in j} q_j$.

The *support* of a strategy s, denoted by $\mathrm{supp}(s)$, is the subset of the action set that is assigned strictly positive probability.

Payoffs and Strategy Profiles. A *strategy profile* is a tuple of strategies $S = (q, t_1, \ldots, t_k)$, where q denotes the defender's strategy and t_j denotes the j-th attacker's strategy, $j \in [k]$. A strategy profile is pure if the support of every strategy has size one. The *payoff* of every attacker is 1 in any pure strategy profile where she does not choose a defended vertex, and 0 in all the rest. The payoff of the defender in a pure strategy profile where she defends $V(H)$, is the number of attackers that choose a vertex in $V(H)$. Under a strategy profile, the *expected payoff* of the defender is the expected number of attackers that she catches, which we call *defense value*, and the expected payoff of the attacker is the probability that she will not get caught. A *best response* strategy for a participant is a strategy that maximizes her expected payoff, given that the strategies of the rest of the participants are fixed. A *Nash equilibrium* is a strategy profile where all the participants are playing a best response strategy. In other words, neither the defender nor any of the attackers can increase their expected payoff by unilaterally changing their strategy.

Definition 2 (Defense Ratio). *For a given graph G we define a measure of the quality of a strategy profile S, called* defense ratio *of G and denoted $DR(G, S)$, as the ratio of the total number of attackers k over the defense value.*

In this work we are only interested in the cases where S is an equilibrium. For a given graph, when in equilibrium, the defender's expected payoff is unique (due to Corollary 1(a)) and achieves the *equilibrium defense ratio* $DR(G, S^*)$, where S^* is an equilibrium. The defense strategy in S^* which achieves this defense ratio will be termed *best-defense strategy*.

Definition 3 (MaxMin Probability, p^*). *We call* MaxMin Probability *of a graph G the maximum, over all defense strategies, minimum vertex probability in G, that is:*

$$p^*(G) := \max_{q \in \Delta_\theta} \min_{i \in [n]} p_i.$$

As we will show in Lemma 1, the equilibrium defense ratio of a graph G turns out to be $DR(G, S^*) = 1/p^*(G)$.

Definition 4 (Price of Defense). *The* Price of Defense, *PoD, for a given parameter λ of the game, is the worst defense ratio, over all graphs, achievable in equilibrium, that is:*

$$PoD(\lambda) = \max_G DR(G, S^*).$$

Definition 5 (Defense-Optimal Graph). *For a given λ, a graph G^* that achieves the minimum equilibrium defense ratio over all graphs, i.e. $G^* \in \arg\min_G DR(G, S^*)$, is called* defense-optimal graph.

In the following, for ease of presentation, whenever we refer to defense optimality, we implicitly assume that λ has a fixed value.

3   Nash Equilibria

In this section, we provide a characterization of Nash equilibria in CSD games, as well as important properties of their structure which prove useful for the development of our algorithms in the remainder of the paper.

Theorem 1 (Equilibrium characterization). *For a given graph G, in any equilibrium with support $S \subseteq [\theta]$ of the defender and support $T_j \subseteq [n]$ of each attacker $j \in [k]$, the following conditions are necessary and sufficient:*

1. $\min_{i \in [n]} p_i$ *is maximized over all defense strategies, and*
2. $\bigcup_{j \in [k]} T_j \subseteq V^*$, *where $V^* := \{i \in [n] \mid \min_{i \in [n]} p_i$ is maximized over all defense strategies\}, and*
3. *every $s \in S$ has the maximum expected total number of attackers on its vertices over all pure strategies.*

Lemma 1. *For any given graph G, the equilibrium defense ratio is $DR(G, S^*) = \frac{1}{p^*(G)}$, where $p^*(G) := \max_{q \in \Delta_\theta} \min_{i \in [n]} p_i$ and S^* is an equilibrium.*

Proof. By Theorem 1, in an equilibrium, every attacker will have in her support only vertices that are defended with probability exactly $p^*(G)$. Therefore, the expected number of attackers that the defender catches is $p^*(G) \cdot k$. By definition of the defense ratio, $\mathrm{DR}(G, S^*) = \frac{k}{p^*(G) \cdot k} = \frac{1}{p^*(G)}$. □

Corollary 1. *The following hold:*

(a) *For a given graph G, in any equilibrium, the expected payoff of the defender and each attacker is unique.*

(b) *For a given graph G, in any equilibrium with support $S \subseteq [\theta]$ of the defender, for every $s \in S$ there exists a vertex $v \in s$ such that $p_v = p^*(G)$.*

(c) *In any CSD game on a graph G, the problem of finding the equilibrium defense ratio (or equivalently, $p^*(G)$) for $k \geq 2$ attackers reduces to the same problem in the game with $k = 1$ attacker, which is a two-player constant-sum game.*

Proof.(a) By Theorem 1, in an equilibrium the defender chooses a strategy that induces probability $p^*(G)$ to some vertex of G (Condition 1). Also, each of the attackers has in her support T only vertices with vertex probability $p^*(G)$. Therefore, all attackers attack only such vertices and the expected payoff of the defender is $k \cdot p^*(G)$. Consider also an attacker with strategy $t = (t_1, t_2, \ldots, t_n)$. Her expected payoff is $\sum_{i \in [n]} t_i(1 - p_i)$, where p_i is the vertex probability of vertex i. This value is equal to $\sum_{i \in T} t_i(1 - p^*(G)) = 1 - p^*(G)$. Since $p^*(G)$ is unique for a graph G, the expected payoffs of the defender and each attacker is unique.

(b) The proof is by contradiction. Consider an equilibrium where the defender's strategy is $q \in [\theta]$ with support S, and there exists a pure strategy $s \in S$ for which every vertex $v \in s$ has $p_v > p^*(G)$. By Condition 2 of Theorem 1, no attacker has in her support a vertex in s. Therefore, the defender can strictly increase her expected payoff by moving all her probability $q_s > 0$ from s to some other pure strategy s' that contains a vertex which is in the support of some attacker.

(c) Observe that for any given graph G, the quantity $p^*(G)$, by definition, only depends on the graph and not the number of attackers k. That is, $p^*(G)$ is the same for every $k \geq 1$. Lemma 1 states that in any equilibrium S^*, it is $\mathrm{DR}(G, S^*) = \frac{1}{p^*(G)}$, therefore the defense ratio in an equilibrium does not depend on k. This means that when we are given G and we are interested in the equilibrium defense ratio, we might as well consider the game with the single defender and a single attacker. By definition of the game (see Sect. 2) the latter is a two-player constant-sum game. □

The following corollary implies that coordination (resp. individual selfishness) of the attackers cannot increase the attackers' (resp. defender's) expected payoff in equilibrium.

Corollary 2. *Every equilibrium with uncoordinated attackers (i.e. as described in Sect. 2) is an equilibrium with coordinated (i.e. centrally controlled) attackers, and vice versa.*

The following theorem provides an algorithm for computing an equilibrium for any CSD game, whose running time is polynomial in n when $\lambda = c$ or $\lambda = n - c$, where c is a constant natural.

Theorem 2. *For some given graph G and parameter λ, there is an algorithm that computes $p^*(G)$ and also finds an equilibrium in time polynomial in $\binom{n}{\lambda}$.*

Proof. Given a graph G, the number of attackers $k \geq 1$, and some $\lambda \in \{1, 2, \ldots, n\}$, the action set D of the defender is constructed by the vertex sets of at most $\binom{n}{\lambda}$ λ-subgraphs, so for D's cardinality θ it holds that $\theta \leq \binom{n}{\lambda}$. Consider now the mixed strategy $q \in \Delta_\theta$ of the defender, where each pure strategy $j \in [\theta]$ is assigned probability q_j. Consider also the vertex probability p_i for each vertex $i \in [n]$. According to Corollary 1(a) and (c), the unique $p^*(G)$ in the case of a single attacker can be used to derive an equilibrium for the case of $k \geq 2$ attackers. Therefore, we will find $p^*(G)$ for a single attacker, find an equilibrium for that case, and then extend this equilibrium to one in the case of $k \geq 2$ attackers. In more detail, after we find the defense strategy q^* that maximizes $\min_{i \in [n]} p_i$ (Condition 1 of Theorem 1), i.e. yields $p^*(G)$ on the set $V^* := \arg\max_{q \in \Delta_\theta} \min_{i \in [n]} p_i$, an equilibrium is achieved if the single attacker assigns probability $1/|V^*|$ to each vertex of V^*; that is because all conditions of Theorem 1 are satisfied. Then, an equilibrium for $k \geq 2$ is achieved if every attacker plays the same strategy as the single attacker; that is because again all conditions of Theorem 1 are satisfied.

The crucial observation that allows us to design such an algorithm is that we can compute $p^*(G)$ via a Linear Program which has $O\left(\binom{n}{\lambda}\right)$ many variables and $O(n)$ constraints, and therefore its running time is in the worst case polynomial in $\binom{n}{\lambda}$, for $\lambda \in \{2, 3, \ldots, n-1\}$. For the trivial cases $\lambda = 1$ and $\lambda = n$, $D = \{\{i\} | i \in V\}$ and $D = V$ respectively, therefore $p^*(G) = 1/n$ and $p^*(G) = 1$ respectively. So in the rest of the proof we will imply that $\lambda \in \{2, 3 \ldots, n-1\}$. It remains to show how $p^*(G)$ is computed.

Let us denote $p^* := p^*(G) := \max_{q \in \Delta_\theta} \min_{i \in [n]} p_i$. The computation of p^* can be done as follows: First, consider each of the $\binom{n}{\lambda}$ subsets of V of size λ, and find if it is a proper λ-subgraphs of G (i.e. connected); this can be done by running a Depth (or Breadth) First Search algorithm for each subset of size λ. If it is not, then continue with the next subset. If it is, we consider it in the action set $[\theta]$, and assign to it a variable q_j which stands for its assigned probability in a general defense strategy. Now, by definition, for some vertex $i \in [n]$, $p_i = \sum_{\substack{j \in [\theta] \\ i \in j}} q_j$. Therefore, we will consider only pure strategies j which are λ-subgraphs to create the p_i's. To compute the minimum p_i over all i's we introduce the variable p' and write the following set of n inequalities as a constraint in our Linear Program:

$$\sum_{\substack{j \in [\theta] \\ i \in j}} q_j \geq p' \quad , \text{ for } i \in \{1, 2, \ldots, n\}.$$

The variable constraints are $p', q_1, q_2, \ldots, q_\theta \geq 0$ and also $\sum_{j=1}^{\theta} q_j = 1$, and all of the aforementioned constraints can be written in canonical form by applying standard transformations. Finally, the objective function of the Linear Program is variable p' and we require its maximization, which is the value p^*. □

3.1 Connections to Other Types of Games

Although CSD games are defined as a normal form game with $k + 1$ players, we can observe that there are equivalent to other well-studied types of games: polymatrix games and Stackelberg games.

A polymatrix game is defined by a graph where every vertex represents a player and every edge represents a two-player game played by the endpoints of the edge. Every player has the same set of pure strategies in every game he is involved and to play the game he plays the same (mixed) strategy in every game. The payoff of every player is the sum they get from every two-player game they participate in. In a CSD game we observe the following. Firstly, the payoff of every attacker depends only on the strategy the defender plays, thus every attacker is involved only in one two-player game. In addition, all the attackers have the same set of pure strategies and they share the same payoff matrix. Similarly, the payoff the defender gets from catching an attacker depends only on the strategy the defender and this specific attacker chose. Hence, the payoff of the defender can be decomposed into a sum of payoffs from k two-player games. So, a CSD game can be seen as a polymatrix game where the underlying graph is a star with k leaves that correspond to the attackers and the defender is the center of the star. Although many-player polymatrix games have exponentially smaller representation size compared to the equivalent normal-form representation, we should note that this polymatrix game is of exponential size in the worst case since the defender can have exponential in n pure strategies to choose from.

A Stackelberg game is an extensive form two-player game. In the first round, one of the players commits to a (mixed) strategy. In the second round, the other player chooses a best response against the committed strategy of her opponent. In a Stackelberg equilirbium the first player is playing a strategy that maximizes her expected payoff, given that the second player plays a best response (mixed strategy). The MaxMin probability $p^*(G)$ for a CSD game on a graph G corresponds to a Stackelberg equilibrium. By Corollary 1(c), any CSD game with $k \geq 1$ attackers has the same p^* as that of the case with $k = 1$. Furthermore, as in a Stackelberg game, in the CSD game with $k = 1$ the defender chooses a mixed strategy that maximizes her expected payoff, given that the attacker plays a best response (mixed strategy). Therefore, when we are interested in the defense-ratio in equilibrium of a CSD game for some arbitrary $k \geq 1$, finding a Stackelberg equilibrium of the corresponding CSD game with $k = 1$ suffices.

4   Defense-Optimal Graphs

We now focus our attention on defense-optimal graphs. We first characterize defense-optimal graphs with respect to the MaxMin probability p^* and then use this characterization to analyze more specific classes of graphs like cycles and trees. We begin by an exact computation of the equilibrium defense ratio of any defense-optimal graph.

Theorem 3. *In any defense-optimal graph G, we have that $DR(G, S^*) = \frac{n}{\lambda}$.*

As an intermediate corollary of Theorem 3 we get the following characterisation of defense-optimal graphs.

Corollary 3. *A graph G is defense-optimal if and only if all of its vertices are defended with probability $\frac{\lambda}{n}$.*

Someone may wonder whether Corollary 3 can be further exploited to prove that, in general, best-defense strategies in defense-optimal graphs are uniform, i.e. every pure strategy s in the support S of the defender is assigned probability $1/|S|$. However, as we demonstrate in Fig. 1 this is not the case. On the other hand, this claim is true for cyclic graphs and trees.

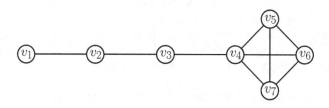

Fig. 1. Here $n = 7$, $\lambda = 3$ and $p^*(G) = 3/7$ is achievable by assigning probability $3/7$ to pure strategy $\{v_1, v_2, v_3\}$ and probability $1/7$ to each of the pure strategies $\{v_4, v_5, v_6\}$, $\{v_4, v_5, v_7\}$, $\{v_4, v_6, v_7\}$, $\{v_5, v_6, v_7\}$, so the graph is defense optimal. However, observe that v_1 cannot participate in more than one pure strategies, so in a uniform defense strategy with support of size r, the vertex probability p_{v_1} has to be $1/r$ (by definition of uniformity), but it also has to be $3/7$. Since $r \in \mathbb{N}$, this is a contradiction.

Observation 1. *All cyclic graphs are defense-optimal.*

Proof. Consider an arbitrary cyclic graph G with n vertices. We will show that the graph can achieve vertex probability $p_i = \frac{\lambda}{n}$ for every $i \in [n]$, thus by Corollary 3 it is defense-optimal. Consider the whole action set D of the defender, i.e. every path starting from a vertex i going clockwise and ending at vertex $i + \lambda - 1$. Observe that there are only n such paths, therefore $\theta := |D| = n$. By assigning probability $\frac{1}{n}$ to each pure strategy $j \in [\theta]$, since each vertex is in exactly λ pure strategies, each vertex $i \in [n]$ has vertex probability $p_i = \lambda \cdot \frac{1}{\theta} = \frac{\lambda}{n}$. $\qquad\square$

4.1   Tree Graphs

In this section we focus on the case where the graph is a tree. We first further refine the characterization of defense-optimal graphs for trees. Then, we
utilise this characterisation to derive a polynomial-time algorithm that decides
in polynomial time whether a given tree is defense-optimal, and if that is the
case, it constructs in polynomial time a defense-optimal strategy for it. On the
other hand, in the case where the tree is not defense-optimal, we show that it is
NP-hard to compute a best-defense strategy for it, namely it is NP-hard to compute $p^*(G)$. We first provide Lemma 2 which will be used in our polynomial-time
algorithm for checking defense-optimality on trees. Henceforth, we write that a
graph is *covered* by a defense strategy if every vertex of the graph is covered by
a λ-subgraph that is in the support of the defense strategy.

Lemma 2. *A tree T is defense-optimal if and only if T can be decomposed into
$\frac{n}{\lambda}$ disjoint λ-subgraphs.*

Proof. ($\Rightarrow$) Let T be defense-optimal. We will show that the support of any
best-defense strategy on T must comprise of pure strategies that are disjoint
λ-subgraphs which altogether cover every $v \in V$. Since those are disjoint and
cover T, it follows that their number is $\frac{n}{\lambda}$ in total.

If $\lambda = 1$ then the above trivially holds. Assume that $\lambda \geq 2$ and consider
a best-defense strategy on T whose support comprises of a collection $\mathcal{L}$ of
λ-subgraphs.

Let $u \in V$ be a leaf of T and let $v \in V$ be its parent. Any λ-subgraph in $\mathcal{L}$
covering u must also cover v, since $\lambda \geq 2$. Also, any λ-subgraph in $\mathcal{L}$ covering
v must also cover u, otherwise p_v would be greater than p_u. Now, consider the
neighbors of v. For those of them that are leaves, the same must hold as holds
for u, namely v and its leaf-children must all be covered by the same exact
λ-subgraph(s).

Consider the case where there is a leaf $u \in V$, such that a *single* λ-subgraph
contains u, its parent v, and all the other leaf-children of v (and, possibly, other
vertices connected to v). Then we can remove this λ-subgraph from $\mathcal{L}$ and the
corresponding tree from T. This leaves the remainder of T being a forest comprising of trees $T_1, \ldots, T_x$, each of which has a (best-) defense strategy comprising
of the corresponding subset of (the remainder of) $\mathcal{L}$ on T_i. Notice that it must
be the case that every tree T_i, $i = 1, 2, \ldots, x$, has size at least λ (otherwise the
initial collection $\mathcal{L}$ would not have covered T). So, if there is always a leaf u in
some tree of the forest, such that a *single* λ-subgraph contains u, its parent v,
and all the other leaf-children of v (and, possibly, other vertices connected to
v), we can proceed in the same fashion for each of the T_i's, always removing a
λ-subgraph from $\mathcal{L}$, and the corresponding vertices from T, until we end up with
an empty tree. This means that $\mathcal{L}$ was indeed a collection of disjoint λ-subgraphs
covering T.

However, assume for the sake of contradiction that at some "iteration" the
assumption does not hold, namely assume that there is a tree in the forest with
no leaf u, such that a single λ-subgraph contains u, its parent v, and all the other

leaf-children of v (and, possibly, other vertices connected to v). This means that there are (at least) two λ-subgraphs in $\mathcal{L}$, namely L_1, L_2, that cover u. Due to our initial observations, u, together with its parent v and all of v's leaf-children are contained in both L_1 and L_2. Since those are different λ-subgraphs, there is a vertex z in the tree which belongs to L_2 but does not belong to L_1. Since $p_z = p_v$ (due to the fact that $\mathcal{L}$ is the support of the defense-optimal strategy and Corollary 3), it must hold that there is a different λ-subgraph, L_3, which covers z but does not cover v or any of its leaf-children. If L_3 also covers a vertex in $L_1 \setminus L_2$[1], then there is a cycle in the tree which is a contradiction. So L_3 must not cover vertices in $L_1 \setminus L_2$. Since L_3 is different to L_2, there must be a vertex z' in the tree which belongs in L_3 but not in L_2 (also not in L_1). Since $p_{z'} = p_z$ (due to the fact that $\mathcal{L}$ is the support of the defense-optimal strategy and Corollary 3), it must hold that there is a different λ-subgraph, L_4, which covers z' but does not cover z or any of the vertices in L_2. Similarly to before, if L_4 covers a vertex in $L_1 \setminus L_2$, then there is a cycle in the tree which is a contradiction. So L_4 must not cover vertices in L_1 or in L_2.

Proceeding in the same way, we result in contradiction since the tree has finite number of vertices and there will need to be an overlap in coverage of some L_j with some L_i, $j > i + 1$, which would mean that there is a cycle in the tree.

Therefore, there cannot be any overlaps between the λ-subgraphs of $\mathcal{L}$, meaning that $\mathcal{L}$ comprises of $\frac{n}{\lambda}$ disjoint λ-subgraphs which altogether cover T.

($\Leftarrow$) Let $\mathcal{L} = \{L_1, \ldots, L_{\frac{n}{\lambda}}\}$ be a collection of $\frac{n}{\lambda}$ disjoint λ-subgraphs that altogether cover T. Let the defender play each L_i, $i \in \{1, \ldots, \frac{n}{\lambda}\}$, equiprobably, that is, with probability $1/\left(\frac{n}{\lambda}\right) = \frac{\lambda}{n}$. Then every vertex $v \in V$ is covered with probability $p_v = \frac{\lambda}{n} = p^*(G)$, meaning that T is defense-optimal. $\square$

With Lemma 2 in hand we can derive a polynomial-time algorithm that decides if a tree is defense-optimal, and if it is, to produce a best-defense strategy.

Theorem 4. *There exists a polynomial-time algorithm that decides whether a tree is defense-optimal and produces a best-defense strategy for it, or it outputs that the tree is not defense-optimal.*

Proof. The algorithm works as follows. Initially, there is a pointer associated with a counter in every leaf of the tree T that moves "upwards" towards an arbitrary root of the tree. For every move of the pointer the corresponding counter increases by one. The pointer moves until one of the following happens: either the counter is equal to λ, or it reaches a vertex with degree greater of equal to 3 where it "stalls". In the case where the counter is equal to λ, we create a λ-subgraph of T, we delete this λ-subgraph from the tree, we move the pointer one position upwards, and we reset the counter back to zero. If a pointer stalls at a vertex of degree $d \geq 3$, it waits until all $d - 1$ pointers reach this vertex.

---

[1] We use $L_i \setminus L_j$ for some λ-subgraphs L_i, L_j to denote the set of vertices which are contained in L_i but not in L_j.

Then, all these pointers are merged to a single one and a new counter is created whose value is equal to the sum of the counters of all d pointers. If this sum is more than λ, then the algorithm returns that the graph is not defense-optimal. If this sum is less than or equal to λ, then we proceed as if there was initially only this pointer with its counter; if the new counter is equal to λ, then we create a λ-subgraph of T and reset the counter to 0; else the pointer moves upwards and the counter increases by one. To see why the algorithm requires polynomial time, observe that we need at most n pointers and n counters and in addition every pointer moves at most n times.

We now argue about the correctness of the algorithm described above. Clearly, if the algorithm does not output that the tree is not defense-optimal, it means that it partitioned T into λ-subgraphs. So, from Lemma 2 we get that T is defense-optimal and the uniform probability distribution over the produced partition covers every vertex with probability $\frac{\lambda}{n}$. It remains to argue that when the algorithm outputs that the graph is not defense-optimal, this is indeed the case. Consider the case where we delete a λ-subgraph of the (remaining) tree. Observe that the λ-subgraph our algorithm deleted should be uniquely covered by this λ-subgraph in any best-defense strategy; any other λ-subgraph would overlap with some other λ-subgraph. Hence, the deletion of such a λ-subgraph was not a "wrong" move of our algorithm and the remaining tree is defense-optimal if and only if the tree before the deletion was defense-optimal. This means that any deletion that occurred by our algorithm did not make the remaining graph non defense-optimal. So, consider the case where after a merge that occurred at vertex v we get that the new counter is $c > \lambda$. Then, we can deduce that all the subtrees rooted at v associated with the counters have strictly less than λ vertices. Hence, in order to cover all the $c > \lambda$ vertices using λ-subgraphs, at least two of these λ-subgraphs cover vertex v. Hence, the condition of Lemma 2 is violated. But since every step of our algorithm so far was correct, it means that v cannot be covered only by one λ-subgraph. Hence, our algorithm correctly outputs that the tree is not defense-optimal. □

In Theorem 4 we showed that it is easy to decide whether a tree is defense-optimal and if this is the case, it is easy to find a best-defense strategy for it. Now we prove that if a tree is not defense-optimal, then it is NP-hard to find a best-defense strategy for it.

Theorem 5. *Finding a best-defense strategy in* CSD *games is* NP-*hard, even if the graph is a tree.*

Proof. We will prove the theorem by reducing from 3-PARTITION. In an instance of 3-PARTITION we are given a multiset with n positive integers $a_1, a_2, \ldots, a_n$ where $n = 3m$ for some $m \in \mathbb{N}_{>0}$ and we ask whether it can be partitioned into m triplets $S_1, S_2, \ldots, S_m$ such that the sum of the numbers in each subset is equal. Let $s = \sum_{i=1}^{n} a_i$. Observe then that the problem is equivalent to asking whether there is a partition of the integers to m triplets such that the numbers in every triplet sum up to $\frac{s}{m}$. Without loss of generality we can assume that $a_i < \frac{s}{m}$ for every $i \in [n]$; if this was not the case, the problem could be trivially answered. So, given an instance of 3-PARTITION, we create a tree $G = (V, E)$ with $s+1$ vertices and $\lambda = \frac{s}{m} + 1$. The tree is created as follows. For every integer a_i, we create a path with a_i vertices. In addition, we create the vertex v_0 and connect it to one of the two ends of each path. We will ask whether $p^*(G) \geq \frac{1}{m}$.

Firstly, assume that the given instance of 3-PARTITION is satisfiable. Then, given S_j we create a $(\frac{s}{m} + 1)$-subgraph of G as follows. If $a_i \in S_j$, then we add the corresponding path of G to the subgraph. Finally, we add vertex v_0 in our $(\frac{s}{m}+1)$-subgraph and the resulting subgraph is connected (by the construction of G). Since the sum of a_i's equals $\frac{s}{m}$, the constructed subgraph has $\frac{s}{m}+1$ vertices. If we assign probability $\frac{1}{m}$ to every $(\frac{s}{m}+1)$-subgraph we get that $p_v \geq \frac{1}{m}$ for every $v \in V$.

To prove the other direction, assume that $p^*(G) \geq \frac{1}{m}$ and observe the following. Firstly, since as we argued it is $a_i < \frac{s}{m}$ for every $i \in [n]$, it holds that every $(\frac{s}{m} + 1)$-subgraph of G contains vertex v_0. Thus, $p_{v_0} = 1$ and $\sum_{v \neq v_0} p_v \geq \frac{s}{m}$, since there are s vertices other than v_0 and for each one of them holds that $p_v \geq \frac{1}{m}$. In addition, observe that $\sum_{v \in V} p_v = \lambda = \frac{s}{m} + 1$. Hence, we get that $p_v = p^*(G) = \frac{1}{m}$ for every vertex $v \neq v_0$. In addition, observe that every pure defense strategy that covers a leaf of this tree, covers all the vertices of the branch. Hence, for every branch of the tree, all its vertices are covered by the same set of pure strategies; if a vertex u that is closer to v_0 is covered by one strategy that does not cover the whole branch, then the leaf u' of the branch is covered with probability less than u. So, in order for $p_v = p^*(G) = \frac{1}{m}$ for every $v \neq v_0$, it means that there exist a $(\frac{s}{m}+1)$-subgraph that *exactly* covers a subset of the paths; this means that if a $(\frac{s}{m} + 1)$-subgraph covers a vertex in a path, then it covers every vertex of the path. Hence, by the construction of the graph, we get that this $(\frac{s}{m} + 1)$-subgraph of G corresponds to a subset of integers in the 3-PARTITION instance that sum up to $\frac{s}{m}$. Since, 3-PARTITION is NP-hard, we get that finding a best-defense strategy is NP-hard. □

4.2   General Graphs

We conjecture that contrary to checking defense-optimality of tree graphs and constructing a corresponding defense-optimal strategy in polynomial time, it is NP-hard to even decide whether a given (general) graph is defense-optimal.

Conjecture 1. It is NP-hard to decide whether a graph is defense-optimal.

5   Approximation Algorithm for $p^*(G)$

We showed in the previous section that, given a graph G, it is NP-hard to find the best-defense strategy, or equivalently, to compute $p^*(G)$. We also presented in Theorem 2 an algorithm for computing the exact value $p^*(G)$ of a given graph G (and therefore its best defense ratio), but this algorithm has running time polynomial in the size of the input only in the cases $\lambda = c$ or $\lambda = n - c$, where c is a constant natural. On the positive side, we present now a polynomial-time algorithm which, given a graph G of n vertices, returns a defense strategy with defense ratio which is within factor $2 + \frac{\lambda - 3}{n}$ of the best defense ratio for G. In particular, it achieves defense ratio $1/p' \leq \left(2 + \frac{\lambda - 3}{n}\right)/p^*(G)$, where $p' = \min_{i \in [n]} p_i$ and every p_i, $i \in [n]$ is the vertex probability determined by the constructed defense strategy. We henceforth write that a collection $\mathcal{L}$ of λ-subgraphs covers a graph $G = (V, E)$, if every vertex of V is covered by some λ-subgraph in $\mathcal{L}$. The algorithm presented in this section returns a collection $\mathcal{L}$ of at most $\frac{2n-3}{\lambda} + 1$ λ-subgraphs that covers G. Therefore, the uniform defense strategy over $\mathcal{L}$ assigns probability at least $1/\left(\frac{2n-3}{\lambda} + 1\right)$ to each λ-subgraph.

For any collection $\mathcal{L}$ of λ-subgraphs and for any $v \in V$, let us denote by $\text{coverage}_{\mathcal{L}}(v)$ the number of λ-subgraphs in $\mathcal{L}$ which v belongs in. Observe that:

$$\sum_{v \in V} \text{coverage}_{\mathcal{L}}(v) = |\mathcal{L}| \cdot \lambda, \tag{1}$$

where $|\mathcal{L}|$ denotes the cardinality of $\mathcal{L}$.

We first prove Lemma 3, to be used in the proof of the main theorem of this Section. We henceforth denote by $V(G)$ and $E(G)$ the vertex set and edge set, respectively, of some graph G.

Lemma 3. *For any tree T of n vertices, and for any $\lambda \leq n$, we can find a collection $\mathcal{L}$ of distinct λ-subgraphs such that for every $v \in V$, it holds that $1 \leq \text{coverage}_{\mathcal{L}}(v) \leq \text{degree}(v)$, except maybe for (at most) $\lambda - 1$ vertices, where for each of them it holds that $\text{coverage}_{\mathcal{L}}(v) = \text{degree}(v) + 1$.*

Proof. We will prove the statement of the lemma by providing Algorithm 1 that takes as input T and λ and outputs the requested collection $\mathcal{L}$ of λ-subgraphs.

The algorithm starts by picking an arbitrary vertex v to serve as the root of the tree. Then it performs a Depth-First-Search (DFS) starting from v. We will distinguish between *visiting* a vertex and *covering* a vertex in the following way. We say that DFS visited a vertex if it considered that vertex as a candidate to be inserted to some λ-subgraph, and we say that DFS covered a vertex if it visited *and* inserted the vertex at some λ-subgraph. By definition, DFS visits in a greedy manner first an uncovered child, and only if there is no such child, it visits its parent (lines 14–17, 21–24). The set-variable that keeps track of the covered vertices is S.

Starting with the root of T, the algorithm simply visits the whole vertex set according to DFS, putting each visited vertex in the same λ-subgraph L_i (starting with $i = 1$) (lines 18–24), and when $|L_i| = \lambda$, a new empty λ-subgraph L_{i+1} is picked to get filled in with λ vertices (lines 26–27) taking care of one extra thing: The first vertex that the algorithm puts in an empty λ-subgraph L_i, $i \in \{1, 2, \dots\}$ is guaranteed to be one that has not been covered by any other λ-subgraph so far (lines 13–17). This ensures that no two λ-subgraphs will eventually be identical.

Algorithm 1. MAIN ALGORITHM

Require: A tree graph $T = (V, E)$ of n vertices, and a natural $\lambda \leq n$.
Ensure: A collection $\mathcal{L}$ of distinct λ-subgraphs that satisfies the statement of Lemma 3.

1: i,   global variable.   % *The index of the λ-subgraph L_i.*
2: *count*,   global variable.   % *Is 0 until the whole tree is covered, then it becomes 1 to allow for the last λ-subgraph to be completed, if it is not already.*
3: S,   global variable.   % *The set of vertices already covered by the algorithm.*
4: *vertex*,   global variable.   % *The vertex considered to be inserted in a λ-subgraph.*

5: $S \leftarrow \emptyset$
6: $i \leftarrow 1$
7: $L_i \leftarrow \emptyset$
8: Pick an arbitrary vertex v of T and consider it the root.
9: $vertex \leftarrow v$
10: $count \leftarrow 0$

11: **while** $count < 2$ **do**
12:     **while** $S \neq V$ **do**
13:         **while** $vertex \in S$ **do**   % *The while-loop to ensure that the first element of L_i is uncovered.*
14:             **if** *vertex* has a child $u \notin S$ **then**
15:                 $vertex \leftarrow u$
16:             **else**
17:                 $vertex \leftarrow$ parent of *vertex*
18:         **while** $|L_i| < \lambda$ **do**   % *The while-loop that fills in the current λ-subgraph L_i.*
19:             $L_i \leftarrow L_i \cup \{vertex\}$
20:             $S \leftarrow S \cup \{vertex\}$
21:             **if** *vertex* has a child $u \notin S$ **then**
22:                 $vertex \leftarrow u$
23:             **else**
24:                 $vertex \leftarrow$ parent of *vertex*
25:         **if** $count < 1$ **then**
26:             $i \leftarrow i + 1$
27:             $L_i \leftarrow \emptyset$
28:         **else**
29:             **break**
30:     $S \leftarrow \emptyset$
31:     $i \leftarrow i - 1$
32:     Pick an arbitrary vertex $v \in L_i$ and consider it the root.
33:     $vertex \leftarrow v$
34:     $count \leftarrow count + 1$

The algorithm will not only visit all vertices in T, but also cover them. That is because there is no point where the algorithm checks whether the currently visited vertex is uncovered and then does not cover it. On the contrary, it covers every vertex that it visits, except for some already covered one in case the current λ-subgraph is empty (lines 13–24). And since DFS by construction visits every vertex, we know that at some point the whole vertex set will be covered, or equivalently, $\text{coverage}_{\mathcal{L}}(v) \geq 1, \forall v \in V$. Therefore, the algorithm will eventually exit the while-loop in lines 12–29.

Now we prove that, after the algorithm terminates, every vertex $v \in V$ is covered at most $degree(v)$ times, except for at most $\lambda - 1$ vertices that are covered $degree(v) + 1$ times. Observe that DFS visits every vertex v at most $degree(v)$ times: (a) v will be visited after its parent u only if v is uncovered (lines 14–15, 21–22), v will get covered (lines 19–20), and will not get visited ever again by its parent since it will be covered (lines 16–17, 23–24). (b) v will be visited at most once by each of its children, say w, only if w does not have an uncovered child (lines 16–17, 23–24), and v will not get ever visited by its parent since v will be covered, and also v cannot be visited a second time by any of its children, since they can never be visited again (they can only be visited through v since T is a tree). Therefore, any vertex v will be visited exactly once after its parent is visited, and at most once by each of its children, having a total of at most $degree(v)$ visits. And since, as argued above, the total number of times a vertex will be covered is at most the number of times it will get visited, when DFS terminates (i.e $S = V$), it will be $\text{coverage}_{\mathcal{L}}(v) \leq degree(v)$, for every $v \in V$.

However, note that the last nonempty λ-subgraph L_i might not consist of λ vertices since the entire V was covered and DFS could not proceed further. In this case, the algorithm empties the set S that keeps track of the covered nodes, takes the current L_i and fills it in with exactly another $\lambda - |L_i|$ vertices. This is done by picking an arbitrary vertex from L_i and setting it as the root of T, and performing one last DFS starting from it until L_i has λ vertices in total (lines 30–33). To ensure that the DFS will continue only until it fills in this current L_i, the algorithm counts the number of times that it runs the while-loop of DFS, namely lines 12–29, via the variable $count$ (line 34), which escapes the while-loop of DFS in case DFS has filled in L_i (lines 28–29) and terminates. Observe that in the last λ-subgraph L_i, a vertex v inserted in the last iteration of DFS ($count = 1$) and was not inserted in L_i by the first run ($count = 0$) might have been covered by the first run of DFS exactly $degree(v)$ times, therefore when the algorithm terminates it has been covered $degree(v) + 1$ times. Since by the end of the first DFS run L_i had at least one vertex, the cardinality of such vertices that are covered more times than their degree are at most $\lambda - 1$.   □

We can now prove the following.

Lemma 4. *For any graph G of n vertices, and for any $\lambda \leq n$, there exist (at most) $\frac{2n-3}{\lambda} + 1$ λ-subgaphs of G that cover G.*

Proof. Consider a spanning tree T of G. Then Lemma 3 applies to T. Observe that a collection $\mathcal{L}$ as described in the statement of the aforementioned lemma has the same qualities for G since $V(T) = V(G)$ and $E(T) \subseteq E(G)$. That is, $\mathcal{L}$ is a collection of distinct λ-subgraphs of G, such that for every $v \in V$, it holds that $1 \leq \text{coverage}_{\mathcal{L}}(v) \leq \text{degree}(v)$, except maybe for (at most) $\lambda - 1$ vertices, for each v of which it is $\text{coverage}_{\mathcal{L}}(v) = \text{degree}(v) + 1$, where by $\text{degree}(v)$ we denote the degree of vertex v in T.

Fix a particular value for λ and consider a collection $\mathcal{L}$ of λ-subgraphs as constructed in the proof of Lemma 3. Then, by Eq. (1),

$$|\mathcal{L}| = \frac{\sum_{v \in V} \text{coverage}_{\mathcal{L}}(v)}{\lambda} \leq \frac{\sum_{v \in V} \text{degree}(v) + (\lambda - 1)}{\lambda} = \frac{2(n-1)}{\lambda} + \frac{\lambda - 1}{\lambda} = \frac{2n - 3}{\lambda} + 1. \quad \Box$$

We conclude with the simple algorithm that achieves a defense strategy with defense ratio which is within factor $2 + \frac{\lambda - 3}{n}$ of the best defense ratio for G.

Algorithm 2. APPROXIMATING THE BEST DEFENSE RATIO
***
Require: Graph $G = (V, E)$ of n vertices, a natural $\lambda \leq n$.
Ensure: A defense strategy that satisfies the statement of Theorem 6.

1: Find a spanning tree T of G.
2: Construct a collection $\mathcal{L}$ of λ-subgraphs of T as described in the proof of Lemma 3.
3: Assign probability $q_i = \frac{1}{|\mathcal{L}|}$ to every λ-subgraph in $\mathcal{L}$, $i = 1, 2, \ldots, |\mathcal{L}|$.

4: **return** The above uniform defense strategy over the collection $\mathcal{L}$.

Theorem 6. *Given any graph $G = (V, E)$, Algorithm 2 computes in time $O(|E|)$ a defense strategy such that, for any combination of attack strategies, the resulting strategy profile S yields defense ratio $\text{DR}(G, S) \leq \left(2 + \frac{\lambda - 3}{n}\right) \cdot \text{DR}(G, S^*)$.*

Proof. As argued in Lemma 4, there is a collection $\mathcal{L}$ of λ-subgraphs with $|\mathcal{L}| \leq \frac{2n}{\lambda} + 1 - \frac{3}{\lambda}$ which covers G. Therefore, the uniform defense strategy returned by Algorithm 2 (which determines the vertex probability p_i for each vertex i) achieves a minimum vertex probability $p' := \min_{i \in [n]} p_i$ for which it holds that:

$$p' = \frac{1}{|\mathcal{L}|} \geq \frac{1}{\frac{2n}{\lambda} + 1 - \frac{3}{\lambda}} = \frac{\frac{\lambda}{n}}{2 + \frac{\lambda - 3}{n}} \geq \frac{1}{2 + \frac{\lambda - 3}{n}} \cdot p^*(G),$$

where the first equality is due to the fact that any leaf $v \in V$ of the spanning tree T of G through which $\mathcal{L}$ was created has $\text{coverage}_{\mathcal{L}}(v) = 1$, and therefore there is such a vertex v in G that is covered by exactly one λ-subgraph; and the last inequality is due to the fact that $p^*(G) \leq \lambda/n$ for any graph G (due to Corollary 3), where $p^*(G)$ is the MaxMin probability of G.

The above inequality implies that if the defender chooses the prescribed strategy the minimum defense ratio cannot be too bad. That is because in the worst case for the defender, each and every attacker will choose a vertex v' on which the aforementioned strategy of the defender results to vertex probability p' (so that the attacker is caught with minimum probability). As a result, the defender will have the minimum possible expected payoff which is $p' \cdot k$. Thus, for the constructed defend strategy and any combination of attack strategies, the resulting strategy profile S yields defense ratio:

$$\mathrm{DR}(G,S) \le \frac{k}{p' \cdot k} \le \left(2 + \frac{\lambda-3}{n}\right) \cdot \frac{1}{p^*(G)} = \left(2 + \frac{\lambda-3}{n}\right) \cdot \mathrm{DR}(G,S^*),$$

where the last equality is due to Lemma 1.

With respect to the running time, notice that Step 1 of Algorithm 2 can be executed in time $O(|V| + |E(G)|) = O(|E(G)|)$. Step 2 can be executed in time $O(|V| + |E(T)|) = O(|V|)$. Finally, Step 3 can be executed in constant time. Therefore, the total running time of Algorithm 2 is $O(|E(G)|)$. □

Corollary 4. *For any graph G there is a polynomial (in both n and λ) time approximation algorithm (Algorithm 2) with approximation factor $1/\left(2 + \frac{\lambda-3}{n}\right)$ for the computation of $p^*(G)$.*

The merit of finding a probability p' that approximates (from below) $p^*(G)$ for a given graph G through an algorithm such as Algorithm 2 is in guaranteeing to the defender that, no matter what the attackers play, she always "catches" at least a portion p' of them in expectation, where the best portion is $p^*(G)$ in an equilibrium. Algorithm 2 guarantees that the defender catches at least $1/\left(2 + \frac{\lambda-3}{n}\right)$ of the attackers in expectation.

6   Bounds on the Price of Defense

In the following theorem we give a lower bound on the PoD for any given n and $2 \le \lambda \le n-1$ by constructing a graph G with particular (very small) $p^*(G)$ (which, by Lemma 1 implies great best defense ratio). Due to lack of space the construction is omitted and can be found in the full version of the paper [1].

Theorem 7. *The* $\mathrm{PoD}(\lambda)$ *is lower bounded by* $\left\lfloor \frac{2(n-1)}{\lambda} \right\rfloor$ *and* $\left\lfloor \frac{2(n-1)}{\lambda+1} \right\rfloor$ *for λ even and odd respectively, when $\lambda \in \{2,3,\dots,n-1\}$.*

Corollary 5. *For any given n and $2 \le \lambda \le n-1$, it holds that* $\left\lfloor \frac{2(n-1)}{\lambda+1} \right\rfloor \le \mathrm{PoD}(\lambda) \le \frac{2(n-1)+\lambda-1}{\lambda}$. *Furthermore, for the trivial cases $\lambda \in \{1,n\}$ it is* $\mathrm{PoD}(1) = n$ *and* $\mathrm{PoD}(n) = 1$.

Proof. The lower bound is established by Theorem 7. The upper bound is due to Theorem 6. For the cases $\lambda = 1$ and $\lambda = n$, observe that the defender's action set is $D = \{\{i\}|i \in V\}$ and $D = \{V\}$ respectively, therefore $p^*(G) = 1/n$ and $p^*(G) = 1$ respectively, and again from Lemma 1 we get the values in the statement of the corollary. □

References

1. Akrida, E.C., Deligkas, A., Melissourgos, T., Spirakis, P.G.: Connected subgraph defense games. CoRR abs/1906.02774 (2019). http://arxiv.org/abs/1906.02774
2. Alon, N., Karp, R.M., Peleg, D., West, D.B.: A graph-theoretic game and its application to the k-server problem. SIAM J. Comput. **24**(1), 78–100 (1995)
3. An, B., Pita, J., Shieh, E., Tambe, M., Kiekintveld, C., Marecki, J.: Guards and protect: Next generation applications of security games. ACM SIGecom Exchanges **10**(1), 31–34 (2011)
4. Aspnes, J., Chang, K.L., Yampolskiy, A.: Inoculation strategies for victims of viruses and the sum-of-squares partition problem. J. Comput. Syst. Sci. **72**(6), 1077–1093 (2006)
5. Attiya, H., Welch, J.: Distributed Computing: Fundamentals Simulations and Advanced Topics. Wiley, Hoboken (2004)
6. Cheswick, W.R., Bellovin, S.M., Rubin, A.D.: Firewalls and Internet Security: Repelling the Wily Hacker, 2nd edn. Addison-Wesley Longman Publishing Co. Inc, Boston (2003)
7. Franklin, M.K., Galil, Z., Yung, M.: Eavesdropping games: a graph-theoretic approach to privacy in distributed systems. J. ACM **47**(2), 225–243 (2000)
8. Jain, M., Conitzer, V., Tambe, M.: Security scheduling for real-world networks. In: Proceedings of the 2013 international conference on Autonomous agents and multiagent systems, pp. 215–222. International Foundation for Autonomous Agents and Multiagent Systems (2013)
9. Kearns, M.J., Ortiz, L.E.: Algorithms for interdependent security games. In: Advances in Neural Information Processing Systems 16 [Neural Information Processing Systems, NIPS], pp. 561–568 (2003)
10. Letchford, J., Conitzer, V.: Solving security games on graphs via marginal probabilities. In: Twenty-Seventh AAAI Conference on Artificial Intelligence (2013)
11. Mavronicolas, M., Michael, L., Papadopoulou, V., Philippou, A., Spirakis, P.: The price of defense. In: Královič, R., Urzyczyn, P. (eds.) MFCS 2006. LNCS, vol. 4162, pp. 717–728. Springer, Heidelberg (2006). https://doi.org/10.1007/11821069_62
12. Mavronicolas, M., Papadopoulou, V., Philippou, A., Spirakis, P.G.: A network game with attackers and a defender. Algorithmica **51**(3), 315–341 (2008)
13. Mavronicolas, M., Papadopoulou, V.G., Persiano, G., Philippou, A., Spirakis, P.G.: The price of defense and fractional matchings. In: 8th International Conference Distributed Computing and Networking, ICDCN, pp. 115–126 (2006)
14. Mavronicolas, M., Papadopoulou, V.G., Philippou, A., Spirakis, P.G.: A graph-theoretic network security game. In: Internet and Network Economics, First International Workshop, WINE, pp. 969–978 (2005)
15. Nash, J.F.: Equilibrium points in n-person games. Proc. Nat. Acad. Sci. U.S.A. **36**(1), 48–49 (1950)
16. Nash, J.: Non-cooperative games. Ann. Math. **54**(2), 286–295 (1951)
17. Spafford, E.H.: The internet worm: crisis and aftermath. Commun. ACM **32**(6), 678–687 (1989)

18. Stallings, W.: Cryptography and network security - principles and practice, 3 edn. Prentice Hall, Upper Saddle River (2003)
19. Vaněk, O., Yin, Z., Jain, M., Bošanský, B., Tambe, M., Pěchouček, M.: Game-theoretic resource allocation for malicious packet detection in computer networks. In: Proceedings of the 11th International Conference on Autonomous Agents and Multiagent Systems-Volume 2, pp. 905–912. International Foundation for Autonomous Agents and Multiagent Systems (2012)
20. Xu, H.: The mysteries of security games: equilibrium computation becomes combinatorial algorithm design. In: Proceedings of the 2016 ACM Conference on Economics and Computation, EC 2016, pp. 497–514. ACM, New York (2016). https://doi.org/10.1145/2940716.2940796, http://doi.acm.org/10.1145/2940716.2940796

Principal-Agent Problems
with Present-Biased Agents

Sigal Oren[1]([⊠]) and Dolav Soker[2]

[1] Ben-Gurion University of the Negev, Beersheba, Israel
sigal3@gmail.com
[2] Google, New York, USA
sdolav@gmail.com

Abstract. We present a novel graph-theoretic principal-agent model in which the agent is present biased (a bias that was well studied in behavioral economics). Our model captures situations in which a principal guides an agent in a complex multi-step project. We model the different steps and branches of the project as a directed acyclic graph with a source and a target, in which each edge has the cost for completing a corresponding task. If the agent reaches the target it receives some fixed reward R. We assume that the present-biased agent traverses the graph according to the framework of Kleinberg and Oren (EC'14) and as such will continue traversing the graph as long as his perceived cost is less than R. We further assume that each edge is assigned a value and if the agent reaches the target the principal's payoff is the sum of values of the edges on the path that the agent traversed. Our goal in this work is to understand whether the principal can efficiently compute a subgraph that maximizes his payoff among all subgraphs in which the agent reaches the target. For this central question we provide both impossibility results and algorithms.

1 Introduction

Present bias is one of the biases known to many on a personal level. It is often highlighted as one of the main reasons making it so difficult to regularly attend the gym, start a diet or complete a paper well ahead of the deadline. Intuitively, individuals with present bias perceive the cost of completing a task today as greater than it really is and as a result postpone it to a later time. This bias has been studied in psychology and behavioral economics since the 50's [1,12, 14,16] and was used to explain different behaviors including procrastination, abandonment of tasks and the benefits of reducing the set of choices [1,5,8,13].

A recent line of work [2–4,6,10,11,17] originating from [9] uses a graph theoretic framework to analyze the behavior of present-biased agents. In this framework, an agent traverses a directed acyclic graph from a source node s to a target node t. We refer to this graph as the *task graph*. The nodes of the task graph represent states of intermediate progress towards some goal and the directed edges between them represent "tasks" the agent must complete to move from one state to the next. Similarly to quasi-hyperbolic discounting [12], we assume

---

The work was done while D. Soker was a student at Ben-Gurion University.

© Springer Nature Switzerland AG 2019
D. Fotakis and E. Markakis (Eds.): SAGT 2019, LNCS 11801, pp. 237–251, 2019.
https://doi.org/10.1007/978-3-030-30473-7_16

that the present-biased agent has a present bias parameter $b > 1$[1] and his cost for executing a task now is multiplied by b. We define the *perceived cost* of an agent currently at v for reaching t as the cost of the min-cost path from v to t in a graph in which the costs of all the outgoing edges from v are multiplied by b. At each node the agent will move to the next node according to the path minimizing his perceived cost. We further assume that there is a reward R on the target and the agent will continue traversing the graph as long as his perceived cost is at most R. The model is formally defined in Sect. 2.

In the classic principal-agent problem [7], the principal should motivate the agent to invest enough effort to succeed at some task. Here we suggest to consider more general situations in which the agent has to complete a complicated multi-step project and there are different subsets of tasks he can work on in order to complete the project. Such situations can be easily modeled using a task graph. Furthermore, as often is the case, we assume that the agent exhibits present bias and traverses the task graph as in [9]. The principal gains a positive payoff only if the agent reaches the target. In this case the payoff of the principal is a function of the path that the agent traversed.

As a motivating example, consider the following scenario: To complete his PhD, a graduate student may select between working on three small-scale projects, each will constitute only a small contribution to the relevant literature, or working on a single large-scale project that will make significant contribution. The cost of completing the large project is 10 while the cost of each small project is 4. This scenario is illustrated in Fig. 1. The student's goal is to obtain a PhD and if he is present-biased with $b > 4/3$ he will follow the lower path. The advisor's goal might be different, as the small projects might have negligible contribution to his CV. Thus, the advisor might take actions to guide the student towards the larger project.

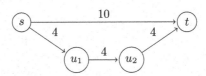

Fig. 1. A task graph for the PhD student and supervisor's principal-agent problem.

Model. Let $G = (V, E)$ be a directed acyclic task graph. Each edge $e \in E$ has a cost $c(e)$ for the agent and a value $v(e)$ for the principal. We consider the following Stackelberg game in which the principal plays first and chooses

---

[1] In quasi-hyperbolic discounting a cost that will be incurred t steps from now is discounted by $\beta \cdot \delta^t$. Where $\beta \in [0, 1]$ is a present bias parameter and $\delta \in [0, 1]$ is an exponential discounting parameter. Our model is equivalent to the (β, δ) model for $\beta = 1/b$ and $\delta = 1$. Similar to [9] and some of its follow-ups we focus on the case that $\delta = 1$ to highlight the effects of present bias.

a subgraph G' of G. Then, the present-biased agent traverses the graph G' as described earlier (see a formal model in Sect. 2). If the agent reaches the target by traversing a path P then the agent receives a reward of R and the principal's payoff is $\sum_{e \in P} v(e)$. Otherwise the principal's payoff is 0.

It is important to note that since the agent is present biased it is often the case that the principal cannot choose a subgraph that is just a simple s-t-path to maximize his utility. To see this, consider the task graph depicted in Fig. 2. Assume that the agent's present bias parameter is $b = 4$, the reward for reaching the target is $R = 6.5$ and the path that maximizes the principal's utility is $P = s, v, u, w, t$. The agent will not traverse P in isolation since its perceived cost at s is $4 \cdot 1 + 3 = 7$. However, if we include the edge (v, t), the agent's perceived cost at s will be reduced to $4 \cdot 1 + 2 = 6$ and it is not hard to see that he will traverse P. We refer to the path $\hat{P} = v, t$ as a *shortcut*. Intuitively, a shortcut is a path from a node in $v \in P$ to t such that at some node in P, prior to v, the agent plans to follow the shortcut and at v the agent decides to continue traversing P instead. In the example in Fig. 2 the agent at s planned to take the shortcut $\hat{P} = v, t$ but when he reached v he decided to take the path $P = v, u, w, t$ instead.

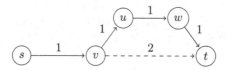

Fig. 2. For $b = 4$, $R = 6.5$ the agent will not traverse $P = s, v, u, w, t$ in isolation but if we include the edge (v, t) the agent will traverse P.

Results. Our goal in this paper is to understand when the principal can *efficiently* compute a subgraph maximizing his payoff among all subgraphs in which the agent reaches the target. We focus on the case that the principal has a unique path P maximizing his payoff. We ask: can the principal motivate the agent to follow P? We name this problem the *P-Motivating Subgraph problem (PMS)*.

We show that in the general case the P-motivating subgraph problem is NP-complete. A natural next step is to look for special cases in which the problem can be solved in polynomial time. One simple case is when all the edges have the same cost. In this case, it is not hard to see that a present-biased agent will behave exactly as a non-biased agent (i.e., $b = 1$). Thus, in this case, all the principal has to do is to check whether the agent will traverse P in isolation or not. In contrast, we show that even if there are two possible values for the costs of the edges the PMS problem is already NP-complete. We refer to this problem as PMS with two weights and formally show:

Theorem 1. *The P-motivating subgraph problem with two weights is NP-complete for any present bias parameter $b > 2$.*

Note that the PMS problem with two weights has a natural interpretation. It is common that tasks can be partitioned according to their difficulty such that the agent attributes a cost w^- to all the easy tasks and a cost w^+ to all the difficult tasks. This is often the case with assignments in courses.

In light of the hardness results we move to consider a parameterized complexity type of result. Our main technical contribution is an algorithm for the PMS problem that runs in time $O(|V|^{O(k)})$ where k is the number of edges of weight w^- on the path P. This assumption is valid in situations in which the principal prefers that the agent will complete more difficult tasks. In such cases the path maximizing the principal's payoff may only include a small number of easy tasks. This is, for instance, the situation in the example of the PhD student and supervisor we gave earlier. Formally, we show

Theorem 2. *For any parameter $b > 1$ PMS is in XP^2 when parameterized by the number of edges of weight w^- in P.*

The technical reason for taking k to be the number of edges of weight w^- on the path is that this is an upper bound on the number of nodes in the path P that are *potential* sources of a shortcut. We prove the theorem by establishing connections between shortcuts and disjoint paths which allow us to build on the work of Shiloach and Perl [15] on computing two disjoint paths between two pairs of source and target nodes in a directed acyclic graph.

Finally, to complete the picture, we show that the PMS problem for general costs is NP-complete even if the path has only one potential source of a shortcut. Together with our result that the PMS problem is hard when the cost of the edges can take two different values this suggests two independent sources of hardness of the PMS problem: (1) a numeric one related to the different costs (2) a graph theoretic one related to computing a set of paths with certain properties.

Related Work. When the principal is indifferent between all paths in the graph we recover the well studied *motivating subgraph problem* presented in [9]: Does a task graph G has a *motivating subgraph* in which the agent will reach the target t? [9] provides a characterization of minimal motivating subgraphs, essentially showing that such subgraphs consist of the path that the agent traverses and some shortcuts. Later, Tang et al. [17] and Albers and Kraft [2] independently proved that the motivating subgraph problem is NP-complete. Albers and Kraft [2] also consider an approximation version of the question, formulated as follows: Consider a graph G and let R^* be the minimal reward for which a motivating subgraph exists. What is the minimal α such that for $R = \alpha R^*$ a motivating subgraph can be found in polynomial time. They present a $\Theta(\sqrt{n})$ approximation algorithm and show that it is NP-hard to get a better approximation ratio. A relaxation of the problem in which it is possible to increase the cost on the edges was also studied [3].

---

[2] XP is the class of parameterized problems that can be solved in time $n^{f(k)}$, where k is the parameter, n is the input's size and f is a computable function.

Paper Outline. In Sect. 2 we define our model and provide some basic observations. In Sect. 3 we present our main technical result: an algorithm for efficiently solving the PMS problem with two weights for a constant number of potential sources for shortcuts. Lastly, in Sect. 4 we provide a reduction showing that the PMS problem with two weights is NP-complete and that for arbitrary weights the PMS problem is NP-complete even if there is a single potential source for a shortcut.

2   Model and Preliminaries

We begin with a detailed definition of the behavior of present-biased agents in the framework of [9]: An agent with present bias parameter $b > 1$ is traversing a directed acyclic task graph $G = (V, E)$ from a source node s to a target node t. Each node in the graph represents a progress point. The edges have weights representing the cost of continuing from one progress point to another. We denote by $c(v, u)$ the cost associated with the edge (v, u) and by $c(P) = \sum_{e \in P} c(e)$ the cost of a path P in G. The perceived cost of an agent currently at v of a path that begins with the edge (v, u) and then continues with the min-cost path connecting u and t is: $b \cdot c(v, u) + d(u, t)$ where $d(u, t)$ is the cost of the min-cost path connecting u and t (when G is unclear from the context we will use $d_G(x, y)$ to denote the cost of the min-cost path connecting x and y in G). At each node the present-biased agent will take the first edge on a path minimizing his perceived cost.

In this paper we focus on a version of the problem where the agent receives a reward R if he reaches the target. Let

$$v_{next} = argmin_{u \in N(v)} b \cdot c(v, u) + d(u, t)$$

where $N(v)$ is the set of v's neighbors. The agent will continue from node v to node v_{next} if the perceived cost of taking the path starting with (v, v_{next}) and continuing with the min-cost path between v_{next} and t is at most R (i.e., $b \cdot c(v, v_{next}) + d(v_{next}, t) \leq R$). Note that after the agent continues to v_{next}, the perceived cost of the agent changes and he might readjust his original plan.

Recall that we focus on the case that the principal has a unique path P maximizing his payoff and ask whether the principal can efficiently compute a subgraph in which the agent will follow the path P. When such a subgraph exists we refer to it as a P-motivating subgraph. Formally, we consider the following problem:

Definition 1 (P-Motivating Subgraph (PMS)). *Given a task graph G, a reward R, an s-t path P and an agent with present bias parameter $b \geq 1$: Does there exist a P-motivating subgraph?*

As discussed in the introduction, the path that a present-biased agent traverses may depend on paths that he will never follow. We refer to such paths as shortcuts:

Definition 2 (Shortcut). *Consider a path $P = u_1, \ldots, u_l$ in a subgraph $H \subseteq G$. For a node $u_i \in P$, consider a path $\hat{P} = u_i, v_2, \ldots, v_{k-1}, t$ such that the part of $\hat{P}$ from v_2 to t is a min-cost path. $\hat{P}$ is a shortcut if*

(1) $d_H(u_i, t) = c(\hat{P})$

(2) $d_H(u_i, t) < c(u_i, u_{i+1}) + d_H(u_{i+1}, t)$

(3) $b \cdot c(u_i, u_{i+1}) + d_H(u_{i+1}, t) < b \cdot c(u_i, v_2) + d_H(v_2, t)$

Notice that conditions (1) and (2) imply that there exists a node on the path P for which the agent will plan to follow the shortcut $\hat{P}$. Condition (3) implies that an agent at u_i will continue to u_{i+1} instead of taking the shortcut $\hat{P}$.

The paper [9] provides a characterization of minimal motivating subgraphs. These are motivating subgraphs that if we remove any edge from the graph the agent will no longer reach t. In the context of P-motivating subgraphs this characterization can be paraphrased as follows:

Proposition 1 ([9]). *A minimal P-motivating subgraph includes the $s - t$ path $P = u_1, \ldots, u_l$ and for each $1 < i < l$ at most a single shortcut $\hat{P}_i = u_i, \ldots, t$. The out-degree of each node on P is at most 2 and the out-degree of each node not on P is exactly 1.*

In this paper we mainly discuss task graphs with two weights: for every $e \in E, c(e) \in \{w^-, w^+\}$ where $w^+ > w^- > 0$. We refer to edges of cost w^- as *light* edges and edges of cost w^+ as *heavy* edges. Observe that when there are two possible costs only nodes that are the source of a light edge can be the source of a shortcut. Furthermore, we have that the cost of the first edge in such a shortcut is w^+:

Observation 3. *Consider an s-t-path $P = u_1, \ldots, u_l$. For graphs with two costs, if $\hat{P} = u_i, v_2, \ldots, v_{k-1}, t$ is a shortcut from u_i to t then $c(u_i, u_{i+1}) = w^-$ and $c(u_i, v_2) = w^+$.*

Proof. Since $\hat{P}$ is a shortcut we have that $c(u_i, v_2) + d(v_2, t) < c(u_i, u_{i+1}) + d(u_{i+1}, t)$ and $b \cdot c(u_i, u_{i+1}) + d(u_{i+1}, t) < b \cdot c(u_i, v_2) + d(v_2, t)$. By adding the two inequalities and rearranging we get that $c(u_i, v_2) > c(u_i, u_{i+1})$. Since we only have two costs this implies that $c(u_i, v_2) = w^+$ and $c(u_i, u_{i+1}) = w^-$. □

As we will see, for graphs with two different costs, the hardness of computing a P-motivating subgraph is tightly related to the number of sources for potential shortcuts. In particular, we will show that when the number of sources for potential shortcuts is constant the problem can be solved in polynomial time.

It is instructive to consider two special cases of the PMS problem with two weights. The first case we consider is when all the edges of the path have cost w^+.

In this case, by Observation 3 we have that the minimal P-motivating subgraph cannot include any shortcuts, thus all we need to check is whether the agent will traverse the path P in isolation or not. For a path of length l this amounts to checking whether $b \cdot w^+ + (l-1)w^+ \le R$. In the full version we provide detailed analysis of another special case: the path P includes exactly one edge of cost w^- and the rest of the edges are of cost w^+. The analysis provides a good introduction to the ingredients and techniques we later use to solve the more general case.

3   An Algorithm for Computing a P-Motivating Subgraph

In this section we present our main technical result: a polynomial time algorithm for PMS with two weights when the number of light edges on P is constant. Our starting point is the minimal motivating subgraph characterization from Proposition 1. We observe that the assertion that in the minimal P-motivating subgraph all nodes, except the sources of the shortcuts, have an out-degree of 1 implies that any two shortcuts $\hat{P}_1$ and $\hat{P}_2$ merge at the first node they intersect. We refer to such paths as converging paths:

Definition 3 (Converging Paths). $\hat{P}_1 = x_1, \ldots, x_k$ and $\hat{P}_2 = y_1, \ldots, y_l$ are converging paths if there exists $1 \le i \le k$ and $1 \le j \le l$ such that $x_1, \ldots, x_{i-1}$ and $y_1, \ldots, y_{j-1}$ are disjoint paths and $x_i, \ldots, x_k = y_j, \ldots, y_l$.

Recall that by Observation 3 we have that for any shortcut $\hat{P} = u_i, v_2, \ldots, v_{k-1}, t$ such that $u_i \in P$, $c(u_i, u_{i+1}) = w^-$. Thus, given a path $P = u_1, \ldots, u_l$ we define the set $S(P) = \{u_i | u_i \in P, \ c(u_i, u_{i+1}) = w^-\}$ which is the set of nodes that are potential sources of shortcuts. This leads us to to the following definition:

Definition 4 (Potential Shortcuts Set). Let $P = u_1, \ldots, u_l$. A set of paths $\hat{\mathcal{P}}$ is a set of potential shortcuts if:

- For every node $u_i \in S(P)$ there exists a single path $\hat{P}(u_i) \in \hat{\mathcal{P}}$ that starts at u_i and ends at t.
- $\forall \hat{P} \in \hat{\mathcal{P}}$, either the cost of the first edge of $\hat{P}$ is w^+ or $P \cup \hat{P} = P \cup (\hat{P} - \hat{P})$.[3]
- All the paths in $\hat{\mathcal{P}}$ are converging.

Consider a set of potential shortcuts $\hat{\mathcal{P}}$. The driving force behind our algorithm is the next proposition showing that in order to determine whether the subgraph $H = P \cup \hat{\mathcal{P}}$ is P-motivating all we need to know is the value of $d_H(u_i, t)$ for each node $u_i \in P$ which is a source of a potential shortcut:

Proposition 2. Let $P = u_1, \ldots, u_l$. There exists a P-motivating subgraph if and only if there exists a set of potential shortcuts $\hat{\mathcal{P}}$ such that for $H = P \cup \hat{\mathcal{P}}$ all the following conditions hold:

---

[3] In order to simplify the analysis we require the set of potential shortcuts to include a path for each node $u_i \in S(P)$. As a result, in some cases the path $\hat{P}(u_i)$ is not a proper shortcut and all its edges are in $P \cup (\hat{\mathcal{P}} - \hat{P}(u_i))$.

- Continuation *Conditions: the agent at u_i continues to u_{i+1}:*
  - *For $u_i \in S(P)$: $b \cdot w^- + d_H(u_{i+1}, t) \leq R$.*
  - *For $u_i \in P \setminus S(P)$: $b \cdot w^+ + d_H(u_{i+1}, t) \leq R$.*
- Shortcuts *Conditions: For each $u_i \in S(P)$ such that the out-degree of u_i in H is 2:*
  - $d_H(u_i, t) = c(\hat{P}(u_i))$.
  - $b \cdot w^- + d_H(u_{i+1}, t) < b \cdot w^+ + (d_H(u_i, t) - w^+)$.

Proof. It is clear from the statement that if there exists a set of potential shortcuts $\hat{P}$ for which the conditions of the proposition hold, then H is P-motivating. For the other direction, let H be a minimal P-motivating subgraph. Since this is a P-motivating subgraph all the *continuation* conditions must hold. Proposition 1 implies that H consists of P and a set of shortcuts $\hat{P}$[4]. Roughly speaking, the shortcuts conditions also hold since they imply that the shortcuts are indeed necessary and that the agent does not traverse some shortcut instead of traversing the path P. More formally, assume toward a contradiction that the shortcuts conditions are violated for some node u_i:

- If the first condition is violated then $d_H(u_i, t) < c(\hat{P}(u_i))$. This is in contradiction to the minimality of H as the graph in which we remove the node following u_i on the shortcut $\hat{P}(u_i)$ is also P-motivating.
- If the first condition holds for all nodes but there exists a node u_i for which the second condition is violated, then $b \cdot w^- + d_H(u_{i+1}, t) \geq b \cdot w^+ + (d_H(u_i, t) - w^+)$. In this case the agent at u_i will take the shortcut $\hat{P}(u_i)$ instead of continuing to u_{i+1}.[5]   □

Observe that these conditions imply that given the path $P = u_1, \ldots, u_l$ and the cost $c(\hat{P}(u_i))$ for every $u_i \in S(P)$, we can compute $d_H(u_j, t)$ for any node $u_j \in P$. Consider a node $u_j \in P \setminus S(P)$ and let u_i be the next node on P such that $u_i \in S(P)$, then $d_H(u_j, t) = (i - j - 1) \cdot w^+ + c(\hat{P}(u_i))$. Thus, we conclude:

Corollary 1. *Consider two sets of potential shortcuts $\hat{P}, \hat{P}'$ such that for every $u_i \in S(P)$: $c(\hat{P}(u_i)) = c(\hat{P}'(u_i))$. The agent will traverse the same path in $H = P \cup \hat{P}$ and in $H' = P \cup \hat{P}'$.*

By the corollary we have that to determine whether a P-motivating subgraph exists it suffices to check all cost combinations of potential shortcuts sets. Let $k = |S(P)|$ denote the number of sources for potential shortcuts. While the number of different sets of potential shortcuts is exponential, the number of cost combinations is at most $|V|^{2k}$ which is polynomial for a constant k. The reason for this is that for every path $\hat{P}$, $c(\hat{P}) = l \cdot w^- + h \cdot w^+$, where $0 \leq l \leq |V|$ is the number of light edges in $\hat{P}$ and $0 \leq h \leq |V|$ is the number of heavy edges in $\hat{P}$. In total there are at most $|V|^{2k}$ options for cost combinations of k paths.

---

[4] We can easily extend this set to a set of potential shortcuts.
[5] We assume that in case that $b \cdot w^- + d_H(u_{i+1}, t) = b \cdot w^+ + (d_H(u_i, t) - w^+)$ the agent will break ties in favor of taking the shortcut.

We use this to develop an algorithm for solving the PMS problem in polynomial time when the number of light edges in P is constant. Formally:

Theorem 4. *PMS is in XP when parameterized by the number of light edges in P.*

Proof. Consider the following algorithm for deciding if a P-motivating subgraph exists. The algorithm gets as an input a task graph G and a path $P = u_1, \ldots, u_l$:

1. Let $S(P) = \{u_i | u_i \in P, c(u_i, u_{i+1}) = w^-\} = \{s_1, \ldots, s_k\}$ be the set of sources of potential shortcuts.
2. Construct a subgraph G' of G by removing from G for each node $u_i \in P \backslash S(P)$ all outgoing edges except for (u_i, u_{i+1}) and for each node $u_i \in S(P)$ all outgoing edges of weight w^- except for the edge (u_i, u_{i+1}).
3. In the graph G' compute a list L such that $(c_1, \ldots, c_k) \in L$ if and only if there exists a set of potential shortcuts $\hat{P}_1, \ldots, \hat{P}_k$ such that $\forall i$, $\hat{P}_i = s_i \to t$ and $c(\hat{P}_i) = c_i$.
4. For each $(c_1, \ldots, c_k) \in L$, check if according to Proposition 2 a set of potential shortcuts with these costs together with P is a P-motivating subgraph. Return *true* if this is the case.
5. Otherwise return *false*.

The correctness of the algorithm is guaranteed by Proposition 2 and Corollary 1. Since the size of L is at most $|V|^{2k}$ it is easy to see that the running time of all steps except of step 3 is polynomial. In Proposition 3 (below) we prove that the running time of step 3 is also polynomial. We note that in case the algorithm returns true then it is possible to compute in polynomial time a P-motivating subgraph. In particular, the table we use to compute the list L in step 3 can be utilized to do so efficiently. □

3.1 Computing All the Cost Combinations of Converging Paths

The next proposition is the heart of the claim that the algorithm in Theorem 4 runs in polynomial time.

Proposition 3. *For a weighted directed acyclic graph $G = (V, E)$ in which all edge costs are in $\{w^-, w^+\}$, source nodes $s_1, \ldots, s_k$ and a target node t, there is an algorithm with running time $O(|V|^{O(k)})$ that computes all the possible cost combinations for a set of k converging paths: $P_1, \ldots, P_k$ such that $\forall 1 \leq i \leq k$, $P_i = s_i \to t$.*

Our algorithm is based on Shiloach and Perl's algorithm for finding two disjoint paths from two different source nodes to two different target nodes ([15]). Given a graph $G = (V, E)$ they construct a product graph $G' = (V', E')$ such that the nodes of the graph G' are pairs of nodes in V. The edges of G' are defined in a way that every path in G' represents two disjoint paths in G. Building on this idea, given a task graph G we define a product graph $\widehat{G^k}$ such that every path in $\widehat{G^k}$ represents k converging paths in G. We then show how to use $\widehat{G^k}$ to compute all cost combinations of converging paths. Note that the problem we consider is more complicated than the one considered in [15] since not only do we consider convergent paths instead of disjoint paths and extend the number of paths to k, but we also need to reason about the costs of the paths and not just whether they exist or not.

Construction of the Product Graph $\widehat{G^k}$. Consider a directed acyclic graph $G = (V, E)^6$, source nodes $s_1, \ldots, s_k$ and a target node t. $\forall v \in V$ let $l(v)$ be the length (i.e., number of edges) of the longest path connecting v and t. Note that since the graph G is a DAG it is possible to compute $l(v)$ in polynomial time[7]. In the product graph $\widehat{G^k} = (\widehat{V^k}, \widehat{E^k})$ the nodes of the graph are k-tuples of nodes of G: $\widehat{V^k} = \{\langle v_1, \ldots, v_k \rangle \mid \forall 1 \leq i \leq k, v_i \in V\}$. An edge $(\langle x_1, \ldots, x_i, \ldots, x_k \rangle, \langle y_1, \ldots, y_i, \ldots, y_k \rangle)$ is in $\widehat{E^k}$ if and only if

1. There exists i such that $x_i \neq y_i$ and $l(x_i) = \max_{1 \leq j \leq k} l(x_j)$.
2. $(x_i, y_i) \in E$.
3. $\forall j$, if $x_j = x_i$ then $y_j = y_i$ else $y_j = x_j$.

We illustrate this construction in Fig. 3.

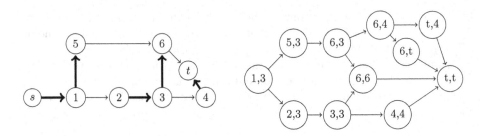

Fig. 3. On the left we have a task graph G with source nodes 1 and 3. Bold edges have weight w^+ and light edges have a weight of w^-. On the right we have a part of the product graph $\widehat{G^2}$ constructed based on G. Consider the node $\langle 1, 3 \rangle$ in $\widehat{G^2}$ since $l(1) = 4$ and $l(3) = 2$ in all the edges leaving $\langle 1, 3 \rangle$ the second coordinate remains fixed. Since $(1, 5), (1, 2) \in E$ we have that $(\langle 1, 3 \rangle, \langle 2, 3 \rangle), (\langle 1, 3 \rangle, \langle 5, 3 \rangle) \in \widehat{E^2}$.

Notice that the number of nodes in the product graph is $O(|V|^k)$ which is polynomial in $|V|$ for a fixed k. Intuitively, every node $\langle x_1, \ldots, x_i, \ldots, x_k \rangle \in \widehat{V^k}$ represents a state of k paths where the current node of the i'th path is $x_i \in V$. An edge $(\langle x_1, \ldots, x_i, \ldots, x_k \rangle, \langle y_1, \ldots, y_i, \ldots, y_k \rangle) \in \widehat{E^k}$ represents an extension of a set of converged paths (i.e., paths that are currently at the same node) by an edge from G, while the rest of the paths remain the same. The next definition formalizes this intuition:

---

[6] In this construction we ignore the costs of the edges.

[7] For example, we can go over the graph in reverse topological order and update for each node the length of the maximum path connecting it to t.

Definition 5 (Induced Path). *A path* $\widehat{P}$ *in the product graph* $\widehat{G^k}$ *defines an induced path* $I(\widehat{P})_i$ *for every coordinate* $1 \leq i \leq k$:

$$I(\widehat{P})_i = \{(x_i, y_i) \in E \mid (\langle x_1, \ldots, x_i, \ldots, x_k \rangle, \langle y_1, \ldots, y_i, \ldots, y_k \rangle) \in \widehat{P} \text{ and } x_i \neq y_i\}$$

As an example, consider the graph in Fig. 3 and let $\widehat{P} = \langle 1,3 \rangle, \langle 2,3 \rangle, \langle 3,3 \rangle,$ $\langle 6,6 \rangle, \langle t,t \rangle$.

The induced paths are $I(\widehat{P})_1 = 1,2,3,6,t$ and $I(\widehat{P})_2 = 3,6,t$.

We prove that every k converging path in G corresponds to a path in $\widehat{G^k}$. Thus, roughly speaking, the complicated problem of computing k converging paths boils down to the simpler problem of finding a path with certain properties in the product graph.

Proposition 4. *Consider a directed acyclic graph G, a set $S = \{s_1, \ldots, s_k\}$ of source nodes and a target node t. G includes k converging paths: $P_1, \ldots, P_k$ such that $\forall i, P_i = s_i \to t$ if and only if there is a path $\langle s_1, s_2, \ldots, s_k \rangle \to \langle t, t, \ldots, t \rangle$ in $\widehat{G^k}$ such that $\forall i, I(\widehat{P})_i = P_i$.*

In the full version we prove each direction of the proposition by a separate induction. To get a feel of the proof that the induced paths of a path in $\widehat{G^k}$ correspond to converging paths, we sketch the proof for $k = 2$. Assume towards contradiction that there exists a path $\widehat{P}$ in $\widehat{G^2}$ that $I(\widehat{P})_1$ and $I(\widehat{P})_2$ are not converging. This implies that there exists a node $v \in V$ in which P_1 and P_2 intersect and there are two different nodes $u, u' \in V$ such that $(v, u) \in P_1$ and $(v, u') \in P_2$. By construction it is easy to see that the path $\widehat{P}$ cannot include the node $\langle v, v \rangle$ since after visiting this node the two induced paths have to converge. The only option is that $\widehat{P}$ first visited a node $\langle a, v \rangle$ for $a \in V$ (not necessarily immediately) afterwards it visited a node $\langle b, u \rangle$ for $b \in V$ and later it visited a node $\langle v, c \rangle$ for $c \in V$. We observe that in such case $l(a) < l(v)$ and thus, by construction, $\widehat{P}$ has to reach from $\langle a, v \rangle$ a node $\langle b, v \rangle$ such that $l(b) = l(v)$ before reaching the node $\langle b, u \rangle$. By the assumption that $\widehat{P}$ does not include $\langle v, v \rangle$ we have that $b \neq v$. However, since $l(b) = l(v)$ and $b \neq v$, it has to be the case that v is not reachable from b, in contradiction to the assumption that $\langle v, c \rangle \in \widehat{P}$.

An Algorithm for Computing all Possible Cost Combinations. We now use the product graph to compute a list of all possible cost combinations. To this end we first observe that the product graph of a DAG is a DAG:

Observation 5. *The product graph of a directed acyclic graph (DAG) is a DAG.*

Proof. Let G be a directed acyclic graph and assume towards contradiction that $\widehat{G}$ contains a cycle of length $d > 1$:

$$\widehat{C} = \langle v_1^1, \ldots, v_k^1 \rangle \cdots \to \cdots \langle v_1^d, \ldots, v_k^d \rangle, \langle v_1^1, \ldots, v_k^1 \rangle.$$

Since the cycle has length greater than 1 there has to be an index j such that $v_j^1 \neq v_j^2$ this implies that $I(\widehat{C})_j = v_j^1, v_j^2, \ldots, v_j^1$. Thus, by Proposition 4 we have that $I(\widehat{C})_j$ is a cycle in G in contradiction to the fact that G is a DAG. $\qquad \square$

Our algorithm goes over the nodes of $\widehat{G}$ in reverse topological order. For each node $\langle v_1, \ldots, v_k \rangle$ it maintains a list of all the potential cost combinations for k converging path from $\langle v_1, \ldots, v_k \rangle$ to $\langle t, \ldots, t \rangle$. As previously discussed, there are at most $|V|^{2k}$ possible cost combinations for the k converging paths and this essentially implies that the following algorithm runs in polynomial time.

1. Initialization: For each node x define an empty list $L(x)$. Process the nodes of the graph $\widehat{G}$ in reverse topological order. For every node $x = \langle x_1, \ldots, x_k \rangle$ and every neighbor of x, $y = \langle y_1, \ldots, y_k \rangle \in N(x)$:
    (a) Let $X = \{j | x_j \neq y_j\}$ and let $e = (x_i, y_i) \in E$ for some $i \in X$.[8]
    (b) For every cost combination $(c_1, \ldots, c_k) \in L(y)$:

    $$- \ \forall 1 \leq j \leq k, \ c'_j = \begin{cases} c_j + c(e), & j \in X \\ c_j, & j \notin X \end{cases}$$

    $-$ If $(c'_1, \ldots, c'_k) \notin L(x)$, add it to $L(x)$.
2. Return $L(\langle s_1, \ldots, s_k \rangle)$.

To see that the algorithm runs in polynomial time, observe that the lists we keep for each node are of polynomial size (their length is at most $|V|^{2k}$). The running time of the algorithms under the most naive implementation in which L is simply a linked list is $O(|V|^k \cdot |V|^k \cdot |V|^{2k} \cdot k \cdot |V|^{2k}) = O(k \cdot |V|^{6k})$.[9]

The correctness of the algorithm is obtained by applying Proposition 4 to prove that $(c_1, \ldots, c_k) \in L(\langle s_1, \ldots, s_k \rangle)$ if and only if there exist k converging paths $P_1, \ldots, P_k$ such that $\forall 1 \leq i \leq k$, $P_i = s_i \rightarrow t$ and $c(P_i) = c_i$.

4 Hardness of Computing a P-Motivating Subgraph

Lastly, we discuss the hardness of the P-motivating subgraph problem. PMS is clearly in NP: given a subgraph we can check in polynomial time whether a present-biased agent will reach t. In the full version we prove:

Proposition 5. *The PMS problem with arbitrary weights is NP-complete for any $b > 1$ even when there is only a single node in P that is a potential source for a shortcut.*

The proof is based on a reduction from the subset sum problem: given a set of integers $X = \{x_1, \ldots, x_n\}$ and an integer L, does there exist a subset $S \subseteq X$ such that $\sum_{x \in S} x = L$? The rough idea is to construct a task graph and a path P such that there exists a P-motivating subgraph if and only if there exists a shortcut of cost exactly x. The costs of the edges that can form the path are chosen such that for each subset $S \subseteq X$ there exists a path of cost $\sum_{x \in S} x$.

---

[8] By construction $\forall i, j \in X$ we have that $x_i = x_j$ and $x_i = y_j$.

[9] By replacing the list with a binary matrix of size $(|V| \times |V|)^k$ in which a cell $(h_1, l_1, \ldots, h_k, l_k)$ represents whether there are k converging paths such that each path i has h_i edges of weight w^+ and l_i edges of weight w^-, we can reduce the running time to $O(k \cdot |V|^{4k})$.

The PMS problem is also hard in the restricted case in which the edge costs can only have two values:

Theorem 6. *The PMS problem with two weights is NP-complete for any $b > 2$.*

The proof of this theorem is composed of two reductions. First, we define the following problem:

Definition 6 (Equal-Weight Converging Paths (EWCP)). *Given a weighted DAG G, k source nodes $\{s_1, \ldots, s_k\}$, a target node t and a real number x, determine whether there exist k converging paths $P_1 = s_1 \to t, \ldots, P_k = s_k \to t$ such that $\forall 1 \le i \le k$, $c(P_i) = x$.*

We reduce from 3-SAT to show that EWCP with two weights is NP-complete. The reduction borrows from a reduction of [2] showing that the k-disjoint connecting paths problem is NP-complete. Formally, in the full version we show:

Proposition 6. *The equal-weight converging path problem where all edges have weights w^- or w^+ is NP-complete for any $w^+ > w^- > 0$.*

Then we reduce from EWCP with two weights to PMS. Roughly speaking, given an instance of EWCP we create a path $P = s, v_1, \ldots, v_{2k+2}, \ldots, t$ such for $1 \le i \le k$, each node v_{2i-1} is connected to an EWCP source node s_i. The idea is to use the k converging paths as shortcuts. Without them, the agent will not be able to traverse P.

5   Discussion

Present bias is a common and central bias inhibiting individuals from completing projects and following their plans in general. In this paper we present a new variant of the principal-agent problem that features present-biased agents. Our results demonstrate that when the optimal path for the principal is unique it is NP-hard for the principal to compute his optimal strategy even when there are only two possible values for the costs of the edges. This is since in such a case if there exists a P-motivating subgraph the algorithm must return it and we proved that the PMS problem with two weights is NP-hard.

Moreover, we conclude that the following approximation variant of the problem does not admit an FPTAS unless P = NP: compute a motivating subgraph in which the payoff of the principal approximates the maximal payoff in any motivating subgraph. To show the hardness of this problem we reduce from PMS. Given the input graph for PMS we set $v(e) = v > 0$ for any $e \in P$ and $v(e) = 0$ for any $e \notin P$. Note that in this instance it is hard to distinguish between the case where the payoff of the principal in the optimal motivating subgraph is $n \cdot v$ and the case where the best motivating subgraph does not use all edges of P and thus the payoff is at most $(n - 1) \cdot v$. Hence, for a sufficiently small value of ε, an FPTAS algorithm will have to return a P-motivating subgraph if one exists. We conclude that there is no FPTAS for our principal-agent problem unless $P = NP$. A fascinating open question is to determine the best approximation ratio that is possible to achieve in polynomial time. Another interesting direction is to identify other definitions for the principal's utility under which computing the motivating subgraph maximizing the payoff of the principal is tractable.

Our model naturally lends itself to a variety of extensions opening up new and exciting open questions. For example, consider the variant in which the reward comes out of the pocket of the principal and the principal has to find a motivating subgraph maximizing $\sum_{e \in P} v(e) - R$. Another direction to explore is allowing the principal to increase the costs on the edges instead of just removing edges, in a manner similar to [3].

References

1. Akerlof, G.A.: Procrastination and obedience. Am. Econ. Rev. Pap. Proc. **81**(2), 1–19 (1991)
2. Albers, S., Kraft, D.: Motivating time-inconsistent agents: a computational approach. In Proceedings 12th Workshop on Internet and Network Economics (2016)
3. Albers, S., Kraft, D.: On the value of penalties in time-inconsistent planning. In: 44th International Colloquium on Automata, Languages, and Programming, ICALP, Warsaw, Poland, pp. 10:1–10:12 (2017)
4. Albers, S., Kraft, D.: The price of uncertainty in present-biased planning. In Proceedings 13th Workshop on Internet and Network Economics (2017)
5. Ariely, D., Wertenbroch, K.: Procrastination, deadlines, and performance: self-control by precommitment. Psychol. Sci. **13**(3), 219–224 (2002)
6. Gravin, N., Immorlica, N., Lucier, B., Pountourakis, E.: Procrastination with variable present bias. In: Proceedings of the 2016 ACM Conference on Economics and Computation, EC 2016, p. 361. ACM (2016)
7. Grossman, S.J., Hart, O.D.: An analysis of the principal-agent problem. Econometrica: J. Econometric Soc. **51**(1), 7–45 (1983)
8. Kaur, S., Kremer, M., Mullainathan, S.: Self-control and the development of work arrangements. Am. Econ. Rev. Pap. Proc. **100**(2), 624–628 (2010)
9. Kleinberg, J., Oren, S.: Time-inconsistent planning: a computational problem in behavioral economics. In: Proceedings of the Fifteenth ACM Conference on Economics and Computation, EC 2014, pp. 547–564. ACM (2014)
10. Kleinberg, J., Oren, S., Raghavan, M.: Planning problems for sophisticated agents with present bias. In: Proceedings of the 2016 ACM Conference on Economics and Computation, EC 2016, pp. 343–360. ACM (2016)
11. Kleinberg, J., Oren, S., Raghavan, M.: Planning with multiple biases. In: Proceedings of the 2017 ACM Conference on Economics and Computation, EC 2017, pp. 567–584. ACM (2017)
12. Laibson, D.: Golden eggs and hyperbolic discounting. Q. J. Econ. **112**(2), 443–478 (1997)
13. O'Donoghue, T., Rabin, M.: Doing it now or later. Am. Econ. Rev. **89**(1), 103–124 (1999)

14. Pollak, R.A.: Consistent planning. Rev. Econ. Stud. **35**(2), 201–208 (1968)
15. Shiloach, Y., Perl, Y.: Finding two disjoint paths between two pairs of vertices in a graph. J. ACM (JACM) **25**(1), 1–9 (1978)
16. Strotz, R.H.: Myopia and inconsistency in dynamic utility maximization. Rev. Econ. Stud. **23**(3), 165–180 (1955)
17. Tang, P., Teng, Y., Wang, Z., Xiao, S., Xu, Y.: Computational issues in time-inconsistent planning. In: AAAI, pp. 3665–3671 (2017)

On a Simple Hedonic Game
with Graph-Restricted Communication

Vittorio Bilò[1], Laurent Gourvès[2(✉)], and Jérôme Monnot[2]

[1] Department of Mathematics and Physics, University of Salento, Lecce, Italy
vittorio.bilo@unisalento.it
[2] Université Paris-Dauphine, Université PSL, CNRS, LAMSADE, 75016 Paris, France
{laurent.gourves,jerome.monnot}@dauphine.fr

Abstract. We study a hedonic game for which the feasible coalitions are pre-scribed by a graph representing the agents' social relations. A group of agents can form a feasible coalition if and only if their corresponding vertices can be spanned with a star. This requirement guarantees that agents are connected, close to each other, and one central agent can coordinate the actions of the group. In our game everyone strives to join the largest feasible coalition. We study the existence and computational complexity of both Nash stable and core stable partitions. Then, we provide tight or asymptotically tight bounds on their quality, with respect to both the price of anarchy and stability, under two natural social functions, namely, the number of agents who are not in a singleton coalition, and the number of coalitions. We also derive refined bounds for games in which the social graph is restricted to be claw-free. Finally, we investigate the complexity of computing socially optimal partitions as well as extreme Nash stable ones.

Keywords: Hedonic games · Price of anarchy/stability · Graphs

1 Introduction

Coalition formation, that is the process by which agents gather into groups, is a fervent research topic at the intersection of Multi-Agent Systems, Computational Social Choice and Algorithmic Game Theory. One of the most studied models of coalition formation is that of *hedonic games* [6,8,13,19], where agents have preferences over all possible coalitions they can belong to. As agents are usually assumed to be self-interested, an acceptable *outcome* for a hedonic game, that is a partition of agents into coalitions, needs to be resistant to agents' deviations. Several notions of stability have been inves-tigated in the literature, such as, individual stability, Nash stability, core stability (see, for instance, [1]).

In a recent paper, Igarashi and Elkind [20] add a further constraint to the definition of acceptable outcomes for hedonic games, by introducing the notion of *feasible coali-tion*: a coalition is feasible if and only if it complies with some prescribed properties. For instance, they assume that the set of agents corresponds to the vertex set $V(G)$ of a social graph G and require a coalition to induce a *connected* subgraph of G.

In this work, we restrict the feasibility constraint of [20] to coalitions inducing a subgraph of G *admitting a spanning star*. This requirement guarantees that agents are

© Springer Nature Switzerland AG 2019
D. Fotakis and E. Markakis (Eds.): SAGT 2019, LNCS 11801, pp. 252–265, 2019.
https://doi.org/10.1007/978-3-030-30473-7_17

connected, close to each other, and one central agent can coordinate the actions of the group. We apply this framework within a basic model, falling within the class of additive separable symmetric hedonic games, in which an agent's utility is defined by the cardinality of the coalition she belongs to.

1.1 Game Model, Definitions and Notation

Given an unweighted and undirected graph $G = (V, E)$, a *coalition* is any non-empty subset of V. A *partition* of V is a set of pairwise disjoint coalitions whose union equals V. We denote by $\mathcal{F} \subseteq 2^V$ the set of *feasible coalitions*. We shall consider $\mathcal{F} = \{C \in 2^V \mid G[C] can be spanned with a star\}$, where $G[C]$ is the subgraph of G induced by C and a star is a tree of depth at most 1. A star on one vertex is called *trivial*.

Given an undirected, unweighted and connected graph G, game $(G, \mathcal{F})$ is defined as follows. Each vertex of G is associated with an agent in the game. Let Π be the set of partitions of V. For a partition $\pi \in \Pi$ and a vertex $i \in V$, denote as $\pi(i)$ the coalition in π containing i. The utility of i in π is defined as

$$u_i(\pi) = \begin{cases} |\pi(i)| & \text{if } \pi(i) \in \mathcal{F}, \\ 0 & \text{otherwise.} \end{cases} \qquad (1)$$

We say that agent i has a *profitable deviation* in π if either $\pi(i) \notin \mathcal{F}$, or there exists a coalition $C \in \pi$ such that $C \cup \{i\} \in \mathcal{F}$ and $|C| \geq |\pi(i)|$. In the first case agent i can form the singleton coalition $\{i\}$ which is feasible because it is spanned by a trivial star and yields a utility equal to $1 > u_i(\pi) = 0$. In the second case, agent i increases her utility by joining C. More generally, a set of agents S has a *joint profitable deviation* in π if there exists a partition π', obtained from π by letting every agent $i \in S$ leave coalition $\pi(i)$ and either join another coalition in π or form a new one, such that $u_i(\pi') > u_i(\pi)$ for each $i \in S$.

A partition π is *Nash stable* (resp. *Strong Nash stable*) if no agent (resp. no set of agents) has a profitable deviation (resp. a joint profitable deviation) in π. Nash and strong Nash stable partitions correspond to (pure) Nash and Strong equilibria respectively. We say that a partition is *feasible* if each of its coalitions belongs to $\mathcal{F}$. It is easy to see that, by definition, any Nash stable partition is feasible. In a *core stable* partition π, there is no coalition C for which all its members are better off by forming C. The set of core stable partitions (simply called the *core*) is a subset of the set of Nash stable ones. Strong Nash stability implies core stability but the converse is not always true. Nevertheless, because every agent only cares about the size of its coalition if it is feasible, like in *anonymous hedonic games* [6], strong Nash stability and core stability coincide in our game. We will use the word core instead of strong Nash for the rest of this article.

A *social function* is a function, defined from Π to $\mathbb{R}_{\geq 0}$, measuring the social value of a partition. A social optimum is a partition $\pi^* \in \mathcal{F}$ optimizing a given social function. We consider the following two social functions: *sociality*, defined as $soc(\pi) = |\{i \in V : |\pi(i)| > 1\}|$, and *fragmentation*, defined as $frag(\pi) = |\pi|$. Sociality needs to be maximized, while fragmentation needs to be minimized.

We evaluate the efficiency of stable partitions by means of the well established notions of *price of anarchy* (PoA), *price of stability* (PoS) and their strong versions

(SPoA and SPoS). The PoA (resp. PoS) of game $(G, \mathcal{F})$ with respect to sociality is defined as $\frac{soc(\pi^*)}{soc(\pi)}$, where π^* is a social optimum and π is a Nash stable partition of minimum (resp. maximum) sociality. The PoA (resp. PoS) of game $(G, \mathcal{F})$ with respect to fragmentation is defined as $\frac{frag(\pi)}{frag(\pi^*)}$, where π^* is a social optimum and π is a Nash stable partition of maximum (resp. minimum) fragmentation. By substituting the notion of Nash stable partition with that of core stable one, we obtain the definition of both the SPoA and SPoS. Observe that, for a given game $(G, \mathcal{F})$ and independently of the chosen social function, $\text{PoS}(G, \mathcal{F}) \leq \text{SPoS}(G, \mathcal{F}) \leq \text{SPoA}(G, \mathcal{F}) \leq \text{PoA}(G, \mathcal{F})$.

1.2 Some Motivations

Requiring subgraphs spanned by a star can be interpreted as restricting the model of [20] to communication patterns of small length. In comparison, unbounded multi-hop communication may be costlier, slower, and prone to errors or misunderstandings. Therefore, distant communication should be avoided. These observations provide both theoretical and practical motivations for the constraint considered in this work. Moreover, our game, complemented with a suitable social function, naturally models several interesting scenarios, some of which are outlined in the following.

Unions. Assume that $V(G)$ models a set of workers of a given company, the edge set $E(G)$ the ideological acquaintance, and that the power of a union is measured by its size. Thus, workers want to join the largest unions. However, a union can survive only if it has a leader who is ideologically close to its partners. For this model, it makes sense considering the fragmentation social function that aims at minimizing the number of unions representing the workers and augmenting their negotiation power.

Group Buying. Assume that $V(G)$ models a set of buyers, all interested in the same product, and $E(G)$ their knowledge/trust relationships. Buyers enjoy flowing into large buying groups, as the larger the group, the better the purchasing conditions they can fetch. However, negotiation with the seller is carried out by one group member only, who then gets also in charge of redistributing what is bought to the others. Thus, this agent needs to be trusted by everybody. If one considers the case in which the product has a fixed price and the share each agent pays is equal to the price divided by the cardinality of her buying group, fragmentation becomes equal to the sum of the costs of all players, i.e., to the utilitarian social function.

Sport Tournaments. Assume that $V(G)$ models a set of teams and there is an edge between two teams if they are close enough to meet and practice a given sport (e.g. football). The participants gather into groups in such a way that a central member can host all teams of its group and organize a tournament. Teams will prefer larger tournaments to small ones in order to maximize the number of opponents against which they can play a match. Sociality, here, aims at involving as many teams as possible into the organization of local tournaments, no matter how big those events are.

1.3 Contribution

We focus on the complexity and efficiency of both Nash stable and core stable partitions. Some proofs are omitted due to space constraints.

As to complexity results, we provide two constructive evidences showing existence of core stable partitions, and so also of Nash stable ones. In particular, any sequence of joint profitable deviations converges to a core stable partition, while Theorem 1 characterizes the core as the set of all possible outputs of a polynomial time greedy algorithm. These two facts complement each other, as the first does not need any coordination among the agents, but provides no guarantees of fast convergence, whereas the second, while requiring centralized coordination (dictated by the greedy choices of the proposed algorithm), guarantees efficient computation. We then provide bounds on the PoA, PoS, SPoA and SPoS under social functions sociality and fragmentation. In particular, we consider games induced by general (unrestricted) graphs and games induced by claw-free graphs. These results are summarized in Fig. 1.

Efficiency of stable partitions	General graphs		Claw-free graphs	
	Sociality	Fragmentation	Sociality	Fragmentation
PoA	$\frac{n}{3}$	$\frac{n-2}{2}$	$\frac{n}{n-\lfloor\frac{n-1}{2}\rfloor}$	$\left\lfloor\frac{2k}{k+1},2\right\rfloor$
SPoA	$\frac{n}{1+\sqrt{n-1}}$	$\left\lfloor\lceil n/2\rceil,\frac{n}{4}+\frac{11}{20}\right\rfloor$	$\frac{n}{n-\lfloor\frac{n-1}{2}\rfloor}$	$\left\lfloor\frac{2k}{k+1},2\right\rfloor$
SPoS	$\left\lfloor\frac{n}{2+\sqrt{n-2}},\frac{n}{1+\sqrt{n-1}}\right\rfloor$	$\left\lfloor\frac{\lceil n/2\rceil}{2},\frac{n}{4}+\frac{11}{20}\right\rfloor$	$\frac{n}{n-\lfloor\frac{n-1}{2}\rfloor}$	$\left\lfloor\frac{2k}{k+1},2\right\rfloor$
PoS	$\left\lfloor\frac{n}{2\sqrt{n-1}},\frac{n}{1+\sqrt{n-1}}\right\rfloor$	$\left\lfloor\frac{\lceil n/2\rceil}{2},\frac{n}{4}+\frac{11}{20}\right\rfloor$	$\frac{n}{n-\lfloor\frac{n-1}{2}\rfloor}$	1

Fig. 1. Bounds on the efficiency of both Nash stable and core stable partitions with respect to social functions sociality and fragmentation. In the last column $k = \sqrt{n+4} - 2$.

It turns out that the presence of claws in the social graph defining the game is a provable source of inefficiency that has to be taken into account, for instance, whenever mechanisms for coping with selfish behavior can be designed and applied.

Finally, we also address the problem of computing outcomes with prescribed welfare guarantees. In particular, we consider the computation of social optima and extreme (i.e., either best or worst) Nash stable partitions under both social functions. We design a polynomial time algorithm to compute a social optimum for sociality and prove that all other problems are **NP**-hard, except for that of computing a worst Nash stable partition under fragmentation whose complexity remains open.

1.4 Related Work

The language for describing which coalitions are feasible, and how agents value them, is a critical feature in hedonic games. Like in [9, 12], feasible coalitions and their values can be described with the help of a (directed) graph. Igarashi and Elkind [20] and Peters [27] have considered hedonic games defined over graphs: agents are the vertices and feasible coalitions satisfy a given graph property. Regarding the worth of a coalition, a simple and compact representation is given by *additively separable functions* [6]: each agent i assigns a value ν_{ij} to agent j and agent i's worth for a coalition C is $\sum_{j\in C}\nu_{ij}$. See, for example, [23] for a simple hedonic game where $\nu_{ij}\in\{0,1\}$. Our work falls in this framework, as everybody wants to be part of the largest coalition

---

Algorithm 1. Greedy Core
___
Input: Game $(G, \mathcal{F})$ where $G = (V, E)$ is a graph.
Output: A core stable partition π.
1: **while** $V \neq \emptyset$ **do**
2:     take $i \in V$ maximizing the degree $d_G(i)$
3:     $\pi(i) \leftarrow N_G[i]$ *$N_G[i]$ is the closed neighbourhood of i*
4:     $G \leftarrow G[V \setminus N_G[i]]$
5: **end while**
6: **return** π

(ν_{ij} is always 1), and a coalition is feasible if and only if the vertices representing the agents can be covered with a star.

Regarding existing games defined over graphs, Panagopoulou and Spirakis [25] and Escoffier *et al.* [14] studied a game where the vertices of a graph have to select a color (each color corresponds to a coalition), and a vertex's payoff is the number of agents with the same color, provided that it constitutes an independent set.

In many works including the famous stable marriage problem, the coalitions form a matching of a graph (see for example [19]).

For bounds on the price of anarchy and the price of stability in some classes of hedonic games, one can see [4,7,15,21,22]. The computation of socially optimal partitions in hedonic games, according to different social functions, has been treated in [5,9–11,16]. Finally, we refer the reader to [2,3,24,26] for an extensive treatment of the computational complexity of both decision and search problems related to stable partitions in hedonic games.

2   On Core Stable Partitions

Given a partition π, a *strong Nash dynamics* of length ℓ starting from π is a sequence of partitions $\langle \pi = \pi_0, \pi_1, \ldots, \pi_\ell \rangle$ such that, for each $j \geq 1$, π_j is obtained as a result of a joint profitable deviation of some set of agents in π_{j-1}.

A game has the *lexicographical improvement property* (LIP) [18], if every joint profitable deviation strictly decreases the lexicographical order of a certain function defined on Π. It is not difficult to see that for any graph G, game $(G, \mathcal{F})$ has the LIP property if one considers the n-dimensional vector consisting of the values $u_i(\pi)$ for each $i \in V$, sorted in non-increasing order. Thus, a core stable partition always exists as the length of every strong Nash dynamics is finite.

We now prove that, if centralized coordination is allowed, a core stable partition can be computed in polynomial time. This is done by proving that the core is completely characterized by the set of all possible outputs of Algorithm 1.

Theorem 1. *A partition is core stable if and only if it is the output of Algorithm 1 (according to some specific tie breaking rule).*

Proof. The algorithm outputs a feasible partition π. First we show that π is core stable. Assume, by way of contradiction, that π is not core stable. Then, there exists a joint profitable deviation in π for a coalition S such that $|S| \geq 1$. Let i be the agent getting the highest utility in π among the ones belonging to S. As i improves, she will end up in a coalition $C \in \mathcal{F}$ such that $|C| > |\pi(i)|$. If C is created by the algorithm before $\pi(i)$, then i should belong to C in π: a contradiction to the greedy choice. Hence, either C is created by the algorithm after $\pi(i)$ or it is a new coalition created by the joint deviation. As i gets the highest utility in S, C only contains vertices belonging to coalitions created by the algorithm at the step $\pi(i)$ is created or after. This implies that, at the step in which $\pi(i)$ is created, C could have been created too, thus contradicting the greedy choice.

Now, we show that any core stable partition π can be the output of the above algorithm. List the coalitions in π by non-increasing cardinality and define a tie breaking rule R that gives priority to the coalitions in π according to the given ordering, and gives higher priority to a coalition in π with respect to any coalition not in π. Run the algorithm according to rule R to obtain a partition π'. Assume, by way of contradiction, that $\pi \neq \pi'$. List the coalitions in π' by non-increasing cardinality, breaking ties according to R. Let j be the first index at which the two sequences become different and denote as C and C' the j-th coalition in the ordering defined on π and π', respectively. By the definition of R, it must be $|C'| > |C|$ which implies that all vertices in C' can perform a joint profitable deviation in π. This contradicts the assumption that π is a core stable partition.   □

3 Efficiency of Core/Nash Stable Partitions

In this section, we focus on the efficiency of Nash or core stable partitions with respect to both social functions sociality and fragmentation. Before characterizing the price of anarchy, we prove some preliminary lemmas.

Lemma 1. *If G admits a spanning star, then any Nash stable partition for game $(G, \mathcal{F})$ is formed by a unique coalition $V(G)$.*

Lemma 2. *If G is connected with $n \geq 3$, then any Nash stable partition for game $(G, \mathcal{F})$ contains either 2 coalitions of size at least 2 or a coalition of size at least 3.*

We are now ready to characterize the PoA. As it is equal to 1 for any game with three players (by Lemma 1), in the remaining of the section we shall assume $n \geq 4$.

Theorem 2. *For any game with n players, the price of anarchy is $n/3$ with respect to sociality and $(n - 2)/2$ with respect to fragmentation. Both bounds are tight.*

Proof. Fix a Nash stable partition π. By Lemma 2, $soc(\pi) \geq 3$. As the sociality of any partition is upper bounded by n, we obtain an upper bound of $n/3$ on the price of anarchy. By Lemma 1, if the fragmentation of the social optimum is 1, then the price of anarchy is 1, hence assume that its fragmentation is at least 2. By Lemma 2, $frag(\pi) \leq n - 2$ which yields the desired upper bound on the price of anarchy.

A matching lower bound for both social functions can be obtained by considering the game induced by the graph G depicted in Fig. 2. The partition π such that $\pi(v_1) = \{v_1, v_2, v_3\}$ and $\pi(v_i) = \{v_i\}$ for $i > 3$ is Nash stable and has $soc(\pi) = 3$ and

$frag(\pi) = n - 2$. On the other hand, the partition π^* such that $\pi^*(v_1) = \{v_1, v_2\}$ and $\pi^*(v_3) = \{v_3, \ldots, v_n\}$ has a sociality of n and a fragmentation of 2. Comparing the two partitions yields the desired lower bounds.   □

Fig. 2. A graph yielding a game with worst-case price of anarchy.

It is also possible to give an upper bound on the price of anarchy with respect to both social functions which depends on the stability number $\alpha(G)$ of graph G, where $\alpha(G)$ is the largest size of an independent set.

Theorem 3. *The price of anarchy of game* $(G, \mathcal{F})$ *is at most* $\frac{n}{n-\alpha(G)+1}$ *with respect to sociality and at most* $\frac{\alpha(G)}{2}$ *with respect to fragmentation.*

Proof. Fix a Nash stable partition π. As the set of centres of the stars spanning the subgraph induced by each coalition in π forms an independent set in G, it follows that $frag(\pi) \le \alpha(G)$. Thus, by Lemma 2, it follows that the price of anarchy with respect to fragmentation is at most $\frac{\alpha(G)}{2}$. Moreover, as G has at least one edge, the number of singleton coalitions in π can be at most $frag(\pi) - 1 \le \alpha(G) - 1$. So, $soc(\pi) \ge n - \alpha(G) + 1$.   □

For the efficiency of core stable outcomes, we have the following results.

Theorem 4. *For any game* $(G, \mathcal{F})$ *with n players* $SPoS(G, \mathcal{F}) \in \left[\frac{n}{2+\sqrt{n-2}}, \frac{n}{1+\sqrt{n-1}} \right]$ *and* $SPoA(G, \mathcal{F}) = \frac{n}{1+\sqrt{n-1}}$ *hold for the sociality function.*

Proof. For the upper bound, consider a coalition C in a core stable partition π with $|C| = k > 2$ and let c be the center of the spanning star of $G[C]$. No vertex i belonging to a singleton coalition can be adjacent to c, otherwise i would have a profitable deviation in π. Any vertex $i \in C$, with $i \ne c$ can be adjacent to at most $k - 2$ vertices belonging to a singleton coalition, otherwise these vertices, together with i and c would have a joint profitable deviation in π. It follows that for any coalition in π with $k > 2$ vertices, there can be at most $(k - 1)(k - 2)$ singleton coalitions in π. Let s_k be the number of coalitions in π with k vertices for $k \ge 1$. We deduce $\sum_{k=3}^{n} \left(s_k(k^2 - 3k + 2) \right) \ge s_1$ which gives $\sum_{k=2}^{n} \left(s_k(k^2 - 2k + 2) \right) \ge n$ by adding $\sum_{k=2}^{n} k s_k$ on both sides.

As the sociality in a social optimum is upper bounded by n and $soc(\pi) = \sum_{k=2}^{n}(s_k k)$, we obtain that the strong price of anarchy is at most $\frac{n}{\sum_{k=2}^{n}(s_k k)} \le \frac{\overline{k}^2 - 2\overline{k} + 2}{\overline{k}}$, where $\overline{k} = \max\{k : s_k > 0\}$. Moreover, as $soc(\pi) \ge \overline{k}$, the strong price of anarchy is trivially upper bounded by $n/\overline{k}$. It follows that the minimum of the two derived bounds is maximized for $\bar{k}^2 - 2\bar{k} + 2 = n \Leftrightarrow \bar{k} = 1 + \sqrt{n-1}$, which yields the desired upper bound on both $SPoA(G, \mathcal{F})$ and $SPoS(G, \mathcal{F})$.

For the lower bound on the strong price of anarchy, consider the game induced by a tree G rooted at vertex x_0 which has ℓ children denoted as $x_1, \ldots, x_\ell$. For each $i \in [\ell]$, x_i has $\ell - 1$ children, so that G has $n = \ell^2 + 1$ vertices. By using the characterization of core stable partitions given in Theorem 1, it follows that the partition π whose unique non-singleton coalition is $\{x_0, x_1, \ldots, x_\ell\}$ is a core stable partition for game $(G, \mathcal{F})$. As $soc(\pi) = \ell + 1$ and there is a partition with sociality n, we get that the strong price of anarchy is lower bounded by $\frac{n}{\ell+1}$. By using $n = \ell^2 + 1$, it follows that $\frac{n}{\ell+1} = \frac{n}{1+\sqrt{n-1}}$ and this provides the desired lower bound.

For the lower bound on the strong price of stability, add a vertex y to the previous instance which is solely connected to x_0, so that G now has $n = \ell^2 + 2$ vertices. In this case, the partition π whose unique non-singleton coalition is $\{x_0, x_1, \ldots, x_\ell, y\}$ is the unique core stable partition for game $(G, \mathcal{F})$. As $soc(\pi) = \ell + 2$, we get that the strong price of stability is lower bounded by $\frac{n}{\ell+2}$. By using $n = \ell^2 + 2$, it follows that $\frac{n}{\ell+2} = \frac{n}{2+\sqrt{n-2}}$ and this provides the desired lower bound. $\quad\square$

Theorem 5. *For any instance $(G, \mathcal{F})$ of the game with n players and the fragmentation function, it holds that $\lceil n/2 \rceil / 2 \leq PoS(G, \mathcal{F}) \leq SPoA(G, \mathcal{F}) \leq n/4 + 11/20$.*

Proof. Let π be a core stable partition and set $k = \Delta(G) + 1$, where $\Delta(G)$ is the maximum degree of G. It follows that the fragmentation of the social optimum is at least $\lceil n/k \rceil$. From Theorem 1, we know that there is a coalition of size k in π. Thus $frag(\pi) \leq 1 + n - k$, and $SPoA(G, \mathcal{F}) \leq \frac{k(1+n-k)}{n}$. If n is even then $\frac{k(1+n-k)}{n} \leq n/4 + 1/2$, otherwise $\frac{k(1+n-k)}{n} \leq \frac{(n+1)^2}{4n} = \frac{n}{4} + \frac{1}{2} + \frac{1}{4n} \leq \frac{n}{4} + \frac{11}{20}$, where the last inequality is due to the hypothesis $n \geq 4$ (the smallest odd n is 5).

To show the lower bound when n is even, consider the game induced by the graph depicted in Fig. 3. We have a social optimum π^* such that $\pi^*(x_1) = \{x_1, \ldots, x_{\lfloor n/2 \rfloor}\}$ and $\pi^*(y_1) = V \setminus \pi^*(x_1)$ and yielding $frag(\pi^*) = 2$. There are only two Nash stable partitions, namely π_1 and π_2, both having fragmentation equal to $\lceil n/2 \rceil$. In particular, π_1 is such that $\pi_1(x_1) = \{x_1, \ldots, x_{\lfloor n/2 \rfloor}, y_1\}$ and $\pi_1(y_i) = \{y_i\}$ for $i > 1$, while π_2 flips the roles of x and y. When n is odd, take the same instance and add a new vertex solely connected to x_1 to get the lower bound of $\lceil n/2 \rceil / 2$. $\quad\square$

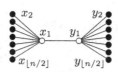

Fig. 3. A graph yielding a game with worst-case price of stability with respect to fragmentation.

We conclude the section with a lower bound on the price of stability for the sociality social function showing that the quality of the best Nash stable partition cannot be better than twice that of the worst core stable one.

Theorem 6. *For any game with n players, the price of stability with respect to sociality is at least $\frac{n}{2\sqrt{n}-1}$.*

Proof. Consider the game yielded by the graph $G = (V, E)$ such that $V = X \cup Y_1 \cup \ldots \cup Y_\ell$, with $X = \{x_1, \ldots, x_\ell\}$ and, for each $i \in [\ell]$, $Y_i = \{y_{i,1}, \ldots, y_{i,\ell-1}\}$. The set of edges E is such that $G[X] = K_\ell$ (i.e. complete graph on ℓ vertices) and, for each $i \in [\ell]$ each vertex in Y_i is connected to x_i only. Note that, by setting $\ell = \sqrt{n}$, we obtain $|V| = n$. We shall prove that, in any Nash stable partition, the sociality is at most $2\sqrt{n} - 1$. Given that there is a partition of sociality n, this will yield the corresponding lower bound. Fix a Nash stable partition π. We claim that π contains a unique non-singleton coalition containing X. This easily follows from the fact that X defines a clique and that, by the topology of G, in any feasible coalition C containing a vertex of $x \in X$, C induces a subgraph of G admitting a spanning star centred at x. Moreover, as π is Nash stable, we shall have that the unique non-singleton coalition in π will also contain all vertices in Y_i for a certain $i \in [\ell]$. No other vertices of G can be added to the coalition without violating the feasibility constraint and no other non-singleton coalition can be constructed as the remaining vertices yield an independent set of G. Hence, the sociality of π is $2\sqrt{n} - 1$.   □

3.1 Claw-Free Graphs

In this subsection, we consider the case in which the graph G is claw-free, i.e., it does not contain an induced $K_{1,3}$ (i.e. complete bipartite graph with 1 and 3 vertices on the respective sides). It will turn out that the presence of claws in graph G is a provable source of inefficiency as the price of anarchy with respect to both social functions (and so also all the other metrics) for games played on claw-free graphs drops to a value which never exceeds 2. Claws (and more generally induced stars with a large number of leaves) are problematic for our social functions when the center c of a claw belongs to a partition $\pi(c)$ which does not admit a spanning star of center c. In this case some leaves of the claw are isolated: they cannot join $\pi(c)$ or group themselves because they are disconnected. For the social function sociality, the following two theorems provide an asymptotically tight characterization.

Theorem 7. *For any game with n players, the price of anarchy with respect to sociality is at most $\frac{n}{n-\lfloor\frac{n-1}{2}\rfloor}$, that is $\frac{2n}{n+2}$ and $\frac{2n}{n+1}$ when n is even and odd, respectively.*

Proof. Fix a Nash stable partition π and let i be a vertex belonging to a singleton coalition in π. Clearly, i cannot be adjacent to a vertex being a center of a spanning star of any subgraph induced by a coalition in π. So, i can only be adjacent to leaves of spanning stars of any subgraph induced by coalitions in π. Assume that there exists a vertex j also belonging to a singleton coalition in π and sharing a neighbour k with i. Let c be the center of a star spanning $G[\pi(k)]$. As $\{i, j, c\}$ is independent, we find that the set of vertices $\{i, j, k, c\}$ induces a claw in G: a contradiction. Two vertices forming a singleton coalition cannot share the same leaf of a star spanning a non-singleton coalition. Thus, denote by α the number of vertices that are centres of a star spanning the subgraph induced by a non-singleton coalition in π, we get that the number of vertices forming singleton coalitions in π, which is integral, is upper bounded by $\lfloor \frac{n-\alpha}{2} \rfloor \leq \lfloor \frac{n-1}{2} \rfloor$. This implies that $soc(\pi) \geq n - \lfloor \frac{n-1}{2} \rfloor$, which yields the desired upper bound because the optimal sociality $soc(\pi^*)$ is n.   □

Theorem 8. *For any game with n players, the price of stability with respect to sociality is at least* $\frac{n}{n - \lfloor \frac{n-1}{2} \rfloor}$.

Proof. Suppose $n = 2p$ and consider the graph $G_{2p} = (V_{2p}, E_{2p})$ such that $V_{2p} = X_p \cup Y_p$, where $X_p = \{x_1, \ldots, x_p\}$, $Y_p = \{y_1, \ldots, y_p\}$, X_p forms a clique, and each vertex y_i is adjacent to vertex x_i. One can see that G_{2p} is claw-free, and $X_p \subseteq \pi(x_i)$ for any Nash stable partition π. Moreover, to have feasibility, a coalition C can contain at most one vertex from Y_p. It follows that the sociality of any Nash stable partition is at most $p + 1$. As there exists a partition with sociality $2p$, the claimed lower bound follows. For $n = 2p + 1$, use the same construction with $Y_p = \{y_1, \ldots, y_{p-1}\}$.   □

For the fragmentation social function, a slightly different situation occurs.

Theorem 9. *For any game with n players, the price of anarchy with respect to fragmentation is at most* 2.

Proof. Fix a Nash stable partition π. If $frag(\pi) \leq 2$, we are done. If $frag(\pi) \geq 3$, consider three distinct coalitions $C_1, C_2, C_3 \in \pi$ and let c_1, c_2 and c_3 be the centres of the spanning stars of $G[C_1]$, $G[C_2]$ and $G[C_3]$, respectively. As π is Nash stable, the set of vertices $U = \{c_1, c_2, c_3\}$ induces an independent set of G. We claim that these vertices cannot belong to a same cluster in some social optimum π^*. Assume, by way of contradiction, that there exists a cluster $C^* \in \pi^*$ containing U. As U induces an independent set of G no vertex in U can be the center of the star spanning $G[C^*]$. So, there exists $c^* \in C^*$ which is adjacent to all vertices in U. But $U \cup \{c^*\}$ induces a claw: a contradiction. Hence, for every two coalitions in π, there must exist a distinct coalition in π^* which yields the desired upper bound.   □

Theorem 10. *For any positive integer k, there exists a game whose strong price of stability with respect to fragmentation is at least* $\frac{2k}{k+1}$.

Proof. For an integer $k \geq 1$, define the following graph G_k. The set of vertices is $V = X_1 \cup \ldots \cup X_k \cup Y \cup Z$, with $Y = \{y_1, \ldots, y_k\}$ and $Z = \{z_1^a, z_1^b, \ldots, z_k^a, z_k^b\}$ so that $|Z| = 2k$. As to sets X_i for $i \in [k]$, we have that $|X_i| = 2i$, $G_k(X_i)$ induces a clique and X_i contains two special vertices, namely $\overline{x}_i$ and $\underline{x}_i$, which are the only vertices adjacent to vertices in $V \setminus X_i$. In particular, $\overline{x}_i$ is adjacent to both z_i^a and z_i^b, while $\underline{x}_i$ is adjacent to y_i and, for $i < k$, to both z_{i+1}^a and z_{i+1}^b. Finally, for each $i \in [k]$, z_i^a and z_i^b are adjacent and, for $i > 1$, they are both adjacent to y_{i-1}. Again, we show two fundamental properties:

Property 1. (i) G_k is claw-free, (ii) $(X_i \cup \{z_i^a, z_i^b\}, \{y_i\} \mid i \in [k])$ is a core stable partition.

Consider the feasible partition π^* containing the following coalitions: $X_k \cup \{y_k\}$, $X_i \cup \{y_i\} \cup \{z_{i+1}^a, z_{i+1}^b\}$ for each $i \in [k-1]$ and $\{z_1^a, z_1^b\}$. As $frag(\pi) = k + 1$, the claimed lower bound follows.   □

Theorem 11. *For any game with n players, the price of stability with respect to fragmentation is* 1.

Proof. Our aim is to show that, given a partition π, it is possible to schedule profitable deviations so as to obtain a Nash dynamics starting from π and ending up to a Nash stable partition π_ℓ such that $frag(\pi_\ell) \le frag(\pi)$. Choosing a social optimum as starting partition will yield the claim. Our scheduling algorithm is defined as follows: given a partition π, if more than one player have a profitable deviation in π, break ties in favour of a player who does not constitute a center for any spanning star of the subgraph induced by the coalition she belongs to. By the LIP property, we are guaranteed that the Nash dynamics defined by this scheduling algorithm always ends to a Nash stable partition π_ℓ for any starting partition π.

Assume, by way of contradiction, that $frag(\pi_\ell) > frag(\pi)$. This implies that there are two partitions $\pi, \pi' \in \Pi$ and a player i such that π' is obtained as a result of a profitable deviation of i in π and $frag(\pi) < frag(\pi')$. The latter condition can happen only if $G[\pi(i) \setminus \{i\}]$ does not admit a spanning star, which implies that $G[\pi(i)]$ admits only one spanning star centred at i. So, there are at least two distinct vertices u and v other than i belonging to $\pi(i)$. Let $C \in \pi$ be the coalition joined by i and let $j \ne i$ be the center of a spanning star for $G[C \cup \{i\}]$. Clearly it must be $\{i, j\} \in E$. If $\{u, j\} \in E$, then $C \cup \{u\} \in \mathcal{F}$. But as $u_u(\pi) = u_i(\pi) = |\pi(i)| < u_i(\pi') = |C \cup \{i\}| = |C \cup \{u\}|$, it follows that u has a profitable deviation in π and, by the definition of the scheduling algorithm, i should have not been chosen. The same argument holds for v. Thus, we have detected a set of vertices $\{i, j, u, v\}$ inducing a claw in G: a contradiction. $\square$

4   Computing Partitions with Prescribed Properties

In this section, we address the complexity of computing partitions with some prescribed properties, such as, for example, being a social optimum or being a Nash stable partition either maximizing or minimizing a given social function.

4.1   Computing a Social Optimum

Proposition 1. *Computing a social optimum with respect to fragmentation is* **NP**-*hard, even for claw-free graphs.*

The following result relies on the efficient computation of a minimum edge-cover.

Proposition 2. *For connected graphs on n vertices, computing a social optimum π^* with respect to sociality is polynomial and* $soc(\pi^*) = n$.

4.2   Computing an Extreme Stable Partition

Using Theorem 11 and Proposition 1, we deduce that computing a best Nash stable partition with respect to fragmentation is **NP**-hard, even for claw-free graphs. We now show that hardness holds also for the sociality social function.

Theorem 12. *Computing the best Nash stable partition with respect to sociality is **NP**-hard when the input graph G has maximum degree equal to 5.*

Proof. We propose a reduction from 3-DIMENSIONAL MATCHING (3-DM in short). An instance of 3-DM consists of a collection $\mathcal{C} = \{s_1, \ldots, s_m\} \subseteq X \times Y \times Z$ of m triples, where $X = \{x_1, \ldots, x_n\}$, $Y = \{y_1, \ldots, y_n\}$ and $Z = \{z_1, \ldots, z_n\}$ are 3 pairwise disjoint sets of size n. A matching is a subset $M \subseteq \mathcal{C}$ such that no two elements in M agree in any coordinate, and the purpose of 3-DM is to answer the question: does there exist a perfect matching M on $\mathcal{C}$, that is, a matching of size n? This problem is known to be **NP**-complete (problem [SP1] p. 221 in [17]), even if each element $t \in X \cup Y \cup Z$ appears in at most 3 triples. We start from such an instance $I = (\mathcal{C}, X \cup Y \cup Z)$ of 3-DM and we build a game $(G, \mathcal{F})$, where $G = (V, E)$ is as follows. Vertex set contains $L \cup R$ where $L = \{l_1, \ldots, l_m\}$ corresponds to the different triples of $\mathcal{C}$ and $R = R^x \cup R^y \cup R^z$ where $R^t = \{r_1^t, \ldots, r_n^t\}$ for $t = x, y, z$, corresponds to elements of $X \cup Y \cup Z$. Moreover for $t = x, y, z$, $l_i r_j^t \in E$ if and only if t_j is an element of triplet s_i. This particular bipartite graph is usually called *representative bipartite graph*. We use gadget $H(l_i)$ for each $l_i \in L$ to characterize triplets of $\mathcal{C}$. This gadget is illustrated in Fig. 4 (on the left of the drawing).

The construction of G is complete and its maximum degree is 5. We claim that I admits a perfect matching if and only if there exists a Nash stable partition π of $(G, \mathcal{F})$ with $soc(\pi) \geq 5m + n$. Actually, we will prove that $5m + n$ is the best value reachable by any Nash stable partition π.

Clearly, if $\mathcal{C}' \subseteq \mathcal{C}$ is a set of n triples forming a perfect matching, then consider the following partition π. For $s_i = (x_{i_1}, y_{i_2}, z_{i_3}) \in \mathcal{C}'$, set $\pi(d_i) = \{d_i\}$, $\pi(e_i) = \{e_i\}$ and $\pi(l_i) = \{b_i, c_i, l_i, r_{i_1}^x, r_{i_2}^y, r_{i_3}^z\}$; an illustration of these 3 coalitions is proposed in Fig. 4 (on the right of the drawing). Otherwise, for $s_i \notin \mathcal{C}'$, set $\pi(l_i) = \{l_i, b_i, c_i, d_i, e_i\}$. It is easy to see that π is a Nash stable partition with $soc(\pi) = 6n + 5(m - n) = 5m + n$.

Fig. 4. On the left: Gadget $H(l_i)$. On the right: a possible Nash stable partition for agents in $H(l_i)$.

Conversely, let π be a Nash stable partition. If $\{l_i, r_j^t\} \subseteq \pi(r_j^t)$ for some $j \in [n]$ and $t = x, y, z$, then:

Property 2. (i) $\forall i' \neq i, l_{i'} \notin \pi(r_j^t)$, (ii) $\{b_i, c_i\} \subseteq \pi(r_j^t)$ and $\{d_i, e_i\} \cap \pi(r_j^t) = \emptyset$.

Let $Cov = \{l_i \in L : |\pi(l_i)| = 6\}$ and $p = |Cov|$. By Property 2, for every $l_i \in Cov$, it must be $\pi(e_i) = \{e_i\}$ and $\pi(d_i) = \{d_i\}$ (see right picture of Fig. 4 for an illustration); actually, these collations correspond to a (3-dimensional) matching of size p. Using Property 2, we know that we also loose in the sociality function, as many trivial stars as the number of vertices of $R \setminus (\cup_{l_i \in Cov} \pi(l_i))$. Hence, $soc(\pi) \leq 5m + 3n - (2p + 3n - 3p) = 5m + p \leq 5m + n$ and the last inequality is tight only when $|Cov| = n$ or equivalently when $C' = \{s_i | l_i \in Cov\}$ is a perfect matching. $\qquad\square$

Corollary 1. *Computing the best Nash stable partition with respect to either sociality or fragmentation is NP-hard even for planar graphs.*

As to the problem of computing a worst Nash stable partition, we give a hardness result with respect to sociality, while the case of the fragmentation social function remains open.

Theorem 13. *Computing the worst Nash stable partition with respect to sociality is NP-hard when the input graph G has maximum degree equal to 11.*

5 Conclusion

Two problems are left open: closing the gap between upper and lower bounds on the PoS with respect to sociality for games played on general graphs, and determining the complexity of computing a worst Nash stable partition with respect to fragmentation. Addressing the problem of computing extreme core stable partition is also worth to be investigated.

Applying our feasibility constraint (i.e. imposing a spanning star) to hedonic games having agents' preferences other than the ones considered in this paper is clearly an interesting research direction. Other graph patterns are appealing in our opinion: the largest distance between any pair of agents, or the distance to some agent of the coalition can be upper bounded by a given number (for the latter the distance to some agent is 1 in this paper).

References

1. Aziz, H., Savani, R.: Hedonic games. In: Handbook of Computational Social Choice, pp. 356–376 (2016)
2. Aziz, H., Brandt, F., Seedig, H.G.: Stable partitions in additively separable hedonic games. In: Proceedings of AAMAS, pp. 183–190 (2011)
3. Ballester, C.: NP-completeness in hedonic games. Games Econ. Behav. **49**(1), 1–30 (2004)
4. Balliu, A., Flammini, M., Melideo, G., Olivetti, D.: Nash stability in social distance games. In: Proceedings of AAAI, pp. 342–348 (2017)
5. Balliu, A., Flammini, M., Olivetti, D.: On Pareto optimality in social distance games. In: Proceedings of AAAI, pp. 349–355 (2017)
6. Banerjee, S., Konishi, H., Sönmez, T.: Core in a simple coalition formation game. Soc. Choice Welf. **18**(1), 135–153 (2001)

7. Bilò, V., Fanelli, A., Flammini, M., Monaco, G., Moscardelli, L.: Nash stable outcomes in fractional hedonic games: existence, efficiency and computation. J. Artif. Intell. Res. **62**, 315–371 (2018)

8. Bogomolnaia, A., Jackson, M.O.: The stability of hedonic coalition structures. Games Econ. Behav. **38**(2), 201–230 (2002)

9. Brânzei, S., Larson, K.: Coalitional affinity games and the stability gap. In: Proceedings of IJCAI 2009, pp. 79–84 (2009)

10. Charikar, M., Guruswami, V., Wirth, A.: Clustering with qualitative information. J. Comput. Syst. Sci. **71**(3), 360–383 (2005)

11. Demaine, E.D., Emanuel, D., Fiat, A., Immorlica, N.: Correlation clustering in general weighted graphs. Theor. Comput. Sci. **36**(3), 360–383 (2005)

12. Deng, X., Papadimitriou, C.H.: On the complexity of cooperative solution concepts. Math. Oper. Res. **19**(2), 257–266 (1994)

13. Drèze, J.H., Greenberg, J.: Hedonic coalitions: optimality and stability. Econometrica **48**(4), 987–1003 (1980)

14. Escoffier, B., Gourvès, L., Monnot, J.: Strategic coloring of a graph. Internet Math. **8**(4), 424–455 (2012)

15. Feldman, M., Lewin-Eytan, L., Naor, J.: Hedonic clustering games. ACM Trans. Parallel Comput. **2**(1), 1–48 (2015)

16. Flammini, M., Monaco, G., Moscardelli, L., Shalom, M., Zaks, S.: Online coalition structure generation in graph games. In: Proceedings of AAMAS, pp. 1353–1361 (2018)

17. Garey, M.R., Johnson, D.S.: Computers and Intractability: A Guide to the Theory of NP-Completeness. W. H. Freeman & Co., New York (1979)

18. Harks, T., Klimm, M., Möhring, R.H.: Strong equilibria in games with the lexicographical improvement property. Int. J. Game Theor. **42**(2), 461–482 (2013)

19. Hoefer, M., Vaz, D., Wagner, L.: Dynamics in matching and coalition formation games with structural constraints. Artif. Intell. **262**, 222–247 (2018)

20. Igarashi, A., Elkind, E.: Hedonic games with graph-restricted communication. In: Proceedings of AAMAS, pp. 242–250 (2016)

21. Monaco, G., Moscardelli, L., Velaj, Y.: Stable outcomes in modified fractional hedonic games. In: Proceedings of AAMAS, pp. 937–945 (2018)

22. Monaco, G., Moscardelli, L., Velaj, Y.: On the performance of stable outcomes in modified fractional hedonic games with egalitarian social welfare. In: Proceedings of AAMAS (to appear)

23. Olsen, M., Baekgaard, L., Tambo, T.: On non-trivial Nash stable partitions in additive hedonic games with symmetric 0/1-utilities. Inf. Process. Lett. **112**(23), 903–907 (2012)

24. Olsen, M.: Nash stability in additively separable hedonic games and community structures. Theor. Comput. Syst. **45**(4), 917–925 (2009)

25. Panagopoulou, P.N., Spirakis, P.G.: A game theoretic approach for efficient graph coloring. In: Hong, S.-H., Nagamochi, H., Fukunaga, T. (eds.) ISAAC 2008. LNCS, vol. 5369, pp. 183–195. Springer, Heidelberg (2008). https://doi.org/10.1007/978-3-540-92182-0_19

26. Peters, D., Elkind, E.: Simple causes of complexity in hedonic games. In: Proceedings of AAAI, pp. 617–623 (2015)

27. Peters, D.: Graphical hedonic games of bounded treewidth. In: Proceedings of AAAI, pp. 586–593 (2016)

Social Choice

Impartial Selection with Additive Approximation Guarantees

Ioannis Caragiannis[1], George Christodoulou[2], and Nicos Protopapas[2(⊠)]

[1] University of Patras, Patras, Greece
caragian@ceid.upatras.gr
[2] University of Liverpool, Liverpool, UK
{G.Christodoulou,N.Protopapas}@liverpool.ac.uk

Abstract. Impartial selection has recently received much attention within the multi-agent systems community. The task is, given a directed graph representing nominations to the members of a community by other members, to select the member with the highest number of nominations. This seemingly trivial goal becomes challenging when there is an additional impartiality constraint, requiring that no single member can influence her chance of being selected. Recent progress has identified impartial selection rules with optimal approximation ratios. Moreover, it was noted that worst-case instances are graphs with few vertices. Motivated by this fact, we propose the study of *additive approximation*, the difference between the highest number of nominations and the number of nominations of the selected member, as an alternative measure of the quality of impartial selection.

Our positive results include two randomized impartial selection mechanisms which have additive approximation guarantees of $\Theta(\sqrt{n})$ and $\Theta(n^{2/3} \ln^{1/3} n)$ for the two most studied models in the literature, where n denotes the community size. We complement our positive results by providing negative results for various cases. First, we provide a characterization for the interesting class of strong sample mechanisms, which allows us to obtain lower bounds of $n - 2$, and of $\Omega(\sqrt{n})$ for their deterministic and randomized variants respectively. Finally, we present a general lower bound of 2 for all deterministic impartial mechanisms.

Keywords: Impartial selection · Voting · Mechanism design

1 Introduction

We study the problem that arises in a community of individuals that want to select a community member that will receive an award. This is a standard social choice problem [7], that is typically encountered in scientific and sports communities but has also found important applications in distributed multi-agent systems. To give an entertaining example, the award for the player of the year[1]

---

[1] https://en.wikipedia.org/wiki/PFA_Players%27_Player_of_the_Year.

© Springer Nature Switzerland AG 2019
D. Fotakis and E. Markakis (Eds.): SAGT 2019, LNCS 11801, pp. 269–283, 2019.
https://doi.org/10.1007/978-3-030-30473-7_18

by the Professional Footballers Association (PFA) is decided by the members of PFA themselves; each PFA member votes the two players they consider the best for the award and the player with the maximum number of votes receives the award. Footballers consider it as one of the most prestigious awards, due to the fact that it is decided by their opponents. In distributed multi-agent systems, *leader election* (e.g., see [2]) can be thought of as a selection problem of similar flavor. Other notable examples include (see [10]) the selection of a representative in a group, funding decisions based on peer reviewing or even (see [1]) finding the most popular user of a social network.

The input of the problem can be represented as a directed graph, which we usually call nomination profile. Each vertex represents an individual and a direct edge indicates a vote (or nomination) by a community member to another.

A *selection mechanism* (or *selection rule*) takes a nomination profile as input and returns a single vertex as the winner. Clearly, there is a highly desirable selection rule: the one which always returns the highest in-degree vertex as the winner. Unfortunately, such a rule suffers from a drawback that is pervasive in social choice. Namely, it is *susceptible to manipulation*.

In particular, the important constraint that makes the selection challenging is *impartiality*. As every individual has a personal interest to receive the award, selection rules should take the individual votes into account but in such a way that no single individual can increase his/her chance of winning by changing his/her vote. The problem, known as *impartial selection*, was introduced independently by Holzman and Moulin [11] and Alon et al. [1]. Unfortunately, the ideal selection rule mentioned above is not impartial. Consider the case with a few individuals that are tied with the highest number of votes. The agents involved in the tie might be tempted to lie about their true preferences to break the tie in their favor.

Impartial selection rules may inevitably select as the winner a vertex that does not have the maximum in-degree. Moulin and Holzman [11] considered minimum axiomatic properties that impartial selection rules should satisfy. For example, a highly desirable property, called negative unanimity, requires that an individual with no votes at all, should never be selected. Alon et al. [1] quantified the efficiency loss with the notion of approximation ratio, defined as the worst-case ratio of the maximum vertex in-degree over the in-degree of the vertex which is selected by the rule. According to their definition, an impartial selection rule should have as low approximation ratio as possible. This line of research was concluded by the work of Fischer and Klimm [10] who proposed impartial mechanisms with the optimal approximation ratio of 2.

It was pointed out in [1,10], that the most challenging nomination profiles for both deterministic and randomized mechanisms are those with small in-degrees. In the case of deterministic mechanisms, the situation is quite extreme as all deterministic mechanisms can be easily seen to have an unbounded approximation ratio on inputs with a maximum in-degree of 1 for a single vertex and 0 for all others; see [1] for a concrete example. As a result, the approximation ratio does not seem to be an appropriate measure to classify deterministic selection mechanisms. Finally, Bousquet et al. [6] have shown that if the maximum in-degree is large enough, randomized mechanisms that return a near optimal impartial winner do exist.

We deviate from previous work and instead propose to use *additive approximation* as a measure of the quality of impartial selection rules. Additive approximation is defined using the *difference* between the maximum in-degree and the in-degree of the winner returned by the selection mechanism. Note that deterministic mechanisms with low additive approximation always return the highest in-degree vertex as the winner when his/her margin of victory is large. When this does not happen, we have a guarantee that the winner returned by the mechanism has a close-to-maximum in-degree.

Our Contribution. We provide positive and negative results for impartial selection mechanisms with additive approximation guarantees. We distinguish between two models. In the first model, which was considered by Holzman and Moulin [11], nomination profiles consist only of graphs with all vertices having an out-degree of 1. The second model is more general and allows for multiple nominations and abstentions (hence, vertices have arbitrary out-degrees).

As positive results, we present two randomized impartial mechanisms which have additive approximation guarantees of $\Theta(\sqrt{n})$ and $\Theta(n^{2/3} \ln^{1/3} n)$ for the single nomination and multiple nomination models, respectively. Notice that both these additive guarantees are $o(n)$ functions of the number n of vertices. We remark that an $o(n)$-additive approximation guarantee can be translated to an $1 - \epsilon$ multiplicative guarantee for graphs with sufficiently large maximum in-degree, similar to the results of [6]. Conversely, the multiplicative guarantees of [6] can be translated to an $O(n^{8/9})$-additive guarantee[2]. This analysis further demonstrates that additive guarantees allow for a more smooth classification of mechanisms that achieve good multiplicative approximation in the limit.

Our mechanisms first select a small sample of vertices, and then select the winner among the vertices that are nominated by the sample vertices. These mechanisms are randomized variants of a class of mechanisms which we define and call *strong sample mechanisms*. Strong sample mechanisms are deterministic impartial mechanisms which select the winner among the vertices nominated by a sample set of vertices. In addition, they have the characteristic that the sample set does not change with changes in the nominations of the vertices belonging to it. For the single nomination model, we provide a characterization, and we show that all these mechanisms should use a fixed sample set that does not depend on the nomination profile. This yields a $n - 2$ lower bound on the additive approximation guarantee of any deterministic strong sample mechanism. For randomized variants, where the sample set is selected randomly, we present an $\Omega(\sqrt{n})$ lower bound which indicates that our first randomized impartial mechanism is best possible among all randomized variants of strong sample mechanisms. Finally,

---

[2] The authors in [6] do not provide additive guarantees, hence we based our calculations on their provided bounds on the multiplicative guarantee $1 - \epsilon$. It is important to note however that they claim that they have not optimized their parameters, so it is possible that this guarantee can be further reduced by a tighter analysis.

for the most general multiple nomination model, we present a lower bound of 2 for all deterministic mechanisms.

Due to space limitations some proofs are omitted. The reader is referred to the full version of the paper.

Related Work. Besides the papers by Holzman and Moulin [11] and Alon et al. [1], which introduced impartial selection as we study it here, de Clippel et al. [9] considered a different version with a divisible award. Alon et al. [1] used the approximation ratio as a measure of quality for impartial selection mechanisms. After realizing that no deterministic mechanism achieves a bounded approximation ratio, they focused on randomized mechanisms and proposed the 2-PARTITION mechanism, which guarantees an approximation ratio of 4 and complemented this positive result with a lower bound of 2 for randomized mechanisms.

Later, Fischer and Klimm were able to design a mechanism that achieves an approximation ratio of 2, by generalizing 2-PARTITION. Their optimal mechanism, called PERMUTATION, examines the vertices sequentially following their order in a random permutation and selects as the winner the vertex of highest degree counting only edges with direction from "left" to "right." They also provided lower bounds on the approximation ratio for restricted inputs (e.g., with no abstentions) and have shown that the worst case examples for the approximation ratio are tight when the input nomination profiles are small.

Bousquet et al. in [6] noticed this bias towards instances with small in-degrees and examined the problem for instances of very high maximum in-degree. After showing that PERMUTATION performs significantly better for instances of high in-degree, they have designed the SLICING mechanism with near optimal asymptotic behaviour for that restricted family of graphs. More precisely, they have shown that, if the maximum in-degree is large enough, SLICING can guarantee that the winner's in-degree approximate the maximum in-degree by a small error. As we discussed in the previous section, the SLICING mechanism can achieve an additive guarantee of $O(n^{8/9})$.

Holzman and Moulin [11] explore impartial mechanisms through an axiomatic approach. They investigate the single nomination model and propose several deterministic mechanisms, including their MAJORITY WITH DEFAULT rule. MAJORITY WITH DEFAULT defines a vertex as a default winner and examines if there is any vertex with in-degree more than $\lceil n/2 \rceil$, ignoring the edge from the default vertex. If such a vertex exists, then this is the winner; otherwise the default vertex wins. While this mechanism has the unpleasant property that the default vertex may become the winner with no incoming edges at all, its additive approximation is at most $\lceil n/2 \rceil$. Further to that, they came up with a fundamental limitation of the problem: no impartial selection mechanism can be simultaneously negative and positive unanimous (i.e., never selecting as a winner a vertex of in-degree 0 and always selecting the vertex of in-degree $n-1$, whenever there exists one).

Mackenzie in [14] characterized symmetric (i.e., name-independent) rules in the single nomination model. Tamura and Ohseto [16] observed that when the demand for only one winner is relaxed, then impartial, negative unanimous and positive unanimous mechanisms do exist. Later on, Tamura [15] characterized them. On the same agenda, Bjelde et al. in [5] proposed a deterministic version of the permutation mechanism that achieves the 1/2 bound by allowing at most two winners. Alon et al. [1] also present results for selecting multiple winners.

Finally, we remark that impartiality has been investigated as a desired property in other contexts where strategic behaviour occurs. Recent examples include peer reviewing [3,12,13], selecting impartially the most influential vertex in a network [4], linear regression algorithms as a means to tackle strategic noise [8], and more.

2 Preliminaries

Let $N = \{1, ..., n\}$ be the set of $n \geq 2$ agents. A *nomination graph* $G = (N, E)$ is a directed graph with vertices representing the agents. The set of outgoing edges from each vertex represents the nominations of each agent; it contains no self-loops (as, agents are not allowed to nominate themselves) and can be empty (as an agent is, in general, allowed to abstain). We write $\mathcal{G} = \mathcal{G}_n$ for the set of all graphs with n vertices and no self-loops. We also use the notation $\mathcal{G}^1 = \mathcal{G}_n^1$ to denote the subset of $\mathcal{G}$ with out-degree exactly 1. For convenience in the proofs, we sometimes denote each graph G by a tuple $\mathbf{x}$, called *nomination profile*, where x_u denotes the set of outgoing edges of vertex u in G. For $u \in N$, we use the notation $\mathbf{x}_{-u}$ to denote the graph $(N, E \setminus (\{u\} \times N))$ and, for the set of vertices $U \subseteq N$, we use $\mathbf{x}_{-U}$ to denote the graph $(N, E \setminus (U \times N))$. We use the terms nomination graphs and nomination profiles interchangeably.

The notation $\delta_S(u, \mathbf{x})$ refers to the in-degree of vertex u in the graph $\mathbf{x}$ taking into account only edges that originate from the subset $S \subseteq N$. When $S = N$, we use the shorthand $\delta(u, \mathbf{x})$ and if the graph is clearly identified by the context we omit $\mathbf{x}$ too, using $\delta(u)$. We denote the maximum in-degree vertex of graph $\mathbf{x}$ as $\Delta(\mathbf{x}) = \max_{u \in N} \delta(u, \mathbf{x})$ and, whenever $\mathbf{x}$ is clear from the context, we use Δ instead.

5. f is not defined without g. Hence, it is ambiguous to use f(x) in isolation. A *selection mechanism* for a set of graphs $\mathcal{G}' \subseteq \mathcal{G}$, is a function $f : \mathcal{G}' \to [0, 1]^{n+1}$, mapping each graph of $\mathcal{G}'$ to a probability distribution over all vertices (which can be potential winners) as well as to the possibility of returning no winner at all. A selection mechanism is *deterministic* in the special case where for all $\mathbf{x}$, $(f(\mathbf{x}))_u \in \{0, 1\}$ for all vertices $u \in N$.

A selection mechanism is *impartial* if for all graphs $\mathbf{x} \in \mathcal{G}'$, it holds $(f(\mathbf{x}))_u = (f(x'_u, \mathbf{x}_{-u}))_u$ for every vertex u. In words, the probability that u wins must be independent of the set of its outgoing edges. Let $\mathbb{E}[\delta(f(\mathbf{x}))]$ be the expected in-degree of f on $\mathbf{x}$, i.e. $\mathbb{E}[\delta(f(\mathbf{x}))] = \sum_{u \in N}(f(\mathbf{x}))_u \delta(u, \mathbf{x})$. We call f $\alpha(n)$-additive if

$$\max_{\mathbf{x} \in \mathcal{G}_n} \{\Delta(\mathbf{x}) - \mathbb{E}[\delta(f(\mathbf{x}))]\} \leq \alpha(n),$$

for every $n \in \mathbb{N}$.

3   Upper Bounds

In this section we provide randomized selection mechanisms for the two best studied models in the literature. First, in Sect. 3.1 we propose a mechanism for the single nomination model of Holzman and Moulin [11], where nomination profiles consist only of graphs with all vertices having an out-degree of 1. Then, in Sect. 3.2 we provide a mechanism for the more general model studied by Alon et al. [1], which allows for multiple nominations and abstentions.

3.1   The RANDOM k-SAMPLE Mechanism

Our first mechanism, RANDOM k-SAMPLE, forms a sample S of vertices by repeating k times the selection of a vertex uniformly at random with replacement. Any vertex that is selected at least once belongs to the sample S. Let $W := \{u \in N \setminus S : \delta_S(u, \mathbf{x}) \geq 1\}$ be the set of vertices outside S that are nominated by the vertices of S. If $W = \emptyset$, no winner is returned. Otherwise, the winner is a vertex in $\arg\max_{u \in W} \delta_{N \setminus W}(u, \mathbf{x})$. We note here the crucial fact that the selection of the sample set S is independent of the nomination profile $\mathbf{x}$.

Impartiality follows since a vertex that does not belong to W (no matter if it belongs to S or not) cannot become the winner and the nominations of vertices in W are not taken into account for deciding the winner among them. We now argue that, for a carefully selected k, this mechanism also achieves a good additive guarantee.

Theorem 1. *For $k = \Theta(\sqrt{n})$, the RANDOM k-SAMPLE mechanism is impartial and $\Theta(\sqrt{n})$-additive in the single nomination model.*

Proof. Consider a nomination graph and let u^* be a vertex of maximum in-degree Δ. In our proof of the approximation guarantee, we will use the following two technical lemmas.

Lemma 1. *If $u^* \in W$, then the winner has in-degree at least $\Delta - k$.*

Proof. This is clearly true if the winner returned by RANDOM k-SAMPLE is u^*. Otherwise, the winner w satisfies

$$\delta(w, \mathbf{x}) \geq \delta_{N \setminus W}(w, \mathbf{x}) \geq \delta_{N \setminus W}(u^*, \mathbf{x}) = \delta(u^*, \mathbf{x}) - \delta_W(u^*, \mathbf{x}) \geq \Delta - k.$$

The first inequality is trivial. The second inequality follows by the definition of the winner w. The third inequality follows since W is created by nominations of vertices in S, taking into account that each vertex has out-degree exactly 1. Hence, $\delta_W(u^*, \mathbf{x}) \leq |W| \leq |S| \leq k$.   □

Lemma 2. *The probability that u^* belongs to the nominated set W is*

$$\mathbf{Pr}\left[\, u^* \in W \,\right] = \left(1 - \left(1 - \frac{\Delta}{n-1}\right)^k\right)\left(1 - \frac{1}{n}\right)^k.$$

Proof. Indeed, u^* belongs to W if it does not belong to the sample S and instead some of the Δ vertices that nominate u^* is picked in some of the k vertex selections. The probability that u^* is not in the sample is

$$\mathbf{Pr}\left[\,u^* \notin S\,\right] = \left(1 - \frac{1}{n}\right)^k, \tag{1}$$

i.e., the probability that vertex u^* is not picked in some of the k vertex selections. Observe that the probability that some of the Δ vertices that nominate u^* is picked in a vertex selection step assuming that u^* is never selected is $\frac{\Delta}{n-1}$. Hence, the probability that some of the Δ vertices nominating u^* is in the sample assuming that $u^* \notin S$ is

$$\mathbf{Pr}\left[\,\delta_S(u^*,\mathbf{x}) \geq 1 | u^* \notin S\,\right] = 1 - \left(1 - \frac{\Delta}{n-1}\right)^k. \tag{2}$$

The lemma follows by the chain rule

$$\begin{aligned}
\mathbf{Pr}\left[\,u^* \in W\,\right] &= \mathbf{Pr}\left[\,u^* \neq S \wedge \delta_S(u^*,\mathbf{x}) \geq 1\,\right] \\
&= \mathbf{Pr}\left[\,\delta_S(u^*,\mathbf{x}) \geq 1 | u^* \notin S\,\right] \cdot \mathbf{Pr}\left[\,u^* \notin S\,\right]
\end{aligned}$$

and Eqs. (1) and (2). □

By Lemmas 1 and 2, we have that the expected degree of the winner returned by mechanism RANDOM k-SAMPLE is

$$\mathbb{E}\left[\,\delta(w,\mathbf{x})\,\right] \geq \mathbf{Pr}\left[\,u^* \in W\,\right] \cdot (\Delta - k) = \left(1 - \left(1 - \frac{\Delta}{n-1}\right)^k\right)\left(1 - \frac{1}{n}\right)^k (\Delta - k)$$

$$\geq \left(1 - \left(1 - \frac{\Delta}{n-1}\right)^k\right)\left(1 - \frac{k}{n}\right)(\Delta - k) > \left(1 - \left(1 - \frac{\Delta}{n-1}\right)^k\right)(\Delta - 2k)$$

$$= \Delta - 2k - \left(1 - \frac{\Delta}{n-1}\right)^k (\Delta - 2k)$$

The second inequality follows by Bernoulli's inequality $(1+x)^r \geq 1 + rx$ for every real $x \geq -1$ and $r \geq 0$ and the third one since $n > \Delta$. Now, the quantity $\left(1 - \frac{\Delta}{n-1}\right)^k (\Delta - 2k)$ is maximized for $\Delta = \frac{n-1+2k^2}{k+1}$ to a value that is at most $\frac{n+1}{k+1} - 2$. Hence,

$$\mathbb{E}\left[\,\delta(w,\mathbf{x})\,\right] \geq \Delta - 2(k-1) - \frac{n+1}{k+1}.$$

By setting $k \in \Theta(\sqrt{n})$, we obtain that $\mathbb{E}\left[\,\delta(w,\mathbf{x})\,\right] \geq \Delta - \Theta(\sqrt{n})$, as desired. □

3.2   The SIMPLE k-SAMPLE Mechanism

In the most general model, we have the randomized mechanism SIMPLE k-SAMPLE, which is even simpler than RANDOM k-SAMPLE. Again, SIMPLE k-SAMPLE forms a sample S of vertices by repeating k times the selection of a vertex uniformly at random with replacement. The winner (if any) is a vertex w in $\arg\max_{u \in N \setminus S} \delta_S(u, \mathbf{x})$. We remark that, for technical reasons, we allow S to be a multi-set if the same vertex is selected more than once. Then, edge multiplicities are counted in $\delta_S(u, \mathbf{x})$. Clearly, SIMPLE k-SAMPLE is impartial. The winner is decided by the vertices in S, which in turn have no chance to become winners. Our approximation guarantee is slightly weaker now.

Theorem 2. *For* $k = \left\lceil 4^{1/3} n^{2/3} \ln^{1/3} n \right\rceil$, *mechanism* SIMPLE k-SAMPLE *is impartial and* $\Theta(n^{2/3} \ln^{1/3} n)$-*additive.*

Proof. Let u^* be a vertex of maximum in-degree Δ. If $\Delta \leq k$, SIMPLE k-SAMPLE is clearly $\Theta(n^{2/3} \ln^{1/3} n)$-additive. So, in the following, we assume that $\Delta > k$. Let C be the set of vertices of in-degree at most $\Delta - k - 1$. We first show that the probability $\mathbf{Pr}[\,\delta(w, \mathbf{x}) \leq \Delta - k - 1\,]$ that some vertex of C is returned as the winner by SIMPLE k-SAMPLE is small.

Notice that if some of the vertices of C is the winner, then either vertex u^* belongs to the sample set S or it does not belongs to S but it gets the same or fewer nominations compared to some vertex u of C. Hence,

$$
\begin{aligned}
&\mathbf{Pr}[\,\delta(w, \mathbf{x}) \leq \Delta - k - 1\,] \\
&\leq \mathbf{Pr}[\,u^* \in S\,] + \mathbf{Pr}[\,u^* \notin S \wedge \delta_S(u^*, \mathbf{x}) \leq \delta_S(u, \mathbf{x}) \text{ for some } u \in C \text{ s.t. } u \notin S\,] \\
&\leq \mathbf{Pr}[\,u^* \in S\,] + \sum_{u \in C} \mathbf{Pr}[\,u^* \notin S \wedge u \notin S \wedge \delta_S(u^*, \mathbf{x}) \leq \delta_S(u, \mathbf{x})\,] \\
&= \mathbf{Pr}[\,u^* \in S\,] + \sum_{u \in C} \mathbf{Pr}[\,u^*, u \notin S\,] \cdot \mathbf{Pr}[\,\delta_S(u^*, \mathbf{x}) \leq \delta_S(u, \mathbf{x})|u^*, u \notin S\,] \qquad (3)
\end{aligned}
$$

We will now bound the rightmost probability in (3). The proof of Claim 1 appears in the full version of the paper.

Claim 1. *For every* $u \in C$, $\mathbf{Pr}[\,\delta_S(u^*, \mathbf{x}) \leq \delta_S(u, \mathbf{x})|u^* \notin S, u \notin S\,] \leq \exp\left(-\frac{k^3}{2n^2}\right)$.

Using the definition of $\mathbb{E}[\delta(w,\mathbf{x})]$, inequality (3), and Claim 1, we obtain

$$\mathbb{E}[\delta(w,\mathbf{x})] \geq (\Delta - k) \cdot (1 - \mathbf{Pr}[\delta(w,\mathbf{x}) \leq \Delta - k - 1])$$

$$\geq (\Delta - k)\left(\mathbf{Pr}[u^* \notin S] - \sum_{u \in C} \mathbf{Pr}[u^*, u \notin S] \cdot \mathbf{Pr}[\delta_S(u^*,\mathbf{x}) \leq \delta_S(u,\mathbf{x})|u^*, u \notin S]\right)$$

$$\geq (\Delta - k)\left(1 - \frac{1}{n}\right)^k - (\Delta - k)\left(\sum_{u \in C}\left(1 - \frac{2}{n}\right)^k \cdot \exp\left(-\frac{k^3}{2n^2}\right)\right)$$

$$\geq (\Delta - k)\left(1 - \frac{k}{n}\right) - (\Delta - k) \cdot n \cdot \exp\left(-\frac{k^3}{2n^2}\right)$$

$$\geq \Delta - 2k - n^2 \cdot \exp\left(-\frac{k^3}{2n^2}\right). \tag{4}$$

The last inequality follows since $n \geq \Delta$. Setting $k = \left\lceil 4^{1/3}n^{2/3}\ln^{1/3} n \right\rceil$, (4) yields $\mathbb{E}[\delta(w,\mathbf{x})] \geq \Delta - \Theta\left(n^{2/3}\ln^{1/3} n\right)$, as desired. $\qquad\square$

4 Lower Bounds

In this section we complement our positive results by providing impossibility results. First, in Sect. 4.1, we provide lower bounds for a class of mechanisms which we call strong sample mechanisms, in the single nomination model of Holzman and Moulin [11]. Then, in Sect. 4.2, we provide a lower bound for the most general model of Alon et al. [1], which applies to any deterministic mechanism.

4.1 Strong Sample Mechanisms

In this section, we give a characterization theorem for a class of impartial mechanisms which we call *strong* sample mechanisms. We then use this characterization to provide lower bounds on the additive approximation of deterministic and randomized mechanisms that belong to this class. Our results suggest that mechanism RANDOM k-SAMPLE from Sect. 3.1 is essentially the best possible randomized mechanism in this class.

For a graph $G \in \mathcal{G}^1$ and a subset of vertices S, let $W := W_S(G)$ be the set of vertices outside S nominated by S, i.e. $W = \{w \in N \setminus S : (v,w) \in E, v \in S\}$. A *sample* mechanism[3] (g,f) firstly selects a subset S using some function $g : \mathcal{G}^1 \to \mathcal{P}(N) \setminus \emptyset$, and then applies a (possibly randomized) selection mechanism f by restricting its range on vertices in W; notice that if $W = \emptyset$, f does not select any vertex. We say that the mechanism is randomized if g uses randomization in the selection of S, otherwise it is deterministic (even if f uses randomization).

---

[3] For simplicity we use the notation (g,f) rather the more precise $(g, f(g))$.

This definition allows for a large class of impartial mechanisms. For example, the special case of sample mechanisms with $|S| = 1$ (in which, the winner has in-degree at least 1), coincides with all negative unanimous mechanisms defined by Holzman and Moulin [11]. Indeed, when $|S| = 1$, the set W in never empty and the winner has in-degree at least 1. This is not however the case for $|S| > 1$, where W could be empty when all vertices in S have outgoing edges destined for vertices in S and no winner can be declared. Characterizing all sample mechanisms is an outstanding open problem. We are able to provide a first step, by providing a characterization for the class of strong sample mechanisms. Informally, in such mechanisms, vertices cannot affect their chance of being selected in the sample set S.

Definition 1. *(Strong sample mechanisms) We call a sample mechanism (g, f) with sample function $g : \mathcal{G}^1 \to \mathcal{P}(N)$ strong, if $g(x'_u, \mathbf{x}_{-u}) = g(\mathbf{x})$ for all $u \in g(\mathbf{x})$, $x'_u \in N \setminus \{u\}$ and $\mathbf{x} \in \mathcal{G}^1$.*

The reader may observe the similarity of this definition with impartiality (function g of a strong sample mechanism satisfies similar properties with function f of an impartial selection mechanism). The following lemma describes a straightforward, yet useful, consequence of the above definition.

Lemma 3. *Let (g, f) be a strong sample mechanism and let $S \subseteq N$. For any nomination profiles $\mathbf{x}, \mathbf{x}'$ with $\mathbf{x}_{-S} = \mathbf{x}'_{-S}$, if $S \setminus g(\mathbf{x}) \neq \emptyset$ then $S \setminus g(\mathbf{x}') \neq \emptyset$.*

Proof. For the sake of contradiction, let us assume that $S \setminus g(\mathbf{x}') = \emptyset$, i.e., the sample vertices in $\mathbf{x}'$ are disjoint from S. Then, by Definition 1, $g(\mathbf{x})$ remains the same as outgoing edges from vertices in S should not affect the sample set. But then, $S \setminus g(\mathbf{x}) = \emptyset$, which is a contradiction. □

In the next theorem, we provide a characterization of the sample function of impartial strong sample mechanisms. The theorem essentially states that the only possible way to choose the sample set must be independent of the graph.

Theorem 3. *Any impartial deterministic strong sample mechanism (g, f) selects the sample set independently of the nomination profile, i.e., for all $\mathbf{x}, \mathbf{x}' \in \mathcal{G}^1$, $g(\mathbf{x}) = g(\mathbf{x}') = S$.*

Proof. Consider any sample mechanism (g, f) and any nomination profile $\mathbf{x} \in \mathcal{G}^1$. It suffices to show that for any vertex u, and any deviation x'_u, the sample set must remain the same, i.e., $g(x'_u, \mathbf{x}_{-u},) = g(x)$. If $u \in g(x)$, this immediately follows by Definition 1. In the following, we prove two claims that this is also true when $u \notin g(\mathbf{x})$; Claim 2 treats the case where u is a winner of a profile, while Claim 3 treats the case where u is a not a winner.

Claim 2. *Let (g, f) be an impartial deterministic strong sample mechanism and let $\mathbf{x}$ be a nomination profile. Then the sample set must remain the same for any other vote of the winner, i.e., $g(\mathbf{x}) = g(x'_{f(\mathbf{x})}, \mathbf{x}_{-f(\mathbf{x})})$ for any $x'_{f(\mathbf{x})} \in N \setminus f(\mathbf{x})$.*

Proof. Let $w = f(\mathbf{x})$ be the winner, for some nomination profile $\mathbf{x}$. We will prove the claim by induction on $\delta(w, \mathbf{x})$.

(**Base case:** $\delta(w, \mathbf{x}) = 1$) Let $S = g(\mathbf{x})$ be the sample set for profile $\mathbf{x}$. Assume for the sake of contradiction that when w changes its vote to x'_w, the sample for profile $\mathbf{x}' = (x'_w, \mathbf{x}_{-w})$ changes, i.e., $g(\mathbf{x}') = S' \neq S$. We first note that impartiality of f implies that $w = f(\mathbf{x}')$. Next, observe that the vertex voting for w in S must be also in S'; otherwise, w becomes a winner without getting any vote from the sample set, which contradicts our definition of sample mechanisms. We will show that this must be the case for all vertices in S.

To do this, we will expand two parallel branches, creating a series of nomination profiles starting from $\mathbf{x}$ and $\mathbf{x}'$ which will eventually lead to contradiction. Figure 1 depicts the situation for $\mathbf{x}$ and $\mathbf{x}'$.

(a) profile $\mathbf{x}$ (b) profile $\mathbf{x}'$

Fig. 1. The starting profiles $\mathbf{x}$ and $\mathbf{x}'$ in Claim 2. Red denotes the winner, while green dashed vertices denote the members of the sample sets S and S', respectively. (Color figure online)

We start with the profile $\mathbf{x}'$. Consider a vertex $s' \in S' \setminus S$. We create a profile $\mathbf{z}'$ in which all vertices in $S' \setminus s'$ vote for s' (i.e., $z_v = s'$, for each $v \in S' \setminus s'$), vertex s' votes for w (i.e., $z_v = w$), while the rest of the vertices vote as in $\mathbf{x}'$ (i.e., $z_v = x_v$, for each $v \notin S'$). For illustration, see Figs. 2a and b. By the definition of a strong sample mechanism, we obtain $g(\mathbf{z}') = g(\mathbf{x}')$, since only votes of vertices in S' have changed. Notice also that $f(\mathbf{z}') = w$, as this is the only vertex outside S' that receives votes from S'. We now move to profile $\mathbf{x}$ and apply the same sequence of deviations, involving all the vertices in S'. These lead to the profile $\mathbf{z}$, which differs from $\mathbf{z}'$ only in the outgoing edge of vertex w.

By Lemma 3, there is a vertex $v \in S'$ such that $v \notin g(\mathbf{z})$. If $v = s'$, then we end up in a contradiction. This is because $f(\mathbf{z}) \neq w$, since s' is the only vertex voting for w in $\mathbf{z}'$ and s' is not in the sample, while $f(\mathbf{z}') = w$, as stated by the other branch and since, when w deviates to x'_w, the created profile is $(x'_w, \mathbf{z}_{-w}) = \mathbf{z}'$ contradicting impartiality (see Figs. 2a and b).

We are now left with the case where $s' \in g(\mathbf{z})$ and $v \neq s'$. Starting from $\mathbf{z}$ and $\mathbf{z}'$, we will create profiles $\mathbf{y}$ and $\mathbf{y}'$ (see Figs. 2c and d) as follows: we construct $\mathbf{y}$ by letting s' vote towards v (i.e., $y_{s'} = v$), v vote towards w (i.e., $y_v = w$) and

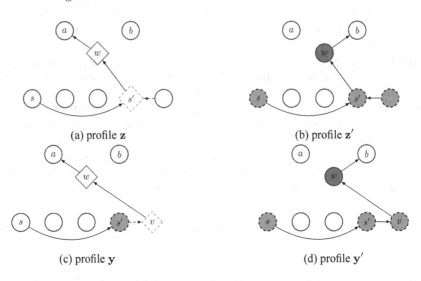

(a) profile $\mathbf{z}$ (b) profile $\mathbf{z}'$

(c) profile $\mathbf{y}$ (d) profile $\mathbf{y}'$

Fig. 2. Profiles $\mathbf{z}$ and $\mathbf{z}'$ in the base case of the proof of Claim 2: Red denotes the winner, while green dashed vertices denote the members of the sample sets S and S'. A red solid diamond denotes a vertex that cannot win and a green/light diamond denotes a vertex that cannot be in the sample. (Color figure online)

$y_i = z_i$ for all other vertices $i \neq v, s'$. By the strong sample property, when s' votes towards v the sample set is preserved, i.e., v cannot get in the sample. Also, when v votes, v cannot get in the sample (by a trivial application of Lemma 3); therefore, $v \notin g(\mathbf{y})$. Hence, w cannot be the winner as its only incoming vote is from v, a vertex that does not belong to the sample set $g(\mathbf{y})$.

Starting from $\mathbf{z}'$, we create similarly $\mathbf{y}'$ by letting s' vote towards v ($y'_{s'} = v$), v to vote towards w ($y'_v = w$) and $y'_i = z_i$ for all other vertices $i \neq v, s'$. In this case, S' will be preserved as sample set in profile $\mathbf{y}'$ (i.e. $g(\mathbf{y}') = S'$). Therefore, w is the only vertex voted by the sample set and must be the winner, leading to a contradiction (see Figs. 2c and d).

(Induction step) Assume as induction hypothesis that, for all profiles $\mathbf{x} \in \mathcal{G}^1$, it holds $g(\mathbf{x}) = g(x'_w, \mathbf{x}_{-w})$ when $\delta(w, \mathbf{x}) \leq \lambda$, for some $\lambda \geq 1$. Now, consider any profile $\mathbf{x}$ where $f(\mathbf{x}) = w$ and $\delta(w, \mathbf{x}) = \lambda + 1$ and assume for the sake of contradiction that there is some graph $\mathbf{x}' = (x'_w, \mathbf{x}_{-w})$ where $g(\mathbf{x}') = S' \neq S$. Without loss of generality, let $\delta_S(w, \mathbf{x}) \leq \delta_{S'}(w, \mathbf{x}')$.

Starting from $\mathbf{x}'$, we create profile $\mathbf{z}'$, by letting all vertices in S' vote for some $s' \in S'$ and s' vote for w, i.e., $z'_v = s'$ for each vertex $v \in S' \setminus \{s'\}$ and $z'_{s'} = w$. The strong sample property implies that $g(\mathbf{z}') = S'$ and $f(\mathbf{z}') = w$. We focus now on profile $\mathbf{x}$, and create the profile $\mathbf{z}$, by performing the same series of deviations, i.e., by letting all vertices in $S' \setminus s'$ vote for s' and s' vote for w. Note, here, that $\mathbf{z}$ differs from $\mathbf{z}'$ only in the outgoing edge of w. Like before, Lemma 3 establishes that there will be some vertex $v \in S'$ such that $v \notin g(\mathbf{z})$, i.e., $g(\mathbf{z}) \neq S'$. Turning our attention back to $\mathbf{z}'$, we let w deviate towards x_w, creating profile $(x_w, \mathbf{z}'_{-w})$. Observe that $(x_w, \mathbf{z}'_{-w}) = \mathbf{z}$. Since $\delta(w, \mathbf{z}') < \delta(w, \mathbf{x})$, by the induction hypothesis we have $g(\mathbf{z}) = S'$, a contradiction. □

The next claim establishes the remaining case, where no vertex $u \notin g(\mathbf{x}), u \neq f(\mathbf{x})$ can change the sample set.

Claim 3. *Let (g, f) be an impartial deterministic strong sample mechanism, $\mathbf{x}$ be a nomination profile and u a vertex with $u \notin g(\mathbf{x}), u \neq f(\mathbf{x})$. Then $g(\mathbf{x}) = g(x'_u, \mathbf{x}_{-u})$ for any other vote $x'_u \in N \setminus u$.*

Proof. For the sake of contradiction, assume that there exists some nomination profile $\mathbf{x}' = (x'_u, \mathbf{x}_{-u})$ with $g(\mathbf{x}') = S' \neq S$. Starting from $\mathbf{x}'$, we define a profile $\mathbf{z}'$ in which all vertices in S' vote for u, and the rest vote as in $\mathbf{x}'$. That is, $z'_v = u$, for all $v \in S'$ and $z'_v = x'_v$ otherwise. Clearly $f(\mathbf{z}') = u$, as all the sample vertices vote for u. By Claim 2, we know that $g(x_u, \mathbf{z}'_{-u}) = g(\mathbf{z}') = S'$.

Starting from $\mathbf{x}$, we define a profile $\mathbf{z}$ in which all vertices in S' vote for u, and the rest vote as in $\mathbf{x}$. Since $S' \neq S = g(\mathbf{x})$, by Lemma 3, we get $g(\mathbf{z}) \neq S'$. Observe that $\mathbf{z} = (x_u, \mathbf{z}'_{-u})$, which leads to a contradiction. □

This completes the proof of Theorem 3. □

We next use Theorem 3 to obtain lower bounds on the additive approximation guarantee obtained by deterministic and randomized strong sample mechanisms respectively.

Corollary 1. *There is no impartial deterministic strong sample mechanism with additive approximation better than $n - 2$.*

Proof. Let S be the sample set which, by Theorem 3, must be selected independently of $\mathbf{x}$, and let $v \in S$. Define $\mathbf{x}$ so that all vertices in $N \setminus v$ vote for v and all other vertices have in-degree either 0 or 1. Then, $\Delta(\mathbf{x}) = n - 1$, but the mechanism selects vertex x_v of in-degree exactly 1. □

We remark that the strong sample mechanism that uses a specific vertex as singleton sample achieves this additive approximation guarantee.

We next provide a lower bound on the additive approximation guarantee of randomized strong sample mechanisms, which shows that RANDOM k-SAMPLE (with $k = \Theta(\sqrt{n})$; see Sect. 3.1) is an optimal mechanism from this class.

Corollary 2. *There is no impartial randomized strong sample mechanism with additive approximation better than $\Omega(\sqrt{n})$.*

Proof. By Theorem 3, any impartial deterministic strong sample mechanism selects its sample S independently of the graph. Hence, any impartial randomized strong sample mechanism decides its sample with a probability distribution over all possible sample sets $S \subseteq N$, independently of the graph. We examine two cases for this probability distribution.

If there is a vertex u^* with $\mathbf{Pr}\,[\,v \in S\,] \geq 1/\sqrt{n}$, then consider a nomination profile consisting of vertex u^* having maximum in-degree $\Delta = n - 1$ (i.e., all other vertices are pointing to it), with all other vertices having in-degree either 1 or 0. Since u^* belongs to the sample (and, hence, cannot be the winner) with probability at least $1/\sqrt{n}$, the expected degree of the winner is at most $1 + (n-1)(1 - 1/\sqrt{n}) = \Delta - \Theta(\sqrt{n})$.

Otherwise, assume that every vertex v has probability at most $1/\sqrt{n}$ of being selected in the sample set. Consider a nomination profile with a vertex u^* having maximum degree $\Delta = \sqrt{n}/2$ and all other vertices having in-degree either 0 or 1. Consider a vertex u pointing to vertex u^*. The probability that u belongs to the sample is at most $1/\sqrt{n}$. Hence, by the union bound, the probability that some of the $\sqrt{n}/2$ vertices pointing to u^* is selected in the sample set is at most $1/2$. Hence, the probability that u^* is returned as the winner is not higher than $1/2$ and the expected in-degree of the winner is at most $1 + \sqrt{n}/2 \cdot 1/2 = \Delta - \Theta(\sqrt{n})$. □

4.2 General Lower Bound

Our last result is a lower bound for all deterministic impartial mechanisms in the most general model of Alon et al. [1], where each agent can nominate multiple other agents or even abstain. We remark that our current proof applies to mechanisms that always select a winner. Due to space limitations we refer the reader to the full version of the paper for the proof of Theorem 4.

Theorem 4. *There is no impartial deterministic α-additive mechanism for $\alpha \leq 2$.*

References

1. Alon, N., Fischer, F., Procaccia, A., Tennenholtz, M.: Sum of us: strategyproof selection from the selectors. In: Proceedings of the 13th Conference on Theoretical Aspects of Rationality and Knowledge, pp. 101–110 (2011)
2. Attiya, H., Welch, J.: Distributed Computing: Fundamental, Simulations, and Advanced Topics, 2nd edn. Wiley, Hoboken (2004)
3. Aziz, H., Lev, O., Mattei, N., Rosenschein, J.S., Walsh, T.: Strategyproof peer selection: mechanisms, analyses, and experiments. In: Proceedings of the 30th AAAI Conference on Artificial Intelligence (AAAI), pp. 397–403 (2016)
4. Babichenko, Y., Dean, O., Tennenholtz, M.: Incentive-compatible diffusion. In: Proceedings of the 27th International Conference on World Wide Web (WWW), pp. 1379–1388 (2018)
5. Bjelde, A., Fischer, F., Klimm, M.: Impartial selection and the power of up to two choices. ACM Trans. Econ. Comput. 5(4), 21 (2017)
6. Bousquet, N., Norin, S., Vetta, A.: A near-optimal mechanism for impartial selection. In: Liu, T.-Y., Qi, Q., Ye, Y. (eds.) WINE 2014. LNCS, vol. 8877, pp. 133–146. Springer, Cham (2014). https://doi.org/10.1007/978-3-319-13129-0_10
7. Brandt, F., Conitzer, V., Endriss, U., Lang, J., Procaccia, A.D. (eds.): Handbook of Computational Social Choice. Cambridge University Press, New York (2016)

8. Chen, Y., Podimata, C., Procaccia, A.D., Shah, N.: Strategyproof linear regression in high dimensions. In: Proceedings of the 19th ACM Conference on Economics and Computation (EC), pp. 9–26 (2018)
9. de Clippel, G., Moulin, H., Tideman, N.: Impartial division of a dollar. J. Econ. Theor. **139**(1), 176–191 (2008)
10. Fischer, F., Klimm, M.: Optimal impartial selection. SIAM J. Comput. **44**(5), 1263–1285 (2015)
11. Holzman, R., Moulin, H.: Impartial nominations for a prize. Econometrica **81**(1), 173–196 (2013)
12. Kahng, A., Kotturi, Y., Kulkarni, C., Kurokawa, D., Procaccia, A.D.: Ranking wily people who rank each other. In: Proceedings of the 32nd AAAI Conference on Artificial Intelligence (AAAI), pp. 1087–1094 (2018)
13. Kurokawa, D., Lev, O., Morgenstern, J., Procaccia, A.D.: Impartial peer review. In: Proceedings of the 24th International Joint Conference on Artificial Intelligence (IJCAI), pp. 582–588 (2015)
14. Mackenzie, A.: Symmetry and impartial lotteries. Games Econ. Behav. **94**, 15–28 (2015)
15. Tamura, S.: Characterizing minimal impartial rules for awarding prizes. Games Econ. Behav. **95**, 41–46 (2016)
16. Tamura, S., Ohseto, S.: Impartial nomination correspondences. Soc. Choice Welfare **43**(1), 47–54 (2014)

The Convergence of Iterative Delegations in Liquid Democracy in a Social Network

Bruno Escoffier[1], Hugo Gilbert[2(✉)], and Adèle Pass-Lanneau[1,3]

[1] Sorbonne Université, CNRS, LIP6 UMR 7606, 4 place Jussieu, 75005 Paris, France
{bruno.escoffier,adele.pass-lanneau}@lip6.fr
[2] Gran Sasso Science Institute, L'Aquila, Italy
hugo.gilbert@gssi.it
[3] EDF R&D, 7 bd Gaspard Monge, 91120 Palaiseau, France

Abstract. Liquid democracy is a collective decision making paradigm which lies between direct and representative democracy. One of its main features is that voters can delegate their votes in a transitive manner so that: A delegates to B and B delegates to C leads to A indirectly delegates to C. These delegations can be effectively empowered by implementing liquid democracy in a social network, so that voters can delegate their votes to any of their neighbors in the network. However, it is uncertain that such a delegation process will lead to a stable state where all voters are satisfied with the people representing them. We study the stability (w.r.t. voters preferences) of the delegation process in liquid democracy and model it as a game in which the players are the voters and the strategies are their possible delegations. We answer several questions on the equilibria of this process in any social network or in social networks that correspond to restricted types of graphs.

Keywords: Computational social choice · Liquid democracy · Algorithmic decision theory · Delegative voting · Games and equilibria

1 Introduction

Liquid Democracy (LD) is a voting paradigm which offers a middle-ground between direct and representative democracy. One of its main features is the concept of *transitive delegations*, i.e., each voter can delegate her vote to some other voter, called representative or proxy, which can in turn delegate her vote and the ones that have been delegated to her to another voter. Consequently, a voter who decides to vote has a voting weight corresponding to the number of people she represents, i.e., herself and the voters who directly or indirectly delegated to her. This voter is called the *guru* of the people she represents. This approach has been advocated recently by many political parties as the German Pirate party or the Sweden's Demoex party and is implemented in several online tools [2,12]. One main advantage of this framework is its flexibility, as it enables voters to vote directly for issues on which they feel both concerned and expert and to delegate for others. In this way, LD provides a middle-ground between direct democracy,

D. Fotakis and E. Markakis (Eds.): SAGT 2019, LNCS 11801, pp. 284–297, 2019.
https://doi.org/10.1007/978-3-030-30473-7_19

which is strongly democratic but which is likely to yield high abstention rates or uninformed votes, and representative democracy which is more practical but less democratic [11,15]. Importantly, LD can be conveniently used in a Social Network (SN), where natural delegates are connected individuals. These choices of delegates are also desirable as they ensure that delegations rely on a foundation of trust. For these reasons, several works studying LD in the context of an SN enforce the constraint that voters may only delegate directly to voters that are connected to them (i.e., their neighbors in the underlying graph) [3,4,13].

Aim of this Paper. We tackle the problem of the stability of the delegation process in the LD setting. Indeed, it is likely that the preferences of voters over possible gurus will be motivated by different criteria and contrary opinions. Hence, the iterative process where each voter chooses her delegate may end up in an unstable situation, i.e., a situation in which some voters would change their delegations. To illustrate this point, consider an election where the voters could be positioned on a line in a way that represents their right-wing left-wing political identity. If voters are ideologically close enough, each voter, starting from the left-side, could agree to delegate to her closest neighbor on her right. By transitivity, this would lead to all voters having an extreme-right voter for guru. These situations raise the questions: "Under what conditions do the iterative delegations of the voters always reach an equilibrium? Does such an equilibrium even exist? Can we determine equilibria that are more desirable than others?".

We assume that voters are part of an SN so that they can only delegate directly to their neighbors. Voters' preferences over possible gurus are given by preference orders. In this setting, the delegation process yields a game where each voter seeks to minimize the rank of her guru in her preference order. We answer the questions raised above with a special emphasize on SN.

2 Related Work

The stability of the delegation process is one of the several algorithmic issues raised by LD. These issues have recently raised attention in the AI literature.

Are votes in an LD setting "more correct"? The idea underlying LD is that its flexibility should allow each voter to make an informed vote either by voting directly, or by finding a suitable guru. Several works have investigated this claim leading to both positive and negative results [11,13]. On the one hand, Green-Armytage [11], proposed a spacial voting setting in which transitive delegations decrease on average a function measuring how well the votes represent the voters attached to them. On the other hand, Kahng et al. [13], studied an election on a binary issue with a ground truth. In their model, no procedure working locally on the SN to organize delegations can guarantee that LD is, at the same time, never less accurate and sometimes strictly more accurate than direct voting.

How much delegations should a guru get? This question raises the concern that a guru could become too powerful. In the setting of Gölz et al. [10], each voter can vote or specify multiple delegation options. Then an algorithm should select delegations to minimize the maximum voting power of a voter. The authors give a $(1+\log(n))$-approximation algorithm (n is the number of voters) and show that approximating the problem within a factor $\frac{1}{2}\log_2(n)$ is NP-hard. Lastly, they gave evidence that allowing multiple possible delegation options (instead of one) leads to a large decrease of the maximum voting power of a voter.

Are votes in an LD setting rational? Christoff and Grossi [7] studied the potential loss of a rationality constraint when voters should vote on different issues that are logically linked and for which they delegate to different representatives. Following this work, Brill and Talmon [6] considered an LD framework in which each voter should provide a linear order over possible candidates. To do so each voter may delegate different binary preference queries to different proxies. The delegation process may then yield incomplete and even intransitive preferences. Notably, the authors showed that it is NP-hard to decide if an incomplete ballot obtained in this way can be completed to obtain complete and transitive preferences while respecting the constraints induced by the delegations.

Are delegations in an LD setting rational? Bloembergen et al. [3], considered an LD setting where voters are connected in an SN and can only delegate to their neighbors in the network. The election is on a binary issue where some voters should vote for the 0 answer and the others should vote for the 1 answer (voters of type τ_0 or τ_1). Each voter i has an accuracy level $q_i \in [0.5, 1]$ representing the probability with which she makes the correct choice. Similarly, each pair (i, j) of voters do not know if they are of the same type which is also modeled by a probability p_{ij}. Hence, a voter i which has j as guru has a probability that j makes the correct vote (w.r.t. to i) which is a formula including the p_{ij} and q_j values. The goal of each voter is to maximize the accuracy of her vote/delegation. This modeling leads to a class of games, called *delegation games*. The authors proved the existence of pure Nash equilibria in several types of delegation games and gave upper and lower bounds on the price of anarchy of such games.

Our approach is closest to this last work as we consider the same type of delegation games. However, our preference model is more general as we assume that each voter has a preference order over her possible gurus. These preferences may be dictated by competence levels and types of voters as in [3]. However, we do not make such hypothesis as the criteria to choose a delegate are numerous: geographic locality, cultural, political or religious identity, et caetera. Considering this more general framework strongly modifies the resulting delegation games.

3   Notation and Settings

3.1   Notation and Nash-Stable Delegation Functions

We denote by $\mathcal{N} = \{1, \dots, n\}$ a set of voters that take part in a vote.[1] These voters are connected in an SN which is represented by an undirected graph

---

[1] Note that similarly to [10], we develop a setting where candidates are not mentioned. Proceeding in this way enables a general approach encapsulating different ways of specifying how candidates structure the preferences of voters over gurus.

$\mathcal{G}_{SN} = (\mathcal{N}, \mathcal{E})$, i.e., the vertices of $\mathcal{G}_{SN}$ are the voters and $(i,j) \in \mathcal{E}$ if voters i and j are connected in the SN. Let $\texttt{Nb}(i)$ denote the set of neighbors of i in $\mathcal{G}_{SN}$. Each voter i can declare intention to either vote herself, delegate to one of her neighbors $j \in \texttt{Nb}(i)$, or abstain. A *delegation function* is a function $d : \mathcal{N} \to \mathcal{N} \cup \{0\}$ such that $d(i) = i$ if voter i wants to vote, $d(i) = j \in \texttt{Nb}(i)$ if i wants to delegate to j, and $d(i) = 0$ if i wants to abstain.

Given a delegation function d, let $\texttt{Gu}(d)$ denote the *set of gurus*, i.e., $\texttt{Gu}(d) = \{j \in \mathcal{N} \mid d(j) = j\}$. The *guru of a voter* $i \in \mathcal{N}$, denoted by $\texttt{gu}(i,d)$, can be found by following the successive delegations starting from i. Formally, $\texttt{gu}(i,d) = j$ if there exists a sequence of voters $i_1, \ldots, i_\ell$ such that $d(i_k) = i_{k+1}$ for every $k \in \{1, \ldots, \ell - 1\}$, $i_1 = i$, $i_\ell = j$ and $j \in \texttt{Gu}(d) \cup \{0\}$. However, it may happen that no such j exists because the successive delegations starting from i end up in a circuit, i.e., $i = i_1$ delegates to i_2, who delegates to i_3, and so on up to i_ℓ who delegates to i_k with $k \in \{1, \ldots, \ell - 1\}$. In this case, we consider that the ℓ voters abstain, as none of them takes the responsibility to vote, i.e., we set $\texttt{gu}(i_k, d) = 0$ for all $k \in \{1, \ldots, \ell\}$. Such a definition of gurus allows to model the transitivity of delegations: if $d(i) = j$, $d(j) = g$, and $d(g) = g$, then the guru of i will be g. Hence a voter i can delegate directly to one of its neighbors in $\texttt{Nb}(i)$, but she can also delegate indirectly to another voter through a chain of delegations. Note that because voters can only delegate directly to their neighbors, such a chain of delegations coincides with a path in $\mathcal{G}_{SN}$.

Given a voter i and a delegation function d, we now consider how voter i may change her delegation to get a guru different from her current guru $\texttt{gu}(i,d)$. She may decide to vote herself, to abstain, or to delegate to a neighbor j with a different guru, and in the latter case she would get $\texttt{gu}(j,d)$ as a guru. We denote by $\texttt{Att}(i,d) = \cup_{j \in \texttt{Nb}(i)}\texttt{gu}(j,d)$ the set of gurus of the neighbors of i in $\mathcal{G}_{SN}$. Then the gurus that i can get by deviating from $d(i)$ is exactly the set $\texttt{Att}(i,d) \cup \{0, i\}$.

Example 1. Consider the SN represented in Fig. 1 with the delegation function d defined by $d(1) = d(3) = d(4) = 4$, $d(2) = d(5) = 5$, $d(6) = d(9) = 6$, $d(7) = 8$ and $d(8) = 9$. In this example, $\texttt{Gu}(d) = \{4, 5, 6\}$, with $\texttt{gu}(1,d) = \texttt{gu}(3,d) = \texttt{gu}(4,d) = 4$, $\texttt{gu}(2,d) = \texttt{gu}(5,d) = 5$ and $\texttt{gu}(6,d) = \texttt{gu}(7,d) = \texttt{gu}(8,d) = \texttt{gu}(9,d) = 6$. If she wants, voter 1 can change her delegation. If she delegates to voter 3, then this will not change her guru as $\texttt{gu}(3,d) = 4$. However, if she delegates to 2, then her new guru will be 5. In any case, she can also decide to modify her delegation to declare intention to vote or abstain. Note that voter 1 cannot change unilaterally her delegation in order to have 6 as guru. Indeed, voter 6 does not belong to $\texttt{Att}(1,d) = \{4,5\}$, as there is no delegation path from a neighbor of 1 to 6.

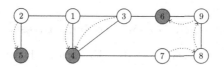

Fig. 1. SN in Ex. 1. Dotted arrows represent the delegations and gurus are in gray.

We assume that each voter i has a preference order $\succ_i$ over who could be her guru in $\mathcal{N} \cup \{0\}$. For every $i \in \mathcal{N}$, and for every $j, k \in \mathcal{N} \cup \{0\}$, we have that $j \succ_i k$ if i prefers to have j as guru (or to vote if $j = i$, or to abstain if $j = 0$) rather than to have k as guru (or to vote if $k = i$, or to abstain if $k = 0$). The collection $\{\succ_i \mid i \in \mathcal{N}\}$ in turn defines a *preference profile* P. Example 2 illustrates these notations on a simple instance with 3 voters.

Example 2 (3-cycle). Consider the following instance in which there are $n = 3$ voters connected in a complete SN and with the following preference profile P:

$$1 : 2 \succ_1 1 \succ_1 3 \succ_1 0$$
$$2 : 3 \succ_2 2 \succ_2 1 \succ_2 0$$
$$3 : 1 \succ_3 3 \succ_3 2 \succ_3 0$$

Put another way, each voter i prefers to delegate to $(i \mod 3) + 1$ rather than to vote directly and each voter prefers to vote rather than to abstain.

As a consequence of successive delegations, a voter might end up in a situation in which she prefers to vote or to abstain or to delegate to another guru that she can reach than to maintain her current delegation. Such a situation is regarded as *unstable* as this voter would modify unilaterally her delegation. This is for instance the case in the previous example if $d(1) = 2$, $d(2) = 3$ and $d(3) = 3$: by successive delegations, the guru of 1 is 3, but 1 would prefer to vote instead. More formally, a delegation function d is *Nash-stable for voter i* if

$$\mathtt{gu}(i, d) \succ_i g \quad \forall g \in (\mathtt{Att}(i, d) \cup \{0, i\}) \setminus \{\mathtt{gu}(i, d)\}.$$

A delegation function d is *Nash-stable* if it is Nash-stable for every voter in $\mathcal{N}$. A Nash-stable delegation function is also called an *equilibrium* in the sequel.

It may seem difficult in practice that voters give a complete linear order over $\mathcal{N} \cup \{0\}$. We now highlight that the computation of equilibria does not require the whole preference profile. We say that voter i is an *abstainer* in P if she prefers to abstain rather than to vote, i.e., if $0 \succ_i i$; she is a *non-abstainer* otherwise. We will denote by $\mathcal{A}$ the set of abstainers. Note that an abstainer (resp. non-abstainer) never votes (resp. abstains) in an equilibrium. Given a preference profile P, we define $\mathtt{Acc}(i) = \{j \in \mathcal{N} | j \succ_i i \text{ and } j \succ_i 0\}$ the set of *acceptable gurus* for i. Note that this set is likely to be small compared to n. A necessary condition for a delegation function to be Nash-stable is that $\mathtt{gu}(i, d) \in \mathtt{Acc}(i)$, or $\mathtt{gu}(i, d) = 0$ and $i \in \mathcal{A}$, or $\mathtt{gu}(i, d) = i$ and $i \notin \mathcal{A}$. Hence when looking for equilibria, the preferences of voter i below 0 (if $i \in \mathcal{A}$) or i (if $i \notin \mathcal{A}$) can be dropped. In the sequel, we may define a preference profile only by giving, for every voter i, if she is an abstainer or not, and her preferences on $\mathtt{Acc}(i)$.

3.2   Existence and Optimization Problems Investigated

Stable situations are obviously desirable. Unfortunately, there are instances for which there is no equilibrium.

Observation 1. *The instance described in Ex. 2 (3-cycle) admits no equilibrium.*

Proof. Assume that there exists an equilibrium d. First, as $\mathcal{G}_{SN}$ is complete, any voter can delegate to any other voter. Second, for any pair of voters, there is always one voter that approves the other as possible guru. Hence, $|\text{Gu}(d)| \leq 1$, otherwise one of the gurus would rather delegate to another guru than vote. On the other hand, there is no voter that is approved as possible guru by all other voters. Hence, $|\text{Gu}(d)| \geq 2$, otherwise one of the voters in $\mathcal{N}\setminus\text{Gu}(d)$ would rather vote than delegate to one of the gurus (here $\mathcal{A} = \emptyset$). We get a contradiction. □

Hence the first problem, called **EX**, that we investigate is the one of the existence of an equilibrium.

EX
INSTANCE: A preference profile P and a social network $\mathcal{G}_{SN}$.
QUESTION: Does there exist an equilibrium?

Note that **EX** is in NP, as given a delegation function, we can easily find the guru of each voter and check Nash-stability in polynomial time.

For instances for which we know that some equilibrium exists, we investigate if we can compute equilibria verifying particular desirable properties. Firstly, given a voter $i \in \mathcal{N} \setminus \mathcal{A}$, we will try to know if there exists a Nash-stable delegation function d for which i is a guru, i.e., $i \in \text{Gu}(d)$. We call this problem **MEMB**.

MEMB
INSTANCE: A preference profile P, a social network $\mathcal{G}_{SN}$ and a voter $i \in \mathcal{N}\setminus\mathcal{A}$.
QUESTION: Is there a Nash-stable delegation function d for which $i \in \text{Gu}(d)$?

Secondly, we will try to find an equilibrium that optimizes some objective function such as (a) minimize the dissatisfaction of the voters (problem **MINDIS** as defined below, where $\text{rk}(i, d)$ is the rank of $\text{gu}(i, d)$ in the preference list of i); (b) minimize the maximum voting power of a guru (problem **MINMAXVP** as defined below, where $\text{vp}(i, d) := |\{j \in \mathcal{N}|\text{gu}(j, d) = i\}|$); or (c) minimize the number of voters who abstain (problem **MINABST**).

Problems **MINDIS, MINMAXVP** and **MINABST**
INSTANCE: A preference profile P and a social network $\mathcal{G}_{SN}$.
SOLUTION: A Nash-stable delegation function d.
MEASURE for **MINDIS**: $\sum_{i \in \mathcal{N}}(\text{rk}(i, d) - 1)$ (to minimize).
MEASURE for **MINMAXVP**: $\max_{i \in \text{Gu}(d)} \text{vp}(i, d)$ (to minimize).
MEASURE for **MINABST**: $|\{i \in \mathcal{N}|\text{gu}(i, d) = 0\}|$ (to minimize).

The last questions investigated capture the dynamic nature of delegations.

3.3   Convergence Problems Investigated

In situations where an equilibrium exists, a natural question is whether a dynamic delegation process necessarily converges towards such an equilibrium. As classically done in game theory (see for instance [14]), we will consider dynamics where iteratively one voter has the possibility to change her delegation.

In a dynamics, we are given a starting delegation function d_0 and a token function $T : \mathbb{N}^* \to \mathcal{N}$ which specifies that voter $T(t)$ has the token at step t: she may change her delegation. This gives a sequence of delegation functions $(d_t)_{t \in \mathbb{N}^*}$ where for any $t \in \mathbb{N}^*$, if $j \neq T(t)$ then $d_t(j) = d_{t-1}(j)$. A dynamics is said to converge if there is a t^* such that for all $t \geq t^*$, $d_t = d_{t^*}$. We assume, as usual, that each voter has the token infinitely many times. A classical way of choosing such a function T is to consider a permutation σ over the voters, and to repeat this permutation over the time to give the token (if $t = r \mod n$ then $T(t) = \sigma(r)$). These dynamics are called *permutation dynamics*. Note that we do not consider moves where a voter delegates to a voter who abstains, or in a way that creates a cycle – in this case, the voter would rather abstain herself.

Given d_0 and T, a dynamics is an *Improved Response Dynamics*[2] (IRD) if for all t, $T(t)$ chooses a move that strictly improves her outcome if any, otherwise does not change her delegation; It is a *Best Response Dynamics* (BRD) if for all t, $T(t)$ chooses $d_t(i)$ so as to maximize her outcome (a BRD is also an IRD). We denote by **IR-CONV** and **BR-CONV** the following problems:

IR-CONV (resp. BR-CONV)
QUESTION: Does a dynamic delegation process under IRD (resp. BRD) necessarily converges whatever the preference profile P and token function T.

3.4   Summary of Results

Our results are presented in Table 1. Section 4 tackles the complexity of **EX**. When $\mathcal{G}_{\mathrm{SN}}$ is complete, we show that **EX** is equivalent to the NP-complete problem of determining if a digraph admits a kernel. We then strengthen this result by showing that **EX** is also NP-complete when the maximum degree of $\mathcal{G}_{\mathrm{SN}}$ is 5 and is W[1]-hard w.r.t. the treewidth of $\mathcal{G}_{\mathrm{SN}}$. Yet, we identify specific SNs that ensure that an equilibrium exists. More precisely, an equilibrium exists whatever the voters' preferences iff $\mathcal{G}_{\mathrm{SN}}$ is a tree. Hence, in Sect. 5, we investigate tree SNs and we design a dynamic programming scheme which solves problems **MEMB**, **MINDIS**, **MINMAXVP** and **MINABST** in polynomial time. Lastly, Sect. 6 studies delegations dynamics. Unfortunately, when an equilibrium exists, we show that a BRD may not converge even if $\mathcal{G}_{\mathrm{SN}}$ is complete or is a path. For a star SN, we obtain that a BRD will always converge, whereas an IRD may not. All missing proofs can be found in the long version of the paper [9].

---

[2] Often termed better response dynamics.

4 Existence of Equilibria: Hardness Results

4.1 Complete Social Networks

We focus in this subsection on the case where $\mathcal{G}_{\mathsf{SN}}$ is a complete graph. We mainly show that determining whether an equilibrium exists or not is an NP-complete problem (Theorem 1), by showing an equivalence with the problem of finding a kernel in a digraph. This equivalence is also helpful to find subcases where an equilibrium always exists.

Table 1. Results (AE: Always Exists; NA: Not Always; NP-C: NP-Complete).

Type of $\mathcal{G}_{\mathsf{SN}}$ or parameter	EX		Problem\ $\mathcal{G}_{\mathsf{SN}}$	Star	Tree		$\mathcal{G}_{\mathsf{SN}}$	Problem	IR	BR
Complete	NP-C		MEMB	$O(n^2)$	$O(n^3)$		with an equilibrium		-CONV	-CONV
Maximum degree = 5	NP-C		MINDIS	$O(n^2)$	$O(n^3)$			Star	NA	Always
Treewidth	W[1]-hard		MINMAXVP	$O(n^2)$	$O(n^4)$			Path	NA	NA
Tree	AE		ABST	$O(n^2)$	$O(n^3)$			Complete	NA	NA

We define the *delegation-acceptability digraph* $G_P = (\mathcal{N} \setminus \mathcal{A}, A_P)$ by its arc-set $A_P = \{(i,j) \mid j \in \mathsf{Acc}(i)\}$. Stated differently, there is one vertex per non-abstainer and there exists an arc from i to j if i accepts j as a guru. For example, in Fig. 2, we give a partial preference profile P involving 5 voters and the corresponding delegation-acceptability digraph G_P. The main result of this subsection, stated in Proposition 1, is a characterization of all sets of gurus of equilibria, as specific subsets of vertices of the delegation-acceptability digraph. Let us introduce additional graph-theoretic definitions. Given a digraph $G = (\mathsf{V}, \mathsf{A})$, a subset of vertices $K \subset \mathsf{V}$ is *independent* if there is no arc between two vertices of K. It is *absorbing* if for every vertex $u \notin K$, there exists $k \in K$ s.t. $(u, k) \in \mathsf{A}$ (we say that k *absorbs* u). A *kernel* of G is an independent and absorbing subset of vertices.

$$1 : 2 \succ_1 1$$
$$2 : 3 \succ_2 2$$
$$3 : 4 \succ_3 2 \succ_3 1 \succ_3 0$$
$$4 : 4$$
$$5 : 2 \succ_5 3 \succ_5 1 \succ_5 5$$

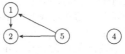

Fig. 2. A partial preference profile P involving 5 voters (left-hand side of the figure) and the corresponding delegation-acceptability digraph G_P (right-hand side of the figure).

Proposition 1. *Assume $\mathcal{G}_{\mathsf{SN}}$ is complete, then given a preference profile P and a subset of voters $K \subseteq \mathcal{N}$, the following propositions are equivalent:*

(i) there exists an equilibrium d s.t. $\mathsf{Gu}(d) = K$;
(ii) K contains no abstainer and K is a kernel of G_P.

Proof. $(i) \implies (ii)$. Let d be an equilibrium for P. It was noted previously that Nash-stability implies the absence of abstainer in $\mathtt{Gu}(d)$. Assume that $\mathtt{Gu}(d)$ is not independent in G_P. Then, there exists $i, j \in \mathtt{Gu}(d)$ such that (i, j) is an arc of G_P, that is, $j \in \mathtt{Acc}(i)$. It implies that i prefers to delegate to j which is in $\mathtt{Nb}(i)$ rather than remaining a guru and hence d is not Nash-stable. Assume now that $\mathtt{Gu}(d)$ is not absorbing for all non-abstainers. Then there exists $i \in \mathcal{N} \backslash (\mathtt{Gu}(d) \cup \mathcal{A})$ such that for every $g \in \mathtt{Gu}(d)$, (i, g) is not an arc of G_P, that is, $g \notin \mathtt{Acc}(i)$. Voter i would prefer to vote herself rather than delegate to any guru in $\mathtt{Gu}(d)$. Thus d is not Nash-stable. This proves that $\mathtt{Gu}(d)$ is a kernel of G_P.

$(ii) \implies (i)(sketched)$. Consider a subset K of non-abstainers such that K is a kernel of G_P. We define a delegation function d by: $d(i) = i$ if $i \in K$; and $d(i) = j$ if $i \notin K$ where j is the voter that i prefers in $K \cup \{0\}$. Note that as $\mathcal{G}_{\mathsf{SN}}$ is complete, each voter in $\mathcal{N} \setminus K$ can delegate directly to any voter in K. It follows that the set of gurus in d is $\mathtt{Gu}(d) = K$. One can easily check that the delegation function d obtained this way is Nash-stable. □

Note that any digraph is the delegation-acceptability digraph of a preference profile P. Indeed, given the digraph, it suffices to build a preference profile P so that every voter prefers to delegate to its out-neighbors, then to vote, then to delegate to other voters, then to abstain. Thus, in a complete SN, determining if a preference profile admits an equilibrium is equivalent to the problem of determining if a digraph admits a kernel, which is an NP-complete problem [8].

Theorem 1. *EX is NP-complete even when the SN is a complete graph.*

Consequently, even when $\mathcal{G}_{\mathsf{SN}}$ is complete, optimization problems **MINDIS**, **MINMAXVP** and **MINABST** are NP-hard as it is NP-hard to decide if their set of admissible solutions is empty or not. We also directly obtain that problem **MEMB** is NP-complete, by a direct reduction from **EX**. Indeed, solving problem **MEMB** for each voter in $\mathcal{N} \setminus \mathcal{A}$ yields the answer to problem **EX**.

As mentioned above, this equivalence is useful to find some interesting sub-cases. Let us consider for instance the case where there is a symmetry in the preferences in the sense that $i \in \mathtt{Acc}(j)$ iff $j \in \mathtt{Acc}(i)$. In this case of symmetrical preference profiles, the delegation-acceptability digraph has the arc (i, j) iff it has the arc (j, i) (it is symmetrical). Then, any inclusion maximal independent set is a kernel. Hence, for any non-abstainer i there exists an equilibrium in which i is a guru (take a maximal independent set containing i).

Proposition 2. *In a complete SN, there always exists an equilibrium when preferences are symmetrical. Moreover, the answer to **MEMB** is always yes.*

4.2   Sparse Social Networks

As the problem **EX** is NP-complete when the SN is a complete graph, it remains NP-complete in any class of graphs that contain cliques, such as interval graphs, split graphs, dense graphs,... Thus, we now focus on classes of graphs that *do not* contain large cliques. We first deal with bounded degree graphs, and show that **EX** remains NP-hard in SNs of degree bounded by 5. We then focus on graphs of bounded treewidth. Interestingly, while we will see in Sect. 5 that **EX** is polynomial if $\mathcal{G}_{SN}$ is a tree (actually, an equilibrium always exists in trees), we obtain that **EX** is W[1]-hard when parameterized by the treewidth of $\mathcal{G}_{SN}$. These results are summarized in Theorem 2.

Theorem 2. *(i)* **EX** *is NP-complete even if the maximum degree of $\mathcal{G}_{SN}$ is at most 5. (ii)* **EX** *is W[1]-hard when parameterized by the treewidth of $\mathcal{G}_{SN}$.*

Another parameter that is worth being considered is the maximal cardinal of a set of acceptable gurus, where the maximum is taken over all voters: $\mathtt{maxa} = \max_{i \in \mathcal{N}} |\mathtt{Acc}(i)|$. This number is likely to be small in practice. Would this assumption help for solving problem **EX**? Unfortunately, in the proof of Theorem 2 (i), $\mathtt{maxa}$ is bounded above by 4, so the problem remains NP-hard when both $\mathtt{maxa}$ and the maximum degree are bounded.

5 Algorithms on Tree Social Networks

5.1 Equilibria and Trees

In this section, we first answer the question of characterizing SNs in which an equilibrium always exists. It turns out that such SNs are exactly trees. More precisely, the existence of an equilibrium is guaranteed whatever the preferences of the voters iff $\mathcal{G}_{SN}$ is a tree.

Theorem 3. *If $\mathcal{G}_{SN}$ is a tree, then for any preference profile there exists an equilibrium. Conversely, if $\mathcal{G}_{SN}$ is a social network such that for any preference profile there exists an equilibrium, then $\mathcal{G}_{SN}$ is a tree.*

Proof. We show the first part of the Theorem (see [9] for the other part). We proceed by induction. The case of a tree of 1 voter is trivial. Consider the result true up to k voters and consider a tree of $k + 1$ voters. Root the tree at some arbitrary vertex. Consider l a leaf and p the parent of this leaf. First if $l \in \mathcal{A}$, or if $l \notin \mathtt{Acc}(p)$, we build an equilibrium in the following way. There exists an equilibrium in the tree without l, add l to this equilibrium by giving her preferred option between abstaining (if $l \in \mathcal{A}$), voting (in this case $l \notin \mathtt{Acc}(p)$) and delegating to p in order to obtain the guru of p as guru.

We now assume $l \notin \mathcal{A}$ and $l \in \mathtt{Acc}(p)$. In this case, we will assume that p can always delegate to l as last resort. Hence, we can consider that the acceptability set of p can be restrained to the voters at least as preferred as l. Secondly, this assumption means that other voters cannot hope to have p as guru. More precisely, the only gurus that they can reach through p are the voters at least as preferred as l, and of course l. To materialize these constraints, we consider the tree without l and where p takes the place of l in all voters preference list (including the one of p). There exists an equilibrium in this tree. If p delegates in the equilibrium (then she prefers her guru to l otherwise p would vote), add l to this equilibrium by giving her preferred option between voting and delegating to the guru of p. Otherwise, make p delegate to l and l votes. □

5.2   Solving Optimization Problems in Trees

Let us address the complexity of problems **MEMB**, **MINDIS**, **MINMAXVP** and **MINABST** in a tree. It will be shown that when $\mathcal{G}_{SN}$ is a tree, a dynamic programming approach is successful in building and optimizing equilibria.

We introduce some additional tools. Let $\mathcal{G}_{SN}$ be a tree $\mathcal{T}$ rooted at some vertex r_0, and let us define accordingly, for every vertex $i \in \mathcal{N}$, $p(i)$ the parent of i and $Ch(i)$ the set of its children. Let $\mathcal{T}_i$ denote the subtree of $\mathcal{T}$ rooted at i.

Let $r \in \mathcal{N}$ and let d be an equilibrium such that the guru of r is some $g_r \in \mathcal{N} \cup \{0\}$. Note that g_r can be in $\mathcal{T}_r \cup \{0\}$: then r delegates downwards, votes or abstains, i.e., $d(r) \in \mathcal{T}_r \cup \{0\}$; or $g_r \notin \mathcal{T}_r \cup \{0\}$: then r delegates to her parent, i.e., $d(r) = p(r)$. Consider the chain of delegations starting from a voter $j \in \mathcal{T}_r \setminus \{r\}$: either it reaches r, and then the guru of j will be g_r, or it does not reach r, hence it is fully determined by the restriction of d to the subtree $\mathcal{T}_r$. Using this remark we will show that an equilibrium can be built inductively by combining delegation functions in subtrees.

Let us now define local equilibria to formalize restrictions of equilibria in a subtree. Let $d : \mathcal{T}_r \longrightarrow \mathcal{N} \cup \{0\}$ be a delegation function over $\mathcal{T}_r$. We define gurus associated with d in a similar way as we defined gurus for delegation function over all voters. Let the guru $\mathbf{gu}(r, d)$ of r be: the first voter i such that $d(i) = i$ reached by the chain of delegations starting from r if r delegates downwards; 0 if $d(r) = 0$; or some $g_r \notin \mathcal{T}_r \cup \{0\}$ if $d(r) = p(r)$. Given a voter $j \in \mathcal{T}_r \setminus \{r\}$, let the guru $\mathbf{gu}(j, d)$ of j be: the first voter i such that $d(i) = i$ reached by the chain of delegations starting from j if the chain does not reach r; 0 if $d(j) = 0$; or $\mathbf{gu}(i, d) = \mathbf{gu}(r, d)$ otherwise.

Given $d : \mathcal{T}_r \longrightarrow \mathcal{N} \cup \{0\}$ and $g_r \in \mathcal{N} \cup \{0\}$, we say that d is a *local equilibrium* on $\mathcal{T}_r$ *with label* g_r if it satisfies:

(i)   either $g_r \notin \mathcal{T}_r \cup \{0\}$ and $d(r) = p(r)$, or there is a chain of delegations in d going from r to g_r.

(ii) for every $i \in \mathcal{T}_r \setminus \{r\}$, voter i does not want to change her delegation, given that the guru of r is g_r, i.e., with $\mathbf{gu}(r, d) := g_r$ it holds that $\mathbf{gu}(i, d) \succ_i g, \forall g \in (\cup_{j \in \mathrm{Nb}(i)} \mathbf{gu}(j, d) \cup \{0, i\}) \setminus \{\mathbf{gu}(i, d)\}$.

(iii) the root r does not want to change her delegation to any of her children, or to vote or to abstain, given that her current guru is g_r, i.e., $g_r \succ_r g, \forall g \in (\cup_{j \in \mathrm{Nb}(r), j \neq p(r)} \mathbf{gu}(j, d) \cup \{0, r\}) \setminus \{g_r\}$.

Note that condition (i) means that the label is consistent with the delegations, i.e., it is indeed possible that g_r is the guru of r in an equilibrium that coincides with d on $\mathcal{T}_r$. Condition (ii) corresponds to Nash-stability for voters in $\mathcal{T}_r \setminus \{r\}$, and condition (iii) is a relaxed Nash-stability for r. This definition slightly generalizes the definition of equilibrium: an equilibrium d is exactly a local equilibrium on $\mathcal{T} = \mathcal{T}_{r_0}$ with label $\mathbf{gu}(r_0, d)$.

Proposition 3. *Let $r \in \mathcal{N}$, $g_r \in \mathcal{N} \cup \{0\}$ and $d: \mathcal{T}_r \longrightarrow \mathcal{N} \cup \{0\}$. Then d is a local equilibrium on $\mathcal{T}_r$ with label g_r iff the following assertions are satisfied:*

(a) $d(r) = r$ (resp. 0, $p(r)$) if $g_r = r$ (resp. $g_r = 0$, $g_r \notin \mathcal{T}_r \cup \{0\}$) and $d(r) = u^*$ if $g_r \in \mathcal{T}_{u^*}$ for some $u^* \in Ch(r)$;

(b) $g_r \succ_r g$ for every $g \in \{0, r\} \setminus \{g_r\}$;

(c) For every $u \in Ch(r)$, (c1) or (c2) is satisfied:

    (c1) $g_r \notin \mathcal{T}_u$ and there exists $g_u \in \mathcal{T}_u \cup \{0\}$ such that $g_r \succ_r g_u$, $g_u \succ_u g_r$, and d is a local equilibrium on $\mathcal{T}_u$ with label g_u;

    (c2) d is a local equilibrium on $\mathcal{T}_u$ with label g_r.

This proposition enables us to build a dynamic programming approach to solve problems on equilibria by reasoning on local equilibria starting from the leaves of the tree to the root. All details are provided in the long version of the paper [9]. This dynamic programming approach yields the following result.

Theorem 4. *In tree SNs, problems **MEMB**, **MINABST**, and **MINDIS** are solvable in $O(n^3)$; problem **MINMAXVP** is solvable in $O(n^4)$.*

The Subcase of Star Social Networks. Interestingly, in the case of a star SN, our dynamic programming approach can be used to solve problems **MEMB**, **MINABST**, **MINDIS** and **MINMAXVP** in $O(n^2)$.

6 Convergence of Delegation Dynamics

In this section, we investigate the convergence of delegation dynamics in several types of graphs. While complete SNs are investigated in Sect. 6.1, our results on two subclasses of tree SNs, paths and stars, are presented in Sect. 6.2.

6.1 Delegation Dynamics in Complete Social Networks

As an equilibrium may not exist in the case of a complete SN (cf. Example 2), it is obvious that IRD or BRD do not always converge. One may wonder whether the convergence is guaranteed in instances where an equilibrium exists. Note that this would not contradict the NP-hardness of problem **EX** in complete SNs (Theorem 1) since the convergence might need an exponential number of steps. Unfortunately, the answer is negative. The game is not even weakly acyclic, as shown in the following Theorem.

Theorem 5. *There exists an instance with a complete SN where an equilibrium exists, but for which there exists a starting delegation function from which no IRD converges to equilibrium.*

Interestingly, the game has a better property if initially all voters declare intention to vote: then, there always exists a permutation dynamics for which BRD converges (though this might not hold for other permutation dynamics).

Theorem 6. *In a complete SN, if an equilibrium exists, then we can find a permutation dynamics which starts when all voters vote, and which always converges under BRD to this equilibrium. However, BRD may not converge even with a permutation dynamics with a starting delegation function in which every voter declares intention to vote.*

6.2 Delegation Dynamics in Tree Social Networks

We now consider the convergence of delegation dynamics in instances with SNs that belong to two subclasses of tree SNs, namely path and star SNs. For these instances, an equilibrium is guaranteed to exist by Theorem 3. Unfortunately, we show that a BRD may not converge if $\mathcal{G}_{SN}$ is a path. Differently, if $\mathcal{G}_{SN}$ is a star, then BRD are guaranteed to converge. However, IRD are not. These results are presented in the two following theorems.

Theorem 7. *There exists a path SN and a BRD that does not converge even if there are no abstainers and if in the starting delegation function, every voter declares intention to vote.*

Theorem 8. *If $\mathcal{G}_{SN}$ is a star and there are no abstainers, an IRD will always converge. If $\mathcal{G}_{SN}$ is a star, a BRD will always converge but there exists an IRD that does not converge.*

7 Conclusion and Future Work

We have proposed a game-theoretic model of the delegation process induced by the liquid democracy paradigm when implemented in a social network. This model makes it possible to investigate several questions on the Nash-equilibria that may be reached by the delegation process of liquid democracy. We have defined and studied several existence and optimization problems defined on these equilibria. Unfortunately, the existence of a Nash-equilibrium is not guaranteed and is even NP-hard to decide even when the social network is complete or has a low maximum degree. In fact, a Nash-equilibrium is only guaranteed to exist whatever the preferences of the voters if the social network is a tree. Hence, we have investigated the case of tree social networks and designed efficient optimization procedures for this special case. Lastly, we have investigated delegation dynamics in several types of social networks highlighting the fact that, in many cases, convergence is not guaranteed.

For future work, similarly to Bloembergen et al. [3] who studied the price of anarchy of delegation games with a more specific model of preferences, it would be interesting to study the price of anarchy of the delegation games developed in this paper. Another direction would be to study a similar game-theoretic analysis of delegations in other frameworks related to liquid democracy such as *viscous democracy* or *flexible representative democracy* [1,5].

Acknowledgments. This work has been supported by the Italian MIUR PRIN 2017 Project ALGADIMAR "Algorithms, Games, and Digital Markets" and the French ANR Project 14-CE24-0007-01 CoCoRICo-CoDec.

References

1. Abramowitz, B., Mattei, N.: Flexible representative democracy: an introduction with binary issues. CoRR abs/1811.02921 (2018)
2. Behrens, J., Kistner, A., Nitsche, A., Swierczek, B.: The principles of LiquidFeedback. Interaktive Demokratie e. V., Berlin (2014)
3. Bloembergen, D., Grossi, D., Lackner, M.: On rational delegations in liquid democracy. In: Proceedings of AAAI 2019 (2019)
4. Boldi, P., Bonchi, F., Castillo, C., Vigna, S.: Voting in social networks. In: Proceedings of the 18th ACM Conference on Information and Knowledge Management, CIKM 2009, Hong Kong, China, 2–6 November 2009, pp. 777–786 (2009)
5. Boldi, P., Bonchi, F., Castillo, C., Vigna, S.: Viscous democracy for social networks. Commun. ACM **54**(6), 129–137 (2011)
6. Brill, M., Talmon, N.: Pairwise liquid democracy. In: Proceedings of the Twenty-Seventh International Joint Conference on Artificial Intelligence, IJCAI 2018, 13–19 July 2018, Stockholm, Sweden, pp. 137–143 (2018)
7. Christoff, Z., Grossi, D.: Binary voting with delegable proxy: an analysis of liquid democracy. In: Proceedings Sixteenth Conference on Theoretical Aspects of Rationality and Knowledge, TARK 2017, Liverpool, UK, 24–26 July 2017, pp. 134–150 (2017)
8. Chvátal, V.: On the computational complexity of finding a kernel. Report CRM-300, Centre de Recherches Mathématiques, Université de Montréal, vol. 592 (1973)
9. Escoffier, B., Gilbert, H., Pass-Lanneau, A.: The convergence of iterative delegations in liquid democracy in a social network. CoRR abs/1904.05775 (2019)
10. Gölz, P., Kahng, A., Mackenzie, S., Procaccia, A.D.: The fluid mechanics of liquid democracy. In: Christodoulou, G., Harks, T. (eds.) WINE 2018. LNCS, vol. 11316, pp. 188–202. Springer, Cham (2018). https://doi.org/10.1007/978-3-030-04612-5_13
11. Green-Armytage, J.: Direct voting and proxy voting. Const. Polit. Econ. **26**(2), 190–220 (2015)
12. Hardt, S., Lopes, L.C.: Google votes: a liquid democracy experiment on a corporate social network. Technical Disclosure Commons (2015)
13. Kahng, A., Mackenzie, S., Procaccia, A.D.: Liquid democracy: an algorithmic perspective. In: Proceedings of the Thirty-Second AAAI Conference on Artificial Intelligence, (AAAI-18), New Orleans, Louisiana, USA, 2–7 February 2018, pp. 1095–1102 (2018)
14. Nisan, N., Schapira, M., Valiant, G., Zohar, A.: Best-response mechanisms. In: Innovations in Computer Science - ICS 2010, Tsinghua University, Beijing, China, 7–9 January 2011. Proceedings, pp. 155–165 (2011)
15. Wiersma, W.: Transitive delegative democracy through facebook for more permanent online movements: How it could address the democratic deficit, and how it could attain critical mass. In: 18th International Conference on Alternative Futures and Popular Protest (2013)

Optimal Truth-Tracking Rules for the Aggregation of Incomplete Judgments

Zoi Terzopoulou$^{(\boxtimes)}$ and Ulle Endriss

Institute for Logic, Language and Computation, University of Amsterdam,
Amsterdam, The Netherlands
{z.terzopoulou,u.endriss}@uva.nl

Abstract. Suppose you need to determine the correct answer to a complex question that depends on two logically independent premises. You can ask several agents to each evaluate either just one of those premises (which they can do with relatively high accuracy) or both premises (in which case their need to multitask will lower their individual accuracy). We first determine the optimal rule to aggregate the individual judgments reported by the agents and then analyse their strategic incentives, depending on whether they are motivated by (*i*) the group tracking the truth, by (*ii*) maximising their own reputation, or by (*iii*) maximising the agreement of the group's findings with their own private judgments. We also study the problem of deciding how many agents to ask for two judgments and how many to ask for just a single judgment.

1 Introduction

Suppose a group of agents need to collectively determine the answer to a binary question that directly depends on the evaluation of several independent criteria. A correct *yes/no* answer—both on the different criteria and on the complex question—exists, but the agents are *a priori* unaware of it. Still, the agents can reflect on the possible answers and obtain a judgment which has a certain probability of being correct. But, most importantly, different agents may assess different parts of the question under consideration. We assume that, under time restrictions and cognitive constraints, the more criteria a given agent tries to assess, the less accurate her judgments are likely to be. This decrease in accuracy might be due to *time pressure* [3,10,16], *multitasking attempts* [1], or *speeded reasoning* [18]. How can the agents then, as a group, maximise the probability of discovering the correct answer to the complex question they are facing?

Example 1. An academic hiring committee needs to decide whether Alice should get the advertised research job. In order to do so, the committee members (professors 1, 2, and 3) have to review two of Alice's papers—Alice will be hired if and only if both these papers are marked as "excellent". Due to an urgent deadline the committee is given only one day to judge the quality of Alice's papers. After the day passes, professor 1 has spent all her time on one of

© Springer Nature Switzerland AG 2019
D. Fotakis and E. Markakis (Eds.): SAGT 2019, LNCS 11801, pp. 298–311, 2019.
https://doi.org/10.1007/978-3-030-30473-7_20

the two papers, while professors 2 and 3 have looked at both, and they express the following "yes" and "no" opinions, related to whether the relevant paper is excellent:

	Paper 1	Paper 2
Professor 1	Yes	–
Professor 2	No	Yes
Professor 3	No	Yes

Assuming that professor 1 has a higher probability to be correct than professor 2 about the first paper, but also taking into account that professor 3 agrees with professor 2, what is the best way to aggregate the given judgments if the committee wants to be as accurate as possible on Alice's evaluation?   △

Judgment aggregation [11,12,15] is a formal framework for group decision making concerned with the aggregation of individual judgments about several logically interconnected propositions into one collective judgment. Along the lines of Example 1, the propositions can be separated into the *premises* (e.g., excellency of Alice's papers) and the *conclusion* (e.g., Alice's hiring), where the conclusion is satisfied if and only if all premises are. Given two independent premises φ and ψ and a group of agents, each of which answers specific questions regarding the premises, in this paper we consider two cases of practical interest:

(i) Free assignment: Each agent chooses with some probability whether to report an opinion only on the first premise, only on the second premise, or on both.

(ii) Fixed assignment: Each agent is asked (and required) to report an opinion only on the first premise, only on the second premise, or on both.

Our main goal is to achieve an aggregate judgment on the conclusion that has— in expectation—high chances to reflect the truth. Notably, under the assumption that the agents are sincere about the judgments they obtain after contemplating their appointed premises, we find that the optimal aggregation rule is always a weighted majority rule assigning each agent a weight that depends on the size of her submitted judgment. We may think of this as a *scoring rule* [17].

But a further problem arises, namely that the agents may behave strategically, trying to manipulate the collective outcome to satisfy their own preferences. We examine the three most natural cases for the preferences of an agent in our context, i.e., preferences that prioritise outcomes that are close to (i) the truth, (ii) the agent's reported judgment, or (iii) the agent's sincere judgment. In addition, we study how an agent's incentives to be insincere relate to the information the agent holds about the judgments reported by her peers.

Finally, knowing in which scenarios the agents are sincere, we ask (from a mechanism-design point of view): Which fixed assignment is the most efficient one, meaning that it achieves the highest probability of producing a correct collective judgment? Our answer here depends heavily on the number of agents in the group as well as on exactly how accurate the agents are individually.

Prior work on judgment aggregation aiming at the tracking of the truth, which can be traced all the way back to the famous Condorcet Jury Theorem [8], has primarily focused on scenarios with two independent premises and one conclusion, like the one investigated in this paper. But such work has solely been concerned with the case of *complete* judgments, that is, the special case where all agents report opinions on all propositions under consideration. Under this assumption, Bovens and Rabinowicz [4] and de Clippel and Eliaz [6] compare two famous aggregation rules (for uniform and varying individual accuracies, respectively): the *premise-based* rule (according to which the collective judgment on the conclusion follows from the majority's judgments on the premises) and the *conclusion-based* rule (which simply considers the opinion of the majority on the conclusion), concluding that most of the time the premise-based rule is superior. Strengthening this result, Hartmann and coauthors [13,14] show that the premise-based rule is optimal across wider classes of aggregation rules too. Generalising the model further, Bozbay et al. [5] study scenarios with any number of premises and agents with incentives to manipulate the collective outcome, and design rules that are optimal truth-trackers, but again assuming complete reported judgments. Also focusing on strategic agents, Ahn and Oliveros [2] wonder "should two issues be decided jointly by a single committee or separately by different committees?" This question differs essentially from the one addressed in our work, since our model accounts for the lower accuracy of the agents who judge a greater number of issues.

This paper proceeds as follows. In Sect. 2 we present our basic model. In Sect. 3 we provide our central result about the optimal aggregation rule for truth-tracking with incomplete judgments and in Sect. 4 we conduct a game-theoretical analysis of our model. We then engage with finding the optimal fixed assignment for sincere agents in Sect. 5, and we conclude in Sect. 6.

2   The Model

Let φ and ψ be two logically independent premises and $c = (\varphi \wedge \psi)$ be the corresponding conclusion, and assume that all three propositions are associated with a correct answer "yes" or "no", where a positive answer on the conclusion is equivalent to a positive answer on both premises. Each agent i in a group $N = \{1, \ldots, n\}$ with $n \geqslant 2$ holds a *sincere* judgment $J_i^* \subseteq \{\varphi, \overline{\varphi}, \psi, \overline{\psi}\}$ that contains at most one formula from each pair of a premise and its negation: $\varphi \in J_i^*$ ($\overline{\varphi} \in J_i^*$) means that agent i judges φ as true (false). Clearly, agent i cannot judge the conclusion without having judged both premises, but her judgment on the conclusion would follow directly from her judgment on the two premises in case she had one. We denote by $\mathcal{J}$ the set of all admissible individual judgments. We say that two judgments J, J' *agree* on their evaluation of a proposition if they both contain either the non-negated or the negated version of that proposition.

An *aggregation rule* F is a function that maps every reported profile $\boldsymbol{J} = (J_1, \ldots, J_n) \in \mathcal{J}^n$ of individual judgments to a set of collective judgments $F(\boldsymbol{J})$. F is resolute if $|F(\boldsymbol{J})| = 1$ for every profile $\boldsymbol{J}$. A collective outcome $J \in F(\boldsymbol{J})$ is a *logically consistent* set $J \subseteq \{\varphi, \overline{\varphi}, \psi, \overline{\psi}, c, \overline{c}\}$ that contains exactly one formula from each pair of a proposition and its negation (namely, it is *complete*). We write $J^{\blacktriangle} \subseteq \{\varphi, \overline{\varphi}, \psi, \overline{\psi}, c, \overline{c}\}$ for the judgment that captures the correct evaluation on all three propositions.

We define N_1^{φ} (N_2^{φ}) to be the sets of agents who report a judgment on one (two) premise(s) and say "yes" (and analogously for "no") on φ. We also define $n_1^{\varphi} = |N_1^{\varphi}|$ and $n_2^{\varphi} = |N_2^{\varphi}|$ to be the relevant cardinalities of these sets.

We denote by p the probability that agent i's judgment J_i^* is correct on a premise when i judges both premises and by q the relevant probability when i only judges a single premise (assuming that the probability of each agent i's judgment being correct on a premise φ is independent (i) of whether φ is true or false and (ii) of what i's judgment on premise ψ is). We assume that the probabilities p and q are the same for all agents, but the agents make their judgments independently of each other. We shall moreover suppose that all agents' judgments are more accurate than a random guess, but not perfect, and that agents judging a single premise are strictly more accurate than those judging both premises, i.e., that $1/2 < p < q < 1$. Then, $P(J^*)$ denotes the probability of the sincere profile J^* to arise and $P(J_{-i}^* \mid J_i^*)$ the probability that the judgments of all agents besides i form the sincere (partial) profile J_{-i}^*, given that i has the sincere judgment J_i^*. Formally, for a fixed assignment:

$$P(J^*) = P(\varphi \text{ true}) \cdot P(J^* \mid \varphi \text{ true}) + P(\varphi \text{ false}) \cdot P(J^* \mid \varphi \text{ false})$$

where $P(J^* \mid \varphi \text{ true}) = q^{n_1^{\varphi}} p^{n_2^{\varphi}} (1-q)^{n_1^{\overline{\varphi}}} (1-p)^{n_2^{\overline{\varphi}}}$, and similarly for $P(J^* \mid \varphi \text{ false})$. The *accuracy* $P(F)$ of a resolute aggregation rule F is defined as:

$$P(F) = \sum_{\substack{J^* \in \mathcal{J}^n \text{ s.t.} \\ F(J^*) \text{ and } J^{\blacktriangle} \text{ agree on } c}} P(J^*)$$

3   Optimal Aggregation

We now define the (irresolute) aggregation rule F_{irr}^{opt}, such that for all profiles J:

$$F_{irr}^{opt}(J) = \underset{\substack{J \\ \text{complete} \\ \text{consistent}}}{\operatorname{argmax}} \sum_{i \in N} w_i \cdot |J \cap J_i|$$

where $w_i = \log \frac{q}{1-q}$ if $|J_i| = 1$, $w_i = \log \frac{p}{1-p}$ if $|J_i| = 2$, and $w_i = 0$ if $|J_i| = 0$. Observe that the base of the logarithm in the definition of w_i is irrelevant.

F_{irr}^{opt} functions as a weighted-majority rule on each premise separately, assigning to the agents weights according to the size of their reported judgments, and subsequently picks that evaluation of the conclusion that is consistent with the collective evaluation of the premises [17]. Then, F^{opt} is a resolute version of F_{irr}^{opt} that, if the obtained collective judgments are more than one, randomly chooses one of them for the collective outcome.

For a resolute aggregation rule F, the probability $P(F)$ depends on the probabilities $P(F$ correct on $\varphi)$ and $P(F$ correct on $\psi)$, which, for simplicity, we call P_φ and P_ψ, respectively. For the remainder of this paper we will further assume that the prior probabilities of the two premises being true or false are equal (and independent of each other). That is, $P(\varphi$ true$) = P(\psi$ true$) = 1/2$. Then:

$$P(F) = \frac{1}{4}\left[(P_\varphi P_\psi) + (P_\varphi P_\psi + (1 - P_\varphi)(1 - P_\psi) + P_\varphi(1 - P_\psi)) + (P_\varphi P_\psi\right.$$
$$\left. + (1 - P_\varphi)(1 - P_\psi) + (1 - P_\varphi)P_\psi) + (P_\varphi P_\psi + P_\varphi(1 - P_\psi) + (1 - P_\varphi)P_\psi)\right]$$
$$= \frac{1}{2} + \frac{P_\varphi P_\psi}{2} \tag{1}$$

Given a (fixed or free) assignment, let us denote by J_F^φ ($J_F^{\overline{\varphi}}$) and J_F^ψ ($J_F^{\overline{\psi}}$) the sets of all possible profiles of reported judgments that lead to a "yes" ("no") collective answer on φ and ψ under the rule F, respectively. Then:

$$P_\varphi = \frac{1}{2} \sum_{J^* \in J_F^\varphi} P(J^* \mid \varphi \text{ true}) + \frac{1}{2} \sum_{J^* \in J_F^{\overline{\varphi}}} P(J^* \mid \varphi \text{ false}) \tag{2}$$

Now, for a fixed assignment and a profile J^*, we have that:

$$P(J^* \mid \varphi \text{ true}) > P(J^* \mid \varphi \text{ false}) \qquad \Leftrightarrow$$
$$q^{n_1^\varphi} p^{n_2^\varphi} (1 - q)^{n_1^{\overline{\varphi}}} (1 - p)^{n_2^{\overline{\varphi}}} > (1 - q)^{n_1^\varphi} (1 - p)^{n_2^\varphi} q^{n_1^{\overline{\varphi}}} p^{n_2^{\overline{\varphi}}} \qquad \Leftrightarrow \tag{3}$$
$$n_1^\varphi \log \frac{q}{1 - q} + n_2^\varphi \log \frac{p}{1 - p} > n_1^{\overline{\varphi}} \log \frac{q}{1 - q} + n_2^{\overline{\varphi}} \log \frac{p}{1 - p}$$

Analogously, we consider a free assignment where agent i makes a sincere judgment on premise φ with probability p_i^φ, on premise ψ with probability p_i^ψ, and on both premises with probability $p_i^{\varphi,\psi}$. Given a sincere profile J^*:

$$P(J^* \mid \varphi \text{ true}) = \sum_{i \in N_1^\varphi \cup N(1,\overline{\varphi})} \sum_{j \in N_2^\varphi \cup N(2,\overline{\varphi})} p_i^\varphi p_j^{\varphi,\psi} q^{n_1^\varphi} p^{n_2^\varphi} (1 - q)^{n_1^{\overline{\varphi}}} (1 - p)^{n_2^{\overline{\varphi}}}$$

Defining $P(J^* \mid \varphi \text{ false})$ similarly, we have as in as in Eq. 3:

$$P(J^* \mid \varphi \text{ true}) > P(J^* \mid \varphi \text{ false}) \qquad \Leftrightarrow$$
$$n_1^\varphi \log \frac{q}{1 - q} + n_2^\varphi \log \frac{p}{1 - p} > n_1^{\overline{\varphi}} \log \frac{q}{1 - q} + n_2^{\overline{\varphi}} \log \frac{p}{1 - p} \tag{4}$$

Theorem 1 states the main results of this section. The proof technique we use is a standard method in research on *maximum likelihood estimators* [9].

Theorem 1. *For any fixed (or free) assignment and sincere judgments, $F^{opt} \in$ $\operatorname{argmax}_F P(F)$. For every other aggregation rule $F' \in \operatorname{argmax}_F P(F)$, F' only differs from F^{opt} on the tie-breaking part.*

Proof. For a fixed assignment, it follows from Eqs. 2 and 3 that P_φ (and P_ψ) will be maximal if and only if F assigns to the agents weights as in F^{opt}. Equation 1 implies that $\max_F P(F) \leqslant \frac{1}{2} + \frac{\max_F P_\varphi \max_F P_\psi}{2}$, so $P(F)$ is maximal if and only if F^{opt} (or a rule that only differs from F^{opt} on the tie-breaking part) is used. The proof is analogous for a free assignment. □

4 Strategic Behaviour

In this section we study the incentives of the agents to report insincere judgments when the most accurate rule F^{opt} is used. We examine in detail both fixed and free assignments. An agent's incentives to be insincere will of course depend on the type of her preferences. We analyse three natural and disjoint cases regarding these preferences, assuming that all agents have the same preference type: First, the agents may want the group to reach a correct judgment—these preferences are called *truth-oriented*. Second, the agents may want to report an opinion that is close to the collective judgment, no matter what that judgment is— these preferences are called *reputation-oriented*. Third, the agents may want the group's judgment to agree with their own sincere judgment—these preferences are called *self-oriented*.[1]

To make things formal, we employ tools from Bayesian game theory. We wish to understand when sincerity by all agents is an equilibrium:

Given that agent i holds the sincere judgment J_i^*, and given that the rest of the agents are going to be sincere no matter what judgments they have, is sincerity (i.e., reporting J_i^*) a best response of agent i?

We examine the *interim* and the *ex-post* case. In both cases agent i already knows her own sincere judgment, but in the former case she is ignorant about the judgments of the rest of the group (only knowing that they have to be probabilistically compatible with her own judgment), while in the latter case she is in addition fully informed about them (this can happen, for example, after some communication action has taken place).

Call $T \in \{$"truth", "reputation", "self"$\}$ the type of the agents' preferences. Let us denote by $U_i^T((J_i, \boldsymbol{J}_{-i}^*), \boldsymbol{J}^*)$ the utility that agent i gets by reporting judgment J_i, when the sincere profile of the group is $\boldsymbol{J}^*$ and all other agents $j \neq i$ report their sincere judgments J_j^*. $EU_i^T((J_i, \boldsymbol{J}_{-i}^*), J_i^*)$ stands for the expected utility that agent i gets by reporting judgment J_i, when her sincere judgment is J_i^* and all other agents j report their sincere judgments for any possible such judgments. More precisely, we have that:

$$U_i^{truth}((J_i, \boldsymbol{J}_{-i}^*), \boldsymbol{J}^*) = |F^{opt}(J_i, \boldsymbol{J}_{-i}^*) \cap J^{\blacktriangle}|$$
$$U_i^{reputation}((J_i, \boldsymbol{J}_{-i}^*), \boldsymbol{J}^*) = |F^{opt}(J_i, \boldsymbol{J}_{-i}^*) \cap J_i|$$
$$U_i^{self}((J_i, \boldsymbol{J}_{-i}^*), \boldsymbol{J}^*) = |F^{opt}(J_i, \boldsymbol{J}_{-i}^*) \cap J_i^*|$$

---

[1] For instance, doctors making judgments about their patients may simply care about the correctness of their collective judgment, participants of an experiment that are paid proportionally to their agreement with the group can be assumed to aim at being seen to agree with their peers, and people who like having their opinions confirmed might manipulate the group to agree with their own privately held judgment.

Also, for any $T \in \{$ "truth", "reputation", "self" $\}$:

$$EU_i^T\big((J_i, \boldsymbol{J}^*_{-i}), J_i^*\big) = \sum_{\boldsymbol{J}^*_{-i}} U_i^T\big((J_i, \boldsymbol{J}^*_{-i}), \boldsymbol{J}^*\big) P(\boldsymbol{J}^*_{-i} \mid J_i^*)$$

We proceed with formally defining *strategyproofness* in our framework, namely the situation where all agents being sincere forms an equilibrium. Given a preference type $T \in \{$ "truth", "reputation", "self" $\}$ and a (fixed or free) assignment, we say that *sincerity always gives rise to an interim equilibrium* if and only if $J_i^* \in \operatorname{argmax}_{J_i \in A_i} EU_i^T\big((J_i, \boldsymbol{J}^*_{-i}), J_i^*\big)$ for all agents i and sincere judgments J_i^*, where $A_i \subseteq \mathcal{J}$ is the set of all judgments that agent i is can potentially report under the given assignment. Similarly, *sincerity always gives rise to an ex-post equilibrium* if and only if the above holds, where EU_i^T is replaced by U_i^T.

Table 1 summarises our results, where "✓" stands for strategyproofness and "✗" designates the existence of a counterexample.

Table 1. Strategyproofness results.

Preferences	Assignment	Fixed		Free	
		interim	*ex-post*	*interim*	*ex-post*
truth-oriented		✓ Thm 4	✓ Thm 4	✓ Thm 4	✓ Thm 4
reputation-oriented		✓ Thm 5	✗ Prp 6	✓ Thm 7	✗ Prp 6
self-oriented		✓ Thm 8	✓ Thm 8	✗ Prp 9	✗ Prp 9

Two fundamental lemmas are in order (the proofs are easy and thus omitted). First, we verify the basic intuition that when an agent holds more information about the reported judgments of the rest of the group, then her incentives to manipulate increase. Second, we stress that whenever we can find a counterexample of *ex-post* strategyproofness under fixed assignments, the same counterexample works for free assignments too.[2]

Lemma 2. *For any assignment and type of preferences, ex-post strategyproofness implies interim strategyproofness.*

Lemma 3. *For any type of preferences, ex-post strategyproofness under free assignments implies ex-post strategyproofness under fixed assignments.*

4.1 Truth-Oriented Preferences

When all agents have truth-oriented preferences and when the rule F^{opt} is used to aggregate their reported judgments, it directly is in everyone's best interest

---

[2] The other direction does not hold. Importantly, a counterexample may go through under free but not fixed assignments because the agents have the option to manipulate by abstaining on some premise they have sincerely thought about.

to be sincere—given that the rest of the group is sincere as well—irrespective of whether the assignment materialised is fixed or free and whether the agents know the judgments of their peers. Intuitively, the agents can trust that the rule F^{opt} will achieve a collective judgment that is as accurate as possible.

Theorem 4. *For any fixed (or free) assignment and truth-oriented preferences:*

(i) sincerity always gives rise to an interim equilibrium
(ii) sincerity always gives rise to an ex-post equilibrium

Proof. By Lemma 2, we only need to prove case *(ii)*. For an arbitrary sincere profile $\boldsymbol{J}^* = (J_i^*, \boldsymbol{J}_{-i}^*)$, we have that $U_i^{truth}((J_i, \boldsymbol{J}_{-i}^*), \boldsymbol{J}^*) = |F^{opt}(J_i, \boldsymbol{J}_{-i}^*) \cap J^{\blacktriangle}|$, where $J^{\blacktriangle}$ captures the true evaluation of the propositions. Now suppose, aiming for a contradiction, that there is an insincere judgment J_i of agent i such that $|F^{opt}(J_i, \boldsymbol{J}_{-i}^*) \cap J^{\blacktriangle}| > |F^{opt}(J_i^*, \boldsymbol{J}_{-i}^*) \cap J^{\blacktriangle}|$. This means that $F^{opt}(J_i, \boldsymbol{J}_{-i}^*) \neq F^{opt}(J_i^*, \boldsymbol{J}_{-i}^*)$. Then, for a suitable aggregation rule $F' \neq F^{opt}$, we can write $F^{opt}(J_i, \boldsymbol{J}_{-i}^*) = F'(J_i^*, \boldsymbol{J}_{-i}^*)$ and derive that $|F(J_i^*, \boldsymbol{J}_{-i}^*) \cap J^{\blacktriangle}| > |F^{opt}(J_i^*, \boldsymbol{J}_{-i}^*) \cap J^{\blacktriangle}|$. But this is impossible, because according to Theorem 1 F^{opt} has to maximise aggrement with $J^{\blacktriangle}$. Hence, it holds that $|F^{opt}(J_i, \boldsymbol{J}_{-i}^*) \cap J^{\blacktriangle}| \leqslant |F^{opt}(J_i^*, \boldsymbol{J}_{-i}^*) \cap J^{\blacktriangle}|$ for all J_i, which implies that $U_i^{truth}((J_i, \boldsymbol{J}_{-i}^*), \boldsymbol{J}^*) \leqslant U_i^{truth}((J_i^*, \boldsymbol{J}_{-i}^*), \boldsymbol{J}^*)$ for all J_i and concludes the proof.   □

4.2   Reputation-Oriented Preferences

When the agents care about the positive reputation they obtain by agreeing with the collective judgment of the group, their incentives to behave insincerely heavily depend on whether they already know the judgments of their peers. Of course: if an agent knows precisely what the collective judgment of the group will be, she can simply change her reported judgment to fully match that collective judgment. On the other hand, we will see that if an agent does not know exactly what the sincere judgments of her peers are, it is more attractive for her to remain sincere (as—knowing that her sincere judgment is more accurate than random—she can reasonably expect the group to agree with her).

Theorem 5. *For any fixed assignment and reputation-oriented preferences, sincerity always gives rise to an interim equilibrium.*

Proof. Given a fixed assignment, an agent i, and a sincere judgment J_i^*, let us call P_{dis} the probability that agent i will disagree with the group on the evaluation of premise φ. Let us assume that agent i's judgment J_i^* concerns both premises φ and ψ (the proof is analogous when J_i^* concerns only premise φ). Now, let us denote by P_g the probability that the group is collectively correct on their evaluation of φ. Recalling that $p > 1/2$ is the probability that agent i is correct on her evaluation of φ, it holds that:

$$P_{dis} \leqslant p(1 - P_g) + (1 - p)P_g = p + P_g(1 - 2p)$$

Now, we have that $p + P_g(1 - 2p) \leqslant 1/2$ if and only if $P_g \geqslant \frac{p-1/2}{2p-1} = 1/2$, which holds since all members of the group are more accurate than random. So, $P_{dis} \leqslant 1/2$, which means that it is more probable for the group's judgment to agree with J_i^* on premise φ than to disagree with it, and the same holds for premise ψ as well. Therefore, agent i has no better option than to report her sincere judgment on the premises that are assigned to her. Formally, $EU_i^{reputation}((J_i, \boldsymbol{J}_{-i}^*), J_i^*)$ is maximised when $J_i = J_i^*$.   □

Proposition 6. *For reputation-oriented preferences, there exists a fixed (and thus a free, from Lemma 3) assignment where sincerity does not always give rise to an ex-post equilibrium.*

Proof. Consider a fixed assignment where all agents in the group are asked about both premises φ and ψ, agent i has the sincere judgment $J_i^* = \{\varphi, \psi\}$, and all other agents j have sincere judgments $J_j = \{\varphi, \overline{\psi}\}$. Agent i would increase her utility by reporting the insincere judgment $J_i = \{\varphi, \overline{\psi}\}$.   □

Theorem 7. *For any free assignment and reputation-oriented preferences, sincerity always gives rise to an interim equilibrium.*

Proof. Consider an arbitrary free assignment and an agent i with sincere judgment J_i^*. Since the given assignment is uncertain, agent i can potentially report any judgment set she wants, that is, $A_i = \mathcal{J}$. Suppose that her sincere judgment J_i^* has size $|J_i^*| = k \in \{1, 2\}$. First, following the same argument as that in the proof of Theorem 5, we see that agent i cannot increase her expected utility by reporting a judgment $J_i \neq J_i^*$ with $|J_i| = k$. We omit the formal details, but the intuition is clear: the group has higher probability to agree with the sincere evaluation of agent i on each premise than to disagree with it.

However, we also need to show that agent i cannot increase her expected utility by reporting a judgment $J_i \neq J_i^*$ with $|J_i| \neq k$. The case where $|J_i| > k$ is straightforward: if agent i has no information about one of the premises, the best she could do is reporting a random judgment on that premise, but this would not increase her expected utility. Thus, we need to consider the case where $|J_i| < k$, and more specifically the only interesting scenario with $|J_i^*| = 2$ and $|J_i| = 1$. Say, without loss of generality, that $J_i = \{\varphi\}$. Let us call $P_{ag,2}^\varphi$ $(P_{ag,2}^\psi)$ the probability that the group will agree with agent i on her evaluation of φ (ψ) given that agent i reports her sincere judgment on both premises, and let us call $P_{ag,1}^\varphi$ the probability that the group will agree with agent i on her evaluation of φ given that she reports her a judgment only on premise φ. We have that:

$$EU_i^{reputation}((J_i^*, \boldsymbol{J}_{-i}^*), J_i^*) = P_{ag,2}^\varphi + P_{ag,2}^\psi \qquad \text{and}$$
$$EU_i^{reputation}((J_i, \boldsymbol{J}_{-i}^*), J_i^*) = P_{ag,1}^\varphi$$

Analogously to the proof of Theorem 5, it holds that $P_{ag,2}^\varphi \geqslant 1/2$ and $P_{ag,2}^\psi \geqslant 1/2$, so $P_{ag,2}^\varphi + P_{ag,2}^\psi \geqslant 1$. This means that $P_{ag,1}^\varphi \leqslant P_{ag,2}^\varphi + P_{ag,2}^\psi$ (because $P_{ag,1}^\varphi \leqslant 1$ is a probability value), so, it is the case that $EU_i^{reputation}((J_i, \boldsymbol{J}_{-i}^*), J_i^*) \leqslant EU_i^{reputation}((J_i^*, \boldsymbol{J}_{-i}^*), J_i^*)$. We conclude that agent i cannot increase her expected utility by reporting J_i instead of J_i^*.   □

4.3   Self-oriented Preferences

Suppose the agents would like the collective outcome to agree with their own sincere judgment. Now, having a fixed or a free assignment radically changes their strategic considerations: under fixed assignments the agents can never increase their utility by lying, while there are free assignments where insincere behaviour is profitable. The critical difference is that when the agents are free to submit judgments of variable size, they can increase the weight that the optimal aggregation rule F^{opt} will assign to their judgment on one of the two premises by avoiding to report a judgment on the other premise, thus having more opportunities to manipulate the outcome in favour of their private judgment.

Theorem 8. *For any fixed assignment and self-oriented preferences:*

(i) sincerity always gives rise to an interim equilibrium
(ii) sincerity always gives rise to an ex-post equilibrium

Proof. By Lemma 2, we only need to show case (ii). Given a fixed assignment, an agent can only report an insincere opinion by flipping her sincere judgment on some of the premises she is asked about. But if she did so, F^{opt} could only favour a judgment different from her own, not increasing her utility. Thus, every agent always maximises her utility by being sincere. $\square$

Proposition 9. *For self-oriented preferences, there exists a free assignment such that sincerity does not always give rise to an interim (and thus also not to an ex-post, from Lemma 2) equilibrium.*

Proof. Consider a group of three agents and a free assignment as follows: Agent 1 reports an opinion on both premises φ, ψ with probability 1/2, and only on premise φ or only on premise ψ with probability 1/4 and 1/4, respectively. Agent 2 reports a judgment on both premises φ, ψ with probability 1, and agent 3 reports a judgment only on premise φ with probability 1. Suppose that agent 1's truthful judgment is $J_1^* = \{\varphi, \psi\}$. Suppose additionally that $q > \frac{p^2}{p^2+(1-p)^2}$. In such a case, if agent 1 decides to report her sincere judgment on both premises, she will always be unable to affect the collective outcome on φ according to the rule F^{opt}, and she will obtain an outcome that agrees with her sincere judgment on ψ with probability $p(p + \frac{1-p}{2}) + (1-p)(1-p+\frac{p}{2}) = p^2 + p + 1 < 1$. However, if agent 1 reports the insincere judgment $J_1 = \{\psi\}$ instead, she will always obtain a collective outcome on ψ that is identical to her own sincere judgment, corresponding to a higher expected utility of value 1. Thus, we can conclude that $J_1^* \notin \operatorname{argmax}_{J_1 \in A_1} EU_1^{\text{self}}\big((J_1, J_2^*, J_3^*), J_1^*\big)$. $\square$

5   Optimal Fixed Assignment

Having a group of n agents, different choices for assigning agents to questions concerning the premises induce different fixed assignments, which in turn yield the correct answer on the conclusion with different probability. In this section we are interested in finding the optimal (viz., the most accurate) such assignment.

Let us denote by $n_1 \leqslant \lfloor \frac{n}{2} \rfloor$ the number of agents that will be asked to report a judgment only on premise φ. For symmetry reasons, we assume that the same number of agents will be asked to report a judgment only on premise ψ, and the remaining $n - 2n_1$ agents will be asked to report a judgment on both premises. Given n_1, $P_{\varphi,n_1}(F^{opt})$ is the probability of the aggregation rule F^{opt} producing a correct evaluation of premise φ, and since we assume that the same number of agents that will judge φ will also judge ψ, it will be the case that $P_{\varphi,n_1}(F^{opt}) = P_{\psi,n_1}(F^{opt})$. As in Sect. 3, the accuracy of F^{opt} regarding the conclusion is $P_{n_1}(F^{opt}) = \frac{1}{2} + \frac{P_{\varphi,n_1}(F^{opt})P_{\psi,n_1}(F^{opt})}{2} = \frac{1}{2} + \frac{P_{\varphi,n_1}(F^{opt})^2}{2}$. Thus, we will maximise $P_{n_1}(F^{opt})$ if and only if we maximise $P_{\varphi,n_1}(F^{opt})$:

$$\underset{0\leqslant n_1 \leqslant \lfloor \frac{n}{2} \rfloor}{\mathrm{argmax}} \ P_{n_1}(F^{opt}) = \underset{0\leqslant n_1 \leqslant \lfloor \frac{n}{2} \rfloor}{\mathrm{argmax}} \ P_{\varphi,n_1}(F^{opt})$$

The optimal assignment depends on the specific number of agents n, but also on the values p and q of the individual accuracy. For small groups of at most four agents, we calculate exactly what the optimal assignment is for any p and q; for larger groups, we provide results for several indicative values of p and q.

Proposition 10. *For $n = 2$, $\mathrm{argmax}_{n_1} P_{\varphi,n_1}(F^{opt}) = 1$. Thus, when there are just two agents, it is optimal to ask each of them to evaluate one of the two premises ($n_1 = 1$) rather than asking both to evaluate both premises ($n_1 = 0$).*

Proof. For $n = 2$ we have two options: $n_1 = 0$ or $n_1 = 1$. We consider them separately. It is the case that $P_{\varphi,0}(F^{opt}) = p^2 + \frac{1}{2}2p(1-p) = p$, while $P_{\varphi,1}(F^{opt}) = q > p$. Thus, $\mathrm{argmax}_{n_1} P_{\varphi,n_1}(F^{opt}) = 1$. $\quad\square$

Proposition 11. *For $n = 3$, $\mathrm{argmax}_{n_1} P_{\varphi,n_1}(F^{opt}) = 1$ if and only if $q \geqslant p^2(3 - 2p)$.*

Proof. For $n = 3$ we have two options: $n_1 = 0$ or $n_1 = 1$. We consider them separately. It is the case that $P_{\varphi,0}(F^{opt}) = p^3 + \binom{3}{2}p^2(1 - p) = p^2(3 - 2p)$, while $P_{\varphi,1}(F^{opt}) = q$ (because the judgment of the agent who reports only on premise φ will always prevail over the judgment of the agent who reports on both premises). Thus, $\mathrm{argmax}_{n_1} P_{\varphi,n_1}(F^{opt}) = 1$ if and only if $q \geqslant p^2(3 - 2p)$. $\square$

Thus, if agents who evaluate both premises are correct 60% of the time, then in case there are three agents, you should ask two of them to focus on a single premise each if and only if their accuracy for doing so is at least 64.8%.

Proposition 12. *For $n = 4$, $\mathrm{argmax}_{n_1} P_{\varphi,n_1}(F^{opt}) = 1$ if $q < \frac{p^2}{(1-p)^2+p^2}$ and $\mathrm{argmax}_{n_1} P_{\varphi,n_1}(F^{opt}) = 2$ otherwise.*

Proof. We provide a sketch. We have three options: $n_1 = 0$, $n_1 = 1$, or $n_1 = 2$.

$$P_{\varphi,0}(F^{opt}) = p^4 + \binom{4}{3}p^3(1-p) + \frac{1}{2}\binom{4}{2}p^2(1-p)^2 = p^2(3-2p)$$

$$P_{\varphi,1}(F^{opt}) = \begin{cases} q(p^2 + 2p(1-p) + \frac{(1-p)^2}{2}) + (1-q)\frac{p^2}{2} & \text{if } q = \frac{p^2}{(1-p)^2+p^2} \\ q & \text{if } q > \frac{p^2}{(1-p)^2+p^2} \\ q(p^2 + 2p(1-p)) + (1-q)p^2 & \text{if } q < \frac{p^2}{(1-p)^2+p^2} \end{cases}$$

$$P_{\varphi,2}(F^{opt}) = q^2 + \frac{1}{2}q(1-q) = q$$

The claim now follows after some simple algebraic manipulations, by distinguishing cases regarding the relation of q to $\frac{p^2}{(1-p)^2+p^2}$. □

Now, for an arbitrary number of agents n and a number of agents n_1 who judge only premise φ (and the same for ψ), we have the following:

$$P_{n_1,\varphi}(F^{opt}) = \sum_{k=0}^{n-2n_1} \left(\sum_{\substack{\ell=0 \\ \text{s.t. } \ell \in W}}^{n_1} P(k,\ell,n,n_1,p,q) + \frac{1}{2} \sum_{\substack{\ell=0 \\ \text{s.t. } \ell \in T}}^{n_1} P(k,\ell,n,n_1,p,q) \right)$$

- k counts how many of the agents that judge both premises are right on φ.
- ℓ counts how many of the agents that judge only φ are right on φ.
- $W = \{\ell \mid \ell \log \frac{q}{1-q} + k \log \frac{p}{1-p} > (n_1 - \ell) \log \frac{q}{1-q} + (n - 2n_1 - k) \log \frac{p}{1-p}\}$.
- $T = \{\ell \mid \ell \log \frac{q}{1-q} + k \log \frac{p}{1-p} = (n_1 - \ell) \log \frac{q}{1-q} + (n - 2n_1 - k) \log \frac{p}{1-p}\}$.
- $P(k,\ell,n,n_1,p,q) = \binom{n-2n_1}{k}\binom{n_1}{\ell}p^k(1-p)^{n-2n_1-k}q^\ell(1-q)^{n_1-\ell}$.

For large groups with $n \geqslant 5$ it is too complex to calculate the optimal assignment analytically in all cases. We instead look at some representative values of p and q. For that purpose, we define a parameter α that intuitively captures the agents' *multitasking ability*, as follows: $\alpha = \frac{p-0.5}{q-0.5}$. Clearly, $0 < \alpha < 1$, and the smaller α is, the worse multitaskers the agents can be assumed to be.

We consider three types for the agents' multitasking ability: *good, average,* and *bad,* corresponding to values for α of 0.8, 0.5, and 0.2, respectively. In addition, we consider four types for the agents' accuracy on a single question: *very high, high, medium,* and *low,* corresponding to values for q of 0.9, 0.8, 0.7, and 0.6, respectively. Table 2 demonstrates our findings regarding the optimal assignment in terms of the number n_1 for these characteristic cases, for groups with at most 15 members.[34] In general, we can observe that the better multitaskers the agents are, the lower the number n_1 that corresponds to the best assignment is. This verifies an elementary intuition suggesting that if the agents are good at multitasking, then it is profitable to ask many of them about both premises, while if the agents are bad at multitasking, it is more beneficial to ask them about a single premise each.

---

[3] For *any* assignment, collective accuracy converges to 1 as the size of the group grows larger. Thus, our analysis is most interesting for groups that are not very large.

[4] The calculations were performed using a computer program in R.

Table 2. Optimal assignment in terms of the number of agents who should be asked about premise φ only, for different group size, individual accuracy, multitasking ability.

n \ q \ α	Low bad	avg.	good	Medium bad	avg.	good	High bad	avg.	good	Very high bad	avg.	good
5	2	2	2	2	2	2	2	2	2	2	2	2
6	3	3	3	3	3	3	3	3	3	3	3	3
7	3	3	2	3	3	2	3	3	3	3	3	3
8	4	4	3	4	3	3	4	4	3	4	4	3
9	4	4	4	4	4	4	4	4	4	4	4	4
10	5	5	5	5	5	5	5	5	5	5	5	5
11	5	5	4	5	5	4	5	5	5	5	5	5
12	6	6	5	6	5	5	6	6	5	6	6	5
13	6	6	6	6	6	6	6	6	6	6	6	6
14	7	7	7	7	7	7	7	7	7	7	7	7
15	7	7	6	7	7	6	7	7	7	7	7	7

6 Conclusion

We have contributed to the literature on the truth-tracking of aggregation rules by considering scenarios where the agents may not all judge the same number of issues that need to be decided by the group. Assuming that multitasking is detrimental to the agents' accuracy, we have found what the optimal method to aggregate the judgments of the agents is, and we have analysed the incentives for strategic behaviour that the agents may exhibit in this new context.

For this first study on the topic a few simplifying assumptions have been made. First, we have assumed that all agents have the same accuracy and that there are only two premises. Our analysis can be naturally extended beyond this case, considering different accuracies and more than two premises: the optimal aggregation rule would still be a weighted majority rule, agents with truth-oriented preferences would still always be sincere, while agents who care about their reputation or their individual opinion would still find reasons to lie—but careful further work is essential here in order to clarify all relevant details. Second, we have assumed that the exact values of the agents' accuracies are known, but this is often not true in practice. Thus, to complement our theoretical work, our results could be combined with existing experimental research that measures the accuracy of individual agents on specific application domains, ranging from human-computer interaction [1] to crowdsourcing [7].

References

1. Adler, R.F., Benbunan-Fich, R.: Juggling on a high wire: multitasking effects on performance. Int. J. Hum. Comput. Stud. **70**(2), 156–168 (2012)
2. Ahn, D.S., Oliveros, S.: The condorcet jur(ies) theorem. J. Econ. Theor. **150**, 841–851 (2014)

3. Ariely, D., Zakay, D.: A timely account of the role of duration in decision making. Acta Psychol. **108**(2), 187–207 (2001)

4. Bovens, L., Rabinowicz, W.: Democratic answers to complex questions–an epistemic perspective. Synthese **150**(1), 131–153 (2006)

5. Bozbay, I., Dietrich, F., Peters, H.: Judgment aggregation in search for the truth. Games Econ. Behav. **87**, 571–590 (2014)

6. de Clippel, G., Eliaz, K.: Premise-based versus outcome-based information aggregation. Games Econ. Behav. **89**, 34–42 (2015)

7. Cohensius, G., Porat, O.B., Meir, R., Amir, O.: Efficient crowdsourcing via proxy voting. In: Proceedings of the 7th International Workshop on Computational Social Choice (COMSOC-2018) (2018)

8. de Condorcet, M.: Essai sur l'application de l'analyse à la probabilité des décisions rendues à la pluralité des voix. Imprimerie Royale (1785)

9. Dawid, A.P., Skene, A.M.: Maximum likelihood estimation of observer error-rates using the EM algorithm. Appl. Stat. **28**(1), 20–28 (1979)

10. Edland, A., Svenson, O.: Judgment and decision making under time pressure. In: Svenson, O., Maule, A.J. (eds.) Time Pressure and Stress in Human Judgment and Decision Making, pp. 27–40. Springer, Boston (1993)

11. Endriss, U.: Judgment aggregation. In: Brandt, F., Conitzer, V., Endriss, U., Lang, J., Procaccia, A.D. (eds.) Handbook of Computational Social Choice, pp. 399–426. Cambridge University Press, Cambridge (2016)

12. Grossi, D., Pigozzi, G.: Judgment Aggregation: A Primer, Synthesis Lectures on Artificial Intelligence and Machine Learning, vol. 8. Morgan & Claypool Publishers, Los Altos (2014)

13. Hartmann, S., Pigozzi, G., Sprenger, J.: Reliable methods of judgement aggregation. J. Logic Comput. **20**(2), 603–617 (2010)

14. Hartmann, S., Sprenger, J.: Judgment aggregation and the problem of tracking the truth. Synthese **187**(1), 209–221 (2012)

15. List, C., Pettit, P.: Aggregating sets of judgments: an impossibility result. Econ. Philos. **18**(1), 89–110 (2002)

16. Payne, J.W., Bettman, J.R., Johnson, E.J.: Adaptive strategy selection in decision making. J. Exp. Psychol. Learn. Mem. Cogn. **14**(3), 534 (1988)

17. Terzopoulou, Z., Endriss, U., de Haan, R.: Aggregating incomplete judgments: axiomatisations for scoring rules. In: Proceedings of the 7th International Workshop on Computational Social Choice (COMSOC-2018) (2018)

18. Wilhelm, O., Schulze, R.: The relation of speeded and unspeeded reasoning with mental speed. Intelligence **30**(6), 537–554 (2002)

The Distortion of Distributed Voting

Aris Filos-Ratsikas[1], Evi Micha[2], and Alexandros A. Voudouris[3]($\boxtimes$)

[1] École polytechnique fédérale de Lausanne, Lausanne, Switzerland
aris.filosratsikas@epfl.ch
[2] University of Toronto, Toronto, Canada
emicha@cs.toronto.edu
[3] University of Oxford, Oxford, UK
alexandros.voudouris@cs.ox.ac.uk

Abstract. Voting can abstractly model any decision-making scenario and as such it has been extensively studied over the decades. Recently, the related literature has focused on quantifying the impact of utilizing only limited information in the voting process on the societal welfare for the outcome, by bounding the *distortion* of voting rules. Even though there has been significant progress towards this goal, all previous works have so far neglected the fact that in many scenarios (like presidential elections) voting is actually a distributed procedure. In this paper, we consider a setting in which the voters are partitioned into disjoint districts and vote locally therein to elect local winning alternatives using a voting rule; the final outcome is then chosen from the set of these alternatives. We prove tight bounds on the distortion of well-known voting rules for such distributed elections both from a worst-case perspective as well as from a best-case one. Our results indicate that the partition of voters into districts leads to considerably higher distortion, a phenomenon which we also experimentally showcase using real-world data.

Keywords: Distributed voting · District-based elections · Distortion

1 Introduction

In a decision-making scenario, the task is to aggregate the opinions of a group of different people into a common decision. This process is often distributed, in the sense that smaller groups first reach an agreement, and then the final outcome is determined based on the options proposed by each such group. This can be due to scalability issues (e.g., it is hard to coordinate a decision between a very large number of participants), due to different roles of the groups (e.g., when each group represents a country in the European Union), or simply due to established institutional procedures (e.g., electoral systems).

This work has been supported by the Swiss National Science Foundation under contract number 200021_165522 and by the European Research Council (ERC) under grant number 639945 (ACCORD).

© Springer Nature Switzerland AG 2019
D. Fotakis and E. Markakis (Eds.): SAGT 2019, LNCS 11801, pp. 312–325, 2019.
https://doi.org/10.1007/978-3-030-30473-7_21

For example, in the US presidential elections, the voters in each of the 50 states cast their votes within their regional district, and each state declares a winner; the final winner is taken as the one that wins a weighted plurality vote over the state winners, with the weight of each state being proportional to its size. Another example is the Eurovision Song Contest, where each participating country holds a local voting process (consisting of a committee vote and an Internet vote from the people of the country) and then assigns points to the 10 most popular options, on a 1–12 scale (with 11 and 9 omitted). The winner of the competition is the participant with the most total points.

The foundation of utilitarian economics, which originated near the end of the 18th century, revolves around the idea that the outcome of a decision making process should be one that maximizes the well-being of the society, which is typically captured by the notion of the *social welfare*. A fundamental question that has been studied extensively in the related literature is whether the rules that are being used for decision making actually achieve this goal, or to what extend they fail to do so. This motivates the following question: *What is the effect of distributed decision making on the social welfare?*

The importance of this investigation is highlighted by the example of the 2016 US presidential election [24]. While 48.2% of the US population (that participated in the election) viewed Hillary Clinton as the best candidate, Donald Trump won the election with only 46.1% of the popular vote. This was due to the district-based electoral system, and the outcome would have been different if there was a single pool of voters instead. A similar phenomenon occurred in the 2000 presidential election as well, when Al Gore won the popular vote, but George W. Bush was elected president.

1.1 Our Setting and Contribution

For concreteness, we use the terminology of voting as a proxy for any distributed decision-making scenario. A set of voters are called to vote on a set of alternatives through a district-based election. In other words, the set of voters is partitioned into *districts* and each district holds a local election, following some voting rule. The winners of the local elections are then aggregated into the single winner of the general election. Note that this setting models many scenarios of interest, such as those highlighted in the above discussion.

We are interested in the effect of the distributed nature of elections on the social welfare of the voters (the sum of their valuations for the chosen outcome). Typically, this effect is quantified by the notion of *distortion* [22], which is defined as the worst-case ratio between the maximum social welfare for any outcome and the social welfare for the outcome chosen through voting. Concretely, we are interested in bounding the distortion of voting rules for district-based elections.

We consider three cases when it comes to the district partition: (a) *symmetric districts*, in which every district has the same number of voters and contributes the same weight to the final outcome, (b) *unweighted districts*, in which the weight is still the same, but the sizes of the districts may vary, and (c) *unrestricted districts*, where the sizes and the weights of the districts are unconstrained. For each case, we show upper and lower bounds on the distortion of voting rules.

First, in Sect. 3, we consider general voting rules (which might have access to the numerical valuations of the voters) and provide distortion guarantees for any voting rule as a function of the worst-case distortion of the voting rule when applied to a single district. As a corollary, we obtain distortion bounds for *Range Voting*, the rule that outputs a welfare-maximizing alternative, and prove that it is optimal among all voting rules for the problem. Then, in Sect. 4, we consider ordinal rules and provide a general lower bound on the distortion of any such rule. For the widely-used Plurality voting rule, we provide *tight* distortion bounds, proving that it is asymptotically the best ordinal voting rule in terms of distortion. In Sect. 5, we provide experiments based on real data to evaluate the distortion on "average case" and "average worst case" district partitions. Finally, in Sect. 6, we explore whether *districting* (i.e., manually partitioning the voters into districts in the best-way possible) can allow to recover the winner of Plurality or Range Voting in the election without districts. We conclude with possible avenues for future work in Sect. 7. Due to space constraints, most proofs are omitted; see the full version of the paper [12].

1.2   Related Work

The distortion framework was first proposed by Procaccia and Rosenschein [22] and subsequently it was adopted by a series of papers; for instance, see [1,2,4,5, 7,8,13]. The original idea of the distortion measure was to quantify the loss in performance due to the lack of *information*, meaning how well an ordinal voting rule (i.e., one that has access only to the preference orderings induced by the numerical values of the voters) can approximate the cardinal objective. In our paper, the distortion will be attributed to two factors: *always* the fact that the election is being done in districts, and *possibly* also the fact that the voting rules employed are ordinal. Our setting follows closely that of Boutilier et al. [7] and Caragiannis et al. [8], with the novelty of introducing district-based elections and measuring their distortion. The worst-case distortion bounds of voting rules in the absence of districts can be found in the aforementioned papers.

The ill effects of district-based elections have been highlighted in a series of related articles, mainly revolving around the issue of *gerrymandering* [23], i.e., the systematic manipulation of the geographical boundaries of an electoral constituency in favor of a particular political party. The effects of gerrymandering have been studied in the related literature before [6,9,19], but never in relation to the induced distortion of the elections. While our district partitions are not necessarily geographically-based, our worst-case bounds capture the potential effects of gerrymandering on the deterioration of the social welfare. Other works on district-based elections and distributed decision-making include [3,10].

Related to our results in Sect. 6 is the paper by Lewenberg et al. [20], which explores the effects of districting with respect to the winner of Plurality, when ballot boxes are placed on the real plane, and voters are partitioned into districts based on their nearest ballot box. The extra constraints imposed by the geological nature of the districts in their setting leads to an NP-hardness result for the districting problem, whereas for our unconstrained districts, making the Plurality winner the winner of the general election is always possible in polynomial time. In contrast, the problem becomes NP-hard when we are interested in the winner of Range Voting instead of Plurality.

2   Preliminaries

A *general election* $\mathcal{E}$ is defined as a tuple $(\mathcal{M}, \mathcal{N}, \mathcal{D}, \mathbf{w}, \mathbf{v}, \mathbf{f})$, where

- $\mathcal{M}$ is a set of m *alternatives*;
- $\mathcal{N}$ is a set of n *voters*;
- $\mathcal{D}$ is a set of $k \geq 2$ districts, with district $d \in \mathcal{D}$ containing n_d voters such that $\sum_{d \in \mathcal{D}} n_d = n$ (i.e., the districts define a partition of the set of voters);
- $\mathbf{w} = (w_d)_{d \in \mathcal{D}}$ is a *weight-vector* consisting of a weight $w_d \in \mathbb{R}_{>0}$ for each district $d \in \mathcal{D}$;
- $\mathbf{v} = (\mathbf{v}_1, \ldots, \mathbf{v}_n)$ is a *valuation profile* for the n voters, where $\mathbf{v}_i = (v_{ij})_{j \in \mathcal{M}}$ contains the *valuation* of voter i for all alternatives, and $\mathcal{V}^n$ is the set of all such valuation profiles;
- $\mathbf{f} = (f_d)_{d \in \mathcal{D}}$ is a set of *voting rules* (one for each district), where $f_d : \mathcal{V}^{n_d} \to \mathcal{M}$ is a map of valuation profiles with n_d voters to alternatives.

For each voter $i \in \mathcal{N}$, we denote by $d(i)$ the district she belongs to. For each district $d \in \mathcal{D}$, a *local* or *district* election between its members takes place, and the winner of this election is the alternative $j_d = f_d((\mathbf{v}_i)_{i:d(i)=d})$ that gets elected according to f_d. The outcome of the general election $\mathcal{E}$ is an alternative

$$j(\mathcal{E}) \in \arg\max_{j \in M} \sum_{d \in \mathcal{D}} w_d \cdot \mathbf{1}\{j_d = j\},$$

where $\mathbf{1}\{X\}$ is equal to 1 if the event X is true, and 0 otherwise. In simple words, the winner $j(\mathcal{E})$ of the general election is the alternative with the highest weighted approval score, breaking ties arbitrarily. For example, when all weights are 1, $j(\mathcal{E})$ is the alternative that wins the most local elections.

Following the standard convention, we adopt the unit-sum representation of valuations, according to which $\sum_{j \in \mathcal{M}} v_{ij} = 1$ for every voter $i \in \mathcal{N}$. For a given valuation profile $\mathbf{v}$, the *social welfare* of alternative $j \in \mathcal{M}$ is defined as the total value the agents have for her:

$$\mathrm{SW}(j|\mathbf{v}) = \sum_{i \in \mathcal{N}} v_{ij}.$$

Throughout the paper, we assume that the same voting rule is applied in every local election (possibly for a different number of voters though, depending on how the districts are defined); we denote this voting rule by f and also let $f(\mathbf{v})$ be the alternative that is chosen by f when the voters have the valuation profile $\mathbf{v}$.

The distortion of a voting rule f *in a local election* with η voters is defined as the worst-case ratio, over all possible valuation profiles of the voters participating in that election, between the maximum social welfare of any alternative and the social welfare of the alternative chosen by the voting rule:

$$\mathtt{dist}(f) = \sup_{\mathbf{v} \in \mathcal{V}^\eta} \frac{\max_{j \in \mathcal{M}} \mathrm{SW}(j|\mathbf{v})}{\mathrm{SW}(f(\mathbf{v})|\mathbf{v})}.$$

The distortion of a voting rule f *in a general election* is defined as the worst-case ratio, over all possible general elections $\mathcal{E}$ that use f as the voting rule within the districts, between the maximum social welfare of any alternative and the social welfare of the alternative chosen by the general election:

$$\mathtt{gdist}(f) = \sup_{\mathcal{E}: f \in \mathcal{E}} \frac{\max_{j \in \mathcal{M}} \mathrm{SW}(j|\mathbf{v})}{\mathrm{SW}(j(\mathcal{E})|\mathbf{v})}.$$

Again, in simple words, the distortion of a voting rule f is the worst-case over all the possible valuations that voters can have and over all possible ways of partitioning these voters into districts. When $k = 1$, we recover the standard definition of the distortion.

Next, we define some standard properties of voting rules.

Definition 1 (Properties of voting rules). *A voting rule f is*

- ordinal, *if the outcome only depends on the preference orderings induced by the valuations and not the actual numerical values themselves. Formally, given a valuation profile $\mathbf{v}$, let $\Pi_\mathbf{v}$ be the ordinal preference profile formed by the values of the agents for the alternatives (assuming some fixed tie-breaking rule). A voting rule is ordinal if for any two valuation profiles $\mathbf{v}$ and $\mathbf{v}'$ such that $\Pi_\mathbf{v} = \Pi_{\mathbf{v}'}$, it holds that $f(\mathbf{v}) = f(\mathbf{v}')$.*
- unanimous, *if whenever all agents agree on an alternative, that alternative gets elected. Formally, whenever there exists an alternative $a \in \mathcal{M}$ for whom $v_{ia} \geq v_{ij}$ for all voters $i \in \mathcal{N}$ and all alternatives $j \in \mathcal{M}$, then $f(\mathbf{v}) = a$.*
- (strictly) Pareto efficient, *if whenever all agents agree that an alternative a is better than b, then b cannot be elected. Formally, if $v_{ia} > v_{ib}$ for all $i \in \mathcal{N}$, then $f(\mathbf{v}) \neq b$.[1]*

Remark. It is not hard to see that we can assume that the best voting rule in terms of distortion is Pareto efficient, without loss of generality. Indeed, for any voting rule f that is not Pareto efficient, we can construct the following Pareto efficient rule f': for every input on which f outputs a Pareto efficient alternative, f' outputs the same alternative; for every input on which f outputs an alternative that is not Pareto efficient, f' outputs a maximal Pareto improvement, that is, a Pareto efficient alternative which all voters (weakly) prefer more than the

---

[1] Pareto efficiency usually requires that there is no other alternative who all voters *weakly prefer* and who one voter *strictly prefers*. We use the strict definition in our proofs, as it is also without loss of generality with respect to distortion.

alternative chosen by f. Clearly, f' is Pareto efficient and achieves a social welfare at least as high as f. Note also that Pareto efficiency implies unanimity. In our proofs, we will use both of these properties without loss of generality. Finally, most of the voting rules that are being employed in practice are ordinal, with the notable exception of *Range Voting*, which is the voting rule that outputs the alternative that maximizes the social welfare.

We consider the following three basic cases for the general elections, depending on the size and the weight of the districts:

- *Symmetric Elections*: all districts consist of n/k voters and have the same weight, i.e., $n_d = n/k$ and $w_d = 1$ for each $d \in \mathcal{D}$.
- *Unweighted Elections*: all districts have the same weight, but not necessarily the same number of voters, i.e., $w_d = 1$ for each $d \in \mathcal{D}$.
- *Unrestricted Elections*: there are no restrictions on the sizes and weights of the districts.

Of course, the class of symmetric elections is a subclass of that of unweighted elections which in turn is a subclass of the class of unrestricted elections.

3 The Effect of Districts for General Voting Rules

Our aim in this section is to showcase the immediate effect of using districts to distributively aggregate votes. To this end, we present tight bounds on the distortion of all voting rules in a general election. We will first state a general theorem relating the distortion $\texttt{gdist}(f)$ of any general election that uses a voting rule f for the local elections, with the distortion $\texttt{dist}(f)$ of the voting rule.

Theorem 1. *Let f be a voting rule with* $\texttt{dist}(f) = \gamma$. *Then, the distortion* $\texttt{gdist}(f)$ *of f in the general election is at most*

(i) $\gamma + \frac{\gamma m k}{2}$ *for symmetric elections;*

(ii) $\gamma + \frac{\gamma m}{2} \left(\frac{n + \max_{d \in \mathcal{D}} n_d}{\min_{d \in \mathcal{D}} n_d} - 1 \right)$ *for unweighted elections;*

(iii) $\gamma + \gamma m \left(\frac{n}{\min_{d \in \mathcal{D}} n_d} - 1 \right)$ *for unrestricted elections.*

We now turn to concrete voting rules and consider perhaps the most natural such rule: *Range Voting* (RV).

Definition 2 (Range Voting (RV)). *Given a valuation profile* $\mathbf{v} = (\mathbf{v}_1, ..., \mathbf{v}_\eta)$ *with η voters, Range Voting elects the alternative that maximizes the social welfare of the voters.*

Note that the rule is both unanimous and Pareto efficient. Immediately from the definition of the rule and Theorem 1, we have the following corollary.

Corollary 1. *The distortion* $\texttt{gdist}(\text{RV})$ *of RV in the general election is at most*

(i) $1 + \frac{mk}{2}$ for symmetric elections;

(ii) $1 + \frac{m}{2} \left(\frac{n + \max_{d \in \mathcal{D}} n_d}{\min_{d \in \mathcal{D}} n_d} - 1 \right)$ for unweighted elections;

(iii) $1 + m \left(\frac{n}{\min_{d \in \mathcal{D}} n_d} - 1 \right)$ for unrestricted elections.

We continue by presenting matching lower bounds on the distortion of any voting rule in a general election. The high-level idea in the proof of the following theorem is that the election winner is chosen arbitrarily among the alternatives with equal weight, which might lead to the cardinal information within the districts to be lost.

Theorem 2. *The distortion of all voting rules in a general election is at least*

(i) $1 + \frac{mk}{2}$ for symmetric elections;

(ii) $1 + \frac{m}{2} \left(\frac{n + \max_{d \in \mathcal{D}} n_d}{\min_{d \in \mathcal{D}} n_d} - 1 \right)$ for unweighted elections;

(iii) $1 + m \left(\frac{n}{\min_{d \in \mathcal{D}} n_d} - 1 \right)$ for unrestricted elections.

4   Ordinal Voting Rules and Plurality

Although Range Voting is quite natural, its documented drawback is that it requires a very detailed informational structure from the voters, making the elicitation process rather complicated. For this reason, most voting rules that have been applied in practice are ordinal (see Definition 1), as such rules present the voters with the much less demanding task of reporting a preference ordering over the alternatives, rather than actual numerical values.

Thus, a very meaningful question, from a practical point of view, is "What is the distortion of ordinal voting rules?" The most widely used such rule is *Plurality Voting*. Besides its simplicity, the importance of this voting rule also comes from the fact that it is used extensively in practice. For instance, it is used in presidential elections in a number of countries like the USA and the UK.

Definition 3 (Plurality Voting (PV)). *Given a valuation profile* **v** *and its induced ordinal preference profile* $\Pi_{\mathbf{v}}$, *PV elects the alternative with the most first position appearances in* $\Pi_{\mathbf{v}}$, *breaking ties arbitrarily.*

It is known that the distortion $\texttt{dist}(PV)$ of Plurality Voting is $O(m^2)$ [8]. Therefore, if we plug-in this number to our general bound in Theorem 1, we obtain corresponding upper bounds for PV. However, in the following we obtain much better bounds, taking advantage of the structure of the mechanism; these bounds are actually tight.

Theorem 3. *The distortion* $\texttt{gdist}(PV)$ *of PV is exactly*

(i) $1 + \frac{3m^2 k}{4}$ for symmetric elections;

(ii) $1 + \frac{m^2}{4} \left(\frac{3n + \max_{d \in \mathcal{D}} n_d}{\min_{d \in \mathcal{D}} n_d} - 1 \right)$ for unweighted elections;

(iii) $1 + m^2 \left(\frac{n}{\min_{d \in \mathcal{D}} n_d} - \frac{1}{2} \right)$, *for unrestricted elections.*

Proof. We prove only the upper bounds for the first two parts here; the upper bound for the third part as well as the matching lower bounds can be found in the full version.

Consider a general unweighted election $\mathcal{E}$ with a set $\mathcal{M}$ of m alternatives, a set $\mathcal{N}$ of n voters, a set $\mathcal{D}$ of k districts such that each district d consists of n_d voters and has weight $w_d = 1$. Let $\mathbf{v}$ be the valuation profile consisting of the valuations of all voters for all alternatives, which induces the ordinal preference profile $\Pi_\mathbf{v}$. To simplify our discussion, let $\mathcal{N}_d(j)$ be the set of voters in district d that rank alternative j in the first position, and also set $|\mathcal{N}_d(j)| = n_d(j)$.

Let $a = j(\mathcal{E})$ be the winner of the election and denote by $A \subseteq \mathcal{D}$ the set of districts in which a wins according to PV. Then, we have that

$$\mathrm{SW}(a|\mathbf{v}) = \sum_{i \in \mathcal{N}} v_{ia} \geq \sum_{i:d(i) \in A} v_{ia}. \tag{1}$$

Since a has the plurality of votes in each district $d \in A$, we have that $n_d(a) \geq n_d(j)$ for every $j \in \mathcal{M}$, and by the fact that $\sum_{j \in \mathcal{M}} n_d(j) = n_d$, we obtain that $n_d(a) \geq \frac{n_d}{m}$. Similarly, for each agent $i \in \mathcal{N}_d(a)$ we have that $v_{ia} \geq v_{ij}$ for every $j \in \mathcal{M}$, and by the unit-sum assumption, we obtain that $v_{ia} \geq \frac{1}{m}$. We also have that $\sum_{d \in A} n_d \geq |A| \cdot \min_{d \in \mathcal{D}} n_d$. Hence,

$$\sum_{i:d(i) \in A} v_{ia} \geq \sum_{d \in A} \sum_{i \in \mathcal{N}_d(a)} v_{ia} \geq \frac{1}{m} \cdot \sum_{d \in A} n_d(a)$$

$$\geq \frac{1}{m^2} \sum_{d \in A} n_d \geq \frac{1}{m^2} \cdot |A| \cdot \min_{d \in \mathcal{D}} n_d. \tag{2}$$

Let b the optimal alternative, and denote by $B \subset \mathcal{D}$ the set of districts in which b is the winner. We split the social welfare of b into three parts:

$$\mathrm{SW}(b|\mathbf{v}) = \sum_{i:d(i) \in A} v_{ib} + \sum_{i:d(i) \in B} v_{ib} + \sum_{i:d(i) \notin A \cup B} v_{ib}. \tag{3}$$

We will now bound each term individually. First consider a district $d \in A$. Then, the welfare of the agents in d for b can be written as

$$\sum_{i:d(i)=d} v_{ib} = \sum_{i \in \mathcal{N}_d(a)} v_{ib} + \sum_{i \in \mathcal{N}_d(b)} v_{ib} + \sum_{i \notin \mathcal{N}_d(a) \cup \mathcal{N}_d(b)} v_{ib}.$$

Since a is the favourite alternative of every agent $i \in \mathcal{N}_d(a)$, $v_{ib} \leq v_{ia}$. By definition, the value of every agent $i \in \mathcal{N}_d(b)$ for b is at most 1. The value of every agent $i \notin \mathcal{N}_d(a) \cup \mathcal{N}_d(b)$ for b can be at most $1/2$ since otherwise b would definitely be the favourite alternative of such an agent. Combining these observations, we get

$$\sum_{i:d(i)=d} v_{ib} \leq \sum_{i \in \mathcal{N}_d(a)} v_{ia} + n_d(b) + \frac{1}{2} \sum_{j \neq a,b} n_d(j)$$

$$\leq \sum_{i:d(i)=d} v_{ia} + \frac{1}{2} n_d(b) + \frac{1}{2} \sum_{j \neq a} n_d(j)$$

$$\leq \sum_{i:d(i)=d} v_{ia} + \frac{1}{2} n_d(a) + \frac{1}{2} \left(n_d - n_d(a) \right)$$

$$= \sum_{i:d(i)=d} v_{ia} + \frac{1}{2} n_d,$$

where the second inequality follows by considering the value of all agent in d for alternative a, while the third inequality follows by the fact that a wins b by plurality. By summing over all districts in A, we can bound the first term of (3) as follows:

$$\sum_{i:d(i) \in A} v_{ib} \leq \sum_{i:d(i) \in A} v_{ia} + \frac{1}{2} \sum_{d \in A} n_d. \tag{4}$$

For the second term of (3), by definition we have that the value of each agent in the districts of B for alternative b can be at most 1, and therefore

$$\sum_{i:d(i) \in B} v_{ib} \leq \sum_{d \in B} n_d.$$

For the third term of (3), observe that the total value of the agents in a district $d \notin A \cup B$ for b must be at most $\frac{3}{4} n_d$; otherwise b would necessarily be ranked first in strictly more than half of the agents' preferences and therefore win in the district. Hence,

$$\sum_{i:d(i) \notin A \cup B} v_{ib} \leq \frac{3}{4} \sum_{d \notin A \cup B} n_d.$$

By substituting the bounds for the three terms of (3), as well as by taking into account the facts that $|B| \leq |A|$ and $|A| \geq 1$, we can finally upper-bound the social welfare of b as follows:

$$SW(b|\mathbf{v}) \leq \sum_{i:d(i) \in A} v_{ia} + \frac{1}{2} \sum_{d \in A} n_d + \sum_{d \in B} n_d + \frac{3}{4} \sum_{d \notin A \cup B} n_d$$

$$= \sum_{i:d(i) \in A} v_{ia} + \frac{1}{4} \left(3n + \sum_{d \in B} n_d - \sum_{d \in A} n_d \right)$$

$$\leq \sum_{i:d(i) \in A} v_{ia} + \frac{1}{4} \left(3n + |B| \cdot \max_{d \in \mathcal{D}} n_d - |A| \cdot \min_{d \in \mathcal{D}} n_d \right)$$

$$\leq \sum_{i:d(i) \in A} v_{ia} + \frac{1}{4} \cdot |A| \cdot \left(3n + \max_{d \in \mathcal{D}} n_d - \min_{d \in \mathcal{D}} n_d \right) \tag{5}$$

By (1), (2) and (5), we can upper-bound the distortion of PV as follows:

$$
\texttt{gdist}(\text{PV}) = \frac{\text{SW}(b|\mathbf{v})}{\text{SW}(a|\mathbf{v})}
$$

$$
\leq \frac{\sum_{i:d(i)\in A} v_{ia} + \frac{1}{4} \cdot |A| \cdot (3n + \max_{d\in\mathcal{D}} n_d - \min_{d\in\mathcal{D}} n_d)}{\sum_{i:d(i)\in A} v_{ia}}
$$

$$
\leq 1 + \frac{m^2}{4}\left(\frac{3n + \max_{d\in\mathcal{D}} n_d}{\min_{d\in\mathcal{D}} n_d} - 1\right).
$$

This completed the proof of part (ii). For part (i), we get the desired bound of $1 + \frac{3m^2 k}{4}$ by simply setting $\min_{d\in\mathcal{D}} n_d = \max_{d\in\mathcal{D}} n_d = n/k$.   □

Our next theorem shows that PV is asymptotically the best possible voting rule among all deterministic ordinal voting rules.

Theorem 4. *The distortion* $\texttt{gdist}(f)$ *of any ordinal Pareto efficient voting rule* f *is*

(i) $\Omega(m^2 k)$, *for symmetric elections;*

(ii) $\Omega\left(m^2 \frac{n + \max_{d\in\mathcal{D}} n_d}{\min_{d\in\mathcal{D}} n_d}\right)$, *for unweighted elections;*

(iii) $\Omega\left(\frac{m^2 n}{\min_{d\in\mathcal{D}} n_d}\right)$, *for unrestricted elections.*

5   Experiments

Thus far, we have studied the worst-case effect of the partition of voters into districts on the distortion of voting rules. In this section, we further showcase this phenomenon experimentally by using real-world utility profiles that are drawn from the Jester dataset [16], which consists of ratings of 100 different jokes in the interval $[-10, 10]$ by approximately 70,000 users; this dataset has been used in a plethora of previous papers, including the seminal work of Boutilier et al. [7]. Following their methodology, we build instances with a set of alternatives that consists of the eight most-rated jokes. For various values of k, we execute 1000 independent simulations as follows: we select a random set of 100 users among the ones that evaluated all eight alternatives, rescale their ratings so that they are non-negative and satisfy the unit-sum assumption, and then divide them into k districts.

For the partition into districts, we consider both *random* partitions as well as *bad* partitions in terms of distortion. For the construction of the latter, for each instance consisting of a specific value of k and a set of voters, we create 100 random partitions of the voters into k districts, simulate the general election (based on the voting rules we consider) and then keep the partition with maximum distortion.

We compare the average distortion of four rules: Range Voting, Plurality, Borda, and Harmonic. Borda and Harmonic are two well-known positional scoring rules defined by the scoring vectors $(m-1, m-2, ..., 0)$ and $(1, 1/2, ..., 1/m)$,

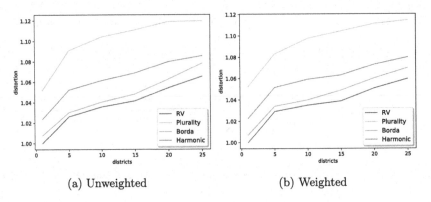

(a) Unweighted   (b) Weighted

Fig. 1. Average distortion from 1000 simulations as a function of the number of districts k with random partitions of voters into districts.

Table 1. Average distortion from 1000 simulations with bad partitions of voters into districts.

	Unweighted					Weighted				
k	1	2	3	4	5	1	2	3	4	5
Range Voting	1	4.82	4.51	4.50	4.60	1	4.46	4.96	5.14	5.14
Plurality	1.05	5.03	4.66	4.71	4.81	1.05	4.77	5.29	5.47	5.49
Borda	1.01	4.83	4.47	4.50	4.61	1.01	4.51	4.98	5.16	5.18
Harmonic	1.02	4.97	4.60	4.62	4.72	1.02	4.64	5.16	5.35	5.36

respectively. According to these rules, each voter assigns points to the alternatives based on the positions she ranks them, and the alternative with the most points is the winner; Plurality can also be defined similarly by the scoring vector $(1, 0, ..., 0)$.

Figure 1 depicts the results of our simulations for unweighted and weighted districts when the partition into districts is random and $k \in \{1, 5, 10, 15, 20, 25\}$; for weighted districts, the weights are drawn uniformly at random from a given interval. As one can observe, the behaviour of the four voting rules is very similar in both cases, and it is evident that as the number of districts increases, the distortion increases as well. For instance, the distortion of Plurality increased by 3.71% for $k = 5$ compared to $k = 1$ (i.e., when there are no districts) and by 6.44% for $k = 25$; these values are similar for the other rules as well, although a bit lower. Table 1 contains the results of our simulations for unweighted and weighted districts when the partition into districts is bad (in terms of distortion) and $k \in \{1, 2, 3, 4, 5\}$. As in the case of random districts, we can again observe that the distortion increases as k increases, but now the difference between the cases with districts ($k \geq 2$) and without districts ($k = 1$) is more clear; the distortion is almost five times higher.

6   Best-Case Partitions via Districting

In this section we turn out attention to a somewhat different setting. We assume that the k districts are not a priori defined, and instead we are free to decide the partition of the voters into the districts so as to minimize their effect on the distortion of the underlying voting rule; we refer to the process of partitioning the voters into k districts as *k-districting*. We consider symmetric districts, and start our analysis with the question of whether it is possible to define the districts so that the optimal alternative (i.e., the one that maximizes the social welfare of the voters) wins the general election when RV is used as the voting rule. Unfortunately, as we show with our next theorem, this is not always possible.

Theorem 5. *For every $k \geq 2$, there exists an instance such that no symmetric k-districting allows the optimal alternative to win the general election when RV is the voting rule.*

Proof. Consider a general election with $n + 1$ alternatives $\mathcal{M} = \{a_1, ..., a_n, b\}$ and let k be such that n/k is an integer for simplicity; then, each district must consist of exactly n/k voters. Let $\varepsilon \in \left(0, \frac{1}{2(n+1)}\right)$ and let $\mathbf{v}$ be the valuation profile according to which voter i has value $\frac{n}{n+k} + \varepsilon$ for alternative a_i and value $\frac{k}{n+k} - \varepsilon$ for alternative b; her value for the remaining alternatives is zero.

Since $\mathrm{SW}(a_i|\mathbf{v}) = \frac{n}{n+k} + \varepsilon$ for every $i \in [n]$ and $\mathrm{SW}(b|\mathbf{v}) = \frac{nk}{n+k} - n\varepsilon$, alternative b is clearly the optimal alternative. However, observe that all possible sets of n/k voters that can be included together in a district cannot make b the winner of the district when the voting rule is RV. Indeed, the welfare of such a set of voters for b is only $\frac{n}{n+k} - \frac{n\varepsilon}{k}$, while their welfare for the alternatives they rank first is $\frac{n}{n+k}$. Therefore, there is no symmetric k-districting that can make b the winner of the general election with RV.   □

In fact, the instance used in the above proof indicates that even the best-case distortion of RV may be at least k. We continue the bad news by showing that the problem of deciding whether it is possible to define the districts such that the optimal alternative wins the general election with RV is NP-hard for $k = 2$.

Theorem 6. *Deciding whether there is a symmetric 2-districting such that the optimal alternative is the winner of the general election with RV is NP-hard.*

In contrast to the above result for the optimal alternative and RV, we next show that we can always find a symmetric k-districting so that the PV winner without districts can be made the winner of the general election when PV is used as the voting rule within the districts. Since the voting rule is PV, we assume that the only knowledge which we can leverage in order to define the districts is about the favourite alternatives of the voters (i.e., for each voter, we know the alternative she approves).

Theorem 7. *For any $k \geq 2$, there always exists a symmetric k-districting that allows the winner of PV without districts to win the general election with k districts, and this districting can be computed in polynomial time.*

We conclude this section by showing that the above result for PV is essentially tight. This follows by the existence of instances where any partition of the voters into any number of districts yields distortion for the general election with PV that is asymptotically equal to the distortion of PV without districts.

Theorem 8. *There exist instances where any symmetric districting yields distortion* $\mathtt{gdist}(\mathrm{PV}) = \Omega(m^2)$.

7 Conclusion and Possible Extensions

In this paper, we have initiated the study of the distortion of distributed voting. We showcased the effect of districting on the social welfare both theoretically from a worst- and a best-case perspective, as well as experimentally using real-world data. Even though we have painted an almost complete picture, our work reveals many interesting avenues for future research.

In terms of our results, possibly the most obvious open question is whether we can strengthen the weak intractability result of Theorem 6 using a reduction from a strongly NP-hard problem, and also extend it to $k \geq 2$. Moving away from the unconstrained normalized setting that we considered here, it would be very interesting to analyze the effect of districts in the case of *metric preferences* [1], a setting that has received considerable attention in the recent related literature on the distortion of voting rules without districts [2,11,14,15,17,21]. Other important extensions include settings in which the partitioning of voters into districts is further constrained by natural factors such as geological locations [20] or connectivity in social networks [18].

References

1. Anshelevich, E., Bhardwaj, O., Elkind, E., Postl, J., Skowron, P.: Approximating optimal social choice under metric preferences. Artif. Intell. **264**, 27–51 (2018)
2. Anshelevich, E., Postl, J.: Randomized social choice functions under metric preferences. J. Artif. Intell. Res. **58**, 797–827 (2017)
3. Bachrach, Y., Lev, O., Lewenberg, Y., Zick, Y.: Misrepresentation in district voting. In: Proceedings of the 25th International Joint Conference on Artificial Intelligence (IJCAI), pp. 81–87 (2016)
4. Benade, G., Nath, S., Procaccia, A.D., Shah, N.: Preference elicitation for participatory budgeting. In: Proceedings of the 31st AAAI Conference on Artificial Intelligence (AAAI), pp. 376–382 (2017)
5. Bhaskar, U., Dani, V., Ghosh, A.: Truthful and near-optimal mechanisms for welfare maximization in multi-winner elections. In: Proceedings of the 32nd AAAI Conference on Artificial Intelligence (AAAI), pp. 925–932 (2018)
6. Borodin, A., Lev, O., Shah, N., Strangway, T.: Big city vs. the great outdoors: voter distribution and how it affects gerrymandering. In: Proceedings of the 27th International Joint Conference on Artificial Intelligence (IJCAI), pp. 98–104 (2018)
7. Boutilier, C., Caragiannis, I., Haber, S., Lu, T., Procaccia, A.D., Sheffet, O.: Optimal social choice functions: a utilitarian view. Artif. Intell. **227**, 190–213 (2015)

8. Caragiannis, I., Nath, S., Procaccia, A.D., Shah, N.: Subset selection via implicit utilitarian voting. J. Artif. Intell. Res. **58**, 123–152 (2017)
9. Cohen-Zemach, A., Lewenberg, Y., Rosenschein, J.S.: Gerrymandering over graphs. In: Proceedings of the 17th International Conference on Autonomous Agents and Multiagent Systems (AAMAS), pp. 274–282 (2018)
10. Erdélyi, G., Hemaspaandra, E., Hemaspaandra, L.A.: More natural models of electoral control by partition. In: Walsh, T. (ed.) ADT 2015. LNCS (LNAI), vol. 9346, pp. 396–413. Springer, Cham (2015). https://doi.org/10.1007/978-3-319-23114-3_24
11. Feldman, M., Fiat, A., Golomb, I.: On voting and facility location. In: Proceedings of the 17th ACM Conference on Economics and Computation (EC), pp. 269–286 (2016)
12. Filos-Ratsikas, A., Micha, E., Voudouris, A.A.: The distortion of distributed voting. CoRR abs/1905.01882 (2019)
13. Filos-Ratsikas, A., Miltersen, P.B.: Truthful approximations to range voting. In: Liu, T.-Y., Qi, Q., Ye, Y. (eds.) WINE 2014. LNCS, vol. 8877, pp. 175–188. Springer, Cham (2014). https://doi.org/10.1007/978-3-319-13129-0_13
14. Goel, A., Hulett, R., Krishnaswamy, A.K.: Relating metric distortion and fairness of social choice rules. In: Proceedings of the 13th Workshop on the Economics of Networks, Systems and Computation (NetEcon), p. 4:1 (2018)
15. Goel, A., Krishnaswamy, A.K., Munagala, K.: Metric distortion of social choice rules: lower bounds and fairness properties. In: Proceedings of the 18th ACM Conference on Economics and Computation (EC), pp. 287–304 (2017)
16. Goldberg, K., Roeder, T., Gupta, D., Perkins, C.: Eigentaste: a constant time collaborative filtering algorithm. Inf. Retrieval **4**, 133–151 (2001)
17. Gross, S., Anshelevich, E., Xia, L.: Vote until two of you agree: mechanisms with small distortion and sample complexity. In: Proceedings of the 31st AAAI Conference on Artificial Intelligence (AAAI), pp. 544–550 (2017)
18. Lesser, O., Naamani-Dery, L., Kalech, M., Elovici, Y.: Group decision support for leisure activities using voting and social networks. Group Decis. Negot. **26**(3), 473–494 (2017)
19. Lev, O., Lewenberg, Y.: "reverse gerrymandering": a decentralized model for multi-group decision making. In: Proceedings of the 33rd AAAI Conference on Artificial Intelligence (AAAI) (2019)
20. Lewenberg, Y., Lev, O., Rosenschein, J.S.: Divide and conquer: using geographic manipulation to win district-based elections. In: Proceedings of the 16th International Conference on Autonomous Agents and Multiagent Systems (AAMAS), pp. 624–632 (2017)
21. Pierczynski, G., Skowron, P.: Approval-based elections and distortion of voting rules. CoRR abs/1901.06709 (2019)
22. Procaccia, A.D., Rosenschein, J.S.: The distortion of cardinal preferences in voting. In: Klusch, M., Rovatsos, M., Payne, T.R. (eds.) CIA 2006. LNCS (LNAI), vol. 4149, pp. 317–331. Springer, Heidelberg (2006). https://doi.org/10.1007/11839354_23
23. Schuck, P.H.: The thickest thicket: partisan gerrymandering and judicial regulation of politics. Columbia Law Rev. **87**(7), 1325–1384 (1987)
24. Wikipedia: 2016 United States presidential election (2016). https://en.wikipedia.org/wiki/2016_United_States_presidential_election

Matchings and Fair Division

On the Existence of Three-Dimensional Stable Matchings with Cyclic Preferences

Chi-Kit Lam and C. Gregory Plaxton$^{(\boxtimes)}$

University of Texas at Austin, Austin, TX 78712, USA
{geocklam,plaxton}@cs.utexas.edu

Abstract. We study the three-dimensional stable matching problem with cyclic preferences. This model involves three types of agents, with an equal number of agents of each type. The types form a cyclic order such that each agent has a complete preference list over the agents of the next type. We consider the open problem of the existence of three-dimensional matchings in which no triple of agents prefer each other to their partners. Such matchings are said to be weakly stable. We show that contrary to published conjectures, weakly stable three-dimensional matchings need not exist. Furthermore, we show that it is NP-complete to determine whether a weakly stable three-dimensional matching exists. We achieve this by reducing from the variant of the problem where preference lists are allowed to be incomplete. Our results can be generalized to the k-dimensional stable matching problem with cyclic preferences for $k \geq 3$.

Keywords: Stable matching · Three-dimensional matching · NP-completeness

1 Introduction

The study of stable matchings was started by Gale and Shapley [9], who investigated a market with two types of agents. The two-dimensional stable matching problem involves an equal number of men and women, each of whom has a complete preference list over the agents of the opposite sex. The goal is to find a matching between the men and the women such that no man and woman prefer each other to their partners. Matchings satisfying this property are said to be stable. Gale and Shapley showed that a solution for the two-dimensional stable matching problem always exists and can be computed in polynomial time. Their result also applies to the variant where preference lists may be incomplete due to unacceptable partners, and the number of men may be different from the number of women.

The problem of generalizing stable matchings to markets with three types of agents was posed by Knuth [13]. In pursuit of an existence theorem and an elegant theory analogous to those of the Gale-Shapley model, the three-dimensional stable matching problem has been studied with respect to a number

© Springer Nature Switzerland AG 2019
D. Fotakis and E. Markakis (Eds.): SAGT 2019, LNCS 11801, pp. 329–342, 2019.
https://doi.org/10.1007/978-3-030-30473-7_22

of preference structures. When each agent has preferences over pairs of agents from the other two types, stable matchings need not exist [1,16]. Furthermore, it is NP-complete to determine whether a stable matching exists [16,18], even if the preferences are consistent with product orders [11]. When two types of agents care primarily about each other and secondarily about the remaining type, a stable matching always exists and can be obtained by computing two-dimensional stable matchings using the Gale-Shapley algorithm in a hierarchical manner [5]. When the types form a cyclic order such that each type of agent cares primarily about the next type and secondarily about the other type, stable matchings need not exist [3].

A prominent problem mentioned in several of the aforementioned papers [3,11,16] is the three-dimensional stable matching problem for the case where the types form a cyclic order such that each type of agent cares only about the next type and not the other type. Following the terminology of the survey of Manlove [15], we call this the three-dimensional stable matching problem with cyclic preferences (3-DSM-CYC), and refer to the three types of agents as men, women, and dogs. A number of stability notions [11] can be considered in 3-DSM-CYC. In this paper, we focus on weak stability, which is the most permissive one and has received the most attention in the literature. It is known that determining whether a 3-DSM-CYC instance has a strongly stable matching is NP-complete [2]. For the variant where ties are allowed, determining the existence of a super-stable matching is also NP-complete [12]. However, it remained an open problem for weakly stable matchings in 3-DSM-CYC.

In 3-DSM-CYC, there are an equal number of men, women, and dogs. Each man has a complete preference list over the women, each woman has a complete preference list over the dogs, and each dog has a complete preference list over the men. A family is a triple consisting of a man, a woman, and a dog. A matching is a set of agent-disjoint families. A family is strongly blocking if every agent in the family prefers each other to their partners in the matching. A matching is weakly stable if it admits no strongly blocking family. This problem is related to applications such as kidney exchange [2] and three-sided network services [4].

The formulation of 3-DSM-CYC first appeared in the paper of Ng and Hirschberg [16], where it is attributed to Knuth. Using a greedy approach, Boros et al. [3] showed that every 3-DSM-CYC instance with at most 3 agents per type has a weakly stable matching. Their result also applies to the k-dimensional generalization of the problem, which we call k-DSM-CYC. For $k \geq 3$, they showed that every k-DSM-CYC instance with at most k agents per type has a weakly stable matching. Using a case analysis, Eriksson et al. [6] showed that every 3-DSM-CYC instance with at most 4 agents per type has a weakly stable matching, and they conjectured that every 3-DSM-CYC instance has a weakly stable matching. In fact, they posed the stronger conjecture that for a certain "strongest link" generalization of 3-DSM-CYC, every instance with at least two agents per type has at least two weakly stable matchings. Eriksson et al. also investigated and ruled out the use of certain arguments based on "effectivity functions" and "balanced games" for proving the 3-DSM-CYC conjecture. Using an efficient greedy procedure, Hofbauer [10] showed that for $k \geq 3$, every k-DSM-CYC instance with at most $k+1$ agents per type has a weakly stable matching. Using a satisfiability problem formulation and an extensive computer-assisted search, Pashkovich and Poirrier [17] showed that every 3-DSM-CYC instance with exactly 5 agents per type has at least two weakly stable matchings. Escamocher and O'Sullivan [7] showed that the number of weakly stable matchings is exponential in the size of the 3-DSM-CYC instance if agents of the same type are restricted to have the same preferences. They also conjectured that for unrestricted 3-DSM-CYC instances, there are exponentially many weakly stable matchings.

Hardness results are known for some related problems. For the variant of 3-DSM-CYC where preference lists are allowed to be incomplete, which we refer to as 3-DSMI-CYC, Biró and McDermid [2] showed that determining whether a weakly stable matching exists is NP-complete. Farczadi et al. [8] showed that determining whether a given perfect two-dimensional matching can be extended to a three-dimensional weakly stable matching in 3-DSM-CYC is also NP-complete. However, the existence of weakly stable matchings in 3-DSM-CYC remained unresolved. Manlove [15] described it as an "intriguing open problem", and Woeginger [19] classified it as "hard and outstanding".

Our Techniques and Contributions. In this paper, we show that there exists a 3-DSM-CYC instance that has no weakly stable matching. This disproves the conjectures of Eriksson et al. [6] and Escamocher and O'Sullivan [7]. Furthermore, we show that determining whether a 3-DSM-CYC instance has a weakly stable matching is NP-complete. We achieve this by reducing from the problem of determining whether a 3-DSMI-CYC instance has a weakly stable matching. Our results generalize to k-DSM-CYC for $k \geq 3$.

Our main technique involves converting each agent in 3-DSMI-CYC to a gadget consisting of one non-dummy agent and many dummy agents. The dummy agents in our gadget give rise to chains of admirers. (See Remark 2 in Sect. 4.3.) By applying the weak stability condition to the chains of admirers, we are able to obtain some control over the partner of the non-dummy agent.

Organization of This Paper. In Sect. 2, we present the formal definitions of k-DSM-CYC and k-DSMI-CYC. In Sect. 3, we show that the NP-completeness result of Biró and McDermid [2] can be extended to k-DSMI-CYC. In Sect. 4, we show that k-DSM-CYC is NP-complete by a reduction from k-DSMI-CYC. In Sect. 5, we conclude by mentioning some potential future work.

2   Preliminaries

In this paper, we use $\langle z \in Z \mid \mathcal{P}(z) \rangle$ to denote the list of all tuples $z \in Z$ satisfying predicate $\mathcal{P}(z)$, where the tuples are sorted in increasing lexicographical order. Given two lists Y and Z, we denote their concatenation as $Y \cdot Z$. For any $k \geq 1$, we use $\oplus_k$ to denote addition modulo k.

2.1   The Models

Let $k \geq 2$. The k-dimensional stable matching problem with incomplete lists and cyclic preferences (k-DSMI-CYC) involves a finite set $A = I \times \{0, \ldots, k-1\}$ of agents, where each agent $\alpha = (i, t) \in A$ is associated with an identifier i and a type t. (When $k = 3$, we can think of the sets $I \times \{0\}$, $I \times \{1\}$, and $I \times \{2\}$ as the sets of men, women, and dogs, respectively.) Each agent $\alpha = (i, t) \in A$ has a strict preference list P_α over a subset of agents of type $t' = t \oplus_k 1$. In other words, every agent in $I \times \{t \oplus_k 1\}$ appears in P_α at most once, and every element in P_α belongs to $I \times \{t \oplus_k 1\}$. For every $\alpha, \alpha', \alpha'' \in A$, we say that α prefers α' to α'' if α' appears in P_α and either agent α'' appears in P_α after α' or agent α'' does not appear in P_α. We denote this k-DSMI-CYC instance as $X = (A, \{P_\alpha\}_{\alpha \in A})$.

Given a k-DSMI-CYC instance $X = (A, \{P_\alpha\}_{\alpha \in A})$, a *family* is a tuple

$$(\alpha_0, \ldots, \alpha_{k-1}) \in A^k$$

such that $\alpha_t \in I \times \{t\}$ and $\alpha_{t \oplus_k 1}$ appears in P_{α_t} for every $t \in \{0, \ldots, k-1\}$. A *matching* μ is a set of agent-disjoint families. In other words, for every $t, t' \in \{0, \ldots, k-1\}$ and $(\alpha_0, \ldots, \alpha_{k-1}), (\alpha'_0, \ldots, \alpha'_{k-1}) \in \mu$, if $\alpha_t = \alpha'_t$, then $\alpha_{t'} = \alpha'_{t'}$. Given a matching μ and an agent $\alpha \in A$, if $\alpha = \alpha_t$ for some $(\alpha_0, \ldots, \alpha_{k-1}) \in \mu$ and $t \in \{0, \ldots, k-1\}$, we say that α is matched to $\alpha_{t \oplus_k 1}$, and we write $\mu(\alpha) = \alpha_{t \oplus_k 1}$. Otherwise, we say that α is unmatched, and we write $\mu(\alpha) = \alpha$.

Given a matching μ, we say that a family $(\alpha_0, \ldots, \alpha_{k-1})$ is *strongly blocking* if α_t prefers $\alpha_{t \oplus_k 1}$ to $\mu(\alpha_t)$ for every $t \in \{0, \ldots, k-1\}$. A matching μ is *weakly stable* if it does not admit any strongly blocking family.

The k-dimensional stable matching problem with cyclic preferences (k-DSM-CYC) is defined as the special case of k-DSMI-CYC in which every agent in $I \times \{t \oplus_k 1\}$ appears exactly once in P_α for every agent $\alpha = (i, t) \in A$.

Notice that when incomplete lists are allowed, the case of an unequal number of agents of each type can be handled within our k-DSMI-CYC model by padding with dummy agents whose preference lists are empty. Hence, the results of Biró and McDermid [2] apply to our 3-DSMI-CYC model. When preference lists are complete, we follow the literature and focus on the case where each type has an equal number of agents. Our result shows that even when restricted to the case of an equal number of agents of each type, a given k-DSM-CYC instance need not admit a weakly stable matching, and determining the existence of a weakly stable matching is NP-complete.

2.2   Polynomial-Time Verification

Given a matching μ of a k-DSMI-CYC instance with n agents per type, it is straightforward to determine whether μ is weakly stable in $O(n^k)$ time by checking that none of the $O(n^k)$ families is strongly blocking. The following theorem shows that when k is large, there is a more efficient method to determine whether a given matching is weakly stable. A proof is provided in [14].

Theorem 1. *There exists a* $\mathrm{poly}(n, k)$*-time algorithm to determine whether a given matching μ is weakly stable for a k-DSMI-CYC instance, where n is the number of agents per type.*

3 NP-Completeness of k-DSMI-CYC

In this section, we show that for every $k \geq 3$, it is NP-complete to determine whether a given k-DSMI-CYC instance has a weakly stable matching. We achieve this by reducing from the problem of determining whether a 3-DSMI-CYC instance has a weakly stable matching.

3.1 The Reduction

Let $k \geq 4$. Consider an input 3-DSMI-CYC instance $X = (A, \{P_\alpha\}_{\alpha \in A})$ where $A = I \times \{0, 1, 2\}$. Our reduction constructs a k-DSMI-CYC instance $\hat{X} = (\hat{A}, \{\hat{P}_{\hat{\alpha}}\}_{\hat{\alpha} \in \hat{A}})$ as follows.

- Let $\hat{I} = I \times I$ and $\hat{A} = I \times I \times \{0, \dots, k-1\}$. For every agent $(i, t) \in A$, we call $\hat{\alpha} = (i, i, t) \in \hat{A}$ the non-dummy agent corresponding to (i, t). We call the agents

$$\{(i, j, t) \in \hat{A} \mid t \notin \{0, 1, 2\} \text{ or } i \neq j\}$$

  dummy agents.
- For every agent $\hat{\alpha} = (i, j, t) \in \hat{A}$, we construct the preference list $\hat{P}_{\hat{\alpha}}$ as follows.
  - If $0 \leq t \leq 1$ and $i = j$, we list in $\hat{P}_{\hat{\alpha}}$ the agents

$$\{(i', j', t') \in I \times I \times \{t+1\} \mid i' = j' \text{ and } (i', t') \text{ is in } P_{(i,t)}\}$$

    in the order in which the corresponding agent (i', t') appears in $P_{(i,t)}$.
  - If $t = 2$ and $i = j$, we list in $\hat{P}_{\hat{\alpha}}$ the agents

$$\{(i', j', t') \in I \times I \times \{3\} \mid i' = i \text{ and } (j', 0) \text{ is in } P_{(i,2)}\}$$

    in the order in which the corresponding agent $(j', 0)$ appears in $P_{(i,2)}$.
  - If $0 \leq t \leq 2$ and $i \neq j$, we define $\hat{P}_{\hat{\alpha}}$ as the empty list.
  - If $3 \leq t \leq k-2$ and $(j, 0)$ is in $P_{(i,2)}$, we define $\hat{P}_{\hat{\alpha}}$ as $\langle (i, j, t+1) \rangle$.
  - If $t = k-1$ and $(j, 0)$ is in $P_{(i,2)}$, we define $\hat{P}_{\hat{\alpha}}$ as $\langle (j, j, 0) \rangle$.
  - If $3 \leq t \leq k-1$ and $(j, 0)$ is not in $P_{(i,2)}$, we define $\hat{P}_{\hat{\alpha}}$ as the empty list.

Figure 1 shows an example of the reduction when $k = 5$ and $I = \{0, 1\}$.

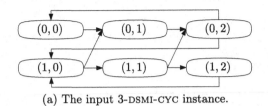

(a) The input 3-DSMI-CYC instance.

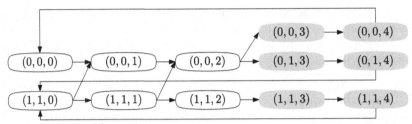

(b) The output 5-DSMI-CYC instance. Agents with empty preference lists are omitted.

Fig. 1. Example of a reduction from 3-DSMI-CYC to 5-DSMI-CYC. An arrow indicates that the target agent appears in the preference list of the source agent.

3.2   Correctness of the Reduction

Proofs of the three claims stated below are provided in [14]. We emphasize that the important special case of Theorem 2 where $k = 3$ is due to Biró and McDermid [2, Lemma 1].

Lemma 1. *Let $k \geq 4$. Consider the reduction given in Sect. 3.1. The output k-DSMI-CYC instance $\hat{X}$ has a weakly stable matching if and only if the input 3-DSMI-CYC instance X has a weakly stable matching.*

Theorem 2. *Let $k \geq 3$. Then there exists a k-DSMI-CYC instance that has no weakly stable matching.*

Theorem 3. *Let $k \geq 3$. Then it is NP-complete to determine whether a k-DSMI-CYC instance has a weakly stable matching.*

4   NP-Completeness of k-DSM-CYC

In this section, we show that for every $k \geq 3$, it is NP-complete to determine whether a k-DSM-CYC instance has a weakly stable matching. We achieve this by reducing from the problem of determining whether a k-DSMI-CYC instance has a weakly stable matching. Since the dimensions of both the input instance and the output instance of the reduction are equal to k, throughout this section, we write $\oplus$ instead of $\oplus_k$ for better readability.

4.1 The Reduction

Let $k \geq 3$. Consider an input k-DSMI-CYC instance $X = (A, \{P_\alpha\}_{\alpha \in A})$ where $A = I \times \{0, \ldots k-1\}$. We may assume that $I = \{0, \ldots, |I| - 1\}$, so agents in A can be compared lexicographically. Our reduction constructs a k-DSM-CYC instance $\hat{X} = (\hat{A}, \{\hat{P}_{\hat{\alpha}}\}_{\hat{\alpha} \in \hat{A}})$ as follows.

- Let $J = \{0, \ldots, (k-1)^2\}$. Let $\hat{I} = J \times A$ and $\hat{A} = J \times A \times \{0, \ldots, k-1\}$. For every agent $\alpha \in A$, we call $J \times \{\alpha\} \times \{0, \ldots, k-1\}$ the gadget corresponding to α.
- For every agent $\hat{\alpha} = (j, \alpha, t) \in \hat{A}$ such that $j = 0$ and $\alpha \in I \times \{t\}$, we call $\hat{\alpha}$ the non-dummy agent corresponding to α. Let $\hat{P}'_\alpha$ be the list obtained by replacing every α' in P_α by $(0, \alpha', t \oplus 1)$. We define the preference list $\hat{P}_{\hat{\alpha}}$ as $\hat{P}'_\alpha \cdot \langle (j', \alpha', t') \in J \times A \times \{t \oplus 1\} \mid \alpha' = \alpha \rangle$ followed by the remaining agents in $J \times A \times \{t \oplus 1\}$ in an arbitrary order.
- For every agent $\hat{\alpha} = (j, \alpha, t) \in \hat{A}$ such that $j = (k-1)^2$, we call $\hat{\alpha}$ a boundary dummy agent, and we define the preference list $\hat{P}_{\hat{\alpha}}$ as

$$\langle (j', \alpha', t') \in J \times A \times \{t \oplus 1\} \mid \alpha' = \alpha \text{ and } j' < (k-1)^2 \rangle$$
$$\cdot \langle (j', \alpha', t') \in J \times A \times \{t \oplus 1\} \mid j' = (k-1)^2 \rangle$$

followed by the remaining agents in $J \times A \times \{t \oplus 1\}$ in an arbitrary order.
- For every agent $\hat{\alpha} = (j, \alpha, t) \in \hat{A}$ such that $(j, \alpha, t) \notin \{0\} \times (I \times \{t\}) \times \{t\}$ and $j < (k-1)^2$, we call $\hat{\alpha}$ a non-boundary dummy agent, and we define the preference list $\hat{P}_{\hat{\alpha}}$ as $\langle (j', \alpha', t') \in J \times A \times \{t \oplus 1\} \mid \alpha' = \alpha \rangle$ followed by the remaining agents in $J \times A \times \{t \oplus 1\}$ in an arbitrary order.

As shown in Fig. 2(a), the gadget corresponding to $\alpha \in I \times \{t\}$ can be visualized as a grid of agents with k rows and $(k-1)^2 + 1$ columns. The non-boundary dummy agents in the same row have essentially the same preferences, which begin with the agents in the next row from left to right. The preferences of the boundary dummy agents are similar to those of the non-boundary dummy agents, except that they incorporate the other boundary dummy agents in a special manner. Meanwhile, the preferences of the non-dummy agent $(0, \alpha, t)$ reflect the preferences of agent α by starting with $\hat{P}'_\alpha$.

Remark 1. The reason our gadget has $(k-1)^2 + 1$ columns will become clearer when we present Lemmas 4 and 5 below. At a high level, Lemma 4 is invoked $k-1$ times within the proof of Lemma 5, and each such invocation leads to an increase in the number of columns of $k-1$.

4.2 Correctness of the Reduction

Lemmas 2 and 3 below show that the reduction in Sect. 4.1 is a correct reduction from k-DSMI-CYC to k-DSM-CYC. The associated proofs are presented in Sects. 4.4 and 4.5.

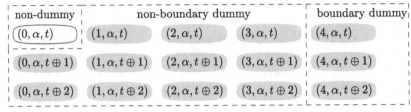

(a) The structure of the gadget.

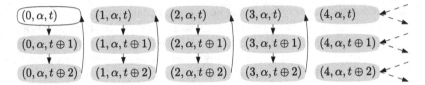

(b) The matching $\hat{\mu}$ induced by μ when α is unmatched in μ.

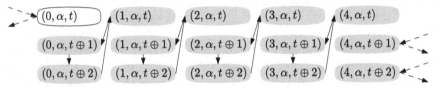

(c) The matching $\hat{\mu}$ induced by μ when α is matched in μ.

Fig. 2. Example of a gadget corresponding to $\alpha \in I \times \{t\}$ when $k = 3$. An arrow indicates that the source agent is matched to the target agent.

Lemma 2. *Let $k \geq 3$. Consider the reduction given in Sect. 4.1. If the input k-DSMI-CYC instance X has no weakly stable matching, then the output k-DSM-CYC instance $\hat{X}$ has no weakly stable matching.*

Lemma 3. *Let $k \geq 3$. Consider the reduction in Sect. 4.1. If the input k-DSMI-CYC instance X has a weakly stable matching, then the output k-DSM-CYC instance $\hat{X}$ has a weakly stable matching.*

Proofs of the next two theorems are provided in [14].

Theorem 4. *Let $k \geq 3$. Then there exists a k-DSM-CYC instance that has no weakly stable matching.*

Theorem 5. *Let $k \geq 3$. Then it is NP-complete to determine whether a k-DSM-CYC instance has a weakly stable matching.*

4.3 Properties of the Gadget

In this subsection, we study the properties of the gadget in the scenario that the non-dummy agent is not matched to a non-dummy agent corresponding to an acceptable partner. In Lemma 4, we show that in this scenario, many agents in the gadget are matched to agents in the same gadget. In Lemma 5, we apply Lemma 4 inductively to show that in the same scenario, every agent in the same family as the non-dummy agent belongs to the same gadget.

Remark 2. In the proof of Lemma 4 below, we can think of $\hat{\alpha}_0, \ldots, \hat{\alpha}_{k-1}$ as a chain of admirers in the gadget corresponding to α, where $\hat{\alpha}_s$ prefers $\hat{\alpha}_{s+1}$ to $\hat{\mu}(\hat{\alpha}_s)$. By applying the weak stability condition to this chain of admirers, we show that $\hat{\alpha}_{k-1}$ is matched to a partner no worse than $\hat{\alpha}_0$.

Lemma 4. *Let $\hat{\mu}$ be a weakly stable matching in $\hat{X}$. Let $t^* \in \{0, \ldots, k-1\}$ and $\alpha^* \in I \times \{t^*\}$ such that $\hat{\mu}(0, \alpha^*, t^*)$ is not in $\hat{P}'_{\alpha^*}$. Let $t \in \{0, \ldots, k-1\}$ and $j \in J$ such that $j \leq (k-1)\cdot(k-2)$. Then $\hat{\mu}(j, \alpha^*, t) \in \{0, \ldots, j+k-1\} \times \{\alpha^*\} \times \{t \oplus 1\}$.*

Proof. Let $\hat{A}_s = \langle (j', \alpha', t') \in J \times \{\alpha^*\} \times \{t \oplus s \oplus 1\} \mid j' \leq j + k - s - 1 \rangle$ for every $s \in \{0, \ldots, k-1\}$. For the sake of contradiction, suppose $\hat{\mu}(j, \alpha^*, t)$ is not in $\hat{A}_0$.

For every $s \in \{0, \ldots, k-2\}$, since the length of $\hat{A}_s$ is greater than the length of $\hat{A}_{s+1}$, there exists $\hat{\alpha}_s$ in $\hat{A}_s$ such that $\hat{\mu}(\hat{\alpha}_s)$ is not in $\hat{A}_{s+1}$. Let $\hat{\alpha}_{k-1} = (j, \alpha^*, t)$. Then $\hat{\alpha}_{k-1}$ is in $\hat{A}_{k-1}$ and $\hat{\mu}(\hat{\alpha}_{k-1})$ is not in $\hat{A}_0$. Since $\hat{\mu}$ is a weakly stable matching of $\hat{X}$, the family $(\hat{\alpha}_{k-t-1}, \ldots, \hat{\alpha}_{(k-t-1)\oplus(k-1)})$ is not strongly blocking. So there exists $s^* \in \{0, \ldots, k-1\}$ such that $\hat{\alpha}_{s^*}$ does not prefer $\hat{\alpha}_{s^* \oplus 1}$ to $\mu(\hat{\alpha}_{s^*})$. Since $\hat{\alpha}_{s^*}$ is in $\hat{A}_{s^*}$, there exists $j^* \leq j + k - s^* - 1$ such that $\hat{\alpha}_{s^*} = (j^*, \alpha^*, t \oplus s^* \oplus 1)$. We consider two cases.

Case 1: $j^* = 0$ and $t \oplus s^* \oplus 1 = t^*$. Then $\hat{\alpha}_{s^*} = (0, \alpha^*, t^*)$ is a non-dummy agent and $\hat{P}'_{\alpha^*} \cdot \hat{A}_{s^* \oplus 1}$ is a prefix of the preference list $\hat{P}_{\hat{\alpha}_{s^*}}$. Since $\mu(\hat{\alpha}_{s^*})$ is not in $\hat{P}'_{\alpha^*} \cdot \hat{A}_{s^* \oplus 1}$ and $\hat{\alpha}_{s^* \oplus 1}$ is in $\hat{A}_{s^* \oplus 1}$, agent $\hat{\alpha}_{s^*}$ prefers $\hat{\alpha}_{s^* \oplus 1}$ to $\mu(\hat{\alpha}_{s^*})$, a contradiction.

Case 2: $j^* \neq 0$ or $t \oplus s^* \oplus 1 \neq t^*$. We consider two subcases.

Case 2.1: $j^* = (k-1)^2$. Since $(k-1)^2 = j^* \leq j + k - s^* - 1 \leq (k-1)^2 - s^*$, we have $s^* = 0$. Hence $\hat{\alpha}_0 = ((k-1)^2, \alpha^*, t \oplus 1)$ is a boundary dummy agent and $\hat{A}_1$ is a prefix of the preference list $\hat{P}_{\hat{\alpha}_0}$. Since $\mu(\hat{\alpha}_0)$ is not in $\hat{A}_1$ and $\hat{\alpha}_1$ is in $\hat{A}_1$, agent $\hat{\alpha}_0$ prefers $\hat{\alpha}_1$ to $\mu(\hat{\alpha}_0)$, a contradiction.

Case 2.2: $j^* < (k-1)^2$. Then $\hat{\alpha}_{s^*}$ is a non-boundary dummy agent and $\hat{A}_{s^* \oplus 1}$ is a prefix of the preference list $\hat{P}_{\hat{\alpha}_{s^*}}$. Since $\mu(\hat{\alpha}_{s^*})$ is not in $\hat{A}_{s^* \oplus 1}$ and $\hat{\alpha}_{s^* \oplus 1}$ is in $\hat{A}_{s^* \oplus 1}$, agent $\hat{\alpha}_{s^*}$ prefers $\hat{\alpha}_{s^* \oplus 1}$ to $\mu(\hat{\alpha}_{s^*})$, a contradiction. $\square$

Lemma 5. *Let $\hat{\mu}$ be a weakly stable matching in $\hat{X}$. Let $j_0, \ldots, j_{k-1} \in J$ and $\alpha_0, \ldots, \alpha_{k-1} \in A$ such that $((j_0, \alpha_0, 0), \ldots, (j_{k-1}, \alpha_{k-1}, k-1)) \in \hat{\mu}$. Let $t^* \in \{0, \ldots, k-1\}$ such that $j_{t^*} = 0$ and $\alpha_{t^*} \in I \times \{t^*\}$. Suppose that $(j_{t^* \oplus 1}, \alpha_{t^* \oplus 1}, t^* \oplus 1)$ is not in $\hat{P}'_{\alpha_{t^*}}$. Then, for every $s \in \{0, \ldots, k-1\}$, we have $\alpha_{t^* \oplus s} = \alpha_{t^*}$ and $j_{t^* \oplus s} \leq (k-1) \cdot s$.*

Proof. We prove the claim by induction on s. When $s = 0$, we have $\alpha_{t^* \oplus s} = \alpha_{t^* \oplus 0} = \alpha_{t^*}$ and $j_{t^* \oplus s} = j_{t^*} = 0 \le (k-1) \cdot s$.

Suppose $\alpha_{t^* \oplus (s-1)} = \alpha_{t^*}$ and $j_{t^* \oplus (s-1)} \le (k-1) \cdot (s-1)$, where $s \in \{1, \ldots, k-1\}$. Since $(j_{t^* \oplus 1}, \alpha_{t^* \oplus 1}, t^* \oplus 1)$ is not in $\hat{P}'_{\alpha_{t^*}}$, agent $\hat{\mu}(0, \alpha_{t^*}, t^*)$ is not in $\hat{P}'_{\alpha_{t^*}}$. Let $t = t^* \oplus (s-1)$. Then $\alpha_t = \alpha_{t^* \oplus (s-1)} = \alpha_{t^*}$ and $j_t = j_{t^* \oplus (s-1)} \le (k-1) \cdot (s-1) \le (k-1) \cdot (k-2)$. So Lemma 4 implies that $\hat{\mu}(j_t, \alpha_{t^*}, t) \in \{0, \ldots, j_t + k - 1\} \times \{\alpha_{t^*}\} \times \{t \oplus 1\}$. Hence $j_{t \oplus 1} \le j_t + k - 1$ and $\alpha_{t \oplus 1} = \alpha_{t^*}$, since $\hat{\mu}(j_t, \alpha_{t^*}, t) = \hat{\mu}(j_t, \alpha_t, t) = (j_{t \oplus 1}, \alpha_{t \oplus 1}, t \oplus 1)$. Thus $\alpha_{t^* \oplus s} = \alpha_{t \oplus 1} = \alpha_{t^*}$ and $j_{t^* \oplus s} = j_{t \oplus 1} \le j_t + k - 1 = j_{t^* \oplus (s-1)} + k - 1 \le (k-1) \cdot (s-1) + k - 1 = (k-1) \cdot s$. $\qquad\square$

4.4   Proof of Lemma 2

The goal of this subsection is to prove Lemma 2. It suffices to show that every weakly stable matching $\hat{\mu}$ in $\hat{X}$ induces a weakly stable matching μ in X.

Recall that each agent in A has a corresponding non-dummy agent in $\hat{A}$, and that a family in X is a tuple of k agents in A such that each agent appears in the preference list of another. Hence we include in μ a family of agents in X whenever the corresponding family of non-dummy agents are matched in $\hat{\mu}$. More formally, we define the matching μ in X induced by $\hat{\mu}$ in $\hat{X}$ as the set of families $(\alpha_0, \ldots, \alpha_{k-1})$ in X satisfying $((0, \alpha_0, 0), \ldots, (0, \alpha_{k-1}, k-1)) \in \hat{\mu}$. Notice that every μ induced by a matching $\hat{\mu}$ in $\hat{X}$ is a valid matching in X since agent-disjoint families in $\hat{X}$ induce agent-disjoint families in X.

Lemma 6 below shows that if $\hat{\mu}$ is weakly stable and matches a non-dummy agent to a non-dummy agent corresponding to an acceptable partner, then μ matches the corresponding agents. Our proof relies on Lemma 5 and the weak stability of $\hat{\mu}$. Notice that if $\hat{\mu}$ is not weakly stable, it may be the case that $\hat{\mu}$ matches a family consisting of $k-1$ non-dummy agents and one dummy agent. In such a case, the corresponding $k-1$ agents are unmatched in the induced matching μ.

Lemma 6. *Let μ be the matching in X induced by a weakly stable matching $\hat{\mu}$ in $\hat{X}$. Let $t \in \{0, \ldots, k-1\}$ and $\alpha \in I \times \{t\}$ such that $\hat{\mu}(0, \alpha, t)$ is in $\hat{P}'_\alpha$. Then $\hat{\mu}(0, \alpha, t) = (0, \mu(\alpha), t \oplus 1)$.*

Proof. For the sake of contradiction, suppose $\hat{\mu}(0, \alpha, t) \ne (0, \mu(\alpha), t \oplus 1)$. Since $\hat{\mu}(0, \alpha, t)$ is in $\hat{P}'_\alpha$, we have $((j_0, \alpha_0, 0), \ldots, (j_{k-1}, \alpha_{k-1}, k-1)) \in \hat{\mu}$ for some $j_0, \ldots, j_{k-1} \in J$ and $\alpha_0, \ldots, \alpha_{k-1} \in A$ such that $(j_t, \alpha_t, t) = (0, \alpha, t)$ and $(j_{t \oplus 1}, \alpha_{t \oplus 1}, t \oplus 1)$ is in $\hat{P}'_\alpha$. Let

$$T = \{t' \in \{0, \ldots, k-1\} \mid \alpha_{t'} \in I \times \{t'\} \text{ and } (j_{t' \oplus 1}, \alpha_{t' \oplus 1}, t' \oplus 1) \text{ is in } \hat{P}'_{\alpha_{t'}}\}.$$

Then $t \in T$. We consider two cases.

Case 1: $T = \{0, \ldots, k-1\}$. Then for every $t' \in T = \{0, \ldots, k-1\}$, we have $\alpha_{t'} \in I \times \{t'\}$ and $(j_{t' \oplus 1}, \alpha_{t' \oplus 1}, t' \oplus 1)$ is in $\hat{P}'_{\alpha_{t'}}$. So $j_{t' \oplus 1} = 0$ and $\alpha_{t' \oplus 1}$ is in $P_{\alpha_{t'}}$ for every $t' \in \{0, \ldots, k-1\}$. Hence $(\alpha_0, \ldots, \alpha_{k-1})$ is a valid family in X. Since μ is induced by $\hat{\mu}$ and $((0,\alpha_0,0),\ldots,(0,\alpha_{k-1},k-1)) \in \hat{\mu}$, we have $(\alpha_0,\ldots,\alpha_{k-1}) \in \mu$. Thus $\mu(\alpha) = \mu(\alpha_t) = \alpha_{t\oplus 1}$, which contradicts $(0,\mu(\alpha),t\oplus 1) \neq \hat{\mu}(0,\alpha,t) = (0,\alpha_{t\oplus 1},t\oplus 1)$.

Case 2: $T \neq \{0,\ldots,k-1\}$. Then there exists a smallest $s^* \in \{1,\ldots,k-1\}$ such that $t \oplus s^* \notin T$. Then $t \oplus (s^*-1) \in T$. Let $t^* = t \oplus s^*$. Since $t^* \oplus (-1) = t\oplus(s^*-1) \in T$, we have $\alpha_{t^*\oplus(-1)} \in I\times\{t^*\oplus(-1)\}$ and $(j_{t^*},\alpha_{t^*},t^*)$ is in $\hat{P}'_{\alpha_{t^*\oplus(-1)}}$. So $j_{t^*} = 0$ and α_{t^*} is in $P_{\alpha_{t^*\oplus(-1)}}$. Hence $\alpha_{t^*} \in I \times \{t^*\}$. Since $\alpha_{t^*} \in I \times \{t^*\}$ and $t^* = t \oplus s^* \notin T$, agent $(j_{t^*\oplus 1},\alpha_{t^*\oplus 1},t^*\oplus 1)$ is not in $\hat{P}'_{\alpha_{t^*}}$. So Lemma 5 implies $\alpha_{t^*\oplus(k-1)} = \alpha_{t^*}$. Hence $\alpha_{t^*\oplus(-1)} = \alpha_{t^*\oplus(k-1)} = \alpha_{t^*} \in I \times \{t^*\}$, which contradicts $\alpha_{t^*\oplus(-1)} \in I \times \{t^* \oplus (-1)\}$. $\qquad\square$

Proof of Lemma 2. For the sake of contradiction, suppose X has no weakly stable matching and $\hat{X}$ has a weakly stable matching $\hat{\mu}$. Let μ be the matching in X induced by $\hat{\mu}$.

Since μ is not a weakly stable matching of X, there exists a strongly blocking family $(\alpha_0,\ldots,\alpha_{k-1})$. Since $\hat{\mu}$ is a weakly stable matching of $\hat{X}$, the family

$$((0,\alpha_0,0),\ldots,(0,\alpha_{k-1},k-1))$$

is not strongly blocking. So there exists $t \in \{0,\ldots,k-1\}$ such that $(0,\alpha_t,t)$ does not prefer $(0,\alpha_{t\oplus 1},t\oplus 1)$ to $\hat{\mu}(0,\alpha_t,t)$. Since $(\alpha_0,\ldots,\alpha_{k-1})$ is a family in X, agent $\alpha_{t\oplus 1}$ is in P_{α_t}. So $(0,\alpha_{t\oplus 1},t\oplus 1)$ is in $\hat{P}'_{\alpha_t}$. Hence $\hat{\mu}(0,\alpha_t,t)$ appears in $\hat{P}'_{\alpha_t}$ no later than $(0,\alpha_{t\oplus 1},t\oplus 1)$, since $\hat{P}'_{\alpha_t}$ is a prefix of the preference list $\hat{P}_{(0,\alpha_t,t)}$.

Since $\hat{\mu}(0,\alpha_t,t)$ is in $\hat{P}'_{\alpha_t}$, Lemma 6 implies $\hat{\mu}(0,\alpha_t,t) = (0,\mu(\alpha_t),t\oplus 1)$. Since $(0,\mu(\alpha_t),t\oplus 1)$ appears in $\hat{P}'_{\alpha_t}$ no later than $(0,\alpha_{t\oplus 1},t\oplus 1)$, agent $\mu(\alpha_t)$ appears in P_{α_t} no later than $\alpha_{t\oplus 1}$. Hence α_t does not prefer $\alpha_{t\oplus 1}$ to $\mu(\alpha_t)$. So $(\alpha_0,\ldots,\alpha_{k-1})$ is not a strongly blocking family of μ, a contradiction. $\qquad\square$

4.5 Proof of Lemma 3

The goal of this subsection is to prove Lemma 3. It suffices to show that every weakly stable matching μ in X induces a weakly stable matching $\hat{\mu}$ in $\hat{X}$. We construct the matching $\hat{\mu}$ induced by μ as follows.

– For every $(\alpha_0,\ldots,\alpha_{k-1}) \in \mu$, we include in $\hat{\mu}$ the family

$$((0,\alpha_0,0),\ldots,(0,\alpha_{k-1},k-1)).$$

– For every agent $\alpha \in A$ and $j \in J$ such that $j < (k-1)^2$, we include in $\hat{\mu}$ the family $((j+\delta_0(\alpha),\alpha,0),\ldots,(j+\delta_{k-1}(\alpha),\alpha,k-1))$, where

$$\delta_t(\alpha) = \begin{cases} 1 & \text{if } \mu(\alpha) \neq \alpha \text{ and } \alpha \in I \times \{t\} \\ 0 & \text{otherwise} \end{cases}$$

– For every $t \in \{0,\ldots,k-1\}$, let R_t be the list

$$\langle (j',\alpha',t') \in \{(k-1)^2\} \times A \times \{t\} \mid \delta_{t'}(\alpha') = 0 \rangle.$$

We include in $\hat{\mu}$ the family $(R_0[s],\ldots,R_{k-1}[s])$ for every $0 \le s < |A| - |\mu|$, where $R_t[s]$ denotes the $(s+1)$th element of R_t.

Figures 2(b) and (c) show the gadget under the matching $\hat{\mu}$.

It is straightforward to check that the families in $\hat{\mu}$ induced by a matching μ are agent-disjoint. Hence $\hat{\mu}$ is a valid matching in $\hat{X}$.

Lemma 7. *Let $\hat{\mu}$ be the matching in $\hat{X}$ induced by a matching μ in X. Let $t \in \{0, \dots, k-1\}$ and $\alpha \in A$ such that $\alpha \in I \times \{t\}$. Let $j' \in J$ and $\alpha' \in A$ such that non-dummy agent $(0, \alpha, t)$ prefers $(j', \alpha', t \oplus 1)$ to $\hat{\mu}(0, \alpha, t)$. Then $(j', \alpha', t \oplus 1)$ is in $\hat{P}'_\alpha$ and α prefers α' to $\mu(\alpha)$.*

Proof. Notice that $\hat{P}'_\alpha \cdot \langle (0, \alpha, t \oplus 1) \rangle$ is a prefix of the preference list $\hat{P}_{(0,\alpha,t)}$ of non-dummy agent $(0, \alpha, t)$. We consider two cases.

Case 1: $\mu(\alpha) \neq \alpha$. Then $\hat{\mu}(0, \alpha, t) = (0, \mu(\alpha), t \oplus 1)$. Since $(0, \alpha, t)$ prefers $(j', \alpha', t \oplus 1)$ to $(0, \mu(\alpha), t \oplus 1)$, agent $(j', \alpha', t \oplus 1)$ appears in $\hat{P}'_\alpha$ before $(0, \mu(\alpha), t \oplus 1)$. Hence α prefers α' to $\mu(\alpha)$.

Case 2: $\mu(\alpha) = \alpha$. Then $\hat{\mu}(0, \alpha, t) = (0, \alpha, t \oplus 1)$. Since $(0, \alpha, t)$ prefers $(j', \alpha', t \oplus 1)$ to $(0, \alpha, t \oplus 1)$, agent $(j', \alpha', t \oplus 1)$ is in $\hat{P}'_\alpha$. Then α' is in P_α, and hence α prefers α' to $\mu(\alpha)$.   □

Lemma 8. *Let $\hat{\mu}$ be the matching in $\hat{X}$ induced by a weakly stable matching μ in X. Let $j_0, \dots, j_{k-1} \in J$ and $\alpha_0, \dots, \alpha_{k-1} \in A$ such that*

$$((j_0, \alpha_0, 0), \dots, (j_{k-1}, \alpha_{k-1}, k-1))$$

is a strongly blocking family of $\hat{\mu}$. Then $j_t - \delta_t(\alpha_t) \geq (k-1)^2$ for every $t \in \{0, \dots, k-1\}$.

Proof. Let $t^* \in \{0, \dots, k-1\}$ such that

$$j_{t^*} - \delta_{t^*}(\alpha_{t^*}) = \min_{t \in \{0, \dots, k-1\}} (j_t - \delta_t(\alpha_t)).$$

For the sake of contradiction, suppose $j_{t^*} - \delta_{t^*}(\alpha_{t^*}) < (k-1)^2$. We consider two cases.

Case 1: $j_{t^*} = 0$ and $\alpha_{t^*} \in I \times \{t^*\}$. Let $T = \{t \mid j_t = 0 \text{ and } \alpha_t \in I \times \{t\}\}$. Then $t^* \in T$. We consider two subcases.

Case 1.1: $T = \{0, \dots, k-1\}$. Then for every $t \in \{0, \dots, k-1\} = T$, since $(0, \alpha_t, t)$ prefers $(0, \alpha_{t \oplus 1}, t \oplus 1)$ to $\hat{\mu}(0, \alpha_t, t)$, Lemma 7 implies that α_t prefers $\alpha_{t \oplus 1}$ to $\mu(\alpha_t)$. Hence $(\alpha_0, \dots, \alpha_{k-1})$ is a strongly blocking family of μ, which contradicts the stability of μ.

Case 1.2: $\{t^*\} \subseteq T \subsetneq \{0, \dots, k-1\}$. Then there exists s^* such that $s^* \in T$ and $s^* \oplus 1 \notin T$. Since $s^* \in T$, we have $j_{s^*} = 0$ and $\alpha_{s^*} \in I \times \{s^*\}$. Since $(0, \alpha_{s^*}, s^*)$ prefers $(j_{s^* \oplus 1}, \alpha_{s^* \oplus 1}, s^* \oplus 1)$ to $\hat{\mu}(0, \alpha_{s^*}, s^*)$, Lemma 7 implies that $(j_{s^* \oplus 1}, \alpha_{s^* \oplus 1}, s^* \oplus 1)$ is in $\hat{P}'_{\alpha_{s^*}}$. Hence $j_{s^* \oplus 1} = 0$ and $\alpha_{s^* \oplus 1} \in I \times \{s^* \oplus 1\}$, which contradicts $s^* \oplus 1 \notin T$.

Case 2: Either $j_{t^*} \neq 0$ or $\alpha_{t^*} \notin I \times \{t^*\}$. Thus $(j_{t^*}, \alpha_{t^*}, t^*)$ is a dummy agent. We consider two subcases.

Case 2.1: $j_{t^*} < (k-1)^2$. Since

$$\hat{\mu}(j_{t^*}, \alpha_{t^*}, t^*) = (j_{t^*} + \delta_{t^* \oplus 1}(\alpha_{t^*}) - \delta_{t^*}(\alpha_{t^*}), \alpha_{t^*}, t^* \oplus 1),$$

and the non-boundary dummy agent $(j_{t^*}, \alpha_{t^*}, t^*)$ prefers $(j_{t^* \oplus 1}, \alpha_{t^* \oplus 1}, t^* \oplus 1)$ to $\hat{\mu}(j_{t^*}, \alpha_{t^*}, t^*)$, we have $j_{t^* \oplus 1} < j_{t^*} + \delta_{t^* \oplus 1}(\alpha_{t^*}) - \delta_{t^*}(\alpha_{t^*})$, which contradicts the definition of t^*.

Case 2.2: $j_{t^*} = (k-1)^2$. Then $\delta_{t^*}(\alpha_{t^*}) = 1$ since $j_{t^*} - \delta_{t^*}(\alpha_{t^*}) < (k-1)^2$. So $\alpha_{t^*} \in I \times \{t^*\}$, and hence $\delta_{t^* \oplus 1}(\alpha_{t^*}) = 0$. Since

$$\hat{\mu}(j_{t^*}, \alpha_{t^*}, t^*) = (j_{t^*} - 1, \alpha_{t^*}, t^* \oplus 1)$$

and the boundary dummy agent $(j_{t^*}, \alpha_{t^*}, t^*)$ prefers $(j_{t^* \oplus 1}, \alpha_{t^* \oplus 1}, t^* \oplus 1)$ to $\hat{\mu}(j_{t^*}, \alpha_{t^*}, t^*)$, we have $j_{t^* \oplus 1} < j_{t^*} - 1 = j_{t^*} + \delta_{t^* \oplus 1}(\alpha_{t^*}) - \delta_{t^*}(\alpha_{t^*})$, which contradicts the definition of t^*. □

Proof of Lemma 3. Suppose X has a weakly stable matching μ. Let $\hat{\mu}$ be the matching in $\hat{X}$ induced by μ. It suffices to show that $\hat{\mu}$ does not admit a strongly blocking family.

For the sake of contradiction, suppose $\hat{\mu}$ admits a strongly blocking family

$$((j_0, \alpha_0, 0), \ldots, (j_{k-1}, \alpha_{k-1}, k-1)).$$

Lemma 8 implies that for every $t \in \{0, \ldots, k-1\}$, we have $j_t - \delta_t(\alpha_t) \geq (k-1)^2$. Since $j_t \leq (k-1)^2$ and $\delta_t(\alpha_t) \geq 0$, we deduce that $j_t = (k-1)^2$ and $\delta_t(\alpha_t) = 0$ for every $t \in \{0, \ldots, k-1\}$. Hence for every $t \in \{0, \ldots, k-1\}$, there exists s_t such that $(j_t, \alpha_t, t) = R_t[s_t]$.

Let $t^* \in \{0, \ldots, k-1\}$ such that

$$s_{t^*} = \min_{t \in \{0,\ldots,k-1\}} s_t.$$

Since $\hat{\mu}(R_{t^*}[s_{t^*}]) = R_{t^* \oplus 1}[s_{t^*}]$ and the boundary dummy agent $R_{t^*}[s_{t^*}]$ prefers boundary dummy agent $R_{t^* \oplus 1}[s_{t^* \oplus 1}]$ to boundary dummy agent $\hat{\mu}(R_{t^*}[s_{t^*}])$, we deduce that $R_{t^* \oplus 1}[s_{t^* \oplus 1}]$ is lexicographically smaller than $R_{t^* \oplus 1}[s_{t^*}]$. Hence $s_{t^* \oplus 1} < s_{t^*}$, which contradicts the definition of t^*. □

5 Concluding Remarks

We have shown that a 3-DSM-CYC instance need not admit a weakly stable matching, and that it is NP-complete to determine whether a given 3-DSM-CYC instance admits a weakly stable matching. It seems that for the three-dimensional stable matching problem, none of the preference structures studied in the literature admits a non-trivial generalization of the existence theorem of Gale and Shapley. (The existence result in Danilov's model [5] follows from applying the Gale-Shapley algorithm in a straightforward manner.) It would be interesting to consider solution concepts such as popular matchings instead of stable matchings in the multi-dimensional matching context.

The 3-DSMI-CYC instance with no weakly stable matching presented by Biró and McDermid [2, Lemma 1] has 6 agents of each type. The reduction of Sect. 4.1 blows up the number of agents by a factor of $k[(k-1)^2+1]$. Thus, for $k=3$, we obtain an explicit construction of a 3-DSM-CYC instance with no weakly stable matching and $6 \cdot 15 = 90$ agents of each type. It would be interesting to identify smaller 3-DSM-CYC instances with no weakly stable matching.

References

1. Alkan, A.: Nonexistence of stable threesome matchings. Math. Soc. Sci. **16**(2), 207–209 (1988)
2. Biró, P., McDermid, E.: Three-sided stable matchings with cyclic preferences. Algorithmica **58**(1), 5–18 (2010)
3. Boros, E., Gurvich, V., Jaslar, S., Krasner, D.: Stable matchings in three-sided systems with cyclic preferences. Discrete Math. **289**(1), 1–10 (2004)
4. Cui, L., Jia, W.: Cyclic stable matching for three-sided networking services. Comput. Netw. **57**(1), 351–363 (2013)
5. Danilov, V.I.: Existence of stable matchings in some three-sided systems. Math. Soc. Sci. **46**(2), 145–148 (2003)
6. Eriksson, K., Sjöstrand, J., Strimling, P.: Three-dimensional stable matching with cyclic preferences. Math. Soc. Sci. **52**(1), 77–87 (2006)
7. Escamocher, G., O'Sullivan, B.: Three-dimensional matching instances are rich in stable matchings. In: van Hoeve, W.-J. (ed.) CPAIOR 2018. LNCS, vol. 10848, pp. 182–197. Springer, Cham (2018). https://doi.org/10.1007/978-3-319-93031-2_13
8. Farczadi, L., Georgiou, K., Könemann, J.: Stable marriage with general preferences. Theory Comput. Syst. **59**(4), 683–699 (2016)
9. Gale, D., Shapley, L.S.: College admissions and the stability of marriage. Am. Math. Mon. **69**(1), 9–15 (1962)
10. Hofbauer, J.: d-dimensional stable matching with cyclic preferences. Math. Soc. Sci. **82**, 72–76 (2016)
11. Huang, C.-C.: Two's company, three's a crowd: stable family and threesome roommates problems. In: Arge, L., Hoffmann, M., Welzl, E. (eds.) ESA 2007. LNCS, vol. 4698, pp. 558–569. Springer, Heidelberg (2007). https://doi.org/10.1007/978-3-540-75520-3_50
12. Huang, C.C.: Circular stable matching and 3-way kidney transplant. Algorithmica **58**(1), 137–150 (2010)
13. Knuth, D.E.: Stable Marriage and Its Relation to Other Combinatorial Problems: An Introduction to the Mathematical Analysis of Algorithms. American Mathematical Society, Providence (1997)
14. Lam, C.K., Plaxton, C.G.: On the existence of three-dimensional stable matchings with cyclic preferences (2019). arXiv:1905.02844
15. Manlove, D.F.: Algorithmics of Matching Under Preferences. World Scientific, Singapore (2013)
16. Ng, C., Hirschberg, D.: Three-dimensional stable matching problems. SIAM J. Discrete Math. **4**(2), 245–252 (1991)
17. Pashkovich, K., Poirrier, L.: Three-dimensional stable matching with cyclic preferences (2018). arXiv:1807.05638
18. Subramanian, A.: A new approach to stable matching problems. SIAM J. Comput. **23**(4), 671–700 (1994)
19. Woeginger, G.J.: Core stability in hedonic coalition formation. In: van Emde Boas, P., Groen, F.C.A., Italiano, G.F., Nawrocki, J., Sack, H. (eds.) SOFSEM 2013. LNCS, vol. 7741, pp. 33–50. Springer, Heidelberg (2013). https://doi.org/10.1007/978-3-642-35843-2_4

Maximum Stable Matching
with One-Sided Ties of Bounded Length

Chi-Kit Lam and C. Gregory Plaxton[(✉)]

University of Texas at Austin, Austin, TX 78712, USA
{geocklam,plaxton}@cs.utexas.edu

Abstract. We study the problem of finding maximum weakly stable matchings when preference lists are incomplete and contain one-sided ties of bounded length. We show that if the tie length is at most L, then it is possible to achieve an approximation ratio of $1 + (1 - \frac{1}{L})^L$. We also show that the same ratio is an upper bound on the integrality gap, which matches the known lower bound. In the case where the tie length is at most 2, our result implies an approximation ratio and integrality gap of $\frac{5}{4}$, which matches the known UG-hardness result.

Keywords: Stable matching · Approximation algorithm · Integrality gap

1 Introduction

The stable matching model of Gale and Shapley [4] involves a two-sided market in which the agents are typically called men and women. Each agent has ordinal preferences over the agents on the opposite side. A matching is said to be stable if no man and woman prefer each other to their partners. Stable matchings always exist and can be computed efficiently by the proposal algorithm of Gale and Shapley. Their algorithm is also applicable when the preference lists are incomplete. In other words, agents are allowed to omit from their preference lists any unacceptable agent on the opposite side. If ties are allowed in the preference lists, the notion of stability can be generalized in several ways [9]. This paper focuses on weakly stable matchings, which always exist and can be obtained by invoking the Gale-Shapley algorithm after breaking all the ties arbitrarily. When incomplete lists are absent, every weakly stable matching is a maximum matching and hence has the same size. When ties are absent, the Rural Hospital Theorem guarantees that all stable matchings have the same size [5,17]. However, when both ties and incomplete lists are present, weakly stable matchings can vary in size.

The problem of finding maximum weakly stable matchings with ties and incomplete lists has been studied in various settings. When ties and incomplete lists are allowed on both sides, there exist polynomial-time algorithms [11,14,15] that achieve an approximation ratio of $\frac{3}{2}$ (=1.5). Meanwhile, it is known [20] that getting an approximation ratio of $\frac{33}{29} - \varepsilon$ ($\approx$1.1379) is NP-hard, and that getting

© Springer Nature Switzerland AG 2019
D. Fotakis and E. Markakis (Eds.): SAGT 2019, LNCS 11801, pp. 343–356, 2019.
https://doi.org/10.1007/978-3-030-30473-7_23

an approximation ratio $\frac{4}{3} - \varepsilon$ (≈ 1.3333) is UG-hard. These hardness results hold in the case of two-sided ties even when the maximum tie length is two. The associated linear programming (LP) formulation has an integrality gap of at least $\frac{3L-2}{2L-1}$, where L is the maximum tie length [10].

For the case where ties appear only on one side of the market, algorithms with better approximation ratios have been developed using an LP-based approach [3,10,12] or the idea of rounding half-integral stable matchings [1,8,16]. The current best approximation ratio of $1 + \frac{1}{e}$ (≈ 1.3679) is attained by the LP-based algorithm that the authors recently presented in [12]. Meanwhile, it is known [7] that getting an approximation ratio of $\frac{21}{19} - \varepsilon$ (≈ 1.1053) is NP-hard, and that getting an approximation ratio of $\frac{5}{4} - \varepsilon$ (≈ 1.25) is UG-hard. These hardness results hold in the case of one-sided ties even when the maximum tie length is two. The associated LP formulation has an integrality gap of at least $1 + (1 - \frac{1}{L})^L$, where L is the maximum tie length [10]. Furthermore, for the case of one-sided ties with unbounded tie length, the integrality gap equals $1 + \frac{1}{e}$ and matches the attainable approximation ratio [12].

For the case of two-sided ties where the maximum tie length is two, Chiang and Pashkovich [2] showed that the algorithm of Huang and Kavitha [8] attains an approximation ratio of $\frac{4}{3}$ (≈ 1.3333), which matches the UG-hardness result [20] and the lower bound of the integrality gap [10]. A couple of results [6,7] are known for the case of one-sided ties with bounded tie length, but they are subsumed by the approximation ratio of $1 + \frac{1}{e}$ for the case of one-sided ties with unbounded tie length.

Our Techniques and Contributions. In this paper, we focus on the problem of finding maximum weakly stable matchings with one-sided ties and incomplete lists when the tie length is bounded. We show that the algorithm of [12] achieves an approximation ratio of $1 + (1 - \frac{1}{L})^L$, where L is the maximum tie length. We also show that the same ratio is an upper bound on the integrality gap, which matches the lower bound of Iwama et al. [10]. For the case where $L = 2$, our result implies an approximation ratio and integrality gap of $\frac{5}{4}$, which matches the UG-hardness result of Halldórsson et al. [7].

Our analysis is based on four key properties established in [12]. Using these key properties, we extend the analysis of the approximation ratio to the case of bounded tie length. Moreover, we present a new, simpler charging argument. The main idea is to decompose the LP solution associated with each man-woman pair into a charge incurred by the man and a charge incurred by the woman based on an exchange function. We derive an upper bound for the charges incurred by a man using the strict ordering of his preferences, and an upper bound for the charges incurred by a woman using the bounded tie length assumption. By choosing a good exchange function, we show that every matched couple incurs a total charge of at most $1 + (1 - \frac{1}{L})^L$, providing an upper bound on the approximation ratio.

In Sect. 2, we review the key properties of the algorithm of [12] after presenting the stable matching model and its LP formulation. In Sect. 3, we present our simpler charging argument and use it to analyze the approximation ratio for the case of bounded tie length.

2   Stable Matching with One-Sided Ties

2.1   The Model

The formal definition of the stable matching problem with one-sided ties and incomplete lists (SMOTI) below follows the notations of [12].

In SMOTI, there are a set I of men and a set J of women. We assume that the sets I and J are disjoint and do not contain the element 0, which we use to denote being unmatched. Each man $i \in I$ has a preference relation $\geq_i$ over the set $J \cup \{0\}$ that satisfies antisymmetry, transitivity, and totality. Each woman $j \in J$ has a preference relation $\geq_j$ over the set $I \cup \{0\}$ that satisfies transitivity and totality. We denote this SMOTI instance as $(I, J, \{\geq_i\}_{i \in I}, \{\geq_j\}_{j \in J})$.

For every man $i \in I$ and woman $j \in J$, man i is said to be acceptable to woman j if $i \geq_j 0$. Similarly, woman j is said to be acceptable to man i if $j \geq_i 0$. The preference lists are allowed to be incomplete. In other words, there may exist $i \in I$ and $j \in J$ such that $0 >_j i$ or $0 >_i j$.

Notice that the preference relations $\{\geq_j\}_{j \in J}$ of the women are not required to be antisymmetric, while the preference relations $\{\geq_i\}_{i \in I}$ of the men are required to be antisymmetric. For every man $i \in I$, we write $>_i$ to denote the asymmetric part of $\geq_i$. For every woman $j \in J$, we write $>_j$ and $=_j$ to denote the asymmetric part and the symmetric part of $\geq_j$, respectively. A *tie* in the preference list of woman j is an equivalence class of size at least 2 with respect to the equivalence relation $=_j$, and the *length* of a tie is the size of this equivalence class.[1] We assume that there is at least one tie in the SMOTI instance, for otherwise every stable matching has the same size. We use L to denote the maximum length of the ties in the preference lists of the women, where $2 \leq L \leq |I| + 1$.

A matching is a subset $\mu \subseteq I \times J$ such that for every $(i, j), (i', j') \in \mu$, we have $i = i'$ if and only if $j = j'$. For every man $i \in I$, if $(i, j) \in \mu$ for some woman $j \in J$, we say that man i is matched to woman j in matching μ, and we write $\mu(i) = j$. Otherwise, we say that man i is unmatched in matching μ, and we write $\mu(i) = 0$. Similarly, for every woman $j \in J$, if $(i, j) \in \mu$ for some man $i \in I$, we say that woman j is matched to man i in matching μ, and we write $\mu(j) = i$. Otherwise, we say that woman j is unmatched in matching μ, and we write $\mu(j) = 0$.

A matching μ is *individually rational* if for every $(i, j) \in \mu$, we have $j \geq_i 0$ and $i \geq_j 0$. An individually rational matching μ is *weakly stable* if for every man $i \in I$ and woman $j \in J$, either $\mu(i) \geq_i j$ or $\mu(j) \geq_j i$. Otherwise, (i, j) forms a *strongly blocking pair*.

---

[1] Some of the literature on stable matching with indifferences does not allow an agent to be indifferent between being matched to an agent and being unmatched. Our formulation of the SMOTI problem allows for this possibility, since we can have $i =_j 0$ for any man i and woman j.

The goal of the maximum stable matching problem with one-sided ties and incomplete lists is to find a maximum-cardinality weakly stable matching for a given SMOTI instance.

2.2   The LP Formulation

The following LP formulation is based on that of Rothblum [18], which extends that of Vande Vate [19].

$$\text{maximize} \qquad \sum_{(i,j)\in I\times J} x_{i,j}$$

$$\text{subject to} \qquad \sum_{j\in J} x_{i,j} \le 1 \qquad \forall i \in I \tag{C1}$$

$$\sum_{i\in I} x_{i,j} \le 1 \qquad \forall j \in J \tag{C2}$$

$$\sum_{\substack{j'\in J \\ j'>_i j}} x_{i,j'} + \sum_{\substack{i'\in I \\ i'\ge_j i}} x_{i',j} \ge 1 \qquad \begin{array}{l}\forall (i,j)\in I\times J \text{ such that} \\ j >_i 0 \text{ and } i >_j 0\end{array} \tag{C3}$$

$$x_{i,j} = 0 \qquad \begin{array}{l}\forall (i,j)\in I\times J \text{ such that} \\ 0 >_i j \text{ or } 0 >_j i\end{array} \tag{C4}$$

$$x_{i,j} \ge 0 \qquad \forall (i,j)\in I\times J \tag{C5}$$

It is known [12,18] that an integral solution $\mathbf{x} = \{x_{i,j}\}_{(i,j)\in I\times J}$ corresponds to the indicator variables of a weakly stable matching if and only if $\mathbf{x}$ satisfies constraints (C1)–(C5).

Given $\mathbf{x}$ which satisfies constraints (C1)–(C5), it is useful to define auxiliary variables

$$w_{i,j} = \begin{cases} 1 & \text{if } j = 0 \\ \displaystyle\sum_{\substack{j'\in J \\ j'>_i j}} x_{i,j'} & \text{if } j \ne 0 \end{cases}$$

for every $(i,j) \in I \times (J \cup \{0\})$, and

$$z_{i,j} = \sum_{\substack{i'\in I \\ i>_j i'}} x_{i',j}$$

for every $(i,j) \in (I \cup \{0\}) \times J$. The following lemma presents some simple properties of the auxiliary variables; see [13] for a proof.

Lemma 1. *The auxiliary variables satisfy the following conditions.*

(1) For every $i \in I$ and $j \in J$, we have $w_{i,j} + x_{i,j} \le 1$.
(2) For every $i \in I$ and $j, j' \in J$ such that $j >_i j'$, we have $w_{i,j} + x_{i,j} \le w_{i,j'}$.
(3) For every $i, i' \in I \cup \{0\}$ and $j \in J$ such that $i =_j i'$, we have $z_{i,j} = z_{i',j}$.
(4) For every $i \in I$ and $j \in J$ such that $j \ge_i 0$ and $i \ge_j 0$, we have $z_{i,j} \le w_{i,j}$.

2.3   The LP-Based Algorithm

Using the LP formulation of Sect. 2.2, the authors have previously established in [12] that there exists a polynomial-time algorithm with an approximation ratio of $1 + \frac{1}{e}$. The algorithm is based on a proposal process in which every man i maintains a priority p_i that gradually increases from 0 to 1. Between two successive increases of the priority of a man i, he attempts to propose to the set of women $\{j \in J \colon j \geq_i 0 \text{ and } p_i \geq w_{i,j}\}$ in decreasing order of his preference, where $w_{i,j}$ is the auxiliary variable corresponding to a fixed optimal fractional solution $\mathbf{x}$ of the LP. Each woman compares the men based on her preferences and breaks the ties by favoring men with higher priorities. The algorithm simulates this process in which the step size of the priority increases is infinitesimally small. More precisely, the algorithm runs in polynomial time and produces a weakly stable matching μ and priority values $\mathbf{p} = \{p_i\}_{i \in I}$ satisfying the following key properties [12, Lemmas 3.1 and 3.3].

(P1) Let $(i,j) \in \mu$. Then $j \geq_i 0$ and $i \geq_j 0$.

(P2) Let $i \in I$ be a man and $j \in J$ be a woman such that $j \geq_i \mu(i)$ and $i \geq_j 0$. Then $\mu(j) \neq 0$ and $\mu(j) \geq_j i$.

(P3) Let $i \in I$ be a man. Then $w_{i,\mu(i)} \leq p_i \leq 1$.

(P4) Let $i \in I$ be a man and $j \in J$ be a woman such that $j \geq_i 0$ and $i \geq_j 0$. Suppose $p_i - \eta > w_{i,j}$. Then $\mu(j) \neq 0$ and $\mu(j) \geq_j i$. Furthermore, if $\mu(j) =_j i$, then $p_{\mu(j)} \geq p_i$.

In [12], a rather complicated charging argument is used to obtain an approximation ratio of $1 + \frac{1}{e}$ by showing that the optimal fractional value of the LP is at most $1 + \frac{1}{e}$ times the size of any matching μ satisfying (P1)–(P4) with respect to some $\mathbf{p}$.

3   Analysis of the Approximation Ratio

In this section, we analyze the approximation ratio of the algorithm of [12] for the case where the maximum tie length is L. Throughout this section, whenever we mention μ and $\mathbf{p}$, we are referring to their values produced by their algorithm. We use $\mathbf{x}$ to refer to the optimal fractional solution of the LP in their algorithm, and we use $\{w_{i,j}\}_{(i,j) \in I \times (J \cup \{0\})}$ and $\{z_{i,j}\}_{(i,j) \in (I \cup \{0\}) \cup J}$ to refer to the auxiliary variables associated with $\mathbf{x}$ as defined in Sect. 2.2.

3.1   The Charging Argument

Our charging argument is based on an exchange function $h \colon [0,1] \times [0,1] \to \mathbb{R}$ that satisfies the following properties.

(H1) For every $\xi_1, \xi_2 \in [0,1]$, we have $0 = h(0, \xi_2) \leq h(\xi_1, \xi_2) \leq 1$.

(H2) For every $\xi_1, \xi_2 \in [0,1]$ such that $\xi_1 > \xi_2$, we have $h(\xi_1, \xi_2) = 1$.

(H3) The function $h(\xi_1, \xi_2)$ is non-decreasing in ξ_1 and non-increasing in ξ_2.

(H4) For every $\xi_1, \xi_2 \in [0, 1]$, we have

$$L \cdot \int_{\xi_2 \cdot (1-1/L)}^{\xi_2} \left(1 - h(\xi_1, \xi)\right) d\xi \leq \max(\xi_2 - \xi_1, 0).$$

Given an exchange function h which satisfies (H1)–(H4), our charging argument is as follows. For every $(i, j) \in I \times J$, we assign to man i a charge of

$$\theta_{i,j} = \int_0^{x_{i,j}} h(1 - p_i, 1 - w_{i,j} - \xi) \, d\xi$$

and to woman j a charge of

$$\phi_{i,j} = \begin{cases} 0 & \text{if } \mu(j) = 0 \text{ or } i >_j \mu(j) \\ x_{i,j} & \text{if } \mu(j) \neq 0 \text{ and } \mu(j) >_j i \\ x_{i,j} - \int_0^{x_{i,j}} h(1 - p_{\mu(j)}, 1 - z_{\mu(j),j} - \xi) \, d\xi & \text{if } \mu(j) \neq 0 \text{ and } \mu(j) =_j i \end{cases}$$

The following lemma shows that the charges are non-negative and cover the value of LP solution.

Lemma 2. *Let $i \in I$ and $j \in J$. Then $\theta_{i,j}$ and $\phi_{i,j}$ satisfy the following conditions.*

(1) $\theta_{i,j} \geq 0$ and $\phi_{i,j} \geq 0$.
(2) $x_{i,j} \leq \theta_{i,j} + \phi_{i,j}$.

Proof. Part (1) is relatively straightforward to establish; see [13] for a proof. We prove part (2) by considering two cases.
    Case 1: $p_i \leq w_{i,j}$. Then (H3) implies

$$0 \leq \int_0^{x_{i,j}} \left(h(1 - p_i, 1 - w_{i,j} - \xi) - h(1 - p_i, 1 - p_i - \xi)\right) d\xi$$

$$= \int_0^{x_{i,j}} \left(h(1 - p_i, 1 - w_{i,j} - \xi) - 1\right) d\xi$$

$$= \theta_{i,j} - x_{i,j}$$

$$\leq \theta_{i,j} + \phi_{i,j} - x_{i,j},$$

where the first equality follows from (H2), the second equality follows from the definition of $\theta_{i,j}$, and the last inequality follows from part (1).
    Case 2: $p_i > w_{i,j}$. We may assume that $x_{i,j} \neq 0$, for otherwise part (1) implies $\theta_{i,j} + \phi_{i,j} \geq 0 = x_{i,j}$. Since $x_{i,j} \neq 0$, constraint (C4) implies $j \geq_i 0$ and $i \geq_j 0$. So (P4) implies $\mu(j) \neq 0$ and $\mu(j) \geq_j i$. We consider two subcases.
    Case 2.1: $\mu(j) >_j i$. Then the definition of $\phi_{i,j}$ implies

$$0 = \phi_{i,j} - x_{i,j} \leq \theta_{i,j} + \phi_{i,j} - x_{i,j}$$

where the inequality follows from part (1).

Case 2.2: $\mu(j) =_j i$. Then (P4) implies $p_i \leq p_{\mu(j)}$. Also, since $\mu(j) =_j i$, parts (3) and (4) of Lemma 1 imply $z_{\mu(j),j} = z_{i,j} \leq w_{i,j}$. Since $p_i \leq p_{\mu(j)}$ and $w_{i,j} \geq z_{\mu(j),j}$, (H3) implies

$$0 \leq \int_0^{x_{i,j}} \left(h(1 - p_i, 1 - w_{i,j} - \xi) - h(1 - p_{\mu(j)}, 1 - z_{\mu(j),j} - \xi) \right) d\xi$$

$$= \theta_{i,j} + \phi_{i,j} - x_{i,j},$$

where the equality follows from the definitions of $\theta_{i,j}$ and $\phi_{i,j}$. $\square$

3.2 Bounding the Charges

To bound the approximation ratio, Lemma 2 implies that it is sufficient to bound the charges. In Lemma 3, we derive an upper bound for the charges incurred by a man using the strict ordering in his preferences. In Lemma 4, we derive an upper bound for the charges incurred by a woman due to indifferences using the bounded tie length assumption. In Lemma 5, we derive an upper bound for the total charges incurred by a matched couple by combining the results of Lemmas 3 and 4.

Lemma 3. *Let $i \in I$ be a man. Then*

$$\sum_{j \in J} \theta_{i,j} \leq \int_0^1 h(1 - p_i, \xi) \, d\xi.$$

Proof. Let $j_1, \ldots, j_{|J|} \in J$ such that $j_1 >_i j_2 >_i \cdots >_i j_{|J|}$. Then parts (1) and (2) of Lemma 1 imply

$$w_{i,j_k} + x_{i,j_k} \leq \begin{cases} w_{i,j_{k+1}} & \text{if } 1 \leq k < |J| \\ 1 & \text{if } k = |J| \end{cases} \tag{1}$$

Hence the definitions of $\{\theta_{i,j_k}\}_{1 \leq k \leq |J|}$ imply

$$\theta_{i,j_k} = \int_0^{x_{i,j_k}} h(1 - p_i, 1 - w_{i,j_k} - \xi) \, d\xi$$

$$= \int_{w_{i,j_k}}^{w_{i,j_k} + x_{i,j_k}} h(1 - p_i, 1 - \xi) \, d\xi$$

$$\leq \begin{cases} \int_{w_{i,j_k}}^{w_{i,j_{k+1}}} h(1 - p_i, 1 - \xi) \, d\xi & \text{if } 1 \leq k < |J| \\ \int_{w_{i,j_{|J|}}}^{1} h(1 - p_i, 1 - \xi) \, d\xi & \text{if } k = |J| \end{cases}$$

where the inequality follows from (1) and (H1). Thus

$$
\begin{aligned}
\sum_{j\in J}\theta_{i,j} &= \sum_{1\leq k\leq |J|}\theta_{i,j_k} \\
&\leq \int_{w_{i,j_{|J|}}}^{1} h(1-p_i,1-\xi)\,d\xi + \sum_{1\leq k<|J|}\int_{w_{i,j_k}}^{w_{i,j_{k+1}}} h(1-p_i,1-\xi)\,d\xi \\
&= \int_{w_{i,j_1}}^{1} h(1-p_i,1-\xi)\,d\xi \\
&\leq \int_{0}^{1} h(1-p_i,1-\xi)\,d\xi \\
&= \int_{0}^{1} h(1-p_i,\xi)\,d\xi,
\end{aligned}
$$

where the second inequality follows from $w_{i,j_1}\geq 0$ and (H1). $\qquad\square$

Lemma 4. Let $j\in J$ be a woman such that $\mu(j)\neq 0$. Then

$$
\sum_{\substack{i\in I\\ \mu(j)=_j i}} \phi_{i,j} \leq \max(p_{\mu(j)}-z_{\mu(j),j},0).
$$

Proof. Let

$$
H(\xi') = \int_{1-z_{\mu(j),j}-\xi'}^{1-z_{\mu(j),j}} \left(1-h(1-p_{\mu(j)},\xi)\right)d\xi
$$

for every $\xi'\in[0,1]$. Then (H1) and (H3) imply that H is concave and non-decreasing. Also (H4) implies

$$
L\cdot H\!\left(\frac{1-z_{\mu(j),j}}{L}\right) = L\cdot\int_{(1-z_{\mu(j),j})(1-1/L)}^{1-z_{\mu(j),j}} \left(1-h(1-p_{\mu(j)},\xi)\right)d\xi
$$
$$
\leq \max(p_{\mu(j)}-z_{\mu(j),j},0). \tag{2}
$$

Let $I'=\{i\in I:\mu(j)=_j i\}$. Then $|I'|\leq L$ since L is the maximum tie-length. Let $i_1,\ldots,i_{|I'|}\in I$ such that $I'=\{i_1,\ldots,i_{|I'|}\}$. Let

$$
\xi_k = \begin{cases} x_{i_k,j} & \text{if } 1\leq k\leq |I'| \\ 0 & \text{if } |I'|<k\leq L \end{cases}
$$

Then the definition of $z_{\mu(j),j}$ implies

$$
1-z_{\mu(j),j} = 1-\sum_{\substack{i\in I\\ \mu(j)>_j i}} x_{i,j} \geq \sum_{i\in I} x_{i,j} - \sum_{\substack{i\in I\\ \mu(j)>_j i}} x_{i,j} \geq \sum_{\substack{i\in I\\ \mu(j)=_j i}} x_{i,j} = \sum_{1\leq k\leq L}\xi_k,
$$

where the first inequality follows from constraint (C2), and the second equality follows from the definitions of $\{\xi_k\}_{1 \le k \le L}$. Hence the monotonicity and concavity of H imply

$$L \cdot H\left(\frac{1 - z_{\mu(j),j}}{L}\right) \ge L \cdot H\left(\frac{1}{L} \sum_{1 \le k \le L} \xi_k\right) \ge \sum_{1 \le k \le L} H(\xi_k). \tag{3}$$

Thus the definitions of $\{\phi_{i,j}\}_{i \in I}$ imply

$$\sum_{\substack{i \in I \\ \mu(j)=_j i}} \phi_{i,j} = \sum_{\substack{i \in I \\ \mu(j)=_j i}} \left(x_{i,j} - \int_0^{x_{i,j}} h(1 - p_{\mu(j)}, 1 - z_{\mu(j),j} - \xi)\, d\xi\right)$$

$$= \sum_{\substack{i \in I \\ \mu(j)=_j i}} \int_{1 - z_{\mu(j),j} - x_{i,j}}^{1 - z_{\mu(j),j}} \left(1 - h(1 - p_{\mu(j)}, \xi)\right) d\xi$$

$$= \sum_{\substack{i \in I \\ \mu(j)=_j i}} H(x_{i,j})$$

$$= \sum_{1 \le k \le L} H(\xi_k)$$

$$\le L \cdot H\left(\frac{1 - z_{\mu(j),j}}{L}\right)$$

$$\le \max(p_{\mu(j)} - z_{\mu(j),j}, 0),$$

where the third equality follows from the definition of H, the fourth equality follows from the definitions of $\{\xi_k\}_{1 \le k \le L}$, the first inequality follows from (3), and the second inequality follows from (2). $\qquad\square$

Lemma 5. *Let $i \in I$ and $j \in J \cup \{0\}$ such that $\mu(i) = j$. Then the following conditions hold.*

(1) If $j \ne 0$, then

$$\sum_{j' \in J} \theta_{i,j'} + \sum_{i' \in I} \phi_{i',j} \le 1 + \int_{1 - p_i}^1 h(1 - p_i, \xi)\, d\xi.$$

(2) If $j = 0$, then $\theta_{i,j'} = 0$ for every $j' \in J$.

Proof.

(1) Suppose $j \ne 0$. Then (P1) implies $j \ge_i 0$ and $i \ge_j 0$. So part (4) of Lemma 1 implies

$$z_{i,j} \le w_{i,j} \le p_i,$$

where the second inequality follows from (P3). So the definitions of $\{\phi_{i',j}\}_{i' \in I}$ imply

$$\sum_{i' \in I} \phi_{i',j} = \sum_{\substack{i' \in I \\ \mu(j)=_j i'}} \phi_{i',j} + \sum_{\substack{i' \in I \\ \mu(j)>_j i'}} x_{i',j} \le \max(p_i - z_{i,j}, 0) + z_{i,j} = p_i, \tag{4}$$

where the first inequality follows from Lemma 4 and the definition of $z_{i,j}$, and the last equality follows from $p_i \geq z_{i,j}$. Also, by Lemma 3, we have

$$
\begin{aligned}
\sum_{j' \in J} \theta_{i,j'} &\leq \int_0^1 h(1 - p_i, \xi)\, d\xi \\
&= \int_0^{1-p_i} h(1 - p_i, \xi)\, d\xi + \int_{1-p_i}^1 h(1 - p_i, \xi)\, d\xi \\
&= \int_0^{1-p_i} 1\, d\xi + \int_{1-p_i}^1 h(1 - p_i, \xi)\, d\xi \\
&= 1 - p_i + \int_{1-p_i}^1 h(1 - p_i, \xi)\, d\xi,
\end{aligned}
\tag{5}
$$

where the second equality follows from (H2). Combining (4) and (5) gives the desired inequality.

(2) Suppose $j = 0$. Let $j' \in J$. Since $\mu(i) = j = 0$, (P3) implies

$$
1 \geq p_i \geq w_{i,0} = 1,
$$

where the last equality follows from the definition of $w_{i,0}$. Hence the definition of $\theta_{i,j'}$ implies

$$
\theta_{i,j'} = \int_0^{x_{i,j'}} h(1 - p_i, 1 - w_{i,j'} - \xi)\, d\xi = \int_0^{x_{i,j'}} h(0, 1 - w_{i,j'} - \xi)\, d\xi = 0,
$$

where the second equality follows from $p_i = 1$, and the third equality follows from (H1).   □

3.3   The Approximation Ratio

To obtain the approximation ratio, it remains to pick a good exchange function h satisfying (H1)–(H4) such that the right hand side of part (1) of Lemma 5 is small. Using a similar technique as in [12], we can formulate this as an infinite-dimensional factor-revealing linear program. More specifically, we can minimize

$$
\sup_{\xi_1 \in [0,1]} \int_{\xi_1}^1 h(\xi_1, \xi)\, d\xi
$$

over the set of all functions h which satisfies (H1)–(H4). Notice that the objective value and the constraints induced by (H1)–(H4) are linear in h. However, the space of all feasible solutions is infinite-dimensional. One possible approach to the infinite-dimensional factor-revealing linear program is to obtaining a numerical solution via a suitable discretization. Using the numerical results as guidance, we obtain the candidate exchange function

$$
h(\xi_1, \xi_2) = \max\left(\{0\} \cup \left\{ \left(1 - \tfrac{1}{L}\right)^k : k \in \{0, 1, 2, \dots\} \text{ and } \xi_1 > \xi_2 \cdot \left(1 - \tfrac{1}{L}\right)^k \right\} \right). \tag{6}
$$

The following lemma provides a formal analytical proof that it satisfies (H1)–(H4) and achieves an objective value of $(1 - \frac{1}{L})^L$.

Lemma 6. *Let h be the function defined by (6). Then the following conditions hold.*

(1) The function h satisfies (H1)–(H4).

(2) For every $\xi_1 \in [0,1]$, we have $\displaystyle\int_{\xi_1}^{1} h(\xi_1, \xi)\, d\xi \leq \left(1 - \frac{1}{L}\right)^L$.

Proof.

(1) It is straightforward to see that (H1)–(H3) hold by inspecting the definition of h. To show that (H4) holds, let $\xi_1, \xi_2 \in [0,1]$. We consider three cases.
Case 1: $\xi_2 \leq \xi_1$. Then

$$L \cdot \int_{\xi_2 \cdot (1-1/L)}^{\xi_2} \left(1 - h(\xi_1, \xi)\right) d\xi = L \cdot \int_{\xi_2 \cdot (1-1/L)}^{\xi_2} (1 - 1)\, d\xi = 0$$
$$= \max(\xi_2 - \xi_1, 0).$$

Case 2: $\xi_2 > \xi_1 = 0$. Then

$$L \cdot \int_{\xi_2 \cdot (1-1/L)}^{\xi_2} \left(1 - h(\xi_1, \xi)\right) d\xi = L \cdot \int_{\xi_2 \cdot (1-1/L)}^{\xi_2} (1 - 0)\, d\xi = \xi_2$$
$$= \max(\xi_2 - \xi_1, 0).$$

Case 3: $\xi_2 > \xi_1 > 0$. Let $k \in \{0, 1, 2, \dots\}$ such that $(1 - \frac{1}{L})^{k+1} < \frac{\xi_1}{\xi_2} \leq (1 - \frac{1}{L})^k$. Then

$$L \cdot \int_{(1-1/L) \cdot \xi_2}^{\xi_2} \left(1 - h(\xi_1, \xi)\, d\xi\right)$$
$$= \xi_2 - L \cdot \int_{(1-1/L) \cdot \xi_2}^{\xi_2} h(\xi_1, \xi)\, d\xi$$
$$= \xi_2 - L \cdot \int_{(1-1/L) \cdot \xi_2}^{\xi_1/(1-1/L)^k} h(\xi_1, \xi)\, d\xi - L \cdot \int_{\xi_1/(1-1/L)^k}^{\xi_2} h(\xi_1, \xi)\, d\xi$$
$$= \xi_2 - L \cdot \int_{(1-1/L) \cdot \xi_2}^{\xi_1/(1-1/L)^k} \left(1 - \frac{1}{L}\right)^k d\xi - L \cdot \int_{\xi_1/(1-1/L)^k}^{\xi_2} \left(1 - \frac{1}{L}\right)^{k+1} d\xi$$
$$= \xi_2 - L \cdot (\xi_1 - \xi_2 \cdot (1 - \tfrac{1}{L})^{k+1}) - L \cdot (\xi_2 \cdot (1 - \tfrac{1}{L})^{k+1} - \xi_1 \cdot (1 - \tfrac{1}{L}))$$
$$= \xi_2 - \xi_1$$
$$= \max(\xi_2 - \xi_1, 0).$$

(2) Let $\xi_1 \in [0,1]$. We may assume that $\xi_1 > 0$, for otherwise

$$\int_{\xi_1}^{1} h(\xi_1, \xi)\, d\xi = \int_{\xi_1}^{1} 0\, d\xi = 0 \leq \left(1 - \frac{1}{L}\right)^L.$$

Let $k \in \{0, 1, 2, \dots\}$ such that $(1 - \frac{1}{L})^{k+1} < \xi_1 \leq (1 - \frac{1}{L})^k$. Then

$$\int_{\xi_1}^1 h(\xi_1, \xi) \, d\xi$$

$$= \int_{\xi_1/(1-1/L)^k}^1 h(\xi_1, \xi) \, d\xi + \sum_{0 \leq k' < k} \int_{\xi_1/(1-1/L)^{k'}}^{\xi_1/(1-1/L)^{k'+1}} h(\xi_1, \xi) \, d\xi$$

$$= \int_{\xi_1/(1-1/L)^k}^1 \left(1 - \frac{1}{L}\right)^{k+1} d\xi + \sum_{0 \leq k' < k} \int_{\xi_1/(1-1/L)^{k'}}^{\xi_1/(1-1/L)^{k'+1}} \left(1 - \frac{1}{L}\right)^{k'+1} d\xi$$

$$= \left(\left(1 - \frac{1}{L}\right)^{k+1} - \xi_1 \cdot \left(1 - \frac{1}{L}\right)\right) + \sum_{0 \leq k' < k} \frac{\xi_1}{L}$$

$$= (1 - \tfrac{1}{L})^{k+1} + \tfrac{\xi_1}{L}(k - L + 1). \tag{7}$$

We consider three cases.
Case 1: $k = L - 1$. Then (7) implies

$$\int_{\xi_1}^1 h(\xi_1, \xi) \, d\xi = (1 - \tfrac{1}{L})^{k+1} + \tfrac{\xi_1}{L}(k - L + 1) = (1 - \tfrac{1}{L})^L.$$

Case 2: $k \geq L$. Then (7) implies

$$\int_{\xi_1}^1 h(\xi_1, \xi) \, d\xi = (1 - \tfrac{1}{L})^{k+1} + \tfrac{\xi_1}{L}(k - L + 1)$$

$$\leq (1 - \tfrac{1}{L})^{k+1} + \tfrac{1}{L}(k - L + 1)(1 - \tfrac{1}{L})^k$$

$$= (1 - \tfrac{1}{L})^L \cdot \tfrac{k}{L} \cdot (1 - \tfrac{1}{L})^{k-L}$$

$$\leq (1 - \tfrac{1}{L})^L \cdot e^{k/L - 1} \cdot e^{(L-k)/L}$$

$$= (1 - \tfrac{1}{L})^L,$$

where the first inequality follows from $\xi_1 \leq (1 - \frac{1}{L})^k$, and the second inequality follows from $e^{k/L-1} \geq \frac{k}{L}$ and $e^{-1/L} \geq 1 - \frac{1}{L}$.
Case 3: $k \leq L - 2$. Then (7) implies

$$\int_{\xi_1}^1 h(\xi_1, \xi) \, d\xi = (1 - \tfrac{1}{L})^{k+1} + \tfrac{\xi_1}{L}(k - L + 1)$$

$$< (1 - \tfrac{1}{L})^{k+1} - \tfrac{1}{L}(L - k - 1)(1 - \tfrac{1}{L})^{k+1}$$

$$= (1 - \tfrac{1}{L})^L \cdot \tfrac{k+1}{L-1} \cdot (1 + \tfrac{1}{L-1})^{L-k-2}$$

$$\leq (1 - \tfrac{1}{L})^L \cdot e^{(k+1)/(L-1)-1} \cdot e^{(L-k-2)/(L-1)}$$

$$= (1 - \tfrac{1}{L})^L,$$

where the first inequality follows from $\xi_1 > (1 - \frac{1}{L})^{k+1}$, and the second inequality follows from $e^{(k+1)/(L-1)-1} \geq \frac{k+1}{L-1}$ and $e^{1/(L-1)} \geq 1 + \frac{1}{L-1}$. $\quad\square$

Lemma 7. $\sum_{(i,j)\in I \times J} x_{i,j} \leq \left(1 + \left(1 - \frac{1}{L}\right)^L\right) \cdot |\mu|.$

Proof. Consider the charging argument with the exchange function h as defined by (6). By part (1) of Lemma 6, the function h satisfies (H1)–(H4). Lemma 2 implies

$$\sum_{(i,j)\in I \times J} x_{i,j} \leq \sum_{(i,j)\in I \times J} (\theta_{i,j} + \phi_{i,j})$$

$$= \sum_{(i,j)\in \mu} \left(\sum_{j'\in J} \theta_{i,j'} + \sum_{i'\in I} \phi_{i',j}\right) + \sum_{\substack{i\in I \\ \mu(i)=0}} \sum_{j\in J} \theta_{i,j} + \sum_{\substack{j\in J \\ \mu(j)=0}} \sum_{i\in I} \phi_{i,j}.$$

$$(8)$$

Part (1) of Lemma 5 implies

$$\sum_{(i,j)\in \mu} \left(\sum_{j'\in J} \theta_{i,j'} + \sum_{i'\in I} \phi_{i',j}\right) \leq \sum_{(i,j)\in \mu} \left(1 + \int_{1-p_i}^1 h(1 - p_i, \xi)\, d\xi\right)$$

$$\leq \sum_{(i,j)\in \mu} \left(1 + \left(1 - \frac{1}{L}\right)^L\right)$$

$$= (1 + (1 - \tfrac{1}{L})^L) \cdot |\mu|, \qquad (9)$$

where the second inequality follows from part (2) of Lemma 6. Part (2) of Lemma 5 implies

$$\sum_{\substack{i\in I \\ \mu(i)=0}} \sum_{j\in J} \theta_{i,j} = 0. \qquad (10)$$

The definitions of $\{\phi_{i,j}\}_{(i,j)\in I \times J}$ imply

$$\sum_{\substack{j\in J \\ \mu(j)=0}} \sum_{i\in I} \phi_{i,j} = 0. \qquad (11)$$

Combining (8)–(11) gives the desired inequality. $\qquad\square$

Using Lemma 7, it is straightforward to establish the following two theorems; see [13] for proof details.

Theorem 1. *There exists a $(1 + (1 - \frac{1}{L})^L)$-approximation algorithm for the maximum stable matching problem with one-sided ties and incomplete lists where the maximum tie length is L.*

Theorem 2. *For the maximum stable matching problem with one-sided ties where the maximum tie length is L, the integrality gap of the LP formulation in Sect. 2.2 is $1 + (1 - \frac{1}{L})^L$.*

References

1. Bauckholt, F., Pashkovich, K., Sanità, L.: On the approximability of the stable marriage problem with one-sided ties (2018). arXiv:1805.05391
2. Chiang, R., Pashkovich, K.: On the approximability of the stable matching problem with ties of size two (2019). arXiv:1808.04510
3. Dean, B.C., Jalasutram, R.: Factor revealing LPs and stable matching with ties and incomplete lists. In: Proceedings of the 3rd International Workshop on Matching Under Preferences, pp. 42–53 (2015)
4. Gale, D., Shapley, L.S.: College admissions and the stability of marriage. Am. Math. Mon. **69**(1), 9–15 (1962)
5. Gale, D., Sotomayor, M.A.O.: Some remarks on the stable matching problem. Discrete Appl. Math. **11**(3), 223–232 (1985)
6. Halldórsson, M.M., Iwama, K., Miyazaki, S., Yanagisawa, H.: Randomized approximation of the stable marriage problem. Theoret. Comput. Sci. **325**(3), 439–465 (2004)
7. Halldórsson, M.M., Iwama, K., Miyazaki, S., Yanagisawa, H.: Improved approximation results for the stable marriage problem. ACM Trans. Algorithms **3**(3), 30 (2007)
8. Huang, C.C., Kavitha, T.: Improved approximation algorithms for two variants of the stable marriage problem with ties. Math. Program. **154**(1), 353–380 (2015)
9. Irving, R.W.: Stable marriage and indifference. Discrete Appl. Math. **48**(3), 261–272 (1994)
10. Iwama, K., Miyazaki, S., Yanagisawa, H.: A 25/17-approximation algorithm for the stable marriage problem with one-sided ties. Algorithmica **68**(3), 758–775 (2014)
11. Király, Z.: Linear time local approximation algorithm for maximum stable marriage. Algorithms **6**(3), 471–484 (2013)
12. Lam, C.K., Plaxton, C.G.: A $(1 + 1/e)$-approximation algorithm for maximum stable matching with one-sided ties and incomplete lists. In: Proceedings of the 30th Annual ACM-SIAM Symposium on Discrete Algorithms, pp. 2823–2840 (2019)
13. Lam, C.K., Plaxton, C.G.: Maximum stable matching with one-sided ties of bounded length. Technical report TR-19-03, Department of Computer Science, University of Texas at Austin, July 2019
14. McDermid, E.: A 3/2-approximation algorithm for general stable marriage. In: Albers, S., Marchetti-Spaccamela, A., Matias, Y., Nikoletseas, S., Thomas, W. (eds.) ICALP 2009. LNCS, vol. 5555, pp. 689–700. Springer, Heidelberg (2009). https://doi.org/10.1007/978-3-642-02927-1_57
15. Paluch, K.: Faster and simpler approximation of stable matchings. Algorithms **7**(2), 189–202 (2014)
16. Radnai, A.: Approximation algorithms for the stable marriage problem. Master's thesis, Department of Computer Science, Eötvös Loránd University (2014)
17. Roth, A.E.: The evolution of the labor market for medical interns and residents: a case study in game theory. J. Polit. Econ. **92**(6), 991–1016 (1984)
18. Rothblum, U.G.: Characterization of stable matchings as extreme points of a polytope. Math. Program. **54**(1), 57–67 (1992)
19. Vande Vate, J.H.: Linear programming brings marital bliss. Oper. Res. Lett. **8**(3), 147–153 (1989)
20. Yanagisawa, H.: Approximation algorithms for stable marriage problems. Ph.D. thesis, Graduate School of Informatics, Kyoto University (2007)

Stochastic Matching on Uniformly Sparse Graphs

Soheil Behnezhad[1(✉)], Mahsa Derakhshan[1], Alireza Farhadi[1], MohammadTaghi Hajiaghayi[1], and Nima Reyhani[2]

[1] University of Maryland, College Park, USA
{soheil,mahsa,farhadi,hajiagha}@cs.umd.edu
[2] Airbnb, San Francisco, USA
nima.reyhani@airbnb.com

Abstract. In this paper, we consider the following *stochastic matching* problem: We are given a graph $G = (V, E)$ where each edge $e \in E$ is *realized* independently with some constant probability p and the goal is to find a constant degree subgraph R of G whose expected realized matching size is close to that of G. This model of stochastic matching has attracted significant attention over the past few years for its various applications in *kidney exchange* and *recommendation systems*.

The main open question of the area is whether a $(1-\epsilon)$ approximation can be achieved. Currently, the best known bounds are close to 0.66 due to algorithms of Assadi and Bernstein [SOSA'19] and Behnezhad et al. [SODA'19]. We show that indeed this bound can be improved to $(1 - \epsilon)$ if the graph G has small *arboricity*. This includes a large family of graphs such as planar or minor-free graphs, bounded treewidth graphs, or arguably any sparse graph that is of interest.

Finally, we also practically study a number of natural algorithms on the dataset of a major online freelancing company.

1 Introduction

We consider the following *stochastic matching* problem. Given a graph $G = (V, E)$ where each edge $e \in E$ is *realized* with some constant probability p, the goal is to compute a constant degree subgraph R of G whose expected matching is close to the expected matching of G. The edges of this subgraph R are also referred to as the *queried* edges of G in the literature. The reason is that one can query (i.e., ask about realization) of only the edges that belong to R and find a large realized matching of G.

This variant of the stochastic matching problem was first formalized by Blum et al. (2015) who gave a $(0.5 - \epsilon)$-approximation for any arbitrarily small constant ϵ. This model has attracted a significant subsequent work since then due to its diverse applications Assadi et al. (2016, 2017); Maehara and Yamaguchi (2018); Behnezhad and Reyhani (2018); Assadi and Bernstein (2019); Behnezhad et al. (2019). The best approximation factor remained to be close to 0.5 until two recent breakthrough papers by Assadi and Bernstein (2019) and Behnezhad et al.

D. Fotakis and E. Markakis (Eds.): SAGT 2019, LNCS 11801, pp. 357–373, 2019.
https://doi.org/10.1007/978-3-030-30473-7_24

(2019)who achieved close to 0.66 approximations (2/3 and $4\sqrt{2} - 5$ respectively). The major question left open in the literature is:

What is the highest approximation factor that is achievable by a subgraph R whose per vertex degree is bounded by a constant? In particular, is it possible to obtain a $(1 - \epsilon)$-approximation for any arbitrarily small constant ϵ?

Our main theoretical result is to show that one can achieve a $(1 - \epsilon)$-approximation for graphs of *bounded arboricity*[1]. Bounded arboricity graphs include a large family of graphs of interest. For instance, all minor-free graphs (such as planar graphs), random graphs within the preferential attachment model (see Barabási and Albert (1999)), graphs with bounded genus, treewidth, or pathwidth, and essentially most interesting family of sparse graphs have constant arboricity. Even if the arboricity α of the input graph is not bounded, our result implies that merely $O(\alpha)$ queries per-vertex suffices to achieve a $(1 - \epsilon)$-approximation. This improves over the trivial bound of $O(n)$ per-vertex queries for the wide range of graphs with $\alpha = o(n)$. We comment that arboricity of a graph can never exceed its maximum degree, therefore we always have $\alpha \leq n$.

On the practical side, we implement a number of natural algorithms to find the subgraph R and evaluate their quality over the internal dataset of a major online freelancing company.

1.1 Applications

Kidney Exchange. Transplant of a kidney from a *living donor* is possible only when the kidney is compatible with the recipient (*patient*), which is not always the case. The simplest way to overcome this problem is to exchange kidneys between two incompatible donor/patient pairs. That is, the donor of the first pair donates a kidney to the patient of the second pair and vice versa. The goal is to identify the maximum number of donor/patient pairs that can exchange kidney. The medical records of patients and donors can be used to rule out a subset of incompatibilities, however, before the transplant a more time consuming and expensive medical test should take place.

The stochastic matching helps finding a large number of compatible donor/patient pairs while make sure that each patient takes the medical test a few number of times. This problem in the stochastic setting has been extensively studied in the literature Akbarpour et al. (2014); Anderson et al. (2015a, 2015b); Awasthi and Sandholm (2009); Dickerson et al. (2012, 2013); Dickerson and Sandholm (2015); Manlove and O'Malley (2014); Ünver (2010).

Recommendation Systems. As online services become more pervasive in everyday life, we face a higher demand for *recommender systems* (henceforth, RSs) that predict the preferences of users over a large number of options. This has resulted in a plethora of studies on RSs for diverse topics from movies Carrer-Neto et al. (2012); Winoto and Tang (2010) to music Lee et al. (2010);

---

[1] The arboricity of a graph equals (asymptotically) the average degree in its densest subgraph. Equivalently, the classic result of Nash-Williams shows that it is equal to the minimum number of forests required to cover the graph.

Nanopoulos et al. (2010); Tan et al. (2011) to television Yu et al. (2006); Basu et al. (1998) to books Núñez-Valdéz et al. (2012); Crespo et al. (2011) to documents Serrano-Guerrero et al. (2011); Porcel et al. (2009); Porcel and Herrera-Viedma (2010); Porcel et al. (2012). Despite this large volume of work, very few papers have addressed this problem in *matching markets* which have applications in online dating, online labor markets, etc. The fundamental difference is that in matching markets, the recommendations have to take into account the preferences of *both* parties whereas in the above-mentioned scenarios, the problem is essentially ranking a set of items for an individual.

For an extreme example, consider a celebrity joining an online dating platform. Naive recommendations that consider only the personal preferences of the users would be disastrous. The celebrity, on one hand, would be overwhelmed with the huge number of requests that s/he considers undesirable and the other users, on the other hand, would get frustrated with their messages left unreplied.

To coupe with this problem, it is natural to evaluate a set of recommendations based on the number (or even the quality) of the matches that it is likely to lead into. Let us denote a set of recommendations by a graph $R = (V, E)$ (R for recommendations) where each vertex in V corresponds to a user and each edge $\{u, v\} \in E$ implies that users u and v are recommended to one another—we call this the *recommendation graph* throughout. It is natural to evaluate a set of recommendations based on the number (or even the quality) of the matches that it is likely to lead into. This is exactly equivalent to the variant of stochastic matching that we study in this paper.

We provide a more detailed discussion on the applications of the stochastic matching model for recommender systems in Sect. 5.

2   The Model

We start by the formal definition of a model of *stochastic matching* that was first introduced for its applications in kidney exchange Blum et al. (2015). We then show how this model can be used in recommendation systems.

The Stochastic Matching Problem. We are given a (not necessarily bipartite[2]) graph $G = (V, E)$, along with a parameter $p : E \to [0, 1]$ in the input. For any subset E' of E, a *realization* E'_p of E' is a subset E'_p of E' where each edge $e \in E'$ appears (or interchangeably is *realized*) in E'_p independently with probability $p(e)$. Note that E'_p is a random variable. Denoting the maximum cardinality matching of a graph on edge set E' by $M(E')$, the *expected matching* $M_p(E')$ of any subset E' of E is defined to be

$$M_p(E') := \mathbb{E}[M(E'_p)]$$

---

[2] Although most matching markets can be modeled by bipartite graphs, some of them, such as dating websites and kidney exchange graphs are not bipartite. Nonetheless, this assumption only makes the model more general since bipartite graphs are special cases of general graphs.

where the expectation is taken over the randomness of E'_p. The goal in the stochastic matching problem is to pick a degree bounded subgraph R of G such that $M_p(R)/M_p(G)$ which is also known as the approximation factor is maximized.

3   Theoretical Analysis

Given a graph $G = (V, E)$ with the arboricity bounded by α and an $0 < \epsilon < 1$, we propose an algorithm that finds a degree bounded subgraph H of G such that $M_p(H) \geq (1 - c\epsilon)M_p(G)$ where c is a constant. The algorithm that we analyze is formally given as Algorithm 1.

Algorithm 1. Algorithm "double marking".

1: **Input:** Input graph $G = (V, E)$ along with realization probabilities $p : E \to [0, 1]$.
2: **Parameter:** Δ
3: For each vertex v, mark $\max\{\deg(v), \Delta\}$ edges incident to it arbitrarily.
4: $R \leftarrow \emptyset$
5: **for** any edge $e \in E$ **do**
6:     Add e to R if and only if it is marked by both of its endpoints.
7: **return** R.

The algorithm above was previously introduced by Solomon (2018) for nonstochastic settings. Particularly, the analysis of Solomon carries over to our setting without any change if $p = 1$. In what follows, we show that with few changes, similar arguments can be used to show that the algorithm also works in the stochastic setting when $p < 1$.

Throughout, we use the following standard observation on bounded arboricity graphs extensively.

Observation 1. *Given a graph $G = (V, E)$ with an arboricity bounded by $\alpha \geq 1$, let $U = V_1 \cup V_2$ be a subset of vertices such that degree of every vertex in V_1 in the graph $G[U]$ is at least $2\alpha(c+1)$ for some constant c, where $G[U]$ is the graph induced by the vertices of U. Then, $|V_1| \leq |V_2|/c$.*

Suppose that each edge in E is realized independently with the probability p, and let $\beta = 12\alpha/\epsilon$ and $\Delta = 3\beta/p$. For each vertex we mark randomly up to Δ of its incident edges, and we add an edge to the graph G_Δ if it is marked by its both end points. It is clear that the degree of each vertex is bounded by Δ in G_Δ. Now we show the size of expected matching in G_Δ is at least $(1 - c\epsilon)M_p(G)$ for some constant c.

We partition the vertices of G into two sets V_l, and V_h where V_l are set of vertices whose degree in G is at most Δ and V_h is the set of all other vertices. In the following lemma we show that the difference between the size of expected matching of G_Δ and G is at most $|V_h|$. The lemma is as follows.

Lemma 1. *Given a graph $G = (V, E)$ and $\Delta > 0$, define $G_\Delta = (V, E_\Delta)$ as above. Let V_h be the set of vertices with a degree greater than Δ. Then, for any realization E_p, we have $|M(E_P \cap E_\Delta)| \geq |M(E_p)| - |V_h|$.*

Proof. We partition the edges of $M(E_p)$ into two sets M_1 and M_2, where M_1 is the set of edges in the matching which are in our sampling, i.e., there are in $E_P \cap E_\Delta$, and M_2 is the set of all other edges. In our sampling, we sample every edge whose both ends are in V_l. Therefore, every edge in M_2 has at least one end in V_h, and the size of M_2 is at most $|V_h|$. Also, we know that

$$|M(E_P \cap E_\Delta)| \geq |M_1| = |M(E_p)| - |M_2| \geq |M(E_p)| - |V_h|,$$

which concludes the proof.

For a realization E_p, we define the heavy and light vertices as follows.

Definition 1. *Given a graph $G = (V, E)$, a sampled graph $G_\Delta = (V, E_\Delta)$ and a realization E_p, we say a vertex $u \in V$ is γ-heavy in the sampled graph G_Δ if at least γ edges from E_Δ which are incident to u are realized. Otherwise, we say that u is γ-light.*

In the following lemma, we show that if in a realization we have a large number of β-heavy vertices, then we can approximately preserve the size of expected matching.

Lemma 2. *Given a graph $G = (V, E)$ with an arboricity bounded by $\alpha \geq 1$, $0 < \epsilon < 1$, $\beta = 12\alpha/\epsilon$ and $\Delta \geq \beta$, let $G_\Delta = (V, E_\Delta)$ defined as above and V_h be the set of vertices with a degree greater than Δ. Then, for any realization E_p such that at least $(1 - \epsilon/2)|V_h|$ vertices in V_h are β-heavy, the followings hold.*

1. $|M(E_p)| \geq |V_h|/30$.
2. $|M(E_p \cap E_\Delta)| \geq (1 - 29\epsilon)|M(E_p)|$.

Proof. Let $E_\Delta^* = E_p \cap E_\Delta$, V_d be the set of β-heavy vertices in V_h, and $V_l = V \setminus V_h$. Then, we have $|V_d| \geq (1 - \epsilon/2)|V_h|$. Let $U \subseteq V_d$ be the set of vertices in V_d which have at least $\beta/2$ neighbors in V_l. We show that size of U is at least $(1 - 2\epsilon/5)|V_d|$. Let $U' = V_d \setminus U$ be the set of vertices in V_d that have less than $\beta/2$ neighbors in V_l. Since the degree of every vertex in V_d is at least β in the realization and all vertices in U' have less than $\beta/2$ neighbors in V_l, every vertex in U' has at least $\beta/2$ neighbors in V_d. It follows from Observation 1 that

$$|U'| \leq \frac{2\alpha|V_d|}{\beta/2 - 1} \leq \frac{2\alpha|V_d|}{5\alpha/\epsilon} \leq \frac{2\epsilon}{5}|V_d|.$$

Therefore,

$$|U| = |V_d| - |U'| \geq (1 - 2\epsilon/5)|V_d| \geq (1 - 2\epsilon/5)(1 - \epsilon/2)|V_h|$$
$$\geq (1 - 9\epsilon/10)|V_h|. \tag{1}$$

Considering the matching $M(E_p)$, we partition its edges into two sets M_1 and M_2, where M_1 is the set of edges in the matching which are in our sampling, i.e., there are in E_Δ^*, and M_2 is the set of all other edges. Let $V_l^{matched}$ and V_l^{free} (resp., $U^{matched}$, U^{free}) be the set of matched and free vertices in V_l (resp., U) using the edges in M_1. We show that there is a almost complete matching between vertices of U^{free} and V_l^{free} using the edges in E_Δ^*. Our claim is formally as follows. The proof is deferred to the appendix.

Claim 1. *Let M^{free} be a maximum matching between the vertices of U^{free} and V_l^{free} using the edges in E_Δ^*, then $|M^{free}| \geq |U^{free}| - 2\epsilon|M(E_p)|$.*

By Claim 1, the size of a maximum matching in $M(E_\Delta^*)$ is at least $|U^{matched}| + |U^{free}| - 2\epsilon|M(E_p)|$. Therefore,

$$
\begin{aligned}
|M(E_p)| &\geq |M(E_\Delta^*)| \\
&\geq |U^{matched}| + |U^{free}| - 2\epsilon|M(E_p)| \\
&= |U| - 2\epsilon|M(E_p)|,
\end{aligned}
$$

which implies that

$$(1 + 2\epsilon)|M(E_p)| \geq |U|. \tag{2}$$

Also, by (1), we have

$$|U| \geq (1 - 9\epsilon/10)|V_h|.$$

By combining this inequality with (2), we get

$$|M(E_p)| \geq \frac{|U|}{1 + 2\epsilon} \geq \frac{(1 - 9\epsilon/10)|V_h|}{1 + 2\epsilon}.$$

Since $\epsilon < 1$, we have $\frac{1 - 9\epsilon/10}{1 + 2\epsilon} \geq 1/30$, therefore,

$$M(E_p) \geq |V_h|/30,$$

which proves the first part of the lemma.

Now we are ready to prove the second part of the lemma. Note that in our sampling, we take every edge whose both ends are in V_l. Therefore, every unmatched edge has at least one end in V_h. Note that we sample every edges between the vertices of V_l. Therefore, every edge in M_2 is incident to at least one vertex in V_h, and since M_2 is the set of edges which are not in our sampling, they are incident to a vertex in V_h which is not matched using the edges in M_1. Therefore, the size of M_2 is at most the number of free vertices in V_h. Thus, $|M_2| \leq |U^{free}| + |V_h \setminus U|$. Let M^{free} be a maximum matching between U^{free} and V_l^{free} in E_Δ^*. By Claim 1, $|M^{free}| \geq |U^{free}| - 2\epsilon|M(E_p)|$. Therefore, $|M(E_\Delta^*)| \geq |M_1| + |M^{free}| \geq |M_1| + |U^{free}| - 2\epsilon|M(E_p)|$. Note that $|M(E_p)| = |M_1| + |M_2|$. Thus,

$$|M(E_p)| - |M(E_\Delta^*)| \leq |M_2| - |U^{free}| + 2\epsilon|M(E_p)|$$

Since $|M_2| \leq |U^{free}| + |V_h \setminus U|$ and $|U| \geq (1 - 9\epsilon/10)|V_h|$ by (1), we have

$$|M(E_p)| - |M(E_\Delta^*)| \leq |V_h \setminus U| + 2\epsilon|M(E_p)|$$
$$= |V_h| - |U| + 2\epsilon|M(E_p)|$$
$$\leq \frac{9\epsilon}{10}|V_h| + 2\epsilon|M(E_p)|.$$

Also, by the first part of the lemma we have $|V_h| \leq 30|M(E_p)|$. Therefore,

$$|M(E_p)| - |M(E_\Delta^*)| \leq \frac{9\epsilon}{10}|V_h| + 2\epsilon|M(E_p)| \leq 29\epsilon|M(E_p)|.$$

Thus,

$$|M(E_\Delta^*)| \geq (1 - 29\epsilon)|M(E_p)|.$$

which proves the second part of the lemma.

Now, we are ready to show that the size of expected matching in G_Δ is at least $(1 - c\epsilon)$ fraction of the size of expected matching in G.

Theorem 1. *Given a graph $G = (V, E)$ such that each edge in E is realized with the probability p, and an $0 < \epsilon < 1$, let suppose that arboricity of G is bounded by $\alpha \geq 1$. Let $\beta = 12\alpha/\epsilon$ and $\Delta = 3\beta/p$, and define $G_\Delta = (V, E_\Delta)$ as above. Then, $M_p(G_\Delta) \geq (1 - c\epsilon)M_p(G)$ for some constant c.*

Proof. Let V_h be the set of vertices whose degree in G is larger than Δ, and $V_l = V \setminus V_h$ be the set of other vertices. Let $v \in V_h$ be a vertex in V_h. v has at least Δ edges and each of its edge is realized with the probability of p. Therefore, the expected number of neighbors of v after the realization is $p\Delta = 3\beta$. We claim that the probability that v has less than β neighbors in a realization is at most $O(\epsilon^3)$. The proof uses the standard form of Chernoff bound, and it can be found in the appendix.

Claim 2. *Let v be a vertex such that its degree in G is at least Δ, then with the probability of at least $1 - \epsilon^3/100$, u has β edges in a realization.*

Now we show that with a probability sufficiently large in a realization at least $(1 - \epsilon/2)$ portion of vertices in V_h are β-heavy. To this purpose, for every vertex $v \in V_h$, we define Y_u to be a random variable which is 1 if v has less than β edges in a realization, and otherwise is 0. By Claim 2, we have

$$\mathbb{E}[Y_u] = \Pr[Y_u = 1] \leq \epsilon^3/100.$$

Let $Y = \sum_{u \in V_h} Y_u$, i.e., Y is a random variable that counts the number of vertices in V_h which have less than β edges in a realization which is the number of β-light vertices in V_h. By the linearity of expectation we have

$$\mathbb{E}[Y] = \sum_{u \in V_h} \mathbb{E}[Y_u] \leq \epsilon^3|V_h|/100.$$

If in our realization less than $(1 - \epsilon/2)|V_h|$ vertices in $|V_h|$ are β-heavy, then the number of β-light vertices is larger than $\epsilon|V_h|/2$, and we have $Y > \epsilon|V_h|/2$. Therefore, by Markov's inequality the probability that at least $(1 - \epsilon/2)|V_h|$ vertices in $|V_h|$ be β-heavy is at least

$$1 - \Pr[Y > \epsilon|V_h|/2\,] \geq 1 - \frac{\mathbb{E}[Y]}{\epsilon|V_h|/2} \geq 1 - \frac{\epsilon^3|V_h|/100}{\epsilon|V_h|/2} = 1 - \epsilon^2/50.$$

For a realization E_p, we use a random variable W to show if in this realization at least $(1 - \epsilon/2)|V_h|$ vertices in $|V_h|$ are β-heavy. Formally, W is 1 if in the realization at least $(1 - \epsilon/2)|V_h|$ vertices in $|V_h|$ are β-heavy and is 0 otherwise. As we showed above, we have

$$\Pr[W = 1] \geq 1 - \epsilon^2/50. \tag{3}$$

Consider the realizations such that at least $(1 - \epsilon/2)|V_h|$ in $|V_h|$ are β-heavy. By Lemma 2, in this case the expected size of the maximum matching in the sampled graph is at least $(1 - 29\epsilon)$ fraction of the size of the maximum matching in the original graph. Particularly, we have

$$\mathbb{E}\Big[M(E_p \cap E_\Delta)|W = 1\Big] \geq (1 - 29\epsilon)\mathbb{E}\Big[M(E_p)|W = 1\Big],$$

which means that

$$\mathbb{E}\Big[M(E_p \cap E_\Delta)|W = 1\Big] - \mathbb{E}\Big[M(E_p)|W = 1\Big] \geq -29\epsilon\mathbb{E}\Big[M(E_p)|W = 1\Big]. \tag{4}$$

Also, we have

$$\mathbb{E}[M(E_p)] \geq \mathbb{E}\Big[M(E_p)\,\big|\,W = 1\Big]\Pr[W = 1]. \tag{5}$$

Combining (4) and (5), we have

$$
\begin{aligned}
\mathbb{E}\Big[M(E_p \cap E_\Delta) &- M(E_p)\,\big|\,W = 1\Big] \\
&= \mathbb{E}\Big[M(E_p \cap E_\Delta)\,\big|\,W = 1\Big] - \mathbb{E}\Big[M(E_p)\,\big|\,W = 1\Big] \\
&\geq -29\epsilon\mathbb{E}\Big[M(E_p)\,\big|\,W = 1\Big] \qquad\qquad \text{By (4)} \\
&\geq \frac{-29\epsilon\mathbb{E}[M(E_p)]}{\Pr[W = 1]}.
\end{aligned}
\tag{6}
$$

Also, by the first part of Lemma 2, we have

$$\mathbb{E}[M(E_p)|W = 1] \geq |V_h|/30.$$

This together with inequality (5), yields,

$$\mathbb{E}[M(E_p)] \geq \mathbb{E}\Big[M(E_p)\,\big|\,W = 1\Big]\Pr[W = 1] \geq \Pr[W = 1]|V_h|/30.$$

Therefore,

$$|V_h| \leq \frac{30\mathbb{E}[M(E_p)]}{\Pr[W=1]}. \tag{7}$$

Also, for the realizations that their number of β-heavy vertices is less than $(1 - \epsilon/2)|V_h|$, by Lemma 1, the difference between the expected size of the matching of the sampled graph and original graph is at most V_h, i.e.,

$$\mathbb{E}\Big[M(E_p \cap E_\Delta)\,|\,W=0\Big] \geq \mathbb{E}\Big[M(E_p)\,|\,W=0\Big] - |V_h|.$$

This together with inequality (7) yields

$$\mathbb{E}[M(E_p \cap E_\Delta)|W=0] \geq \mathbb{E}[M(E_p)|W=0] - \frac{30\mathbb{E}[M(E_p)]}{\Pr[W=1]}.$$

This further implies that,

$$
\begin{aligned}
\mathbb{E}\Big[M(E_p \cap E_\Delta) &- M(E_p)\,\big|\,W=0\Big] \\
&= \mathbb{E}\Big[M(E_p \cap E_\Delta)\,\big|\,W=0\Big] - \mathbb{E}\Big[M(E_p)\,\big|\,W=0\Big] \\
&\geq -\frac{30\mathbb{E}[M(E_p)]}{\Pr[W=1]}.
\end{aligned}
\tag{8}
$$

By (6) and (8) we have

$$
\begin{aligned}
\mathbb{E}\Big[M(E_p \cap E_\Delta) &- M(E_p)\Big] \\
&= \mathbb{E}\Big[M(E_p \cap E_\Delta) - M(E_p)\,\big|\,W=0\Big]\Pr[W=0] \\
&\quad + \mathbb{E}\Big[M(E_p \cap E_\Delta) - M(E_p)\,\big|\,W=1\Big]\Pr[W=1] \\
&\geq -\Pr[W=0]\frac{30\mathbb{E}[M(E_p)]}{\Pr[W=1]} - 29\epsilon\mathbb{E}[M(E_p)]
\end{aligned}
\tag{9}
$$

In the following claim we show that $\frac{\Pr[W=0]}{\Pr[W=1]} \leq \epsilon$. The proof is deferred to the appendix.

Claim 3. $\frac{\Pr[W=0]}{\Pr[W=1]} \leq \epsilon$.

By Claim 3 and inequality (9) we have

$$
\begin{aligned}
\mathbb{E}[M(E_p &\cap E_\Delta) - M(E_p)] \\
&\geq -\Pr[W=0]\frac{30\mathbb{E}[M(E_p)]}{\Pr[W=1]} - 29\epsilon\mathbb{E}[M(E_p)] \\
&\geq -30\epsilon\mathbb{E}[M(E_p)] - 29\epsilon\mathbb{E}[M(E_p)] \qquad\qquad \text{By Claim 3} \\
&\geq -59\epsilon\mathbb{E}[M(E_p)].
\end{aligned}
$$

Therefore, we have

$$\mathbb{E}[M(E_p \cap E_\Delta)] \geq (1 - 59\epsilon)\mathbb{E}[M(E_p)].$$

Note that $M_p(G_\Delta) = \mathbb{E}[M(E_p \cap E_\Delta)]$ and $M_p(G) = \mathbb{E}[M(E_p)]$. Hence,

$$M_p(G_\Delta) \geq (1 - 59\epsilon)M_p(G),$$

which completes the proof for the theorem.

4 Empirical Results

In this section, we analyze Algorithm 1 experimentally on real-world datasets and compare it to a well-studied algorithm of the literature.

4.1 The Studied Algorithms

We study two variations of Algorithm 1. Recall that in Algorithm 1, each vertex marks Δ of its incident edges arbitrarily and an edge will be part of the queried subgraph iff it is marked by both sides. The fact that our theoretical analysis goes through, even when these edges are *arbitrarily* chosen is perhaps surprising and only strengthens our theoretical result. However, in practice, since we are not just considering the worst case scenario, the actual way we choose these incident edges of every vertex turns out to be very important. In this regard, we consider two natural implementations of Algorithm 1. In implementation 1, which we denote by "Algorithm 1 (I1)", every vertex marks Δ of its incident edges uniformly at random and an edge is chosen iff both sides mark it. In implementation 2, every vertex initially chooses Δ tentative incident edges, then if a vertex v is connected to k chosen edges with $k > \Delta$ (note that these edges might be chosen by neighbors of v), it discards $k - \Delta$ edges arbitrarily. We refer to this implementation of Algorithm 1 as "Algorithm 1 (I2)" in our experiments. The main difference between the two implementations, is that the latter tends to choose many more edges per vertex.

We also consider Algorithm 2 that has received a significant attention in the literature Blum et al. (2015); Assadi et al. (2016, 2017); Behnezhad and Reyhani (2018); Maehara and Yamaguchi (2018). The algorithm, roughly, selects a maximum matching from the input graph, adds it to the set of sampled edges and removes it from the graph. This process is continued for Δ steps, which clearly guarantees the degree of every vertex in the sampled subgraph does not exceed Δ.

An advantage of Algorithm 1 is that it is very simple and can be easily implemented in a distributed manner. Algorithm 2, on the other hand, requires Δ iterations of computing a maximum matching. For this, we use the standard $O(m\sqrt{n})$ time algorithm for maximum matchings in bipartite graphs.

4.2   Datasets

The graphs on which we run our experiments, are from the internal dataset of Upwork[3], a major online freelancing company. We have two sets of vertices which model *freelancers* and *clients* (the users hiring freelancers). There is an edge between a freelancer f and a client c iff there is a possibility of a contract between the two. This is inferred directly by the initial job descriptions. For instance, we put an edge between freelancers that have marked themselves as Android developers, to clients that seek to hire an Android developer. Moreover, on each edge $e = (f, c)$, we introduce a success probability p_e that may differ from one edge to another and denotes the probability that freelancer f signs a contract with client c given that they interview each other. The algorithm and parameters with which these probabilities are obtained cannot be revealed, however, we use them as they are to indicate the realization probabilities of the edges.

Algorithm 2. Iterative matching.

1: **Input:** Input graph $G = (V, E)$ along with realization probabilities $p : E \to [0, 1]$.
2: **Parameter:** Δ
3: $E' \leftarrow E$
4: $R \leftarrow \emptyset$
5: **for** Δ rounds **do**
6:    Take a maximum matching M in E'.
7:    Add all the edges of M to R.
8:    $E' \leftarrow E' \backslash M$
9: **return** R.

4.3   Results

Our first set of experiments measure approximation-factor of the proposed algorithms as a function of Δ, i.e., the upper bound on the number of per-vertex queries. The result is highlighted in Fig. 1. As indicated in the figure, there is a huge difference between the performance of the two implementations of Algorithm 1. In fact, perhaps surprisingly, the performance of Algorithm 1 (I2) almost matches that of the standard Algorithm 2, which is much more complicated to implement and requires computing a global maximum matching for Δ iterations.

In our second set of experiments (Fig. 2), we fix a target approximation factor of 0.8, and measure how many per-vertex queries each algorithm requires in order to achieve this approximation factor, as a function of different realization probabilities. More precisely, we change the realization probabilities of all edges and see how the number of required per-vertex queries changes as a function

---

[3] http://upwork.com.

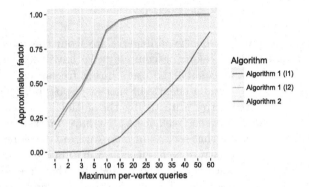

Fig. 1. The obtained approximation-factors of the algorithms, given a fixed upper bound on the number of per-vertex queries.

of that. The fact that Algorithm 1 (I1) requires to set $\Delta \sim 27$ even when the realization probabilities are as high as 0.9, shows the fact that the main issue with this algorithm is indeed the fact that for many vertices, we do not query any edges at all. As indicated in Fig. 2, the other implementation of Algorithm 1 resolves this issue and has a performance, again, very close to the standard algorithm of the literature.

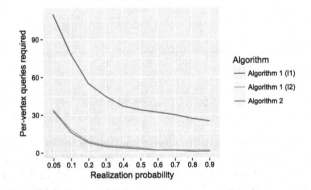

Fig. 2. The number of per-vertex queries required to reach a fixed approximation factor, as we change the realization probability of the edges.

As mentioned before, the fact that on our datasets, Algorithm 1 (I2) and Algorithm 2 have a very similar performance is surprising and far from what we expect in the worst case. A bad input example for Algorithm 1 is illustrated in the figure below. Here, a part of the graph (A) is composed of a large perfect matching, and there is a significantly smaller pool of vertices (B) that are connected to all the vertices in A.

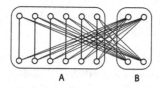

Fig. 3. A bad example for Algorithm 1.

Since Algorithm 1 is very local (i.e., each vertex only looks at its direct edges), most of the sampled vertices will the ones in between sets A and B and very few edges of the actual large perfect matching in A will be sampled. This makes the approximation factor of Algorithm 1 arbitrarily bad (roughly $1/|B|$). Note that, interestingly, the bad example of Fig. 3 crucially has to have a large arboricity of at least $\Theta(|B|)$ since the average degree of the graph is $\Theta(|B|)$ and the arboricity is always larger than average degree.

We construct a similar graph to the one illustrated above and run our algorithms on it (Fig. 4).

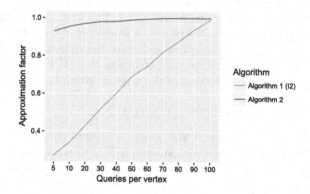

Fig. 4. Performance algorithms on the graph of Fig. 3 with $|A| = 10^3$ and $|B| = 10^2$.

Observe that for Algorithm 1 to achieve a good approximation factor of say 0.9 on this bad example, we have to query almost all the edges whereas Algorithm 2 achieves this with as few as 5 queries per vertex. Fortunately, our prior experiments show that these sort of worst case inputs do not happen in real-world datasets.

5   Discussion: Stochastic Matching and Recommender Systems

From Stochastic Matching to Recommendations. We start by outlining some of the most prominent concerns that recommendations in matching markets need to address. We then proceed to formally describe our measure of effectiveness. These concerns are as follows:

1. The recommendations need to be globally acceptable. That is, a recommendation profile needs to optimize the overall number of successful matches that it is expected to lead into.
2. The recommendations need to take into account the users' preferences. More precisely, the edges picked in R should be more likely to lead into successful matches.
3. Recall that the main purpose of any recommendation system is to recommend only a *small* subset out of a large pool of options. Therefore, there should be a relatively small upper bound on the number of recommendations made to each user.

Suppose, for our target application of recommendations in matching markets, we are given the pairwise success probabilities for each of the potential pairs. To remain as general as possible, we do not pose any restrictions on how these probability functions are computed and define the objective value as a function of these probabilities. Consider a graph $G = (V, E)$ with each vertex $v \in V$ corresponding to a user and each edge $\{u, v\} \in E$ denoting a potential pair. Furthermore, let us denote by function $p : E \to [0, 1]$ the given success probabilities. Any set of recommendations can be seen as a subgraph $R = (V, E_R)$ of G where an edge $\{u, v\}$ is in E_R if and only if u and v are recommended to one another. The score that we assign to each recommendation $R = (V, E_R)$ is the size of its expected matching compared to that of the original graph G. More precisely, we define

$$\text{score}(R) = M_p(R)/M_p(G). \tag{10}$$

Therefore, the score is between 0 and 1 and a higher score is more desirable. The goal is to find a degree bounded subgraph R of G with a high score. Note that if the degrees in R can be arbitrarily large, then the graph G itself would achieve a score of 1. This, however, would map to recommending every pair of users to one another which is not applicable in practice. Therefore for reasons discussed above we want the graph R to be degree bounded while also achieving a good score. This is precisely the objective in the stochastic matching problem and the score corresponds to the guaranteed approximation factor of a stochastic matching algorithm.

Note that all the algorithms that we considered in this setting are called *non-adaptive* which is of practical importance for recommendations. This is in contrast to *adaptive* algorithms (see e.g., Blum et al. (2015)) that may take several *rounds of adaptivity*, with queries conducted at each round depending on the results of the previous round. Synchronizing the queries is impractical for our application since: (1) The delay in interviewing process between even one pair results in a long waiting time for other users. (2) The users are not obligated to follow our recommendations.

References

Akbarpour, M., Li, S., Gharan, S.O.: Dynamic matching market design. In: ACM Conference on Economics and Computation, EC 2014, Stanford, CA, USA, 8–12 June 2014, p. 355 (2014). https://doi.org/10.1145/2600057.2602887

Anderson, R., Ashlagi, I., Gamarnik, D., Kanoria, Y.: A dynamic model of Barter exchange. In: Proceedings of the Twenty-Sixth Annual ACM-SIAM Symposium on Discrete Algorithms, SODA 2015, San Diego, CA, USA, 4–6 January 2015, pp. 1925–1933 (2015a). https://doi.org/10.1137/1.9781611973730.129

Anderson, R., Ashlagi, I., Gamarnik, D., Roth, A.E.: Finding long chains in kidney exchange using the traveling salesman problem. Proc. Natl. Acad. Sci. 112(3), 663–668 (2015b)

Assadi, S., Bernstein, A.: Towards a unified theory of sparsification for matching problems. In: 2nd Symposium on Simplicity in Algorithms, SOSA@SODA, San Diego, CA, USA (OASICS), 8–9 January 2019 (2019)

Fineman, J.T., Mitzenmacher, M. (eds.) Schloss Dagstuhl - Leibniz-Zentrum fuer Informatik, vol. 69, pp. 11:1–11:20 (2019). https://doi.org/10.4230/OASIcs.SOSA.2019.11

Assadi, S., Khanna, S., Li, Y.: The stochastic matching problem with (very) few queries. In: Proceedings of the 2016 ACM Conference on Economics and Computation, EC 2016, Maastricht, The Netherlands, 24–28 July 2016, pp. 43–60 (2016). https://doi.org/10.1145/2940716.2940769

Assadi, S., Khanna, S., Li, Y.: The stochastic matching problem: beating half with a non-adaptive algorithm. In: Proceedings of the 2017 ACM Conference on Economics and Computation, EC 2017, Cambridge, MA, USA, 26–30 June 2017, pp. 99–116 (2017). https://doi.org/10.1145/3033274.3085146

Awasthi, P., Sandholm,T.: Online stochastic optimization in the large: application to kidney exchange. In: IJCAI 2009, Proceedings of the 21st International Joint Conference on Artificial Intelligence, Pasadena, California, USA, 11–17 July 2009, pp. 405–411 (2009). http://ijcai.org/Proceedings/09/Papers/075.pdf

Barabási, A.-L., Albert, R.: Emergence of scaling in random networks. Science 286(543), 509–512 (1999)

Basu, C., Hirsh, H., Cohen, W.W.: Recommendation as classification: using social and content-based information in recommendation. In: Proceedings of the Fifteenth National Conference on Artificial Intelligence and Tenth Innovative Applications of Artificial Intelligence Conference, AAAI 1998, IAAI 1998, Madison, Wisconsin, USA, 26–30 July 1998, pp. 714–720 (1998). http://www.aaai.org/Library/AAAI/1998/aaai98-101.php

Behnezhad, S., Farhadi, A., Hajiaghayi, M.T., Reyhani, N.: Stochastic matching with few queries: new algorithms and tools. In: Proceedings of the Thirtieth Annual ACM-SIAM Symposium on Discrete Algorithms, SODA 2019, San Diego, California, USA, 6–9 January 2019, pp. 2855–2874 (2019). https://doi.org/10.1137/1.9781611975482.177

Behnezhad, S., Reyhani, N.: Almost optimal stochastic weighted matching with few queries (2018)

Blum, A., Dickerson, J.P., Haghtalab, N., Procaccia, A.D.., Sandholm, T., Sharma, A.: Ignorance is almost bliss: near-optimal stochastic matching with few queries. In: Proceedings of the Sixteenth ACM Conference on Economics and Computation, EC 2015, Portland, OR, USA, 15–19 June 2015, pp. 325–342 (2015). https://doi.org/10.1145/2764468.2764479

372   S. Behnezhad et al.

Carrer-Neto, W., Hernández-Alcaraz, M.L., Valencia-García, R., Sánchez, F.G.: Social knowledge-based recommender system. Application to the movies domain. Expert Syst. Appl. 39(12), 10990–11000 (2012). https://doi.org/10.1016/j.eswa.2012.03.025

Crespo, R.G., et al.: Recommendation System based on user interaction data applied to intelligent electronic books. Comput. Hum. Behav. 27(4), 1445–1449 (2011). https://doi.org/10.1016/j.chb.2010.09.012

Dickerson, J.P., Procaccia, A.D., Sandholm, T.: Dynamic matching via weighted myopia with application to kidney exchange. In: Proceedings of the Twenty-Sixth AAAI Conference on Artificial Intelligence, Toronto, Ontario, Canada, 22–26 July 2012 (2012). http://www.aaai.org/ocs/index.php/AAAI/AAAI12/paper/view/5031

Dickerson, J.P., Procaccia, A.D., Sandholm, T.: Failure-aware kidney exchange. In: ACM Conference on Electronic Commerce, EC 2013, Philadelphia, PA, USA, 16–20 June 2013, pp. 323–340 (2013). https://doi.org/10.1145/2482540.2482596

Dickerson, J.P., Sandholm, T.: FutureMatch: combining human value judgments and machine learning to match in dynamic environments. In: Proceedings of the Twenty-Ninth AAAI Conference on Artificial Intelligence, Austin, Texas, USA, 25–30 January 2015, pp. 622–628 (2015). http://www.aaai.org/ocs/index.php/AAAI/AAAI15/paper/view/9497

Lee, S.K., Cho, Y.H., Kim, S.H.: Collaborative filtering with ordinal scale-based implicit ratings for mobile music recommendations. Inf. Sci. 180(11), 2142–2155 (2010). https://doi.org/10.1016/j.ins.2010.02.004

Maehara, T., Yamaguchi, Y.: Stochastic packing integer programs with few queries. In: Proceedings of the Twenty-Ninth Annual ACM-SIAM Symposium on Discrete Algorithms. SIAM (2018)

Manlove, D.F., O'Malley, G.: Paired and altruistic kidney donation in the UK: algorithms and experimentation. ACM J. Exp. Algorithmics 19(1) (2014). https://doi.org/10.1145/2670129

Nanopoulos, A., Rafailidis, D., Symeonidis, P., Manolopoulos, Y.: Musicbox: personalized music recommendation based on cubic analysis of social tags. IEEE Trans. Audio Speech Lang. Process. 18(2), 407–412 (2010)

Núñez-Valdéz, E.R., Lovelle, J.M.C., Martínez, O.S., García-Díaz, V., de Pablos, P.O., Marín, C.E.M.: Implicit feedback techniques on recommender systems applied to electronic books. Comput. Hum. Behav. 28(4), 1186–1193 (2012). https://doi.org/10.1016/j.chb.2012.02.001

Porcel, C., Herrera-Viedma, E.: Dealing with incomplete information in a fuzzy linguistic recommender system to disseminate information in university digital libraries. Knowl.-Based Syst. 23(1), 32–39 (2010). https://doi.org/10.1016/j.knosys.2009.07.007

Porcel, C., Moreno, J.M., Herrera-Viedma, E.: A multi-disciplinar recommender system to advice research resources in University Digital Libraries. Expert Syst. Appl. 36(10), 12520–12528 (2009). https://doi.org/10.1016/j.eswa.2009.04.038

Porcel, C., Tejeda-Lorente, Á., Martínez, M.A., Herrera-Viedma, E.: A hybrid recommender system for the selective dissemination of research resources in a Technology Transfer Office. Inf. Sci. 184(1), 1–19 (2012). https://doi.org/10.1016/j.ins.2011.08.026

Serrano-Guerrero, J., Herrera-Viedma, E., Olivas, J.A., Cerezo, A., Romero, F.P.: A Google wave-based fuzzy recommender system to disseminate information in University Digital Libraries 2.0. Inf. Sci. 181(9), 1503–1516 (2011). https://doi.org/10.1016/j.ins.2011.01.012

Solomon, S.: Local algorithms for bounded degree sparsifiers in sparse graphs. In: 9th Innovations in Theoretical Computer Science Conference, ITCS 2018, Cambridge, MA, USA, 11–14 January 2018, pp. 52:1–52:19. https://doi.org/10.4230/LIPIcs.ITCS.2018.52

Tan, S., Bu, J., Chen, C., He, X.: Using rich social media information for music recommendation via hypergraph model. In: Hoi, S., Luo, J., Boll, S., Xu, D., Jin, R., King, I. (eds.) Social Media Modeling and Computing, pp. 213–237 (2011). https://doi.org/10.1007/978-0-85729-436-4_10

Utku Ünver, M.: Dynamic kidney exchange. Rev. Econ. Stud. **77**(1), 372–414 (2010)

Winoto, P., Tang, T.Y.: The role of user mood in movie recommendations. Expert Syst. Appl. **37**(8), 6086–6092 (2010). https://doi.org/10.1016/j.eswa.2010.02.117

Yu, Z., Zhou, X., Hao, Y., Gu, J.: TV program recommendation for multiple viewers based on user profile merging. User Model. User-Adapt. Interact. **16**(1), 63–82 (2006). https://doi.org/10.1007/s11257-006-9005-6

Fair Division with Subsidy

Daniel Halpern$^{(\boxtimes)}$ and Nisarg Shah

University of Toronto, Toronto, Canada
daniel.halpern@mail.utoronto.ca

Abstract. When allocating a set of goods to a set of agents, a classic fairness notion called *envy-freeness* requires that no agent prefer the allocation of another agent to her own. When the goods are indivisible, this notion is impossible to guarantee, and prior work has focused on its relaxations. However, envy-freeness can be achieved if a third party is willing to subsidize by providing a small amount of money (divisible good), which can be allocated along with the indivisible goods.

In this paper, we study the amount of subsidy needed to achieve envy-freeness for agents with additive valuations, both for a given allocation of indivisible goods and when we can choose the allocation. In the former case, we provide a strongly polynomial time algorithm to minimize subsidy. In the latter case, we provide optimal constructive results for the special cases of binary and identical valuations, and make a conjecture in the general case. Our experiments using real data show that a small amount of subsidy is sufficient in practice.

Keywords: Fair division · Indivisible goods · Envy-freeness · Subsidy

1 Introduction

How to fairly divide goods among people has been a subject of interest for millennia. However, formal foundations of fair division were laid less than a century ago with the work of Steinhaus [29], who proposed the cake-cutting setting where a *divisible* good is to be allocated to n agents with heterogeneous preferences. In the subsequent decades, allocation of divisible goods received significant attention [4,16,25,32,33]. When goods are divisible, one can provide strong fairness guarantees such as *envy-freeness* [17], which requires that no agent prefer the allocation of another agent to her own.

Most real-world applications of fair division, such as divorce settlement or inheritance division, often involve *indivisible* goods. In this case, envy-freeness is impossible to guarantee. For example, if the only available good is a ring, and two agents—Alice and Bob—want it, giving it to either agent would cause the other to envy. Recent research on fair allocation of indivisible goods has focused on achieving relaxed fairness guarantees [2,11,20,27]. For example, *envy-freeness*

Full version of this paper is available at www.cs.toronto.edu/~nisarg/papers/subsidy.pdf.

© Springer Nature Switzerland AG 2019
D. Fotakis and E. Markakis (Eds.): SAGT 2019, LNCS 11801, pp. 374–389, 2019.
https://doi.org/10.1007/978-3-030-30473-7_25

up to one good requires that no agent prefer the allocation of another agent to her own after removing at most one good from the envied agent's bundle. This has lately been a subject of intensive research [7,8,26]. While giving the ring to Alice would satisfy this fairness guarantee, who can blame Bob for thinking that the allocation was unfair? After all, he received nothing!

Intuitively, it seems that if we have money at our disposal, it should help settle the differences and eliminate envy. But can it always help? Suppose that Alice values the ring at $100 while Bob values it at $150. If we give the ring to Alice, then Bob would require at least $150 compensation to not envy Alice. But giving so much money to Bob would make Alice envy Bob. Upon some thought, it becomes clear that the only way to achieve envy-freeness is to give the ring to Bob and give Alice at least $100 (but no more than $150). Is this always possible? When can it be done?

In this paper, we study a setting where we allocate a set of indivisible goods along with some amount of a divisible good (a.k.a. money). The money can either be provided by a third party as a subsidy, or it could already be part of the set of goods available for allocation. Our primary research questions are:

Which allocations of indivisible goods allow elimination of envy using money? And how much money is required to achieve envy-freeness?

1.1   Our Results

Suppose n agents have additive valuations (i.e., the value of a bundle is the sum of the values of the individual items) over m indivisible goods. Without loss of generality, we assume that the value of each agent for each good is in $[0,1]$. We refer to an allocation of indivisible goods as *envy-freeable* if it is possible to eliminate envy by paying each agent some amount of money.

In Sect. 3, we characterize envy-freeable allocations and show how to efficiently compute the minimum payments to agents that are required to eliminate envy in a given envy-freeable allocation.

In Sect. 4, we study the size of the minimum subsidy (total payment to agents) required to achieve envy-freeness. When an (envy-freeable) allocation is given to us, we show that the minimum subsidy required is $\Theta(nm)$ in the worst case, even in the special cases of binary and identical valuations.

The picture gets more interesting when we are allowed to choose the allocation of indivisible goods. In this case, the minimum subsidy is at least $n - 1$ in the worst case. For the special cases of binary and identical valuations, we show that this optimal bound can be achieved through efficient algorithms. For general valuations, we show that it can be achieved for two agents, and conjecture this to be true for more than two agents.

Our experiments in Sect. 5 using synthetic and real data show that the minimum subsidy required in practice is much less than the worst-case bound.

1.2   Related Work

The use of money in fair allocation of indivisible goods has been well-explored. Much of the literature focuses on a setting where the number of goods is at most the number of agents. This is inspired from the classic rent division problem, where the goal is to allocate n indivisible goods to n agents and divide a total cost (rent) among the agents in an envy-free manner [30,31]. In this case, Demange and Gale [13] show that the set of envy-free allocations have a lattice structure; we provide a similar result. Maskin [21] shows that envy-free allocations are guaranteed to exist given a sufficient amount of money; this is easy to show in our setting, so we focus on minimizing the amount of money required. Klijn [19] shows that envy-free allocations can be computed in polynomial time. Several papers focus on concepts other than (or stronger than) envy-freeness. For example, Quinzii [28] shows that the core coincides with competitive equilibria. Bikhchandani and Mamer [6] study the existence of competitive equilibria, which is a stronger requirement than envy-freeness. Ohseto [24] studies the existence of algorithms that are not only envy-free but also strategyproof. This restricted setting with *one good per agent* is substantially different from our general setting with potentially more goods than agents. Svensson [31] shows that in the restricted setting, envy-free allocations are automatically Pareto optimal. This is not true in our setting; and only a weaker condition is implied (Theorem 1).

Among the papers that consider more goods than agents, several consider settings which effectively reduce to one good per agent. For example, Haake et al. [18] consider a fixed partition of the goods into n bundles, so each bundle can be treated as a single good. In contrast, a large portion of our paper (Sect. 4.2) is devoted to finding the optimal bundling of goods. Further, they consider dividing a total cost of C among the agents, whereas we consider paying a non-negative amount of money to each agent. A natural reduction of our problem to their setting would set $C = 0$, compute the payments to agents (which could be negative), and increase all payments equally until they are non-negative. However, it is easy to check that under this reduction, our method requires less subsidy than theirs even for a fixed bundling, and significantly less if we optimize the bundling. Alkan et al. [1] allow more goods than agents, but add fictitious agents until the number of goods and agents are equal. As noted by Meertens et al. [22], their algorithm allocates at most one good to each real agent, throwing away the remaining goods (i.e. assigning them to fictitious agents).

Meertens et al. [22] study a setting more general than ours. They allow agents to have general preference relations over their allocated bundle of indivisible goods and amount of money. In this case, they show that envy-freeness and Pareto optimality may be incompatible regardless of the amount of money available. In contrast, in our setting with quasi-linear preferences, allocations that are both envy-free and Pareto optimal exist given a sufficient amount of money (see the discussion following Proposition 1). Beviá et al. [5] study a setting where each agent arrives at the market with a bundle of goods and an amount of money, and is interested in exchanging the goods and money with other agents. They assume that each agent brings at least as much money as her total value for the goods brought by all the agents, and induce budget-balanced transfers among the agents, making their results incomparable to ours.

To the best of our knowledge, no prior work studies the asymptotic amount of subsidy required to achieve envy-freeness, which is the focus of our work.

2 Preliminaries

For $k \in \mathbb{N}$, let $[k] = \{1, \ldots, k\}$. Let $\mathcal{N} = [n]$ denote the set of *agents*, and let $\mathcal{M}$ denote the set of m indivisible *goods*. Each agent i is endowed with a *valuation* function $v_i : 2^{\mathcal{M}} \to \mathbb{R}_{\geq 0}$ such that $v_i(\emptyset) = 0$. We assume that the valuation is *additive*: $\forall S \subseteq \mathcal{M}, v_i(S) = \sum_{g \in S} v_i(\{g\})$. To simplify notation, we write $v_i(g)$ instead of $v_i(\{g\})$. We denote the vector of valuations by $\mathbf{v} = (v_1, \ldots, v_n)$. We define an *allocation problem* to be the tuple $\mathcal{A} = (\mathcal{N}, \mathcal{M}, \mathbf{v})$.

For a set of goods $S \subseteq \mathcal{M}$ and $k \in \mathbb{N}$, let $\Pi_k(S)$ denote the set of ordered partitions of S into k bundles. Given an allocation problem $\mathcal{A}$, an *allocation* $\mathbf{A} = (A_1, \ldots, A_n) \in \Pi_n(\mathcal{M})$ is a partition of the goods into n bundles, where A_i is the bundle allocated to agent i. Under this allocation, the *utility* to agent i is $v_i(A_i)$, and the *utilitarian welfare* is $\sum_{i=1}^{n} v_i(A_i)$. The following fairness notion is central to our work.

Definition 1 (Envy-Freeness). *An allocation* $\mathbf{A}$ *is called envy-free (EF) if* $v_i(A_i) \geq v_i(A_j)$ *for all agents* $i, j \in \mathcal{N}$.

Envy-freeness requires that no agent prefer another agent's allocation over her own allocation. This cannot be guaranteed when goods are indivisible. Prior literature focuses on its relaxations, such as envy-freeness up to one good [10,20], which can be guaranteed.

Definition 2 (Envy-Freeness up to One Good). *An allocation* $\mathbf{A}$ *is called envy-free up to one good (EF1) if, for all agents* $i, j \in \mathcal{N}$, *either* $v_i(A_i) \geq v_i(A_j)$ *or there exists* $g \in A_j$ *such that* $v_i(A_i) \geq v_i(A_j \setminus \{g\})$. *That is, it should be possible to remove envy between any two agents by removing a single good from the envied agent's bundle.*

We want to study whether (exact) envy-freeness can be achieved by additionally giving each agent some amount of a divisible good, which we refer to as *money*. We denote by $p_i \in \mathbb{R}$ the amount of money received by agent i, and by $\mathbf{p} = (p_1, \ldots, p_n)$ the vector of payments. Throughout most of the paper, we require that $p_i \geq 0$ for each agent i. This corresponds to the *subsidy model*, where a third party subsidizes the allocation problem by donating money. In Sect. 6, we discuss the implications of our results for other models of introducing monetary payments. One other obvious model is one in which there is no outside subsidy and envy is dealt with by agents paying each other. We show these models are essentially equivalent in the sense that any payments in one model can be translated to equivalent payments in the other. In our ring example, Bob giving Alice $50 is equivalent to Alice receiving a $100 subsidy with respect to relative utilities, which is all that matters for envy-freeness.

Given an allocation $\mathbf{A}$ and a payment vector $\mathbf{p}$, we refer to the tuple $(\mathbf{A}, \mathbf{p})$ as the *allocation with payments*. Under $(\mathbf{A}, \mathbf{p})$, the utility of agent i is $v_i(A_i) + p_i$. That is, agents have quasi-linear utilities (equivalently, they express their values for other goods with money as reference). With money, there is a common good to which agents can scale their utilities. Thus, unlike in settings without money, interpersonal comparisons of utilities make sense in our framework. Note that allocation $\mathbf{A}$ is equivalent to allocation with payments $(\mathbf{A}, \mathbf{0})$, where each agent receives zero payment. We can now extend the definition of envy-freeness to allocations with payments.

Definition 3 (Envy-Freeness). *An allocation with payments $(\mathbf{A}, \mathbf{p})$ is envy-free (EF) if $v_i(A_i) + p_i \geq v_i(A_j) + p_j$ for all agents $i, j \in \mathcal{N}$.*

We say that payment vector $\mathbf{p}$ is *envy-eliminating* for allocation $\mathbf{A}$ if $(\mathbf{A}, \mathbf{p})$ is envy-free. Let $\mathcal{P}(\mathbf{A})$ be the set of envy-eliminating payment vectors for $\mathbf{A}$.

Definition 4 (Envy-Freeable). *An allocation $\mathbf{A}$ is called* envy-freeable *if there exists a payment vector $\mathbf{p}$ such that $(\mathbf{A}, \mathbf{p})$ is envy-free, that is, if $\mathcal{P}(\mathbf{A}) \neq \emptyset$.*

Given an allocation problem $\mathcal{A}$, let $\mathcal{E}(\mathcal{A})$ denote the set of envy-freeable allocations. We drop $\mathcal{A}$ from the notation when it is clear from context.

Given an allocation $\mathbf{A}$, its *envy graph* $G_{\mathbf{A}}$ is the complete weighted directed graph in which each agent is a node, and for each $i, j \in \mathcal{N}$, edge (i, j) has weight $w(i, j) = v_i(A_j) - v_i(A_i)$. This is the amount of envy that agent i has for agent j, which can be negative if agent i strictly prefers her own allocation to the allocation of agent j. Note that by definition, $w(i, i) = 0$ for each $i \in \mathcal{N}$. A path P is a sequence of nodes $(i_1, \ldots, i_k)$, and its weight is $w(P) = \sum_{t=1}^{k-1} w(i_t, i_{t+1})$. The path is a *cycle* if $i_1 = i_k$. Given $i, j \in \mathcal{N}$, let $\ell(i, j)$ be the maximum weight of any path which starts at i and ends at j, and let $\ell(i) = \max_{j \in \mathcal{N}} \ell(i, j)$ be the maximum weight of any path starting at i.

3 Envy-Freeable Allocations

In this section, our goal is to characterize envy-freeable allocations of indivisible goods and, given an envy-freeable allocation, to find an envy-eliminating payment vector.

Looking more closely at $G_{\mathbf{A}}$, we can see that $\mathbf{A}$ being envy-free is equivalent to all edge weights of $G_{\mathbf{A}}$ being non-positive. We can extend this connection to the (potentially) larger set of envy-freeable allocations. Note that a permutation of $[n]$ is a bijection $\sigma : [n] \to [n]$.

Theorem 1. *For an allocation $\mathbf{A}$, the following statements are equivalent.*

(a) $\mathbf{A}$ is envy-freeable.
(b) $\mathbf{A}$ maximizes the utilitarian welfare across all reassignments of its bundles to agents, that is, for every permutation σ of $[n]$, $\sum_{i \in \mathcal{N}} v_i(A_i) \geq \sum_{i \in \mathcal{N}} v_i(A_{\sigma(i)})$.
(c) $G_{\mathbf{A}}$ has no positive-weight cycles.

Proof. We show $(a) \Rightarrow (b)$, $(b) \Rightarrow (c)$, and $(c) \Rightarrow (a)$.

$(a) \Rightarrow (b)$: Suppose **A** is envy-freeable. Then, there exists a payment vector **p** such that for all agents $i, j \in \mathcal{N}$, $v_i(A_i) + p_i \geq v_i(A_j) + p_j$, that is, $v_i(A_j) - v_i(A_i) \leq p_i - p_j$. Consider any permutation σ of $[n]$. Then, $\sum_{i \in \mathcal{N}} v_i(A_{\sigma(i)}) - v_i(A_i) \leq \sum_{i \in \mathcal{N}} p_i - p_{\sigma(i)} = 0$.

$(b) \Rightarrow (c)$: Suppose condition (b) holds. Consider a cycle $C = (i_1, \ldots, i_k)$ in $G_\mathbf{A}$. Consider the corresponding permutation σ_C under which $\sigma(i_t) = i_{t+1}$ for each $t \in [k-1]$, and $\sigma(i) = i$ for all $i \notin C$. Then,

$$
\begin{aligned}
w(C) &= \sum_{t=1}^{k-1} w(i_t, i_{t+1}) = \sum_{t=1}^{k-1} v_{i_t}(A_{i_{t+1}}) - v_{i_t}(A_{i_t}) \\
&= \sum_{t=1}^{k-1} \left(v_{i_t}(A_{i_{t+1}}) - v_{i_t}(A_{i_t}) \right) + \sum_{i \notin C} \left(v_i(A_i) - v_i(A_i) \right) \\
&= \sum_{i \in \mathcal{N}} v_i(A_{\sigma(i)}) - v_i(A_i) \leq 0.
\end{aligned}
$$

$(c) \Rightarrow (a)$: Suppose $G_\mathbf{A}$ has no positive-weight cycles. Then, $\ell(i)$, which is the maximum weight of any path starting at i in $G_\mathbf{A}$, is well-defined and finite. Let $p_i = \ell(i)$ for each $i \in \mathcal{N}$. Note that $p_i \geq \ell(i, i) \geq w(i, i) = 0$ for each $i \in \mathcal{N}$. Hence, **p** is a valid payment vector. Also, by definition of longest paths, we have that for all $i, j \in \mathcal{N}$, $p_i = \ell(i) \geq \ell(j) + w(i, j) = p_j + v_i(A_j) - v_i(A_i)$. Hence, $(\mathbf{A}, \mathbf{p})$ is envy-free, and thus, **A** is envy-freeable. □

Theorem 1 provides a way to efficiently check if a given allocation **A** is envy-freeable. This can be done using the maximum weight bipartite matching algorithm [15] to check condition (b) or the Floyd-Warshall algorithm to check condition (c). The proof is provided in the full version.

Proposition 1. *Given an allocation* **A**, *it is possible to check whether* **A** *is envy-freeable in* $O(mn + n^3)$ *time.*

Given Proposition 1, finding an envy-freeable allocation is easy: we can start from an arbitrary allocation **A** and use the maximum weight bipartite matching algorithm to find the reassignment of its bundles that maximizes utilitarian welfare, or we could simply compute the allocation that globally maximizes utilitarian welfare in $O(nm)$ time by assigning each good to the agent who values it most.

But simply knowing an envy-freeable allocation **A** is not enough. We need to find a payment vector **p** such that $(\mathbf{A}, \mathbf{p})$ is envy-free. We would further like to minimize the subsidy required ($\sum_{i \in \mathcal{N}} p_i$). Such a payment vector can easily be computed in polynomial time through a linear program (provided in full version). However, the next result shows that we can compute it in strongly polynomial time (polynomial in the number of inputs, rather than their size). In fact, this payment vector is precisely the one we constructed in the proof of Theorem 1.

Theorem 2. *For an envy-freeable allocation* **A**, *let* $\mathbf{p}^*(\mathbf{A})$ *be given by* $p_i^*(\mathbf{A}) = \ell(i)$ *for all* $i \in \mathcal{N}$, *where* $\ell(i)$ *is the maximum weight of any path starting at* i *in* $G_\mathbf{A}$. *Then,* $\mathbf{p}^*(\mathbf{A}) \in \mathcal{P}(\mathbf{A})$, *and for every* $\mathbf{p} \in \mathcal{P}(\mathbf{A})$ *and* $i \in \mathcal{N}$, $p_i^*(\mathbf{A}) \leq p_i$. *Further,* $\mathbf{p}^*(\mathbf{A})$ *can be computed in* $O(nm + n^3)$ *time.*

Proof. For simplicity, we denote $\mathbf{p}^*(\mathbf{A})$ as $\mathbf{p}^*$. When proving that condition (c) implies condition (a) in Theorem 1, we already showed that $\mathbf{p}^* \in \mathcal{P}(\mathbf{A})$. Thus, we simply need to argue that for every $\mathbf{p} \in \mathcal{P}(\mathbf{A})$, we have that $p_i^* \leq p_i$ for all $i \in \mathcal{N}$.

Fix $\mathbf{p} \in \mathcal{P}(\mathbf{A})$ and $i \in \mathcal{N}$. Consider the longest path starting at i in $G_\mathbf{A}$. Suppose it is $(i_1, \ldots, i_k)$. Hence, $i_1 = i$ and $w(i_1, \ldots, i_k) = \sum_{t=1}^{k-1} w(i_t, i_{t+1}) = p_i^*$. Because $(\mathbf{A}, \mathbf{p})$ is envy-free, we have that for each $t \in [k-1]$,

$$v_{i_t}(A_{i_t}) + p_{i_t} \geq v_{i_t}(A_{i_{t+1}}) + p_{i_{t+1}}$$
$$\Rightarrow p_{i_t} - p_{i_{t+1}} \geq v_{i_t}(A_{i_{t+1}}) - v_{i_t}(A_{i_t}) = w(i_t, i_{t+1}).$$

Summing this over all $t \in [k-1]$, we get

$$p_{i_1} - p_{i_k} \geq w(i_1, \ldots, i_k) = p_i^* \Rightarrow p_i \geq p_i^* + p_{i_k} \geq p_i^*,$$

where the final transition holds because $i_1 = i$ and payments are non-negative.

Finally, $\mathbf{p}^*$ can be computed as follows. We first run the Floyd-Marshall (all-pairs shortest path) algorithm on the graph obtained by negating all edge weights in $G_\mathbf{A}$ to compute $\ell(i, j)$ for all $i, j \in \mathcal{N}$ in $O(nm + n^3)$ time. Then, we compute $\mathbf{p}^*$ in $O(n^2)$ time.   □

We refer to $\mathbf{p}^*(\mathbf{A})$ as the *optimal payment vector* for **A**. When clear from the context, we drop **A** from the notation.

We can also show that for an envy-freeable allocation **A**, $\mathcal{P}(\mathbf{A})$ has a lattice structure and $\mathbf{p}^*$ is its unique minimum element; the proof is provided in the full version. In this lattice, the greatest lower bound (resp., the least upper bound) of two payment vectors is given by the coordinate-wise minimum (resp., maximum).

4   Minimizing and Bounding Subsidy

In this section, we investigate the minimum subsidy required to achieve envy-freeness. We are interested in both the computational complexity of computing the minimum subsidy required in a given allocation problem, and in the minimum subsidy required in the worst case over allocation problems. We consider cases where the (envy-freeable) allocation is given to us, and where we can choose such an allocation to minimize subsidy.

For an envy-freeable allocation **A**, let $\text{sub}(\mathbf{A}) = \sum_{i \in \mathcal{N}} p_i^*(\mathbf{A})$ be the minimum subsidy required to make **A** envy-free. Then, in the former case, we want to compute $\sup_\mathcal{A} \max_{\mathbf{A} \in \mathcal{E}(\mathcal{A})} \text{sub}(\mathbf{A})$ and, in the latter case, we want to compute $\sup_\mathcal{A} \min_{\mathbf{A} \in \mathcal{E}(\mathcal{A})} \text{sub}(\mathbf{A})$.[1]

---

[1] Note that $\mathcal{E}(\mathcal{A}) \neq \emptyset$ because the allocation maximizing utilitarian welfare is always envy-freeable due to Theorem 1.

Without loss of generality, we assume that $v_i(g) \in [0, 1]$ for each agent i and good g. If the valuations lie in $[0, T]$, the worst-case minimum subsidy and the bounds we provide would simply be multiplied by T, the largest value for any single good. We say that valuations are *binary* if $v_i(g) \in \{0, 1\}$ for all agents i and goods g, and *identical* if $v_i(g) = v_j(g)$ for all agents i, j and goods g.

4.1   When the Allocation is Given

In cases where an envy-freeable allocation is already implemented, or if we desire to implement a specific allocation for reasons other than achieving envy-freeness, we may be given an allocation and asked to eliminate envy.

Theorem 2 already shows that we can efficiently compute the minimum amount of subsidy required. To study how much subsidy is needed in the worst case, we begin with the following simple observation.

Lemma 1. *For an envy-freeable allocation* **A**, *no path in* $G_\mathbf{A}$ *has weight more than* m.

Proof. Since $G_\mathbf{A}$ has no positive-weight cycles, we only need to consider simple paths on which no agent appears twice. Consider a simple path $(i_1, \ldots, i_k)$. For $t \in [k-1]$, note that $w(i_t, i_{t+1}) = v_{i_t}(A_{i_{t+1}}) - v_{i_t}(A_{i_t}) \leq |A_{i_{t+1}}|$. Thus, the weight of the path is $\sum_{t=1}^{k-1} w(i_t, i_{t+1}) \leq \sum_{t=1}^{k-1} |A_{i_{t+1}}| = |\cup_{t=2}^{k} A_{i_t}| \leq m$, as desired.   □

We can now pinpoint the subsidy required in the worst case. The upper bound uses Lemma 1 along with the fact that some agent must receive zero payment under the optimal payment vector.

Theorem 3. *When an envy-freeable allocation is given, the minimum subsidy required is* $(n-1)m$ *in the worst case.*

Proof. For the lower bound, consider the instance where $v_i(g) = 1$ for all agents i and goods g. Consider the allocation **A** which assigns all goods to a single agent i^*. It is easy to see that this is envy-freeable, and its optimal payment vector **p** has $p_i = m$ for $i \neq i^*$ and $p_{i^*} = 0$. Hence, we need $(n-1)m$ subsidy.

To prove the upper bound, note that the minimum subsidy required is the sum of weights of longest paths starting at different agents (Theorem 2). Using Lemma 1 and the fact that one agent must receive zero payment (otherwise all payments can be reduced while preserving envy-freeness, which would contradict the minimality of payments), this is at most $(n-1)m$.   □

The lower bound uses an instance with identical binary valuations. Hence, Theorem 3 also holds for the special cases of binary and identical valuations.

4.2   When the Allocation Can Be Chosen

When we are allowed to choose the allocation, computing the minimum subsidy required is NP-hard. This is because checking whether zero subsidy is required

382   D. Halpern and N. Shah

is equivalent to checking whether an envy-free allocation exists, which is NP-hard even for identical valuations [9]. That said, it is possible to compute the minimum subsidy required using a simple integer linear program (details are in the full version).

Recall that when an envy-freeable allocation is *given*, in the worst case we need a subsidy of $(n-1)m$ (Theorem 3). But what if we were able to *choose* the allocation? We show that this does not help improve the bound by a factor larger than m.

Theorem 4. *When the allocation can be chosen, the minimum subsidy required is at least $n-1$ in the worst case, even in the special cases of binary valuations and identical valuations.*

Proof. Consider the instance with identical binary valuations where each agent values a special good at 1 and other goods at 0. Every allocation gives the special good to one of the agents. To achieve envy-freeness, each other agent must be paid at least 1. Hence, a subsidy of at least $n-1$ is needed. □

This raises a natural question: *Can we always find an envy-freeable allocation that requires a subsidy of at most $n-1$?* We answer this question affirmatively for the special cases of binary and identical valuations as well as any valuations with two agents. In addition, we make an interesting conjecture in the general case. First, we take a slight detour.

One promising approach to reducing the subsidy requirement is to start with an allocation that already has limited envy, for example, an allocation that is envy-free up to one good [10,20]. For an envy-freeable EF1 allocation **A**, each edge in $G_\mathbf{A}$ has weight at most 1, so each (simple) path has weight at most $n-1$. Using this improvement over Lemma 1 in Theorem 3, we get the following.

Lemma 2. *For an envy-freeable allocation **A** that is envy-free up to one good, no path in $G_\mathbf{A}$ has weight more than $\min(n-1,m)$. Hence, $\mathrm{sub}(\mathbf{A}) \le (n-1) \cdot \min(n-1,m)$.*

With an envy-freeable EF1 allocation, the subsidy requirement becomes independent of the number of goods, at the expense of becoming quadratic in the number of agents. However, it is not even clear that an envy-freeable EF1 allocation always exists. For the special cases of binary and identical valuations, we show that it does, and in fact, picking a specific EF1 allocation that satisfies other properties allows achieving the optimal subsidy requirement of $n-1$.

Binary Valuations. Recall that with binary valuations, we have $v_i(g) \in \{0,1\}$ for all $i \in \mathcal{N}$ and $g \in \mathcal{M}$. We say that agent i *likes* good g if $v_i(g)=1$. An allocation **A** is *non-wasteful* if each good is allocated to an agent who likes it. Note that because the valuations are binary, non-wasteful is equivalent to Pareto efficiency. For binary valuations, it is easy to see that every non-wasteful allocation is envy-freeable as it satisfies condition (b) of Theorem 1.

Algorithms such as the round-robin method and maximum Nash welfare (MNW) are known to produce non-wasteful EF1 allocations [11]. The round-robin method, given an agent ordering, allows agents to pick goods one-by-one according to the ordering in a cyclic fashion. The MNW algorithm finds the largest set of agents that can simultaneously receive positive utility and returns an allocation maximizing the product of their utilities.

Using a non-wasteful EF1 allocation, we can reduce the $O(mn)$ subsidy requirement to $O(n^2)$. This is the best we can do using the round-robin method with an arbitrary agent ordering (an example is provided in the full version). However, we show that the non-wasteful EF1 allocation returned by the MNW algorithm is special as it requires a subsidy of at most $n - 1$, meeting the lower bound from Theorem 4.

Theorem 5. *For binary valuations, an allocation produced by the maximum Nash welfare algorithm is envy-freeable and requires at most $n - 1$ subsidy.*

Proof. Let $\mathbf{A}$ be an allocation returned by the MNW algorithm. It is easy to see that $\mathbf{A}$ is non-wasteful, and hence, envy-freeable. Next, we show that any path in $G_{\mathbf{A}}$ has weight at most 1. This implies a subsidy requirement of at most $n - 1$ using the same argument as in the proof of Theorem 3.

First, without loss of generality, we assume that each good is liked by at least one agent; if there are goods that are not liked by any agent, we could disregard them in the steps below and allocate them arbitrarily. We already argued that the non-wasteful allocation produced by the MNW algorithm is envy-freeable. Since it assigns each good to an agent who likes it, we have $v_i(A_i) = |A_i|$ for all $i \in \mathcal{N}$ and $v_i(A_j) \leq |A_j|$ for all $i, j \in \mathcal{N}$. It follows that $w(i, j) = v_i(A_j) - v_i(A_i) \leq |A_j| - |A_i|$ for all $i, j \in \mathcal{N}$.

Suppose for a contradiction that there exists a path P^* in $G_{\mathbf{A}}$ such that $w(P^*) > 1$. Because weights are integral, this implies $w(P^*) \geq 2$. Now, we make the following claim; the proof is given in the full version.

Claim. There exists a subpath P of P^* with no negative-weight edges and $w(P) \geq 2$.

Without loss of generality, we further assume that the first edge of P has a positive weight (otherwise we could consider the subpath of P starting at its first positive-weight edge). Let $P = (i_1, \ldots, i_k)$. We want to prove two claims: (a) $|A_{i_k}| \geq |A_{i_1}| + 2$, and (b) for each $t \in [k - 1]$, there exists a good $g \in A_{i_{t+1}}$ which agent i_t likes.

For claim (a), recall that for each $t \in [k - 1]$, we have $w(i_t, i_{t+1}) \leq |A_{i_{t+1}}| - |A_{i_t}|$. Summing over $t \in [k-1]$, we get that $|A_{i_k}| - |A_{i_1}| \geq w(P) \geq 2$, as desired.

Claim (b) holds for $t = 1$ because the first edge has weight $w(i_1, i_2) = v_{i_1}(A_{i_2}) - v_{i_1}(A_{i_1}) > 0$, implying $v_{i_1}(A_{i_2}) > 0$. For $t \in \{2, \ldots, k - 1\}$, using the argument above, we have $|A_{i_t}| - |A_{i_1}| \geq w(i_1, \ldots, i_t) \geq w(i_1, i_2) > 0$. Hence, $v_{i_t}(A_{i_t}) = |A_{i_t}| > 0$. This, along with $w(i_t, i_{t+1}) = v_{i_t}(A_{i_{t+1}}) - v_{i_t}(A_{i_t}) \geq 0$, implies $v_{i_t}(A_{i_{t+1}}) > 0$.

Given the two claims, we derive a contradiction to the fact that $\mathbf{A}$ is returned by the MNW algorithm. Suppose we take a good from $A_{i_{t+1}}$ that agent i_t likes— it exists due to claim (b)—and add it to A_{i_t} for each $t \in [k-1]$. In the resulting allocation, the utility to agent i_k decreases by 1, the utility to agent i_1 increases by 1, and the utility to every other agent remains constant. Since agent i_k had at least 2 more utility than agent i_1 due to claim (a), it is easy to see that the resulting allocation would either give positive utility to strictly more agents (if i_1 had zero utility in the beginning) or strictly increase the product of utilities to the agents with positive utility. Both of these contradict the fact that $\mathbf{A}$ was returned by the MNW algorithm. Hence, every path in $G_{\mathbf{A}}$ has weight at most 1, which implies the desired result.   □

Note that in this proof, along with non-wastefulness, the only property of the MNW algorithm that we used was the following: given allocations $\mathbf{A}^1$ and $\mathbf{A}^2$ such that for some agents $i, j \in \mathcal{N}$, $v_i(A_i^1) \geq v_j(A_j^1) + 2$, $v_i(A_i^2) = v_i(A_i^1) - 1$, $v_j(A_j^2) = v_j(A_j^1) + 1$, and $v_k(A_k^1) = v_k(A_k^2)$ for all $k \in \mathcal{N} \setminus \{i, j\}$, the algorithm cannot return $\mathbf{A}^1$. This property as well as non-wastefulness are implied by the Pigou-Dalton principle [23]. Hence, the result holds for every algorithm which satisfies this principle, including the leximin rule.[2]

This proof leverages several ideas from the literature. Claim (a) shares similarities with a property of MNW allocations established by Darmann and Sch [12], while the trick of passing goods along a path using claim (b) was also used by Barman et al. [3] to show that an MNW allocation can be computed efficiently for binary valuations. Thus, for binary valuations, we can efficiently compute an allocation which needs at most $n - 1$ subsidy.

While the MNW algorithm achieves the optimal worst-case subsidy bound, it does not minimize the subsidy required on every instance. It is easy to construct instances where envy-free allocations exist but the MNW algorithm produces an allocation which requires as much as $n - 2$ subsidy (an example is provided in the full version).

What is the complexity of computing the minimum subsidy required in a given allocation problem? As argued before, we can reduce the problem of checking the existence of an envy-free allocation to the problem of computing the minimum subsidy required. It is not difficult to see that the converse holds too. We can compute the minimum subsidy required by adding a unit subsidy at a time, and checking the existence of an envy-free allocation. The proof of the next result is given in the full version.

Proposition 2. *For binary valuations, the problems of computing the minimum subsidy required and checking the existence of an envy-free allocation are Turing-equivalent.*

Unfortunately, to the best of our knowledge, it is an open question whether existence of an envy-free allocation can be checked efficiently for binary valuations. However, the complexity of a closely related problem is known.

---

[2] The leximin rule finds an allocation that maximizes the minimum utility, subject to that maximizes the second minimum utility, and so on.

Bouveret and Lang [9] show that checking the existence of a non-wasteful envy-free allocation with binary valuations is an NP-complete problem. Using the same argument as before, we have the following.

Corollary 1. *For binary valuations, it is NP-hard to compute the minimum subsidy required to achieve envy-freeness using a non-wasteful allocation.*

Identical Valuations. With identical valuations, we denote the common valuation function of the agents by v. In this case, the utilitarian welfare $\sum_{i \in \mathcal{N}} v(A_i) = v(\mathcal{M})$ is constant. This implies that every allocation is Pareto efficient. Hence, by condition (b) of Theorem 1, every allocation $\mathbf{A}$ is envy-freeable.

Given an allocation $\mathbf{A}$, the optimal payment vector is given by $p_i^*(\mathbf{A}) = \max_{j \in \mathcal{N}} v(A_j) - v(A_i)$ for all $i \in \mathcal{N}$. To see this, note that each agent i requires payment at least $p_i^*(\mathbf{A})$ to not envy the agent with the highest utility. Conversely, $(\mathbf{A}, p_i^*(\mathbf{A}))$ is envy-free as every agent has the same value for all agents' allocations. Thus, $\text{sub}(\mathbf{A}) = n \cdot \max_{j \in \mathcal{N}} v(A_j) - v(\mathcal{M})$. Therefore, minimizing subsidy is equivalent to minimizing the maximum value of any bundle, which is the well-known NP-complete multiprocessor scheduling problem.

Proposition 3. *With identical valuations, every allocation is envy-freeable. An allocation minimizes the subsidy required if and only if it minimizes the maximum utility to any agent. Computing such an allocation is an NP-hard problem.*

What if we simply wanted to achieve the optimal worst-case upper bound of $n - 1$ instead of minimizing the subsidy on every instance? For binary valuations, we achieved this by efficiently choosing a *specific* envy-freeable EF1 allocation— namely, the one produced by the MNW algorithm. For identical valuations, it is easy to see that *any* envy-freeable EF1 allocation $\mathbf{A}$ suffices as $p_i^*(\mathbf{A}) = \max_{j \in \mathcal{N}} v(A_j) - v(A_i)$ is at most 1 for each $i \in \mathcal{N}$ and is zero for some agent. Since we can compute an EF1 allocation efficiently, we have the following.

Proposition 4. *With identical valuations, we can efficiently compute an allocation which requires at most $n - 1$ subsidy.*

Returning to General Valuations. Recall that in the worst case, we need at least $n - 1$ subsidy (Theorem 4). For the special cases of binary and identical valuations, we achieved this optimal bound by finding a special envy-freeable and EF1 allocation, respectively. For general valuations, the problem is that it is not clear if an envy-freeable EF1 allocation is even guaranteed to exist.

Most of the algorithms known in the literature that achieve EF1 are scale-free [11,20], that is, multiplying an agent's valuation by a scalar does not affect the allocation returned. It is easy to see that such algorithms cannot always return an envy-freeable allocation.

Of these algorithms, the round-robin method is of special interest. With a fixed agent ordering, it is scale-free. But what if we chose the *right* agent ordering in a non-scale-free way? We show that this indeed works for two agents. The proof of the next result is provided in the full version.

Theorem 6. *When $n = 2$, there exists an agent ordering such that the allocation returned by the round-robin method with that ordering is envy-freeable and requires at most 1 subsidy.*

Note that this achieves the optimal bound of $n - 1$ for $n = 2$ agents. Unfortunately, this method does not work for $n \geq 3$ agents. In our counterexample (provided in the full version), while the round-robin method fails to produce an envy-freeable EF1 allocation with any agent ordering, there still exists an envy-freeable EF1 allocation. This leads us to the following conjecture.

Conjecture 1. There always exists an envy-freeable allocation that is envy-free up to one good.

If this conjecture is true, then by Lemma 2, we know that the minimum subsidy required in the worst case is $O(n^2)$ (thus independent of m). We conjecture further that the lower bound of $n - 1$ can be achieved.

Conjecture 2. There always exists an envy-freeable allocation which requires at most $n - 1$ subsidy.

In fact, it may be possible that a subsidy of at most $n - 1$ can always be achieved through an envy-freeable EF1 allocation.

5   Experiments

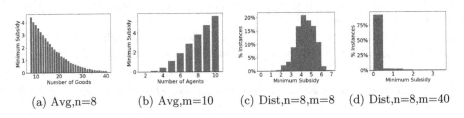

(a) Avg,n=8   (b) Avg,m=10   (c) Dist,n=8,m=8   (d) Dist,n=8,m=40

Fig. 1. The minimum subsidy required in our simulations. Figures (a) and (b) show the minimum subsidy averaged across instances as functions of m and n, respectively. Figures (c) and (d) show the distribution of minimum subsidy for fixed n and m.

In this section, we empirically study the minimum subsidy required in the average case. We compute the minimum subsidy required across all allocations by solving an integer linear program using CPLEX.

To generate synthetic data, we consider instances with $2 \leq n \leq 8$ and $n \leq m \leq 5n$. For each (n, m), we sample $1,000$ instances as follows: For each good g we sample $v^*(g)$ from an exponential distribution with mean 30 and $\sigma^*(g)$ from an exponential distribution with mean 5. Then, for each agent i and good g, we draw $v_i(g)$ from a truncated normal distribution, which has mean $v^*(g)$ and standard deviation $\sigma^*(g)$, and is truncated below at 0.

In addition, we obtained 3,535 real-world fair division instances from a popular fair division website Spliddit.org. These instances have divisible as well as indivisible goods, from 2 to 15 agents, and from 2 to 96 goods. While Spliddit data does not match our model as agents are forced to report valuations that sum to a constant, we believe that it still provides a valuable empirical perspective.

We begin by noting that none of the 114,000 synthetic instances or 3,535 real-world instances required a subsidy of more than $n - 1$, which is evidence in support of Conjecture 2.

In our synthetic experiments, we see that fixing the number of agents, the minimum subsidy required reduces on average as the number of goods increases (Fig. 1(a)). On the other hand, fixing the number of goods, the minimum subsidy required (almost linearly) increases on average as the number of agents increases (Fig. 1(b)). These results are in part due to the fact that the probability of existence of an envy-free allocation (i.e., of requiring no subsidy) increases with more goods but decreases with more agents [14]. Next, we dive into the distribution of the minimum subsidy required, presented in Fig. 1(c) for $n = m = 8$ and in Fig. 1(d) for $n = 8$ and $m = 40$. Again, with more goods, the distribution quickly skews towards requiring little to no subsidy.

Finally, on the real-world data obtained from Spliddit, 68% of the instances required no subsidy (i.e., admitted envy-free allocations), while 93% of the instances required a subsidy of at most 1. Thus, in practice, the amount of subsidy needed to eliminate envy is most likely no greater than the maximum value that any agent places on a single good.

6   Discussion

We have examined the minimum subsidy required both in cases when an allocation is given to us and when it can be chosen. In the former case, we have shown how to compute the minimum subsidy exactly; in both cases, we have provided several useful bounds for cases of interest. However, a number of directions remain open for further research. Perhaps the most immediate question is to settle our two conjectures from Sect. 4.2. Specifically, it may be possible to adapt the iterative algorithm of Lipton et al. [20] to select the good to be allocated in each iteration in a non-scale-free way and achieve the optimal bound of $n - 1$ subsidy. Settling the complexity of checking the existence of an envy-free allocation for binary valuations is also an important open question. Finally, it would be interesting to extend this framework to non-additive valuations.

More broadly, while we modeled the divisible good as external subsidy throughout the paper, our results also have implications for other models of introducing monetary payments. For example, when no subsidy is available but monetary transfers among agents are possible, we would like to find *budget-balanced transfers*, $\mathbf{p}$ where $\sum_{i \in \mathcal{N}} p_i = 0$. It is easy to show that computing the optimal payment vector from Theorem 2 and then reducing the payment to each agent by the average payment finds budget-balanced transfers which minimize the maximum amount that any agent has to pay. Alternatively, one could consider a model where each agent pays to receive goods ($p_i \leq 0$ for each i).

It is again easy to show that we can efficiently minimize the total payment collected in a manner similar to Theorem 2. It would be interesting to study other natural objective functions (e.g., minimizing the number of agents that have a non-zero payment) in such models.

References

1. Alkan, A., Demange, G., Gale, D.: Fair allocation of indivisible goods and criteria of justice. Econometrica **59**(4), 1023–1039 (1991)
2. Barman, S., Krishnamurthy, S.K., Vaish, R.: Finding fair and efficient allocations. In: Proceedings of the 19th ACM Conference on Economics and Computation (EC), pp. 557–574 (2018)
3. Barman, S., Krishnamurthy, S.K., Vaish, R.: Greedy algorithms for maximizing Nash social welfare. In: Proceedings of the 17th International Conference on Autonomous Agents and Multi-Agent Systems (AAMAS), pp. 7–13 (2018)
4. Berliant, M., Dunz, K., Thomson, W.: On the fair division of a heterogeneous commodity. J. Math. Econ. **21**, 201–216 (1992)
5. Beviá, C., Quinzii, M., Silva, J.A.: Buying several indivisible goods. Math. Soc. Sci. **37**(1), 1–23 (1999)
6. Bikhchandani, S., Mamer, J.W.: Competitive equilibrium in an exchange economy with indivisibilities. J. Econ. Theory **74**(2), 385–413 (1997)
7. Bilo, V., et al.: Almost envy-free allocations with connected bundles. In: Proceedings of the 10th Innovations in Theoretical Computer Science Conference (ITCS), pp. 1–21, 14 (2019)
8. Bouveret, S., Cechlárová, K., Elkind, E., Igarashi, A., Peters, D.: Fair division of a graph. In: Proceedings of the 26th International Joint Conference on Artificial Intelligence (IJCAI), pp. 135–141 (2017)
9. Bouveret, S., Lang, J.: Efficiency and envy-freeness in fair division of indivisible goods: logical representation and complexity. J. Artif. Intell. Res. **32**, 525–564 (2008)
10. Budish, E.: The combinatorial assignment problem: approximate competitive equilibrium from equal incomes. J. Polit. Econ. **119**(6), 1061–1103 (2011)
11. Caragiannis, I., Kurokawa, D., Moulin, H., Procaccia, A.D., Shah, N., Wang, J.: The unreasonable fairness of maximum Nash welfare. In: Proceedings of the 17th ACM Conference on Economics and Computation (EC), pp. 305–322 (2016)
12. Darmann, A., Schauer, J.: Maximizing Nash product social welfare in allocating indivisible goods. Eur. J. Oper. Res. **247**(2), 548–559 (2015)
13. Demange, G., Gale, D.: The strategy structure of two-sided matching markets. Econometrica **53**, 873–888 (1985)
14. Dickerson, J.P., Goldman, J., Karp, J., Procaccia, A.D., Sandholm, T.: The computational rise and fall of fairness. In: Proceedings of the 28th AAAI Conference on Artificial Intelligence (AAAI), pp. 1405–1411 (2014)
15. Edmonds, J., Karp, R.M.: Theoretical improvements in algorithmic efficiency for network flow problems. J. ACM (JACM) **19**(2), 248–264 (1972)
16. Eisenberg, E., Gale, D.: Consensus of subjective probabilities: the pari-mutuel method. Ann. Math. Stat. **30**(1), 165–168 (1959)
17. Foley, D.: Resource allocation and the public sector. Yale Econ. Essays **7**, 45–98 (1967)

18. Haake, C.J., Raith, M.G., Su, F.E.: Bidding for envy-freeness: a procedural approach to n-player fair-division problems. Soc. Choice Welfare **19**(4), 723–749 (2002)
19. Klijn, F.: An algorithm for envy-free allocations in an economy with indivisible objects and money. Soc. Choice Welfare **17**(2), 201–215 (2000)
20. Lipton, R.J., Markakis, E., Mossel, E., Saberi, A.: On approximately fair allocations of indivisible goods. In: Proceedings of the 6th ACM Conference on Economics and Computation (EC), pp. 125–131 (2004)
21. Maskin, E.S.: On the fair allocation of indivisible goods. In: Feiwel, G.R. (ed.) Arrow and the Foundations of the Theory of Economic Policy, pp. 341–349. Palgrave Macmillan, London (1987). https://doi.org/10.1007/978-1-349-07357-3_12
22. Meertens, M., Potters, J., Reijnierse, H.: Envy-free and pareto efficient allocations in economies with indivisible goods and money. Math. Soc. Sci. **44**(3), 223–233 (2002)
23. Moulin, H.: Fair Division and Collective Welfare. MIT Press, Cambridge (2004)
24. Ohseto, S.: Characterizations of strategy-proof and fair mechanisms for allocating indivisible goods. Econ. Theory **29**(1), 111–121 (2006)
25. Pazner, E., Schmeidler, D.: Egalitarian equivalent allocations: a new concept of economic equity. Q. J. Econ. **92**(4), 671–687 (1978)
26. Plaut, B., Rougligarden, T.: Almost envy-freeness with general valuations. In: Proceedings of the 29th Annual ACM-SIAM Symposium on Discrete Algorithms (SODA), pp. 2584–2603 (2018)
27. Procaccia, A.D., Wang, J.: Fair enough: guaranteeing approximate maximin shares. In: Proceedings of the 14th ACM Conference on Economics and Computation (EC), pp. 675–692 (2014)
28. Quinzii, M.: Core and competitive equilibria with indivisibilities. Int. J. Game Theory **13**(1), 41–60 (1984)
29. Steinhaus, H.: The problem of fair division. Econometrica **16**, 101–104 (1948)
30. Su, F.E.: Rental harmony: sperner's lemma in fair division. Am. Math. Monthly **106**(10), 930–942 (1999)
31. Svensson, L.G.: Large indivisibles: an analysis with respect to price equilibrium and fairness. Econometrica **51**(4), 939–954 (1983)
32. Varian, H.: Equity, envy and efficiency. J. Econ. Theory **9**, 63–91 (1974)
33. Weller, D.: Fair division of a measurable space. J. Math. Econ. **14**(1), 5–17 (1985)

· **Abstract**

Prophet Inequalities on the Intersection of a Matroid and a Graph

Jackie Baek[ID][1] and Will Ma[ID][2(✉)]

[1] Operations Research Center, Massachusetts Institute of Technology,
Camridge, MA 02139, USA
baek@mit.edu
[2] Operations Research Team, Google Research, Camridge, MA 02139, USA
willma@google.com

Abstract. We consider prophet inequalities in a setting where agents correspond to both elements in a matroid and vertices in a graph, a set of agents is feasible if they form both an independent set in the matroid and an independent set in the graph. Our main result is an ex-ante $\frac{1}{2(d+1)}$-prophet inequality, where d is a graph parameter upper-bounded by the maximum size of an independent set in the neighborhood of any vertex.

We establish this result through a framework that sets both dynamic prices for elements in the matroid (using the method of balanced thresholds), and static but discriminatory prices for vertices in the graph (motivated by recent developments in approximate dynamic programming). The threshold for accepting an agent is then the sum of these two prices.

We show that for graphs induced by a certain family of interval-scheduling constraints, the value of d is 1. Our framework thus provides the first constant-factor prophet inequality when there are both matroid-independence constraints and interval-scheduling constraints. It also unifies and improves several results from the literature, leading to a $\frac{1}{2}$-prophet inequality when agents have XOS valuation functions over a set of items and use them for a finite interval duration, and more generally, a $\frac{1}{d+1}$-prophet inequality when these items each require a bundle of d resources to procure.

The full version of this paper can be found online at: https://arxiv.org/abs/1906.04899.

Keywords: Prophet inequalities · Posted-price mechanisms · Approximate dynamic programming

---

The authors would like to thank the anonymous reviewers for the 12th *Symposium on Algorithmic Game Theory* (SAGT) who made many suggestions that improved the paper.

© Springer Nature Switzerland AG 2019
D. Fotakis and E. Markakis (Eds.): SAGT 2019, LNCS 11801, p. 393, 2019.
https://doi.org/10.1007/978-3-030-30473-7

Author Index

Printed in the United States
By Bookmasters